Proceedings of the 20th International Meshing Roundtable

William Roshan Quadros (Ed.)

Proceedings of the 20th International Meshing Roundtable

Editor

William Roshan Quadros
Sandia National Laboratories
P.O. Box 5800
Albuquerque, NM 87185
USA
E-mail: wrquadr@sandia.gov

ISBN 978-3-642-24733-0   e-ISBN 978-3-642-24734-7

DOI 10.1007/978-3-642-24734-7

Library of Congress Control Number: 2011939091

© 2011 Springer-Verlag Berlin Heidelberg

This work is subject to copyright. All rights are reserved, whether the whole or part of the material is concerned, specifically the rights of translation, reprinting, reuse of illustrations, recitation, broadcasting, reproduction on microfilm or in any other way, and storage in data banks. Duplication of this publication or parts thereof is permitted only under the provisions of the German Copyright Law of September 9, 1965, in its current version, and permission for use must always be obtained from Springer. Violations are liable to prosecution under the German Copyright Law.

The use of general descriptive names, registered names, trademarks, etc. in this publication does not imply, even in the absence of a specific statement, that such names are exempt from the relevant protective laws and regulations and therefore free for general use.

Typeset & Cover Design: Scientific Publishing Services Pvt. Ltd., Chennai, India.

Printed on acid-free paper

9 8 7 6 5 4 3 2 1

springer.com

Preface

The papers in this volume were selected for presentation at the 20th International Meshing Roundtable (IMR), held October 23–26, 2011 in Paris, France. The conference was started by Sandia National Laboratories in 1992 as a small meeting of organizations striving to establish a common focus for research and development in the field of mesh generation. Now after 20 consecutive years, the International Meshing Roundtable has become recognized as an international focal point annually attended by researchers and developers from dozens of countries around the world.

The 20th International Meshing Roundtable consists of technical presentations from contributed papers, research notes, keynote and invited talks, short course presentations, and a poster session and competition. The Program Committee would like to express its appreciation to all who participate to make the IMR a successful and enriching experience.

The papers in these proceedings were selected by the Program Committee from among 54 paper submissions. Based on input from peer reviews, the committee selected these papers for their perceived quality, originality, and appropriateness to the theme of the International Meshing Roundtable. We would like to thank all who submitted papers. We would also like to thank the colleagues who provided reviews of the submitted papers. The names of the reviewers are acknowledged in the following pages.

We extend special thanks to Jacqueline Hunter and Elizabeth Lucero for their time and effort to make the 20th IMR another outstanding conference.

20th IMR Program Committee
August 2011

20th IMR Conference Organization

Committee: **William Roshan Quadros (Committee Chairman)**
Sandia National Laboratories, Albuquerque, NM
wrquadr@sandia.gov

Laurent Anne
Distene, Bruyeres le Chatel, FR
laurent.anne@distene.com

Geoffrey Butlin
TranscenData Europe Ltd., UK
Geoffrey.butlin@transcendata.com

John Dannenhoffer
Syracuse University, Syracuse, NY
jfdannen@ecs.syr.edu

Xiangmin (Jim) Jiao
Stony Brook University, Stony Brook, NY
xjiao@sunysb.edu

Nilanjan Mukherjee – Research Notes
Siemens PLM Software, Milford, OH
mukherjee.nilanjan@siemens.com

Jean-Christophe Weill
CEA, Bruyeres le Chatel, FR
Jean-christophe.weill@cea.fr

Yongjie (Jessica) Zhang – Papers
Carnegie Mellon University, Pittsburgh, PA
jessicaz@andrew.cmu.edu

Coordinator/Web: **Jacqueline A. Hunter**
Sandia National Laboratories, Albuquerque, NM
jafinle@sandia.gov

Elizabeth Lucero
Sandia National Laboratories, Albuquerque, NM
elucero@sandia.gov

Web Site: http://www.imr.sandia.gov

Reviewers

Name	Affiliation
Alauzet, Frederic	INRIA, France
Alliez, Pierre	Inria Sophia-Antipolis, France
Alter, Stephen	NASA Langley Research Center, USA
Anne, Laurent	Distene S.A.S., France
Armstrong, Cecil	Queen's University – Belfast, UK
Benzley, Steven	Brigham Young Univ., USA
Betro, Vincent	U. of Tennessee, Chattanooga, USA
Blacker, Ted	Sandia National Laboratories, USA
Cabello, Jean	Siemens, USA
Chernikov, Andrey	C. of William and Mary, VA, USA
Chevalier, Cedric	CEA/DAM, France
Clark, Brett	Sandia Nat. Labs, USA
Ebeida, Mohamed	Sandia National Laboratories, USA
Francois, Vincent	Université du Québec à Trois-Rivières, Canada
Garimella, Rao	Los Alamos Nat. Labs, USA
Gillette, Andrew	University of Texas – Austin, USA
Guoy, Damron	U. of Illionois, Urbana-Champaign, USA
Gurumoorthy, Balan	Indian Institute of Science, India
Hassan, Oubay	Swansea University, UK
Ito, Yasushi	U. of Alabama, Birmingham, USA
Jiao, Xiangmin	State U. of New York, Stony Brook, USA
Jones, William	NASA Langley Research Center, USA
Karman, Steve	U. of Tennessee, Chattanooga, USA
Labbe, Paul	IREQ, Canada
Ledoux, Franck	CEA, France
Leon, Jean-Claude	Institut National Polytechnique de Grenoble, France
Levin, Joshua	Scientific Computing and Imaging Institute, USA
Mitchell, Scott	Sandia National Laboratories, USA
Mukherjee, Nilanjan	Siemens PLM Software, USA
Ollivier-Gooch, Carl	The University of British Columbia, Canada
Owen, Steve	Sandia Nat. Labs, USA
Park, Mike	NASA Langley Research Center, USA
Qin, Ning	The University of Sheffield, UK
Quadros, William	Sandia National Laboratories, USA
Rand, Alexander	University of Texas – Austin, USA

Sarrate, Jose	Universitat Politecnica de Catalunya, Spain
Shaw, Scott	Airbus Deutschland GMBH, Germany
Shephard, Mark	Rensselaer Polytechnic Institute, USA
Shepherd, Jason	Sandia National Laboratories, USA
Shivanna, Kiran	University of Iowa, USA
Shontz, Suzanne	Pennsylvania State Univ., USA
Si, Hang	Weierstrass Institute of Applied Analysis and Stochastics, Germany
Staten, Matthew	Sandia Nat. Labs, USA
Tautges, Timothy J.	Argonne National Laboratory, USA
Thompson, David	Mississippi State University, USA
Wang, Yuanli	CD-Adaco, USA
Weill, Jean-Christophe	CEA, France
Yamakawa, Soji	Carnegie Mellon U., PA, USA
Yan, Dongming	Geometric Modeling and Scientific Visualization Center, SA
Yu, Zeyun	University of Wisconsin – Milwaukee, USA
Yvinec, Mariette	Inria-Prisme, France
Zhang, Hanzhou	Autodesk-ALGOR, USA
Zhang, Yongjie (Jessica)	Carnegie Mellon University, USA

Contents

TRI MESHING (Monday Morning)

Dynamic Parallel 3D Delaunay Triangulation 3
Panagiotis Foteinos, Nikos Chrisochoides

Quality Surface Meshing Using Discrete Parametrizations ... 21
Emilie Marchandise, Jean-François Rainaud, Christophe Geuzaine

Dual Contouring for Domains with Topology Ambiguity 41
Jin Qian, Yongjie Zhang

MESH SIZE, ORIENTATION, and ANISOTROPY (Monday Afternoon)

High Quality Geometric Meshing of CAD Surfaces 63
Patrick Laug, Houman Borouchaki

Isotropic 2D Quadrangle Meshing with Size and Orientation Control 81
Bertrand Pellenard, Pierre Alliez, Jean-Marie Morvan

Anisotropic Goal-Oriented Mesh Adaptation for Time Dependent Problems 99
Frederic Alauzet, Anca Belme, Alain Dervieux

HEX MESHING (Monday Afternoon)

Automatic All-Hex Mesh Generation of Thin-Walled Solids via a Conformal Pyramid-Less Hex, Prism, and Tet Mixed Mesh 125
Soji Yamakawa, Kenji Shimada

Hexahedral Mesh Refinement Using an Error Sizing
Function 143
Gaurab Paudel, Steven J. Owen, Steven E. Benzley

Parallel Hex Meshing from Volume Fractions 161
Steven J. Owen, Matthew L. Staten, Marguerite C. Sorensen

Volumetric Decomposition via Medial Object and
Pen-Based User Interface for Hexahedral Mesh
Generation 179
Jean Hsiang-Chun Lu, Inho Song, William Roshan Quadros, Kenji Shimada

TET, HYBRID, and POLYHEDRAL MESHING
(Tuesday Morning Parallel Session)

Automatic Decomposition and Efficient Semi-structured
Meshing of Complex Solids 199
Jonathan E. Makem, Cecil G. Armstrong, Trevor T. Robinson

The Cutting Pattern Problem for Tetrahedral Mesh
Generation 217
Xiaotian Yin, Wei Han, Xianfeng Gu, Shing-Tung Yau

Parametrization of Generalized Primal-Dual
Triangulations 237
Pooran Memari, Patrick Mullen, Mathieu Desbrun

Geometrical Validity of Curvilinear Finite Elements 255
Amaury Johnen, Jean-François Remacle, Christophe Geuzaine

Uniform Random Voronoi Meshes 273
Mohamed S. Ebeida, Scott A. Mitchell

MESH OPTIMIZATION
(Tuesday Morning Parallel Session)

A Comparison of Mesh Morphing Methods for $3D$ Shape
Optimization 293
Matthew L. Staten, Steven J. Owen, Suzanne M. Shontz, Andrew G. Salinger, Todd S. Coffey

Families of Meshes Minimizing P_1 Interpolation Error 313
A. Agouzal, K. Lipnikov, Y. Vassilevski

A Log-Barrier Method for Mesh Quality Improvement 329
Shankar P. Sastry, Suzanne M. Shontz, Stephen A. Vavasis

A Novel Geometric Flow-Driven Approach for Quality Improvement of Segmented Tetrahedral Meshes 347
Juelin Leng, Yongjie Zhang, Guoliang Xu

Defining Quality Measures for High-Order Planar Triangles and Curved Mesh Generation................................. 365
Xevi Roca, Abel Gargallo-Peiró, Josep Sarrate

CAD REPAIR and DECOMPOSITION
(Tuesday Afternoon Parallel Session)

Hybrid Approach for Repair of Geometry with Complex Topology 387
Amitesh Kumar, Alan M. Shih

A Surface-Wrapping Algorithm with Hole Detection Based on the Heat Diffusion Equation 405
Franjo Juretić, Norbert Putz

Mesh and CAD Repair Based on Parametrizations with Radial Basis Functions....................................... 419
Cécile Piret, Jean-François Remacle, Emilie Marchandise

Automated Two-Dimensional Multi-block Meshing Using the Medial Object 437
Jeremy Gould, David Martineau, Robin Fairey

QUAD MESHING (Tuesday Afternoon Parallel Session)

A Frontal Delaunay Quad Mesh Generator Using the L^∞ Norm 455
J.-F. Remacle, F. Henrotte, T. Carrier Baudouin, C. Geuzaine, E. Béchet, Thibaud Mouton, E. Marchandise

L_p Lloyd's Energy Minimization for Quadrilateral Surface Mesh Generation 473
Tristan Carrier Baudouin, Jean-François Remacle, Emilie Marchandise, Jonathan Lambrechts

CSALF-Q: A Bricolage Algorithm for Anisotropic Quad Mesh Generation 489
Nilanjan Mukherjee

Jaal: Engineering a High Quality All-Quadrilateral Mesh Generator 511
Chaman Singh Verma, Tim Tautges

CROSS-CUTTING TOPICS
(Wednesday Morning Parallel Session)

Design, Implementation, and Evaluation of the Surface_mesh Data Structure 533
Daniel Sieger, Mario Botsch

The Meccano Method for Isogeometric Solid Modeling 551
José María Escobar, José Manuel Cascón, Eduardo Rodríguez, Rafael Montenegro

A Volume Flattening Methodology for Geostatistical Properties Estimation 569
Mathieu Poudret, Chakib Bennis, Jean-François Rainaud, Houman Borouchaki

Zipper Layer Method 587
Ning Qin, Yibin Wang, Greg Carnie, Shahrokh Shahpar

Fitting Polynomial Surfaces to Triangular Meshes with Voronoi Squared Distance Minimization 601
Vincent Nivoliers, Dong-Ming Yan, Bruno Lévy

Dendritic Meshing: LA-UR 11-04075 619
Brian A. Jean, Rodney W. Douglass, Guy R. McNamara, Frank A. Ortega

Author Index 637

TRI MESHING (Monday Morning)

Dynamic Parallel 3D Delaunay Triangulation

Panagiotis Foteinos[1,2] and Nikos Chrisochoides[2]

[1] Department of Computer Science, College of William and Mary
pfot@cs.wm.edu
[2] Department of Computer Science, Old Dominion University
nikos@cs.odu.edu

Summary. Delaunay meshing is a popular technique for mesh generation. Usually, the mesh has to be refined so that certain fidelity and quality criteria are met. Delaunay refinement is achieved by dynamically inserting and removing points in/from a Delaunay triangulation. In this paper, we present a robust parallel algorithm for computing Delaunay triangulations in three dimensions. Our triangulator offers fully dynamic parallel insertions and removals of points and is thus suitable for mesh refinement. As far as we know, ours is the first method that parallelizes point removals, an operation that significantly slows refinement down. Our shared memory implementation makes use of a custom memory manager and light-weight locks which greatly reduce the communication and synchronization cost. We also employ a contention policy which is able to accelerate the execution times even in the presence of high number of rollbacks. Evaluation on synthetic and real data shows the effectiveness of our method on widely used multi-core SMPs.

Keywords: dynamic Delaunay triangulation, parallel, mesh generation.

1 Introduction

1.1 Motivation

Mesh generation is a fundamental step for finite element analysis or visualization. A popular meshing technique is the *Delaunay mesh generation*, since it is able to mesh domains bounded by polyhedral surfaces [10, 12, 26, 35], curved surfaces [19, 32], or non-manifold surfaces consisted of a single [8, 34] or even multiple materials [18, 33].

The *quality* and *fidelity* of the mesh elements affect the speed and accuracy of the subsequent finite element analysis. Fidelity measures how well the mesh boundary describes the surface of the object to be modelled and quality regards the shape of the elements. Poor fidelity or poor quality meshes undermine the stability of the numerical solvers.

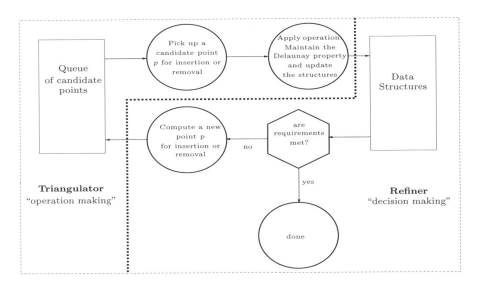

Fig. 1. The mesh refinement procedure illustrating the relationship between the refiner and the triangulator. The refiner computes the points that will potentially improve the mesh, while the triangulator is responsible for the actual operations, i.e, it maintains the Delaunay property and the data structures.

Therefore, after having computed an initial discretization of the input domain, meshers ought to refine it, until the specified quality and fidelity criteria are met. In the literature, refinement is achieved by incrementally inserting in or deleting points from an underlying *Delaunay triangulation* [8, 18, 19, 33, 34]. The triangulator, the backbone of any *Delaunay refiner* (see Figure 1), is responsible for updating the interconnectivity of mesh elements after the insertion or removal of points while maintaining the so called *Delaunay property*. The triangulator has to be able to support *dynamic* (i.e., *on line*) operations, simply because the sequence of the point insertions/removals (that improve the mesh) is not known a priori. In contrast, existing state of the art parallel Delaunay triangulators [6, 7] assume that the points are known before the parallel algorithm actually starts.

Figure 1 illustrates how the refiner and the triangulator cooperate with each other in a mesh refinement algorithm. Although this is the design we adopted for our own sequential mesh refinement algorithm [19], the high-level concept remains valid for all the other Delaunay mesh refinement algorithms in the literature. The key concept is that the refiner computes the point that will have to be inserted or removed, i.e., decides the sequence of the points that will improve the quality and the fidelity of the mesh, while the actual insertion or removal is performed by the triangulator. As stated above, that sequence of points is dynamically revealed by the refiner; decisions made in the past determine the ones that will be made in the future.

In [19], we developed a High Quality mesh Refinement algorithm (abbreviated to HQR hereafter) for medical images. Image-to-Mesh (I2M) conversion software [18, 19, 22, 33, 38] is essential for visualization or medical simulations. HQR at any time maintains the triangulation of points that have already been inserted inside, outside, or precisely on the boundary of the manifold (i.e., curved object) to be meshed. Therein, we prove that after the end of the refinement, a subset of the final triangulation forms a good topological and geometric approximation of the object and is of good quality. That subset (which is uniquely defined and extracted) constitutes the actual mesh of the object (this is a common technique used in the literature for meshing volumes bounded by surfaces [8, 18, 32, 33, 34]). Note that mesh generators for images need to also recover the surface of the object; the surface of the object is unknown and not given as a polyhedral domain. HQR meshes the surface and the volume of the object at the same time.

In this paper, towards the parallelization of our sequential I2M algorithm [19], we present a dynamic *Parallel Delaunay Triangulator* (PDT). Unlike all the other parallel implementations in the literature, we support Delaunay removal of points, an operation that has been shown to significantly improve the quality [26] and fidelity of the final meshes [19]. It will become obvious in the next sections that removing points is not only a slow operation (slower than insertions) but it is also a challenging task.

PDT follows the *master-workers* approach, because in this way we facilitate the integration of the triangulator with any refiner for manifold surfaces. In fact, the triangulator sees the refiner as a black box, which offers great flexibility since no assumptions for the location of the points to be inserted/removed are made. The integration of our parallel triangulator to a parallel refiner is the next step and is outside the scope of this paper.

1.2 Related Work

In the literature, there has been extensive work on developing parallel Delaunay methods in two and three dimensions. Their main limitation is the fact that they do not support removal of points, and therefore, they are not suitable for mesh refinement codes where deletions are needed (see the work by Klinger and Shewchuk [26] and Foteinos et al. [19] for few such refinement schemes).

Blelloch et al. [7], Hardwick [23], and Amato et al. [3] compute the Delaunay triangulation of a given point set by solving the corresponding convex hull problem using a parallel divide and conquer scheme. Teng et al. [36] construct the triangulation by expanding faces on every vertex in parallel. Cignoni et al. [14] present a divide and conquer and a construction parallel algorithm and compare their performance on uniformly distributed points. Blandford et al. [6] present a 3D incremental triangulator that first associates the uninserted vertices to the tetrahedra that contains them. Kohout et al. [27] give two incremental randomized insertion schemes (i.e., the points to be inserted have to be reordered) for computing the Delaunay triangulation in 2D. It is worth noting that all the techniques described so far are not dynamic, since the point set has to be known before the algorithm starts; therefore, they are not suitable for mesh refinement because the list of uninserted vertices constantly changes.

A dynamic parallel refinement algorithm for distributed memory platforms is given by Okusanya and Peraire [31]. The observed speed up, however, is very low: the execution time of 8 processes on uniformly distributed data is higher than that of 2. Dynamic parallel mesh refinement methods are also proposed by Chernikov and Chrisochoides [11]. Therein, the synchronization between processors is greatly reduced by choosing to insert points in parallel only if the insertions do not cause conflicts. Kadow [25] extends the work of Blelloch et al. [7] in order to support dynamic insertions in 2D. Galtier and George [20] propose a domain decomposition scheme. The domain is subdivided by partitioning the polyhedral surface mesh. That partitioning is performed by computing a suitable separator of the domain boundary. Linardakis and Chrisochoides [28] present a 2D decoupling parallel method. The communication between workers is eliminated by inserting extra points on the medial axis of the domain and meshing each resulting subdomain in parallel. Note that the dynamic refinement algorithms mentioned so far target exclusively polyhedral domains, that is, the surface of the object is already represented as a set of polyhedral facets. As already explained in Section 1.1, our parallel triangulator prepares the ground for a parallel guaranteed quality and fidelity mesh generator for objects whose surface and volumes are meshed at the same time, an essentially different problem.

Nave and Chrisochoides [30] parallelize the Bowyer-Watson kernel. The new inserted points, however, are restricted to be the circumcenters of poorly shaped triangles. That implies that *locating* the new point is not necessary, since the first element in the cavity (i.e., the poor element) is known and it needs not to be found. In the literature, meshing curved objects necessitates the insertion of points that are not circumcenters of poor elements. See for example the work in [8, 18, 32, 33, 34]. Therein, points on the voronoi edges are also inserted to guarantee that the mesh boundary is a good approximation of the object. As another example, our sequential refinement algorithm [19] sometimes inserts points outside the circumball of poor elements. For these reasons, our parallel triangulator makes no assumptions about the location of the points to be inserted, and therefore it supports (in fact, it has to support) *parallel locating* as well. Another difference is that Nave and Chrisochoides [30] start their parallel Bowyer-Watson kernel after the sequential construction of an initial mesh. Therefore, there is enough parallelism (i.e., no contention) in the early stages of the refinement. We follow a different approach. Our parallel Bowyer-Watson kernel tries to exploit parallelism in the early stages of the triangulation via the help of a contention manager policy. Furthermore, our implementation supports parallel removals.

1.3 Our Method

In this paper, we present a parallel dynamic Delaunay triangulation algorithm. Its main features include support for dynamic removal of points and ease of integration with any refinement schemes, especially refinement schemes that work directly on manifold surfaces. This makes our implementation suitable for mesh refinement schemes that (a) do not rely only on insertion of points for quality/fidelity improvement but also on Delaunay deletions [19, 26] and (b) decide to insert non-trivial points, i.e., points other than circumcenters or midpoints [18, 32, 33, 34]. Indeed, our parallel triangulator imposes no restrictions on the location of the points to be inserted or removed.

As stated above, we make use of a master-worker approach. More precisely, the triangulator, instead of applying the operation for each point in its global queue sequentially, launches the master thread. The master thread inspects the global queue and as long as there are points left, it moves them to the appropriate private queue of the workers. Each worker is responsible for only the points that are in its private queue. If a worker, during the insertion or removal of a point, encounters a locked vertex, then it aborts the operation (*rollback*) [7, 30] and tries to operate on another vertex in its private queue.

Rollbacks, however, may cause livelocks which result in system-wide starvation [24], especially when there is high contention. Kohout *et al.* [27] overlook this fact and that is the reason the resulting Delaunay triangulation is not always valid: they report that few elements are not Delaunay. Blandford *et al.* [6] deal with livelocks by bootstrapping: they insert the first 500,000 vertices sequentially and therefore the chances for high contention are minimized. As we have already mentioned, however, we are not given any point a priori, and therefore bootstrapping cannot be applied. Generally, dynamic triangulators cannot rely on any pre-processing strategy. We solve the starvation issue by employing a contention management policy. It guarantees correctness even in cases of extreme contention (i.e., when the number workers is very large with respect to the size of the triangulation), without compromising speed. On the contrary, the contention manager always yields faster execution times than these achieved by our sequential algorithm.

The rest of the paper is organized as follows: Section 2 describes our sequential implementation and Section 3 elaborates on the key aspects of our parallel implementation. Section 4 evaluates our parallel code on synthetic and real data, and Section 5 concludes the paper.

2 Sequential Implementation

The Delaunay triangulation $\mathcal{D}(V)$ of a set of vertices $V \subset \mathcal{R}^3$ is a triangulation that satisfies the *Delaunay property*. More precisely, let $\mathcal{B}(t)$ denote the open circumscribing ball (a.k.a *circumball*) of tetrahedron t. Then, t belongs in $\mathcal{D}(V)$ if $\mathcal{B}(t)$ contains no vertex of V. The insertion of a new vertex $v \notin V$

Algorithm 1. The Bowyer-Watson insertion kernel.

1 **Algorithm:** Insert(V, $\mathcal{D}(V)$, v)

  Input : V is the current set of vertices,
  $\mathcal{D}(V)$ is the Delaunay triangulation of V,
  v is the new vertex.
  Output: The Delaunay triangulation of set $V \cup v$.

2 Compute $\mathcal{C}(v) = \{t \in \mathcal{D}(V) \mid v \in \mathcal{B}(t)\}$;
3 Delete all tetrahedra in $\mathcal{C}(v)$;
4 Compute $\partial \mathcal{C}(v)$, the set of triangles incident to a tetrahedron that belongs to $\mathcal{C}(v)$ and to a tetrahedron that does not belong to the cavity;
5 Connect v with all the vertices of $\partial \mathcal{C}(v)$;
6 $V = V \cup v$;

or the removal of an existing vertex $v \in V$ necessitates local transformations such that the Delaunay property is maintained and the triangulation is still valid (i.e., the tetrahedra form a partition of the convex hull of the vertices).

2.1 Sequential Insertion

Let v be the new vertex inserted into V. The triangulation is updated using the well known *Bowyer-Watson* kernel [9, 37]. See Algorithm 1 for an illustration. First, the cavity $\mathcal{C}(v)$ of v is computed. The cavity contains all the tetrahedra in $\mathcal{D}(V)$ whose circumball contains v. Clearly, the elements composing the cavity have to be deleted because they violate the Delaunay property. Let us denote with $\partial \mathcal{C}(v)$ the boundary of the the cavity. As noted in [9, 37], $\partial \mathcal{C}(v)$ is a convex polyhedron and therefore, every vertex on the cavity's boundary is visible from v. Hence, connecting the vertices of $\partial \mathcal{C}(v)$ to v constitutes a valid triangulation. It can also be shown that the new elements created in this way respect the Delaunay property [21].

Computing the cavity of v is trivial, as long as we know one tetrahedron in $\mathcal{D}(V)$ which actually contains v (or one tetrahedron that belongs to the cavity). Thus, we first have to traverse part of the triangulation to locate that tetrahedron before we proceed to cavity computation. For this reason, our algorithm implements the *visibility walk* as described in [16]: starting from an element, we perform orientation checks which will dictate the next element of the walk, until we find the tetrahedron the contains v. In order to launch the location from a good starting element, we implement the *jump and walk* technique as described in [29]. Specifically, a small subset of the vertices in the triangulation is sampled, and the starting element is an element incident to the sample which is closest to v. Although, the jump and walk technique is slower than more elaborate schemes [15], it is an ideal candidate because it achieves fairly good complexity and its parallelization introduces no global synchronization.

Algorithm 2. The removal kernel.

1 **Algorithm:** Remove(V, $\mathcal{D}(V)$, v)

   Input : V is the current set of vertices,
               $\mathcal{D}(V)$ is the Delaunay triangulation of V,
               v is the vertex to be removed.
   Output: The Delaunay triangulation of set $V - v$.

2 Compute $\mathcal{H}(v) = \{t \in \mathcal{D}(V) \mid t \text{ is incident to } v\}$;
3 Delete all tetrahedra in $\mathcal{H}(v)$;
4 Compute $\partial\mathcal{H}(v)$, the set of triangles incident to a tetrahedron that belongs to $\mathcal{H}(v)$ and to a tetrahedron that does not belong to the hole;
5 Compute the small triangulation of the vertices of $\partial\mathcal{H}(v)$;
6 Merge the small triangulation with $\mathcal{D}(V)$;
7 $V = V - v$;

It is worth noting that the Bowyer-Watson kernel never creates flat tetrahedra (which must not exist in legal triangulations). And this is due to the fact that v cannot be coplanar with any of the facets of $\partial\mathcal{C}(v)$.

2.2 Sequential Removal

Removing a vertex $v \in V$ from $\mathcal{D}(V)$ involves re-triangulating the hole $\mathcal{H}(v)$ created by the tetrahedra incident to v. See Algorithm 2. First, we compute the Delaunay triangulation of the vertices on the boundary $\partial\mathcal{H}(v)$ of the hole. We shall refer to that triangulation as "small" triangulation to distinguish it from $\mathcal{D}(V)$ (the "big" triangulation). Then, we sew the small triangulation back to $\mathcal{D}(V)$.

Extra care has to be taken, however, to ensure that the small and big triangulations actually match. The problem is that $\partial\mathcal{H}(v)$ may not appear as a set of facets in the small triangulation due to degenerate cases. As explained in [17], if there are more than 3 cospherical and coplanar vertices in $\partial\mathcal{H}(v)$, then their triangulation in the plane is not unique, and therefore, parts of $\partial\mathcal{H}(v)$ might fail to appear in the small triangulation. Hence, we need a way to resolve ties, so that the small triangulation always matches the boundary of the hole. We do so by keeping track of the order in which the vertices were inserted in $\mathcal{D}(V)$. We use that information when triangulate the vertices of $\partial\mathcal{H}(v)$: they are inserted into the small triangulation according to the order they were inserted earlier in $\mathcal{D}(V)$. In this way, we can guarantee that sewing always gives valid triangulations. The proof of correctness is omitted in this paper.

Removals are more expensive than insertions. During the evaluation of our sequential triangulator, we observed that the removal of 3000 uniformly distributed points is about 6 to 7 times slower than their insertion. See Table 1 for a comparison. (Removals are 6 to 7 times slower than insertions in the CGAL triangulator [2] as well, the fastest dynamic triangulator we are aware of.) This result is counter-intuitive because insertions involve visibility walks, while removals do not. Indeed, each inserted point v stores a pointer to an incident element. Therefore, if v is removed, the hole can be found without triangulation traversals. Although removals do not require triangulations walks, the cost associated with memory management increases. And the reason is that the small triangulation does not contain and maintain only the elements needed to fill the hole. If the hole $\mathcal{H}(v)$ is not convex, then all the elements of the small triangulation that are outside $\mathcal{H}(v)$ but inside the convex hull of the vertices of $\partial\mathcal{H}(v)$ will never be a part of the big triangulation. Therefore, more bookeeping is introduced for maintaining and sewing the small triangulation.

3 Parallel Implementation

Our parallel implementation makes use of the C++ *Boost threads* [1]. When a thread wants to lock an element, it does so by locking its 4 vertices. If the operation is a read-only operation, then the thread asks for a shared ownership. For example, locating the element that contains a vertex does not modify the triangulation. Therefore, multiple location operations might overlap without any blocking.

3.1 Master-Workers Scheme

Each worker is responsible for a specific region. It inserts or removes a vertex as long as it lies inside its region. A worker either inserts or removes vertices. In the former case, it is referred to as an *inserter*, and in the latter as a *remover*. The master thread keeps scanning the global queue and moves vertex v (if any) to a worker's private queue (implemented as a thread safe single linked list) only if v lies inside the worker's region. In order to assign each worker a region, we assume that we are given the positions of the extreme points that will come in the future. This is a reasonable assumption, since in most cases the domain of interest is known. Then, the domain is logically divided into 3D structured blocks, each of which is assigned to a specific worker. Note that there is no global synchronization for the private queues. The master thread places new vertices at the head of the queues and the workers draw them from the tail of their queues, locking each time 2 exactly queue nodes and not all the queue.

3.2 Parallel Operations

When a cavity $\mathcal{C}(v)$ is explored by a worker W_1, all of the cavity's elements are locked exclusively. If another worker W_2 visits any of $\mathcal{C}(v)$'s elements, then W_2 aborts the operation (*rollback*) and moves on to the next point in its private queue [7, 30].

When a worker attempts to remove a vertex v, it acquires an exclusive lock on the elements of the hole $\mathcal{H}(v)$. Similarly to insertions, if another worker happens to be on an element of $\mathcal{H}(v)$, it aborts and tries to operate on another point in its private queue. Also, note that a remover might try to remove a vertex which belongs to an inserter's sample list (recall that the jump and walk location technique implemented by the inserters requires some processing of a small subset of already inserted vertices). Therefore, we require that the inserters exclusively lock their samples. If an inserter finds a sample exclusively locked (by either an another inserter or a remover), then it tries to find another one to start the locating from.

In order to decrease the synchronization (as a result of locking) and communication cost (as a result of reading and writing the shared memory) among the threads, we developed our own custom memory manager and light-weight locks. According to the findings presented by Antonopoulos et al. [4], efficient memory utilization and locking greatly reduces the overhead of maintaining multiple threads.

- *Memory manager*: Allocating and deallocating cells and vertices in multi-threaded implementation is costly, because now the kernel calls a thread safe implementation of the *new/delete* operator. This overhead can be reduced if each thread has its own (private) memory pool from which it asks blocks. If a block is to be deleted, then it throws it back to its memory pool. When the pool does not contain any blocks, only then the thread expands its pool by calling the kernel's *new* operator. In order to further expedite things and exploit localization, when the pool needs to be expanded, a whole chunk of blocks is allocated. We set the chunk size equal to 12 pages for cells and 3 pages for vertices (each page is 4KB). Higher chunk sizes yielded similar results.
- *Light-weight locks*: Locking mechanisms using POSIX mutexes can waste hundreds of cycles. On the contrary, the built-in atomic operations implemented by the C++ GNU compiler just add few stall cycles. Therefore, we decided to implement our own shared and exclusive try-locks using the atomic *fetch_and_add* operation. More precisely, each vertex has a specific flag which we atomically increment or decrement. That flag denotes the number of readers (associated with the vertex) if the value is non-negative. A non-negative flag implements a shared lock. If the flag is negative, then it is exclusively owned by one and only one worker. By replacing the pthread mutexes (that we used in the early stages of the development of our parallel code) with our light-weight locks, the removals sped up significantly. For example, the single-threaded removal of 10,000 random points was faster by 24%. Interestingly, the insertions were not substantially affected. We think that this happens because the number of cells incident to a point (to be removed) tends to be higher than the number of cells of a point's cavity (to be inserted); hence, removing involves more locking and it is more sensitive to the locking mechanism

employed. In uniform data, for example, we counted that on the average the insertion of a point requires 21 locks, while its removal 30 locks.

3.3 Contention Management

When the number of workers is high and the number of inserted vertices is small, increased contention is introduced which hampers the performance of the triangulator. What is even worse, the workers will most likely suffer from *livelocks*, a common pitfall of *non-blocking parallel* algorithms [24]. Livelocks are caused by continuous rollbacks: workers try to lock overlapping sets of vertices for an undefined period of time. Kohout *et al.* [27] overlook that problem and, for that reason, invalid Delaunay elements are reported. Blandford *et al.* [6] solve the problem by bootstrapping: they insert the first 500,000 vertices sequentially. We believe that even in the beginning (i.e., when there are not many vertices inserted) there is parallelism that can be exploited. Therefore, instead of bootstrapping, we decided to follow a more dynamic (yet simple to implement and with little overhead; see Section 4.1) contention policy. This is also one of the differences with the previous work of Nave and Chrisochoides [30]: therein, an initial mesh is first constructed sequentially and then the work is distributed among the workers.

We experimentally found that livelocks are always present in our algorithm when the number of vertices already inserted into the triangulation is less than $750 \times N$, where N is the number of parallel workers. And that fact did not only cause specific threads to starve but also it prevented system-wide throughput: none of the threads did useful work for a long (and thus undefined) time. To solve the problem, each worker W_i keeps track of its progress by calculating its *progress ratio* $u_i = \frac{C_i}{A_i}$, where C_i is the number of completed operations (i.e. amount of useful work) and A_i is the number of attempted operations. If A_i is large compared to C_i, then that means that the worker spends most of its time rolling back. When u_i drops below a specified threshold u^-, then W_i goes to sleep, releasing all its resources. A low ratio implies that W_i finds it difficult to cooperate with other threads, and therefore, it becomes inactive in order to help the other workers do useful work. Conversely, if u_i exceeds a specified threshold u^+ ($\geq u^-$), then it signals a sleeping worker, say W_j (if any). A high ratio implies that W_i does not conflict with other threads often, which indicates that a inactive worker might be able to do some useful work now. The awaken worker W_j can now inspect its private queue to find a point to work on. (Note that when a worker W_j awakes, its counters C_j and A_i are reset to 0, clearing in this way its progress history. We find no good reason why the contention manager should not be memoryless.) In cases of high contention, only one worker will be active, simulating a sequential algorithm.

Although the contention manager's primary goal is to insure correctness (that is, absence of livelocks), we observed that it speeds up the triangulator in its very early stages, i.e., when the triangulation does not contain

Table 1. Contention management on extreme cases

		Insertions		Removals	
#Threads		1	12	1	12
Time (secs)		0.06	0.04	0.39	0.12
Speedup		1	1.5	1	3.3
Max $\frac{\text{\#Rollbacks}}{\text{\#completed operations}}$ (%)		0	80	0	3500

many vertices. Table 1 shows the speed up achieved by 12 inserters and 12 removers. For that experiment, only 3000 points are about to be inserted into and removed from the triangulation, and therefore, we push our algorithm to extremes. The points are normally distributed. (Note that without the contention manager, the presence of livelocks prevented the insertion of any vertex for more than 1 hour.) The thresholds u^- and u^+ were set to 0.7 and 0.9 respectively. Different configurations of the thresholds yielded the same behavior. We observed similar results on different distributions (e.g., uniform distribution, line distribution, points on a box, and grid points) too. We can see from the last row that the contention is very intense. For example, a value of 3500% for $\frac{\text{\#Rollbacks}}{\text{\#completed operations}}$ means that a remover had to roll back 35 times on the average for every single point it removed. Despite that fact, the parallel implementation not only did not encounter any livelocks but also, it yielded faster executions.

4 Experimental Evaluation

In this section, we evaluate the performance of our Parallel Delaunay Triangulator (PDT) on both synthetic and real data. Throughout the evaluation, we used an Intel and an AMD machine. The Intel machine is equipped with a 12-core Xeon X5690 CPU at 3.47GHz and 96GB of memory, and the AMD machine with a 48-core Opteron 6174 CPU at 2.2GHz and 96GB of memory. Both the sequential and the parallel code is written in C++.

4.1 Synthetic Data

We first evaluated our parallel implementation on synthetic data. More precisely, we dynamically feed the global queue with points to be inserted or removed according to three distributions: uniform, normal, and line as described in [7]. For all distributions, we simulate the dynamic insertion of 12M points and the dynamic removal of the first 1.2M points (which account for 10% of the inserted points).

Table 2, Table 3, and Table 4 illustrate the results. The experiments were run on the Intel machine. The reported Speedup1 is the speedup with respect to our single-threaded parallel implementation. The $\frac{\text{\#Rollbacks}}{\text{\#compl. ops}}$ row shows

Table 2. Uniform

#Threads	Insertions					Removals				
	1	2	4	8	12	1	2	4	8	12
Time (secs)	743	355	186	89	61	167	85	46	24	17
Speedup1	1.00	2.09	3.99	8.35	12.18	1.00	1.96	3.63	6.96	9.82
Max $\frac{\#\text{Rollbacks}}{\#\text{compl. ops}}$ (%)	0.000	0.001	0.008	0.013	0.032	0.000	0.000	0.000	0.000	0.003
CGAL time (secs)	402	-	-	-	-	131	-	-	-	-
Speedup2	0.54	1.13	2.16	4.52	6.59	0.78	1.54	2.85	5.46	7.71

Table 3. Normal

#Threads	Insertions					Removals				
	1	2	4	8	12	1	2	4	8	12
Time (secs)	739	356	185	90	87	166	85	46	24	24
Speedup1	1.00	2.08	3.99	8.21	8.49	1.00	1.95	3.61	6.92	6.92
Max $\frac{\#\text{Rollbacks}}{\#\text{compl. ops}}$ (%)	0.000	0.003	0.006	0.018	0.022	0.000	0.000	0.000	0.000	0.001
CGAL time (secs)	400	-	-	-	-	131	-	-	-	-
Speedup2	0.54	1.12	2.16	4.44	4.60	0.79	1.54	2.85	5.46	5.46

Table 4. Line

#Threads	Insertions					Removals				
	1	2	4	8	12	1	2	4	8	12
Time (secs)	1,182	507	240	110	72	169	86	47	25	17
Speedup1	1.00	2.33	4.93	10.75	16.42	1.00	1.97	3.60	6.76	9.94
Max $\frac{\#\text{Rollbacks}}{\#\text{compl. ops}}$ (%)	0.000	0.001	0.035	0.097	0.252	0.000	0.000	0.006	0.030	2.641
CGAL time (secs)	612	-	-	-	-	133	-	-	-	-
Speedup2	0.52	1.21	2.55	5.56	8.50	0.79	1.55	2.83	5.32	7.82

the number of rollbacks with respect to the number of points that were inserted/removed (see Section 3.3). High values imply that the worker spent most of its time rolling back instead of doing useful work (i.e., instead of actually completing an operation). In fact, that row reports the higher value among the workers.

First of all, notice that the penalty introduced by the contention manager is negligible. In most cases, the maximum percentage of rollbacks with respect to the total number of inserted/removed points is less than 0.1%. Also, the total number of seconds that the workers sleep (not shown in the tables) were less than 3 seconds for all the experiments. Therefore, it is safe to look at the $\frac{\#\text{Rollbacks}}{\#\text{compl. ops}}$ to determine whether the synchronization cost is high or not.

For the uniform (Table 2) and line distribution (Table 4), we observe excellent speedups. With 12 workers, the uniform insertions are more than 12 times faster than the single-threaded execution; also, to our surprise, the line insertions are more than 16 times faster. The reason for such superlinear improvements is memory locality and memory reuse achieved by our custom memory manager. For the normal distribution (Table 3), we observe superlinear speedups up to 8 threads. Increasing the number of threads do not result in considerably less execution, although the number of rollbacks is still very small. This behavior is attributable to load imbalance. In fact, only 9 out of the 12 launched workers have work to do (the same applies for the parallel removals too). Load balancing ought to be considered on its own merit (see for example the work by Barker and Chrisochoides [5]), and it is left as future work.

For the parallel removals, notice that the speedups are smaller than those achieved by the inserters. For example, the improvement with 12 uniform removers is 9.82, while the improvement with 12 inserters is superlinear. We believe that this happens because removals are more memory intensive than insertions. During a removal the number of cells to be updated is higher than in the case of an insertion (more on this shortly). Therefore, memory management overhead per removal is higher than that per insertion. When the number of removers increases, that communication cost hampers the speedup of the inserters.

The increased number of cells to be updated during a removal is not only due to the fact that the cells in a cavity of a point to be inserted are fewer than the cells incident to a point to be removed, as reported in Section 3.2. It is also due to the fact that the hole has to be retriangulated which introduces extra memory overhead. For example, during the single threaded experiment of uniform data, we counted that on the average 211 cells are touched per removal, but only 19 cells are touched per insertion.

For comparison, the last row (Speedup2) of the tables above report the speedup of our parallel algorithm over the *Computational Geometry Algorithms Library* (CGAL) [2]. To our knowledge, CGAL triangulator is the fastest sequential dynamic algorithm and it is also open-source software. Note that CGAL triangulator has been highly optimized and tested over the last years and it is the state of the art implementation to date.

Observe that although the time of our single-threaded removal of points is comparable to that of CGAL, our single-threaded insertion is a little bit less than 2 times slower. The reason for such a difference is not only the extra cost associated with locking the elements in the cavity, but also the fact that the jump and walk location technique we employed is asymptotically slower than the optimal *Delaunay hierarchy* used by CGAL. (The worst-case cost of the jump and walk technique is $O(n^{\frac{2}{3}})$, while the cost of the hierarchy is $O(\log n)$.) As explained in Section 2, we used the jump and walk algorithm because it can be parallelized without extra synchronization cost, has a good expected complexity, and requires little memory. Despite the slower single-threaded execution times of our algorithm, it soon outperforms CGAL. The uniform insertions and removals by 12 inserters and 12 removers are 6.59 and 7.71 times faster, respectively. For the line insertions and removals, the speed up over CGAL is 8.5 and 7.82. Lastly, the normal insertions and removals by 8 inserters and 8 removers are 4.6 and 5.46 times faster, respectively.

The last experiment on synthetic data is performed on the AMD machine. In this experiment, we wanted to see the speedup of our algorithm when a larger number of cores is used. We insert 1M uniformly distributed points per

Table 5. Results for the AMD machine

#Threads	Insertions							Removals						
	1	2	4	8	16	32	48	1	2	4	8	16	32	48
Time (secs)	69	86	107	105	128	181	217	24	27	32	33	34	34	35
Max $\frac{\#\text{Rollbacks}}{\#\text{compl. ops}}$ (%)	0.00	0.00	0.01	0.03	0.09	0.20	0.20	0.00	0.00	0.00	0.00	0.00	0.00	0.00
CGAL time(secs)	40	87	194	418	898	1873	2869	16	37	66	132	264	526	800
Speedup	0.58	1.01	1.81	3.98	7.02	10.35	13.22	0.67	1.37	2.06	4.00	7.76	15.47	22.86

inserter and remove 0.1M points per remover. Table 5 depicts the results. The last row shows the speedup of our implementation over CGAL on the same input. Notice that the number of rollbacks is negligible, and therefore synchronization cost is not a problem. However, we observe that after 8 workers both inserters and removers stop scaling well. For example, with 48 inserters the speedup over CGAL is 13.22, while if the scaling was perfect it should be $48 \times 0.58 = 27.84$. Recall that we increase the problem size per worker and therefore one would expect a good scaling, since the number of rollbacks is negligible. Note, however, that the application is memory intensive and that when we increase the problem size, the memory management overhead also increases (the memory reads/writes with 48 workers is at least 48 times more than the memory reads/writes with 1 worker). For these reasons, we believe that after a certain number of workers, the communication cost dominates and the scaling is suboptimal.

4.2 Real Data

We also simulated the sequence of points produced by our high quality sequential I2M algorithm HQR (see Section 1 for some information on HQR) [19]. The quality and fidelity guarantees that HQR proves can be found in [19]. Note that in [19], we developed only our own refiner; the (sequential) triangulator used there was built on top of a third party library (CGAL [2]). We ran HQR on a segmented medical image and we traced the sequence of points that were inserted and removed. Then, we fed that data to our parallel triangulator to simulate a real case. Figure 2 shows the data produced by HQR.

Table 6 depicts the timings. During the simulation, 219,031 points were inserted and 63,356 were removed in parallel. Although the insertions scale well, the removals hit a wall after 4 workers. This is expected because there is a little concurrency to be exploit, i.e., the number of points to be removed per remover is low. (Note that load imbalance is not a problem in this simulation, since we verified that each worker attempted to make roughly the same number of operations, that is, the number of points to be removed is approximately the same for all removers.) This also agrees with the huge rollback percentages: for example, when 12 removers are launched, we noticed that a thread had to try 1,069,400 times just to remove 5,319 points. Clearly, the computation for such a little work of removals is not enough to compensate for the synchronization cost.

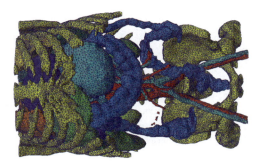

Fig. 2. The result after the termination of our mesh refinement algorithm. This sequence of points that were inserted/removed was fed into our parallel triangulator.

Table 6. Evaluation on real data. Notice the high rollback percentages associated with the removals.

		Insertions					Removals				
#Threads		1	2	4	8	12	1	2	4	8	12
Time (secs)		7.3	4.2	2.5	1.4	0.9	32.0	16.7	10.8	7.1	11.6
Speedup1		1.00	1.74	2.92	5.21	8.11	1.00	1.92	2.96	4.51	2.76
Max $\frac{\#\text{Rollbacks}}{\#\text{compl. ops}}$ (%)		0.0	0.2	1.0	4.7	6.1	0.0	0.1	16.8	42.9	19837.0
CGAL time (secs)		3.9	-	-	-	-	16	-	-	-	-
Speedup2		0.53	0.93	1.56	2.79	4.33	0.50	0.96	1.48	2.25	1.38

5 Conclusions and Future Work

In this paper, we presented a dynamic parallel Delaunay triangulator. Its main feature is its ability to support parallel removal of points, an operation that is much slower than the insertion as we have already explained (Section 2.2). For synthetic data, the execution time of our parallel insertions on the 12-core Intel machine is $4.6 - 8.5$ times faster than the fastest sequential triangulator (CGAL [2]) we are aware of. The corresponding speedup over CGAL for our parallel removals is 5.46 to 7.82. The overall speedup (taking into account both insertions and removals) over CGAL ranges from 4.78 to 8.37 when 12 workers are launched. Removals do not scale as well as insertions, but the reason is not the synchronization cost. In fact, our lightweight lock implementation (see Section 3.2) greatly reduced the time spent for locking by 24%. Indeed, as Table 2, Table 3, and Table 4 show, the time spent on rollbacks is negligible when compared with the time of useful work. We noticed, however, that removals are 11 times more memory intensive than insertions (see Section 4.1). Therefore, removals exhibit higher communication cost which limits scalability. We believe that communication cost is also the reason that our algorithm stop scaling well after 8 workers on the 48-core AMD machine (see Section 4.1).

A case of increased synchronization cost is shown in Table 6. The high $\frac{\#\text{Rollbacks}}{\#\text{compl. ops}}$ numbers indicate exactly that. For example, the value 19837%

implies that a thread had to roll back 198.37 times on average for every point it tried to remove. As explained in Section 4.2, the reason for so much contention is because of the little concurrency that can be exploit. This experiment shows that our contention manager (see Section 3.3) is able to remove livelocks under extreme circumstances.

Our next goal is to combine our parallel triangulator with a parallel refiner for medical images and exploit parallelism in two levels: in the level of the triangulation (operation making) and in the level of the refiner (decision making, see Figure 1). To ease the integration, the triangulator employs a master-workers scheme and sees the refiner (i.e., the master or the masters if the refiner is parallel) as a black box. Note that if the parallel refiners are synchronized, they will never attempt to improve an element that was previously removed.

The experimental evaluation presented in this paper focuses on shared-memory multi-core machines. We are also planning to parallelize our algorithm for larger distributed-memory machines according to the work of Chrisochoides et al. [13]. Therein, a hybrid mesh generation framework is developed that takes advantage of both the smaller shared memory layer and the larger distributed memory layer.

Acknowledgements

The authors would like to thank Andrey Chernikov and Andriy Kot for the fruitful discussions and constructive comments and the anonymous reviewers who helped us improve the paper. This work is supported in part by NSF grants: CCF-1139864, CCF-1136538, and CSI-1136536 and by the John Simon Guggenheim Foundation and the Richard T. Cheng Endowment.

References

1. Boost C++ libraries, http://www.boost.org/
2. Cgal, Computational Geometry Algorithms Library, http://www.cgal.org
3. Amato, N.M., Goodrich, M.T., Ramos, E.A.: Parallel algorithms for higher-dimensional convex hulls. In: IEEE Symposium on Foundations of Computer Science, November 1994, pp. 683–694 (1994)
4. Antonopoulos, C., Blagojevic, F., Chernikov, A., Chrisochoides, N., Nikolopoulos, D.: Algorithm, software, and hardware optimizations for delaunay mesh generation on simultaneous multithreaded architectures. Journal on Parallel and Distributed Computing 69(7), 601–612 (2009)
5. Barker, K., Chrisochoides, N.: Practical performance model for optimizing dynamic load balancing of adaptive applications. In: IEEE International Parallel and Distributed Processing Symposium (IPDPS). IEEE Computer Society Press (2005)
6. Blandford, D.K., Blelloch, G.E., Kadow, C.: Engineering a compact parallel delaunay algorithm in 3d. In: Proceedings of the 22nd Symposium on Computational Geometry, SCG 2006, pp. 292–300. ACM, New York (2006)

7. Blelloch, G.E., Miller, G.L., Hardwick, J.C., Talmor, D.: Design and implementation of a practical parallel delaunay algorithm. Algorithmica 24(3), 243–269 (1999)
8. Boissonnat, J.-D., Oudot, S.: Provably good sampling and meshing of surfaces. Graphical Models 67(5), 405–451 (2005)
9. Bowyer, A.: Computing Dirichlet tesselations. Computer Journal 24, 162–166 (1981)
10. Chernikov, A., Chrisochoides, N.: Three-Dimensional Semi-Generalized Point Placement Method for Delaunay Mesh Refinement. In: Proceedings of the 16th International Meshing Roundtable, Seattle, WA, October 2007, pp. 25–44. Elsevier, Amsterdam (2001)
11. Chernikov, A., Chrisochoides, N.: Three-dimensional delaunay refinement for multi-core processors. In: ACM International Conference on Supercomputing, Island of Kos, Greece, June 2008, vol. 22, pp. 214–224 (2008)
12. Paul Chew, L.: Guaranteed-quality Delaunay meshing in 3D. In: Proceedings of the 13th ACM Symposium on Computational Geometry, Nice, France, pp. 391–393 (1997)
13. Chrisochoides, N., Chernikov, A., Fedorov, A., Kot, A., Linardakis, L., Foteinos, P.: Towards exascale parallel delaunay mesh generation. In: International Meshing Roundtable, Salt Lake City, Utah, October 2009, vol. 18, pp. 319–336 (2009)
14. Cignoni, P., Montani, C., Perego, R., Scopigno, R.: Parallel 3d delaunay triangulation. Computer Graphics Forum 12(3), 129–142 (1993)
15. Devillers, O.: The delaunay hierarchy. Internat. J. Found. Comput. Sci. 13, 163–180 (2002)
16. Devillers, O., Pion, S., Teillaud, M.: Walking in a triangulation. In: Proceedings of the 17th Annual Symposium on Computational Geometry, SoCG 2001, pp. 106–114. ACM, New York (2001)
17. Devillers, O., Teillaud, M.: Perturbations and vertex removal in a 3d delaunay triangulation. In: Proceedings of the 14th ACM-SIAM Symposium on Discrete algorithms, SODA 2003, Philadelphia, PA, USA, pp. 313–319. Society for Industrial and Applied Mathematics (2003)
18. Boltcheva, D., Yvinec, M., Boissonnat, J.-D.: Mesh Generation from 3D Multimaterial Images. In: Yang, G.-Z., Hawkes, D., Rueckert, D., Noble, A., Taylor, C. (eds.) MICCAI 2009. LNCS, vol. 5762, pp. 283–290. Springer, Heidelberg (2009)
19. Foteinos, P., Chernikov, A., Chrisochoides, N.: Guaranteed Quality Tetrahedral Delaunay Meshing for Medical Images. In: Proceedings of the 7th International Symposium on Voronoi Diagrams in Science and Engineering, Quebec City, Canada, June 2010, pp. 215–223 (2010)
20. Galtier, J., George, P.-L.: Prepartitioning as a way to mesh subdomains in parallel. Special Symposium on Trends in Unstructured Mesh Generation, pp. 107–122. ASME/ASCE/SES (1997)
21. George, P.-L., Borouchaki, H.: Delaunay triangulation and meshin. In: Application to finite elements. HERMES (1998)
22. Goksel, O., Salcudean, S.E.: Image-based variational meshing. IEEE Transactions on Medical Imaging 30(1), 11–21 (2011)
23. Hardwick, J.C.: Implementation and evaluation of an efficient parallel delaunay triangulation algorithm. In: Proceedings of the 9th ACM Symposium on Parallel Algorithms and Architectures, pp. 239–248. ACM, New York (1997)

24. Scherer III, W.N., Scott, M.L.: Advanced contention management for dynamic software transactional memory. In: Proceedings of the 24th Annual ACM Symposium on Principles of Distributed Computing, PODC 2005, pp. 240–248. ACM, New York (2005)
25. Kadow, C.M.J.: Parallel Delaunay Refinement Mesh Generation. PhD Thesis, Carnegie Mellon University (2004)
26. Klingner, B.M., Shewchuk, J.R.: Aggressive tetrahedral mesh improvement. In: Proceedings of the International Meshing Roundtable, pp. 3–23. Springer, Heidelberg (2007)
27. Kohout, J., Kolingerová, I., Žára, J.: Practically oriented parallel delaunay triangulation in E^2 for computers with shared memory. Computers & Graphics 28(5), 703–718 (2004)
28. Linardakis, L., Chrisochoides, N.: Graded delaunay decoupling method for parallel guaranteed quality planar mesh generation. SIAM Journal on Scientific Computing 30(4), 1875–1891 (2008)
29. Mücke, E.P., Saias, I., Zhu, B.: Fast randomized point location without preprocessing in two- and three-dimensional delaunay triangulations. In: Proceedings of the 12th ACM Symposium on Computational Geometry, pp. 274–283 (1996)
30. Nave, D., Chew, P., Chrisochoides, N.: Guaranteed quality parallel delaunay refinement for restricted polyhedral domains. In: ACM Symposium on Computational Geometry (SoCG), July 2002, pp. 135–144 (2002)
31. Okusanya, T., Peraire, J.: 3d parallel unstructured mesh generation (1997), http://citeseerx.ist.psu.edu/viewdoc/summary?doi=10.1.1.48.7898
32. Oudot, S., Rineau, L., Yvinec, M.: Meshing volumes bounded by smooth surfaces. In: Proceedings of the International Meshing Roundtable, San Diego, California, USA, September 2005, pp. 203–219. Springer, Heidelberg (2005)
33. Pons, J.-P., Ségonne, F., Boissonnat, J.-D., Rineau, L., Yvinec, M., Keriven, R.: High-Quality Consistent Meshing of Multi-label Datasets. In: Information Processing in Medical Imaging, pp. 198–210 (2007)
34. Rineau, L., Yvinec, M.: Meshing 3d domains bounded by piecewise smooth surfaces. In: Proceedings of the International Meshing Roundtable, pp. 443–460 (2007)
35. Shewchuk, J.R.: Tetrahedral mesh generation by delaunay refinement. In: Proceedings of the 14th ACM Symposium on Computational Geometry, Minneapolis, MN, pp. 86–95 (1998)
36. Ansel Teng, Y., Sullivan, F., Beichl, I., Puppo, E.: A data-parallel algorithm for three-dimensional delaunay triangulation and its implementation. In: ACM Conference on Supercomputing, pp. 112–121. ACM, New York (1993)
37. Watson, D.F.: Computing the n-dimensional Delaunay tesselation with application to Voronoi polytopes. Computer Journal 24, 167–172 (1981)
38. Zhang, Y., Hughes, T.J.R., Bajaj, C.L.: An automatic 3d mesh generation method for domains with multiple materials. Computer Methods in Applied Mechanics and Engineering 199(5-8), 405–415 (2010); Computational Geometry and Analysis

Quality Surface Meshing Using Discrete Parametrizations

Emilie Marchandise[1], Jean-François Remacle[1], and Christophe Geuzaine[2]

[1] Université catholique de Louvain, Institute of Mechanics, Materials and Civil Engineering (iMMC), Place du Levant 1, 1348 Louvain-la-Neuve, Belgium
emilie.marchandise@uclouvain.be, jean-francois.remacle@uclouvain.be
[2] Université de Liège, Department of Electrical Engineering and Computer Science, Montefiore Institute B28, Grande Traverse 10, 4000 Liège, Belgium
cgeuzaine@ulg.ac.be

Summary. We present 3 mapping/flattening techniques for triangulations of poor quality triangles. The implementation of those mappings as well as the boundary conditions are presented in a very comprehensive manner such that it becomes accessible to a wider community than the one of computer graphics. The resulting parameterizations are used to generate new triangulations or quadrilateral meshes for the model that are of high quality.

1 Introduction

There are two kind of applications for which it might be desirable to remesh a $3D$ surface (see Fig. 1).

The first application concerns medical geometries that are often described only by a triangulation (in stereolitography STL format). This triangulation is the result of a segmentation procedure from the CT scan or MRI dicom-images. Those triangulations can be oversampled and have triangles of poor quality with small elementary angles. Those low quality meshes are not suitable for finite element simulations since the quality of the mesh will impact both on the accuracy and efficiency of the numerical method [35, 2]. In this case, it is desirable to build a high quality mesh from those low quality meshes before performing any numerical simulation.

The other application is about CAD models. CAD models are often made of a huge amount of patches that have no physical significance and a straight-forward meshing of those patches often leads to meshes that are not suitable for finite element simulations. Indeed, as most surface mesh algorithms mesh model faces individually, mesh points are generated on the bounding edges of those patches and if thin CAD patches exist in the model they will result in the creation of small distorted triangles with very small angles (Fig.2). Those low quality elements present in the surface mesh will often hinder the convergence of the FE simulations on those surface meshes. Besides, they also

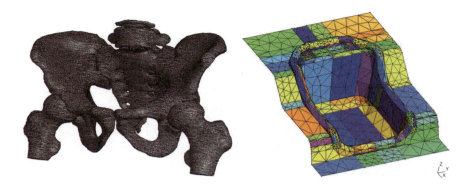

Fig. 1. Examples of geometries for which a remeshing procedure is desirable. Left figure shows an example of an oversampled STL triangulation resulting from a mesh segmentation of a human pelvis and right figure shows the straightforward meshing of a CAD geometry of a maxi-cosi.

prevent the generation of quality volumetric meshes for three-dimensinal finite element computations (CFD, structure mechanics, etc.). An efficient manner to build a high quality mesh for those CAD models is then to build from the initial CAD mesh a cross-patch parametrization that enables the remeshing of merged patches.

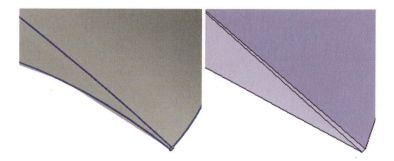

Fig. 2. Example of 2 patches of a CAD geometry (left) for which the mesh (right) contains a very small triangle of poor quality.

There are mainly two approaches for surface remeshing: mesh adaptation strategies [18, 3, 38] and parametrization techniques [6, 40, 26, 36, 19, 22]. Mesh adaptation strategies use local mesh modifications in order both to improve the quality of the input surface mesh and to adapt the mesh to a given mesh size criterion. In parametrization techniques, the input mesh serves as a support for building a continuous parametrization of the surface. (In the case of CAD geometries, the initial mesh can be created using any off the shelf surface mesher for meshing the individual patches.) Surface parametrization techniques originate mainly from the computer graphics community: they have been used extensively for applying textures onto surfaces [5, 23] and have become a very useful and efficient tool for many mesh processing applications [9, 15, 21, 33, 12]. In the context of remeshing procedures, the initial surface is parametrized onto a surface in $\mathcal{R}^2$, the surface is meshed using any standard planar mesh generation procedure and the new triangulation is then mapped back to the original surface [8, 27].

The existing methods for discrete parametrization can be classified as follows: linear, non-linear and hybrid methods. Non-linear methods based on discrete or differential-geometric non-linear distortion measure [17, 34, 41] offer strong guarantees on the absence of triangle folding and flipping at the cost of a generally higher computational effort. Some authors have also suggested hybrid techniques that linearize those non-linear measures at the cost of only a few linear solves [4, 39]. Linear methods require only the resolution of a single linear system. Most methods require to map the vertex of the patch boundary to a given polygon (usually convex) in the parametric plane. This is for example the case of the discrete harmonic map introduced by Eck [8] or the more robust convex combination map of Floater [9]. Some authors also suggested extensions to free boundaries by pinning down only two vertices. This is the case for example in the least square conformal maps (LSCM) introduced by Levy et al. [21] and the discrete conformal parametrizations (DCP) of Desbrun et al. [1]. These mappings could achieve lower angle distortion than previous results. However, as the quality of the parametrization can depend significantly on the choice of the constrained vertices, Mullen et al. [28] suggested to spread the constraints throughout the mesh by constraining that the barycenter of the mapping must be at $(0,0)$ and that the moment of inertia of the boundary must be unit. In [28], those spread constraints are taken into account through recourse to spectral theory.

In this paper, we present and compare three different types of linear harmonic maps for the discrete parametrization of triangulated surfaces. The implementation of those mappings as well as the boundary conditions are presented in a very comprehensive manner such that it becomes accessible to a wider community than the one of computer graphics. The discrete parametrization aims at computing the discrete mapping $\mathbf{u}(\mathbf{x})$ that maps every triangle of the three dimensional surface $\mathcal{S}$ to another triangle of $\mathcal{S}'$ that has a well known parametrization:

$$\mathbf{x} \in \mathcal{S}_\mathcal{T} \subset \mathcal{R}^3 \mapsto \mathbf{u}(\mathbf{x}) \in \mathcal{S}'_\mathcal{T} \subset \mathcal{R}^2 \tag{1}$$

We restrict ourselves to the parametrization of non-closed triangulated surfaces since we have already presented in previous papers [25, 24] efficient techniques to split a closed object into a series of different patches in the context of surface remeshing.

The overall procedure is implemented in the open-source mesh generator Gmsh [14].

2 Harmonic Energy Minimizations

A harmonic map minimizes distortion in the sense that it minimizes the Dirichlet energy of the mapping $\mathbf{u}(\mathbf{x})$:

$$E_D(\mathbf{u}) = \int_\mathsf{M} \frac{1}{2} |\nabla \mathbf{u}|^2 \, ds. \tag{2}$$

subject to Dirichlet boundary conditions $\mathbf{u} = \mathbf{u}_D$ on $\partial \mathsf{M}^i$. Harmonic maps are not in general conformal and do not preserve angles but they are popular since they are very easy to compute and are guaranteed to be one-to-one for convex regions [29, 7].

3 Convex Combination Map

In contrast to the continuous harmonic map (2), Floater et al. showed in [10, 11] that the discrete version of (2) is not always one-to-one. To ensure a discrete maximum principle, the authors introduced a convex combination. One particular type of convex combination maps is for example the barycentric mapping by Tutte [37] that asks every interior vertex $\mathbf{u}_i$ be the barycenter of its neighbors.

$$\mathbf{u}_i = \sum_{k=1}^{d_i} \lambda_k \mathbf{u}_j, \quad \sum_{k=1}^{d_i} \lambda_k = 1, \tag{3}$$

where d_i denotes the number of vertices that are neighbors to node i.

4 Least Square Conformal Map

The least square conformal map as introduced by Levy at al. [21] asks that the gradient if u and the gradient of v be as orthogonal as possible in the parametrization and have the same norm. This can bee seen as an approximation of the Cauchy-Riemann equations. For a piecewise linear mapping, the least square conformal map can be obtained by minimizing the energy:

$$E_{\mathrm{LSCM}}(\mathbf{u}) = \int_\mathsf{M} \frac{1}{2} \left| \nabla u^\perp - \nabla v \right|^2 ds, \tag{4}$$

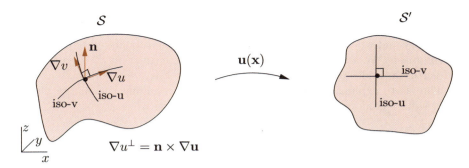

Fig. 3. Definitions for a conformal mapping. $\nabla u^\perp$ denotes the counterclockwise 90° rotation of the gradient ∇u for a 3D surface.

where $\perp$ denotes a counterclockwise 90° rotation in $\mathcal{S}$. For a 3D surface defined by a normal vector $\mathbf{n}$, the counterclockwise rotation of the gradient can be written as: $\nabla u^\perp = \mathbf{n} \times \nabla \mathbf{u}$ (see Fig.3).

Equation (4) can be simplified and rewritten as follows:

$$\begin{aligned} E_{\text{LSCM}}(\mathbf{u}) &= \int_M \frac{1}{2} \left(\nabla u^\perp \cdot \nabla u^\perp + \nabla v \cdot \nabla v - 2\nabla u^\perp \cdot \nabla v \right) ds \\ &= \int_M \frac{1}{2} \left(\nabla u \cdot \nabla u + \nabla v \cdot \nabla v - 2\left(\mathbf{n} \times \nabla u\right) \cdot \nabla v \right) ds. \end{aligned} \quad (5)$$

Recalling the idenity that a "dot" and a "cross" can be interchanged without changing the result, we have

$$E_{\text{LSCM}}(\mathbf{u}) = \int_\mathcal{S} \frac{1}{2} \left(\nabla u \cdot \nabla u + \nabla v \cdot \nabla v - 2\mathbf{n} \cdot (\nabla u \times \nabla v) \right) ds. \quad (6)$$

5 Discrete Harmonic Maps with Finite Elements

We now derive the finite element formulation of the quadratic minimisation problems (2)-(4). We denote by the functional J either the Derichlet energy E_D or the least-square conformal energy E_{LSCM} to be minimized:

$$\min_{\mathbf{u} \in U(\mathsf{M})} J(\mathbf{u}), \quad \text{with} \quad \mathbf{U}(\mathcal{S}) = \{\mathbf{u} \in H^1(\mathcal{S}), \mathbf{u} = \mathbf{u}_D(\mathbf{x}) \text{ on } \partial \mathsf{M}^i\}. \quad (7)$$

We assume the following finite expansions for $\mathbf{u}$

$$\mathbf{u}_h(\mathbf{x}) = \sum_{i \in I} \mathbf{u}_i \phi_i(\mathbf{x}) + \sum_{i \in J} \mathbf{u}_D(\mathbf{x}_i) \phi_i(\mathbf{x}) \quad (8)$$

where I denotes the set of nodes of M that do not belong to the Dirichlet boundary, where J denotes the set of nodes of M that belong to the Dirichlet boundary and where ϕ_i are the nodal shape functions associated to the nodes of the mesh. We assume here that the nodal shape function ϕ_i is equal to 1 on vertex $\mathbf{x}_i$ and 0 on any other vertex: $\phi_i(\mathbf{x}_j) = \delta_{ij}$.

Thanks to expansion Eq. (8), the functional J defining the energy of the least square conformal map Eq. (6) can be rewritten as

$$\begin{aligned}
J(\mathbf{u}_1,\ldots,\mathbf{u}_N) = & \frac{1}{2}\sum_{i\in I}\sum_{j\in I} u_i u_j \int_M \nabla\phi_i \cdot \nabla\phi_j\, ds + \sum_{i\in I}\sum_{j\in J} u_i u_D(\mathbf{x}_j) \int_M \nabla\phi_i \cdot \nabla\phi_j\, ds + \\
& \frac{1}{2}\sum_{i\in I}\sum_{j\in I} v_i v_j \int_M \nabla\phi_i \cdot \nabla\phi_j\, ds + \sum_{i\in I}\sum_{j\in J} v_i v_D(\mathbf{x}_j) \int_M \nabla\phi_i \cdot \nabla\phi_j\, ds + \\
& \sum_{i\in I}\sum_{j\in J} u_D(\mathbf{x}_i) u_D(\mathbf{x}_j) \int_M \nabla\phi_i \cdot \nabla\phi_j\, ds + \sum_{i\in I}\sum_{j\in J} v_D(\mathbf{x}_i) v_D(\mathbf{x}_j) \int_M \nabla\phi_i \cdot \nabla\phi_j\, ds - \\
& \sum_{i\in I}\sum_{j\in J} u_i v_j \int_M \mathbf{n} \cdot (\nabla\phi_i \times \nabla\phi_j)\, ds - \sum_{i\in I}\sum_{j\in J} u_D(\mathbf{x}_i) v_j \int_M \mathbf{n} \cdot (\nabla\phi_i \times \nabla\phi_j)\, ds - \\
& \sum_{i\in I}\sum_{j\in J} u_i v_D(\mathbf{x}_i) \int_M \mathbf{n} \cdot (\nabla\phi_i \times \nabla\phi_j)\, ds - \sum_{i\in I}\sum_{j\in J} u_D(\mathbf{x}_i) v_D(\mathbf{x}_i) \int_M \mathbf{n} \cdot (\nabla\phi_i \times \nabla\phi_j)\, ds.
\end{aligned} \tag{9}$$

In order to minimize J, we can simply cancel the derivative of J with respect to u_k

$$\begin{aligned}
\frac{\partial J}{\partial u_k} = & \sum_{j\in I} u_j \underbrace{\int_M \nabla\phi_k \cdot \nabla\phi_j\, ds}_{A_{kj}} + \sum_{j\in J} u_D(\mathbf{x}_j) \underbrace{\int_M \nabla\phi_k \cdot \nabla\phi_j\, ds}_{A_{kj}} - \\
& \sum_{j\in I} v_j \underbrace{\int_M \mathbf{n} \cdot (\nabla\phi_k \times \nabla\phi_j)\, ds}_{C_{kj}} - \sum_{j\in I} v_D(\mathbf{x}_j) \underbrace{\int_M \mathbf{n} \cdot (\nabla\phi_k \times \nabla\phi_j)\, ds}_{C_{kj}} \\
= & \, 0 \, , \quad \forall k \in I.
\end{aligned} \tag{10}$$

The same can be done for the derivative with respect to v_k.

There are as many equations (10) as there are nodes in I. This system of equations can be written as:

$$\begin{pmatrix} \bar{\bar{A}} & \bar{\bar{C}} \\ \bar{\bar{C}}^T & \bar{\bar{A}} \end{pmatrix} \begin{pmatrix} \bar{U} \\ \bar{V} \end{pmatrix} = \begin{pmatrix} \bar{0} \\ \bar{0} \end{pmatrix} \tag{11}$$

where $\bar{\bar{A}}$ is a symmetric positive definite matrix and $\bar{\bar{C}}$ is an antisymmetric matrix that are both build by assembling the elementary matrices A_{kj} and C_{kj}. Hence the resulting matrix in Eq. 11 is symmetric definite positive and efficient direct sparse symmetric-positive-definite solvers such as TAUCS can be used. The vectors $\bar{U}$ and $\bar{V}$ denote respectively the vector of unknowns u_k and v_k.

In the case of simple Laplacian harmonic maps the matrix C vanishes, which makes the system of equations (11) uncoupled:

$$\bar{\bar{A}}\bar{U} = \bar{0}, \quad \bar{\bar{A}}\bar{V} = \bar{0}. \tag{12}$$

Finally, in the case of a convex combination map the system of equations is also uncoupled as in (12), the matrix A being now the assembly of the following elementary matrices:

$$A_{ij} = \begin{pmatrix} 1 & -0.5 & -0.5 \\ -0.5 & 1 & -0.5 \\ -0.5 & -0.5 & 1 \end{pmatrix} \tag{13}$$

6 Boundary Conditions

It is necessary to impose appropriate boundary conditions to guarantee that the discrete minimization problem has a unique solution and that this unique solution defines a one-to-one mapping (and hence avoids the degenerate solution $\mathbf{u}$ =constant). Dirichlet boundary conditions are often used for the Laplacian harmonic map and the convex combination map to map the boundary nodes of $\partial \mathsf{M}_1$ to a unit circle:

$$u_D(\mathbf{x}_i) = \cos\left(\frac{2\pi l_i(\mathbf{x}_i)}{L}\right), \quad v_D(\mathbf{x}_i) = \sin\left(\frac{2\pi l_i(\mathbf{x}_i)}{L}\right). \tag{14}$$

We have decided to map to a unit circle but all kind of convex fixed boundaries could be considered since the mapping is proven to be one-to-one if the mapped surface is convex [29, 7].

Instead of fixing all the boundary nodes ∂S_1 to a convex polygon, one might fix two (u,v) coordinates, thus pinning down two vertices in the parameter plane with Dirichlet boundary conditions. Indeed, for least square conformal maps, the mapping (11) has full rank only when the number of pinned vertices is greater or equal to 2 [21]. Pinning down two vertices will set the translation, rotation and scale of the solution when solving the linear system $\mathbf{L_C U} = \mathbf{0}$ and will lead to what is called a free-boundary parametrization. It was independendty found by the authors of the LSCM [21] and the DCP [1] that picking two boundary vertices the farthest from each other seems to give good results in general. However, the quality of the conformal parametrization depend drastically on the choice of these constraint vertices. Indeed, global distortion can ensue and a degradation of conformality can be observed around the pinned vertices. Figure 4 compares a LSCM with two pinned vertices with a less distorted LSCM that spreads the constraints throughout the mesh (we call this approach the constrained LSCM or CLSCM)

How can we define a less distorted least square conformal map (CLSCM) without pinning down two vertices ? The idea is to add the two following

Fig. 4. Initial triangulation $\mathcal{S}_\mathcal{T}$ of a boudda statue a) that has been parametrized by computing b) the LSCM with two constrained vertices (shown in red) c) the constrained LSCM solved with a spectral method.

constraints to the minimization problem that that set the translation, rotation and scale of the solution: (i) the barycenter of the solution must be at zero and (ii) the moment of inertia of the boundary $\partial \mathcal{S}_{\mathcal{T}1}$ must be unit. Those constraints can be taken into account through recourse to spectral theory. This idea was derived also by Mullen et al. [28] and named after **spectral conformal parametrization**. In what follows, we try to present the spectral conformal map in a more comprehensive manner than the way it is presented in [28]. The constrained least square conformal map corresponds to the following discrete constrained minimization problem. Find $\mathbf{U}^*$ such that

$$\mathbf{U}^* = \arg\min_{\mathbf{U}} \frac{1}{2} \mathbf{U}^\mathbf{T} \mathbf{L_C} \mathbf{U}, \quad \text{subject to} \quad \mathbf{U}^\mathbf{T} \mathbf{E} = 0, \mathbf{U}^\mathbf{T} \mathbf{B} \mathbf{U} = 1. \quad (15)$$

The first constraint in (15) $\mathbf{U}^\mathbf{T}\mathbf{E} = 0$ states that the barycenter of the solution must be at zero. Indeed, as $\mathbf{E}$ denotes the $2I \times 2$ matrix that is such that $E_{i1} = 1, i = 1, ..., I$ and $E_{i2} = 1, i = I+1, ..., 2I$ (the other entries of $\mathbf{E}$ being zero), the second constraint $\mathbf{U}^\mathbf{T}\mathbf{B}\mathbf{U} = 1$ indicates that the moment of inertia of the boundary must be unit, the $\mathbf{B}$ matrix being a $2I \times 2I$ diagonal matrix with 1 at each diagonal element corresponding to boundary vertices and 0 everywhere else. There are two different ways to solve this constrained minimization problem. The first method tries to find the optimum of the following Lagrangian function $\mathcal{L}(\mathbf{U}, \boldsymbol{\mu})$ with Lagrange multipliers $\mu_i \geq 0$:

$$\mathcal{L}(\mathbf{U}, \lambda \mu) = \frac{1}{2} \mathbf{U}^\mathbf{T} \mathbf{L_C} \mathbf{U} - \mu(\mathbf{U}^\mathbf{T}\mathbf{E}) - \lambda(\mathbf{U}^\mathbf{T}\mathbf{B}\mathbf{U} - 1). \quad (16)$$

The second method is based on spectral theory that shows that the solution of the constrained minimization problem Eq. (15) is the generalized eigenvector $\mathbf{U}^*$ associated to the smallest non-zero eigenvalue of the matrix $\mathbf{L_C}$, i.e the vector satisfying

$$\mathbf{L_C U} = \lambda \mathbf{B U}, \tag{17}$$

where λ is the smallest non-zero eigenvalue of $\mathbf{L_C}$. This generalized eigenvector $\mathbf{U}^*$ is called the Fiedler vector of $\mathbf{L_C}$. It can be shown that optimizing (16) is equivalent to finding the Fiedler vector $\mathbf{U}^*$ (17). From a numerical point of view, there exists very efficient eigensolvers that find the Fiedler vector $\mathbf{U}^*$ of the sparse generalized eigenvalue problem we need to solve (17). Those methods usually proceed through Choleski decomposition (to turn the problem into a conventional eigenvalue problem) and Lanczos iterations, particulary fast in our case since we deal with sparse matrices. Softwares libraries such as Slepc [16] or Arpack [20] provide all those methods for solving the generalized eigenproblem efficiently.

7 Results

In this section, we present several remeshing examples in order to compare the three different mapping techniques. We compare timings as well as mesh qualities for the new triangulations or quadrilateral meshes. The quality of the isotropic meshes is evaluated by computing the aspect ratio of every mesh triangle as follows [14]:

$$\kappa = \alpha \frac{\text{inscribed radius}}{\text{circumscribed radius}} = 4 \frac{\sin \hat{a} \sin \hat{b} \sin \hat{c}}{\sin \hat{a} + \sin \hat{b} + \sin \hat{c}}, \tag{18}$$

$\hat{a}, \hat{b}, \hat{c}$ being the three inner angles of the triangle. With this definition, the equilateral triangle has $\kappa = 1$ and a flat degenerated triangle has $\kappa = 0$. The quality of the quadrangular meshes are evaluated by computing the quality η of every quadrangle as follows:

$$\eta = \max \left(1 - \frac{2}{\pi} \max_k \left(\left| \frac{\pi}{2} - \alpha_k \right| \right), 0 \right), \tag{19}$$

where $\alpha_k, k = 1, .., 4$ are the four angles of the quadrilateral. This quality measure is $\eta = 1$ if the element is a perfect quadrilateral and is $\eta = 0$ if one of those angles is either ≤ 0 or $\geq \pi$.

In the first example, we compare the convex combination map and the harmonic map. The convex combination seems attractive from a mathematical point of view and is widely used in the computer graphics community. However, we show in Fig. 5 why in the context of surface remeshing, this mapping should not be not used as default mapping. Indeed, as the metric tensor $\mathbf{M_u} = (\mathbf{x_{,u}})^T \mathbf{x_{,u}}$ associated with this mapping $\mathbf{u}(\mathbf{x})$ is much more distorted than the one obtained with the harmonic mapping, there is a negative impact on the quality of the resulting mesh. This is illustrated in Fig. 5 where an initially low quality triangulation $\mathcal{S}_\mathcal{T}$ has been remeshed with a parametrization computed with either the convex combination map or the Laplacian harmonic map. For this example, the total time for the parametrization and

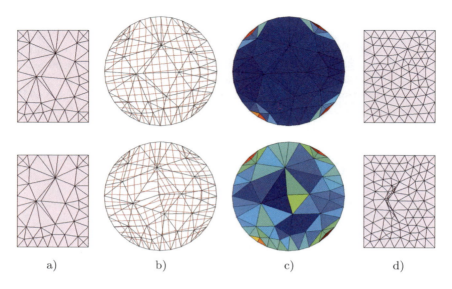

Fig. 5. Poor quality initial triangulation $\mathcal{S}_\mathcal{T}$ (a) that has been remeshed using a harmonic map (top figures) and a convex combination map (bottom figures): b) mapping of the initial mesh onto the unit disk $\mathcal{S}_\mathcal{T}'$ with iso-x and iso-v values in red, c) the determinant of the mesh metric tensor $\mathbf{M_u}$ that defines the area distorstion map and d) the final mesh obtained using the 2D Delaunay algorithms.

the remeshing is $0.008s$. A Delaunay planar mesher has been taken for the remeshing in the parametric space.

In the next example, we compare the remeshing of a human aorta with both the Laplacian and the conformal map. As the geometrical aspect ratio of the triangulation is high, the initial mesh has been automatically split by our algorithm into two different mesh patches. The splitting has been performed with our max-cut mesh partitioner based on a multiscale Laplacian map [24]. As can be seen from Fig. 6, the mapped meshes computed with the Laplacian map present much more distortion close to the boundaries. Again, as most of the planar meshers are more efficient with less distorted meshes, we have that the qualities of the resulting meshes are higher for the conformal map. Indeed, for a radius dependent isotropic remeshing of the aorta, we obtain a minimum quality of $\kappa_{min} = 0.04$ for the harmonic map and $\kappa_{min} = 0.39$ for the conformal map. The mean quality is $\bar{\kappa} = 0.91$ for the harmonic map and $\bar{\kappa} = 0.96$ for the conformal map. Here, a Frontal planar mesher was chosen for the remeshing in the parametric space. The initial triangulation of the aorta contains 12000 triangles and the remeshing procedure for a new mesh of 5500 triangles took us less than $4s$.

We now compare our three mapping techniques for the remeshing of a tooth of very low quality. Fig. 7 shows the remeshing of the tooth and compares

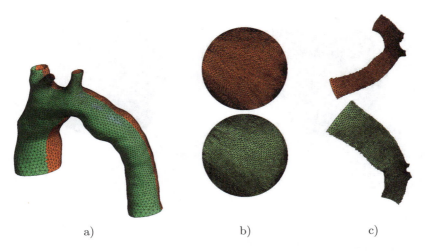

Fig. 6. Remeshing of an STL triangulation of a human aorta that has been split into two mesh patches (a). (b) Harmonic mapping and conformal mapping (c) for those two patches.

the quality of the remeshing procedure using successively a Laplacian map, a conformal map and a convex combination map and choosing a Frontal mesher for the remeshing procedure. As can be seen from Fig.7, the conformal parametrization gives rise to the highest quality mesh while the worst is found to be the convex combination map. The Laplacian mapping has a slightly lower quality that can be explained by a loss in conformality at the boundaries that gives rise to a less smoother mesh metric. The initial triangulation contains 1800 triangles and the remeshed tooth contains about 9000 triangles. The total time for the remeshing is less than $0.8s$ for all of the three mappings. The specific time for the computation of the convex map is $0.13s$, the harmonic map is $0.14s$ and the conformal map is $0.19s$.

An important element in the surface remeshing algorithm is the choice of the planar mesh generator to remesh the parametrized surface. In table 1, we compare the quality of the tooth surface meshes using three different planar mesh generators implemented in Gmsh: a Frontal-Delaunay algorithm [30], a planar Delaunay algorithm [13] and an algorithm based on local mesh adaptation (called MeshAdapt, see [14] for more details). Table 1 shows clearly that the best planar mesh generator is the Gmsh's Frontal-Delaunay algorithm. This is not a surprise: frontal techniques tend to produce meshes that are aligned with principal directions. If the planar domain that has to be meshed is equipped with a smooth metric that conserves angles (i.e., when the mapping is conformal), then the angle between the principal directions is conserved. Frontal algorithms also tend to produce excellent meshes for harmonic maps since harmonic maps are almost conformal except close to the

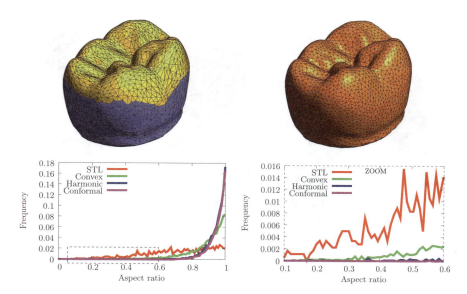

Fig. 7. Remeshing of a tooth with different mappings and with a planar Delaunay mesh generator. Top figures show the initial STL triangulation and new mesh based on the conformal mapping and the bottom figures show the quality histogram obtained when remeshing the STL file with different mappings. The bottom right figure shows a zoom of the quality histogram.

boundaries. The use of Frontal meshers enables us to obtain higher quality mesher from the conformal or harmonic maps. It should be noted that in this case the Frontal mesher was not able to build a mesh given the convex combination map.

Table 1. Quality of the surface mesh of the tooth using different planar mesh generators for the remeshing of the parametric space, where the parametric space has been computed with the 3 presented mappings. The qualities we look at are the the minimum aspect ratio κ_{min} and the mean aspect ratio $\bar{\kappa}$.

Mesh generator	Convex map		Harmonic map		Conformal map	
	κ_{min}	$\bar{\kappa}$	κ_{min}	$\bar{\kappa}$	κ_{min}	$\bar{\kappa}$
MeshAdapt	0.05	0.83	0.17	0.94	0.57	0.95
Delaunay	0.002	0.87	0.18	0.94	0.54	0.94
Frontal	-	-	0.36	0.95	0.65	0.96

Next, we compare the proposed method with other remeshing packages presented in the literature. We consider the well-known standford bunny mesh model of $70k$ triangles[1] (see Fig.8). The original mesh has 5 holes and is of zero genus. For the remeshed bunny of $25k$ triangles presented in Fig.8, we have a minimum quality of $\kappa_{min} = 0.56$ and a mean quality is $\bar{\kappa} = 0.97$. We compare the timings for the conformal map. Prior to computing the parametrization, two different mesh partitioners have been called: a multilevel mesh partitioner (Metis) and a max-cut mesh partitioner based on a multiscale Laplacian map [24]. The timings for the harmonic and convex map are almost similar. We compare in Table 2 some statistics and timings of our algorithm with the least square conformal map (LSCM) of Levy et al. [21], with the multiresolution remeshing of Eck et al. [8] and with the angle based parametrization (ABF) of Zayer [39]. We can see from Table 2 that our method is quite competitive in terms of computationl time with the other methods presented in the literature.

Fig. 8. Remeshing of the bunny mesh model of $70k$ triangles that has been split into 2 mesh partitions. Left figure shows the two partitions, middle figure shows the conformal harmonic parametrization that has been computed for both mesh partitions and right figure shows the remeshed bunny with about $25k$ triangles.

In the last example, we show that one advantage of conformal mappings is that they can be used to produce quadrilateral meshes given an efficient planar quad meshing algorithm. Indeed, as quadrangular meshes are by definition not isotropic but aligned in some directions, one should have a parametrization that preserve the angles between those directions. Indeed, with such a parametrization, the quads that are created by the planar

---

[1] The model can be downloaded at the following web site:
http://www.sonycsl.co.jp/person/nielsen/visualcomputing/
programs/bunny-conformal.obj

Table 2. Remeshing statistics and timings for the remeshing of the bunny mesh model of 70k triangles (new mesh of 25k triangles) with a conformal map. Comparison (when available) of the presented method with other techniques presented in the Computer Graphics community.

Remeshing	Number of partitions	Partition time (s)	Parametrization time (s)	Total remeshing time (s)
LSCM Levy [21] (1.3Ghz)	23	30	95	–
Eck [8] (1.3Ghz)	88	-	-	33.5
ABF++ Zayer [39]	2	-	13	-
LinABF Zayer [39]	2	-	2	-
Presented method (2.4Ghz)				
* laplacian partitioner	2	16.7	1.4	25
* multilevel partitioner	10	7	1.4	14

quad mesh generator will preserve their angles in the 3D space. The only parametrization that preserve angles (in a least square sense) is the conformal parametrization. Figure 9 shows the quadrangular remeshing of half a Falcon aircraft. An initial triangular surface mesh has been generated using a standard frontal surface mesher. Triangles of the surfaces have been patched together to create 12 *compounds of surfaces* to be parametrized (only 7 of them are visible in Fig. 9). The colors of the triangles indicate the different compounds of surfaces of the model.

The images of the different surfaces on their parameter plane can be seen on Figure 10.

The quadrilateral mesh has been obtained with the new Delquad algorithm [31] that generates nearly right triangles combined with the Blossom-quad recombination algorithm [32] that recombines all triangles into quads. The resulting mesh is presented on Figure 9 (bottom). The mesh is composed of 53297 quadrangles. The total time for the surface meshing is 22 seconds. This includes

- The reparametrization of the 12 surfaces (3 sec.),
- The Delquad algorithm applied to the 12 surfaces (10 sec.),
- The Blossom-quad recombination algorithmithm applied to the 12 surfaces (9 sec.).

For the example of the quad mesh of the Falcon aircraft, the worst and average quality of the mesh are $\eta_{min} = 0.17$ and $\bar{\eta} = 0.86$ which can be considered as excellent.

Fig. 9. Initial triangular mesh of half of the Falcon aircraft that has been split into 12 different colored mesh patches (only 7 of those patches are visible) (top) and final quadrangular mesh (bottom).

Fig. 10. Spectral conformal parametrization of the 7 visible surfaces patches (see the colored patches on top Fig. 9) of the Falcon aircraft in the parametric plane.

8 Conclusion

We presented three different ways to compute discrete mappings. The implementation of those mappings as well as the boundary conditions are presented in a very comprehensive manner such that it becomes accessible to a wider community than the one of computer graphics. We showed that the conformal mapping is the best input for our planar meshes and that this mapping allows to generate high quality meshes both triangular and quadrilateral. The overall remeshing technique based on discrete linear parametrization is efficient and renders high quality meshes.

References

1. Alliez, P., Meyer, M., Desbrun, M.: Interactive geometry remeshing. In: Computer graphics (Proceedings of the SIGGRAPH 2002), pp. 347–354 (2002)
2. Batdorf, M., Freitag, L.A., Ollivier-Gooch, C.: A computational study of the effect of unstructured mesh quality on solution efficiency. In: Proc. 13th AIAA Computational Fluid Dynamics Conf. (1997)

3. Bechet, E., Cuilliere, J.-C., Trochu, F.: Generation of a finite element mesh from stereolithography (stl) files. Computer-Aided Design 34(1), 1–17 (2002)
4. Ben-Chen, M., Gotsman, C., Bunin, G.: Conformal flattening by curvature prescription and metric scaling. Computer Graphics Forum 27(2) (2008)
5. Bennis, C., Vézien, J.-M., Iglésias, G.: Piecewise surface flattening for non-distorted texture mapping. In: ACM SIGGRAPH Computer Graphics, pp. 237–246 (1991)
6. Borouchaki, H., Laug, P., George, P.L.: Parametric surface meshing using a combined advancing-front generalized delaunay approach. International Journal for Numerical Methods in Engineering 49, 223–259 (2000)
7. Choquet, C.: Sur un type de représentation analytique généralisant la représentation conforme et défininie au moyen de fonctions harmoniques. Bull. Sci. Math. 69(156-165) (1945)
8. Eck, M., DeRose, T., Duchamp, T., Hoppe, H., Lounsbery, M., Stuetzle, W.: Multiresolution analysis of arbitrary meshes. In: SIGGRAPH 1995: Proceedings of the 22nd Annual Conference on Computer Graphics and Interactive Techniques, pp. 173–182 (1995)
9. Floater, M.S.: Parametrization and smooth approximation of surface triangulations. Computer Aided Geometric Design 14(231-250) (1997)
10. Floater, M.S.: Parametric tilings and scattered data approximation. International Journal of Shape Modeling 4, 165–182 (1998)
11. Floater, M.S.: One-to-one piecewise linear mappings over trinagulations. Math. Comp. 72(685-696) (2003)
12. Floater, M.S., Hormann, K.: Surface parameterization: a tutorial and survey. In: Advances in Multiresolution for Geometric Modelling (2005)
13. George, P.-L., Frey, P.: Mesh Generation. Hermes (2000)
14. Geuzaine, C., Remacle, J.-F.: Gmsh: a three-dimensional finite element mesh generator with built-in pre- and post-processing facilities. International Journal for Numerical Methods in Engineering 79(11), 1309–1331 (2009)
15. Greiner, G., Hormann, K.: Interpolating and approximating scattered 3d data with hierarchical tensor product splines. In: Surface Fitting and Multiresolution Methods, pp. 163–172 (1996)
16. Hernandez, V., Roman, J.E., Vidal, V.: Slepc: A scalable and flexible toolkit for the solution of eigenvalue problems. ACM Trans. Math. Soft. 31(3), 351–362 (2005)
17. Hormann, K., Greiner, G.: Mips: An efficient global. parametrization method. In: Curve and Surface Design, Vanderbilt University Press (2000)
18. Ito, Y., Nakahashi, K.: Direct surface triangulation using stereolithography data. AIAA Journal 40(3), 490–496 (2002)
19. Laug, P., Boruchaki, H.: Interpolating and meshing 3d surface grids. International Journal for Numerical Methods in Engineering 58, 209–225 (2003)
20. Lehoucq, R.B., Sorensen, D.C., Yang, C.: ARPACK Users Guide: Solution of Large Scale Eigenvalue Problems by Implicitly Restarted Arnoldi Methods. Society for Industrial and Applied Mathematics (1997)
21. Lévy, B., Petitjean, S., Ray, N., Maillot, J.: Least squares conformal maps for automatic texture atlas generation. In: Computer Graphics (Proceedings of SIGGRAPH 2002), pp. 362–371 (2002)

22. Spagnuolo, M., Attene, M., Falcidieno, B., Wyvill, G.: A mapping-independent primitive for the triangulation of parametric surfaces. Graphical Models 65(260-273) (2003)
23. Maillot, J., Yahia, H., Verroust, A.: Interactive texture mapping. In: Proceedings of ACM SIGGRAPH 1993, pp. 27–34 (1993)
24. Marchandise, E., Carton de Wiart, C., Vos, W.G., Geuzaine, C., Remacle, J.-F.: High quality surface remeshing using harmonic maps. Part II: Surfaces with high genus and of large aspect ratio. International Journal for Numerical Methods in Engineering 86(11), 1303–1321 (2011)
25. Marchandise, E., Compère, G., Willemet, M., Bricteux, G., Geuzaine, C., Remacle, J.-F.: Quality meshing based on stl triangulations for biomedical simulations. International Journal for Numerical Methods in Biomedical Engineering 83, 876–889 (2010)
26. Marcum, D.L.: Efficient generation of high-quality unstructured surface and volume grids. Engrg. Comput. 17, 211–233 (2001)
27. Marcum, D.L., Gaither, A.: Unstructured surface grid generation using global mapping and physical space approximation. In: Proceedings, 8th International Meshing Roundtable, pp. 397–406 (1999)
28. Mullen, P., Tong, Y., Alliez, P., Desbrun, M.: Spectral conformal parameterization. In: In ACM/EG Symposium of Geometry Processing (2008)
29. Rado, T.: Aufgabe 41. Math-Verien, p. 49 (1926)
30. Rebay, S.: Efficient unstructured mesh generation by means of delaunay triangulation and bowyer-watson algorithm. Journal of Computational Physics 106, 25–138 (1993)
31. Remacle, J.-F., Henrotte, F., Carrier-Baudouin, T., Bechet, E., Marchandise, E., Geuzaine, C., Mouton, T.: A frontal delaunay quad mesh generator using the l_∞ norm. International Journal for Numerical Methods in Engineering (2011)
32. Remacle, J.-F., Lambrechts, J., Seny, B., Marchandise, E., Johnen, A., Geuzaine, C.: Blossom-quad: a non-uniform quadrilateral mesh generator using a minimum cost perfect matching algorithm. International Journal for Numerical Methods in Engineering (submitted 2011)
33. Sheffer, A., Praun, E., Rose, K.: Mesh parameterization methods and their applications. Found. Trends. Comput. Graph. Vis. 2(2), 105–171 (2006)
34. Sheffer, A., Lévy, B., Lorraine, I., Mogilnitsky, M., Bogomyakov, E.: Abf++: fast and robust angle based flattening. ACM Transactions on Graphics 24(311-330) (2005)
35. Szczerba, D., McGregor, R.H.P., Székely, G.: High quality surface mesh generation for multi-physics bio-medical simulations. In: Shi, Y., van Albada, G.D., Dongarra, J., Sloot, P.M.A. (eds.) ICCS 2007. LNCS, vol. 4487, pp. 906–913. Springer, Heidelberg (2007)
36. Tristano, J.R., Owen, S.J., Canann, S.A.: Advancing front surface mesh generation in parametric space using riemannian surface definition. In: Proceedings of 7th International Meshing Roundtable, Sandia National Laboratory, pp. 429–455 (1998)
37. Tutte, W.T.: How to draw a graph. In: Proceedings of the London Mathematical Society, vol. 13, pp. 743–768 (1963)

38. Wang, D., Hassan, O., Morgan, K., Weatheril, N.: Enhanced remeshing from stl files with applications to surface grid generation. Commun. Numer. Meth. Engng 23, 227–239 (2007)
39. Zayer, R., Lévy, B., Seidel, H.-P.: Linear angle based parameterization. In: ACM/EG Symposium on Geometry Processing Conference Proceedings (2007)
40. Zheng, Y., Weatherill, N.P., Hassan, O.: Topology abstraction of surface models for three-dimensional grid generation. Engrg. Comput. 17(28-38) (2001)
41. Zigelman, G., Kimmel, R., Kiryati, N.: Texture mapping using surface flattening via multi-dimensional scaling. IEEE Trans. on Visualisation and Computer Graphics 8(198-207) (2002)

Dual Contouring for Domains with Topology Ambiguity

Jin Qian and Yongjie Zhang*

Department of Mechanical Engineering, Carnegie Mellon University,
5000 Forbes Ave, Pittsburgh, PA 15213, USA
Tel.: (412) 268-5332; Fax: (412) 268-3348
jessicaz@andrew.cmu.edu

Summary. This paper describes an automatic and robust approach to generate quality triangular meshes for complicated domains with topology ambiguity. In previous works, we developed an octree-based Dual Contouring (DC) method to construct triangular meshes for complicated domains. However, topology ambiguity exists and causes non-conformal meshes. In this study, we discuss all possible topology configurations and develop an extension of DC which guarantees the correct topology. We first generate one base mesh with the previous DC method. Then we analyze all the octree leaf cells and categorize them into 31 topology groups. In order to discriminate these cells, we compute the values of their face and body saddle points based on a tri-linear representation inside the cells. Knowing the correct categorization, we are able to modify the base mesh and introduce more minimizer points within the same cell. With these minimizer points we update the mesh connectivities to preserve the correct topology. Finally we use a Laplacian smoothing technique to improve the mesh quality. Our main contribution is the topology categorization and mesh modification. We have applied our algorithm to three complicated domains and obtained good results.

Keywords: Dual Contouring, topology ambiguity, triangular mesh, tri-linear representation, saddle points.

1 Introduction

Accurate representation of an iso-surface is one important problem in scientific visualization and mesh generation. Given 3D scanned images, we aim to generate quality triangular surfaces with correct topology. The Marching Cubes (MC) algorithm [14] visits each cell in the volume and performs local triangulation based on the sign configuration of the eight grid points. Accelerated algorithms [2, 21] were further developed to reduce the running time by avoiding visiting unnecessary cells. The iso-surface inside the cubic cells may have complicated shapes or topology ambiguities. To handle them, the

---

* Corresponding author.

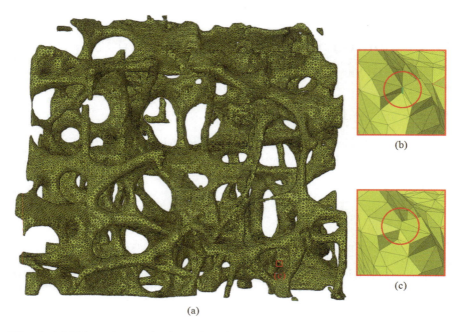

Fig. 1. (a) Triangular mesh of a trabecular bone structure with complicated topology. (b) and (c) show zoom-in details of a local region with topology ambiguity (see the red circle). The green line in (b) denotes one non-manifold edge, which is resolved in (c).

function values of the face and body saddles in the cell are used to decide the correct topology and to generate consistent triangulation [15, 13]. Main drawbacks of MC include uniform and large-size mesh, badly shaped triangles as well as the loss of sharp features. To generate an adaptive iso-surface, people developed ways to triangulate cells with different levels. However, when adjacent cells have different resolution levels, cracks are introduced and a fan of triangles have to be inserted around the gravity center of the coarse triangles [20]. Octree based Dual Contouring (DC) method [10] combines SurfaceNets [8] and the extended Marching Cubes [12] algorithms, and it is able to generate adaptive iso-surfaces with good aspect ratios and sharp feature preservation. Despite being adaptive and feature-preserving, DC has the drawback that one cell can contain only one minimizer, leaving possible non-manifold meshes. To address this deficiency, the vertex clustering algorithm [22, 19] together with topology constraints [17, 3, 11] was developed. However, for datasets with very complicated topology (see the trabecular bone structure in Fig. 1), the existing DC methods fail to recognize all the topology ambiguities and may lead to non-conformal meshes.

To distinguish the topology ambiguities, we need to study the function properties inside the cubic cell. Since the function value is only given at the eight grid points of each cell, we model the interior region with a trilinear interpolation. Depending on the function values of these eight grid points, there are 14 unique configurations for one cell. For some of these configurations, topology ambiguities arise. These ambiguities are either on the face of the cell or interior to the cell. To discriminate them, we compute the function values at body and face saddle points [15]. With respect to the values at the saddle points, we decide whether the iso-surface forms a tunnel linking the two grid points or it is separated into several parts. With all these ambiguities resolved, we develop a new algorithm that modifies the mesh generated using the standard DC cell by cell. In this algorithm, multiple minimizers may be introduced to one cell, eliminating the drawback that the DC method has only one minimizer within one cell. To improve the mesh quality, we relocate the vertex positions via a Laplacian smoothing technique [25]. Our algorithm keeps the mesh conformal across cell faces and attains quality meshes with correct topology.

The remainder of this paper is organized as follows: Section 2 reviews related previous work. Section 3 talks about the standard DC method. Section 4 and 5 discusses the resolving and triangulation of 2D and 3D ambiguities. Section 6 presents results and discussion, and finally Section 7 draws conclusions.

2 Previous Work

Marching Cubes. As one of the most popular iso-contouring techniques, the MC algorithm [14] classifies cubic cells into 256 configurations, depending on whether the eight vertices are positive or negative. After considering symmetry and complementary, the 256 cases can be reduced to 14 unique ones. If the two endpoints of any edge have different signs, then the edge is intersected by the iso-surface. The intersection point can be estimated via a linear interpolation. For each of the 14 cases, the approximation of the iso-surface can be created by the triangulation of multiple intersection points. These configurations are incorporated into a lookup table, and each entry in the table contains a triangulation pattern. Then the MC visits one cell at a time until all the cells are treated.

The MC is straight-forward and easy to implement; however, it has several drawbacks. The main problem of MC is that it requires uniform cell structure, which may lead to huge mesh size. Meanwhile, the vertices in MC are restricted on the cell edges, and this introduces many elements with small angles. Furthermore, sharp features in the data are not well preserved. Another drawback is that there is a possibility for discrepancy in the connection of the shared face of two adjacent cells, which are caused by un-categorized ambiguities.

Extended Marching Cubes. The MC has been the focus of much further research to improve its quality of iso-surface representation [13]. To improve the mesh quality, additional information in the volume, such as surface normals, are utilized to reduce sharp angles [12]. To make the iso-surface adaptive, people developed ways to triangulate over cells with different levels. However, when the resolution levels of adjacent cells are different, there will be cracks and a fan of triangles needs to be inserted around the gravity center of the coarse triangle [20]. To handle the topology ambiguities, face ambiguity [16, 5] and interior ambiguity [15, 4] have been discussed. In order to distinguish these ambiguities, a strategy based on saddle points, where the first partial derivative of the tri-linear function equals to zero, was proposed. After considering all the ambiguities, an enhanced look-up table consisting of 31 cases were built [13]. These improved MC methods provide better quality meshes and handle ambiguities; however, they still lack the capability to mesh large datasets adaptively.

Dual Contouring. The Dual Contouring (DC) method [10, 24], which combines SurfaceNets [8] and the extended Marching Cubes [12] algorithm, generates a dual mesh from an adaptive octree instead of a uniform one. The adaptation of the octree can be controlled by certain error functions. It was first designed for the triangulation from Hermit data [10] and then extended to tetrahedral and hexahedral mesh generation [24, 23], as well as domains with heterogeneous materials [26]. The DC has several advantages compared with MC. Most importantly it is able to handle large datasets using an adaptive octree and preserves sharp features well. However, a problem of DC is the restriction that each leaf cell only holds one minimizer. Such restriction leads to non-manifold meshes and thus makes the accurate representation of topology ambiguities impossible. As a follow-up, vertex clustering, which inserts multiple minimizers inside one leaf cell, was introduced [22, 19]. It was further applied with topology constraints so that the resulting contours are always 2-manifold [17, 3, 11]. However, these techniques still lack the capability of recognizing all the topology ambiguities.

3 An Overview of Standard Dual Contouring

The DC method [10] was developed based on an adaptive octree data structure, which achieves denser cells along the boundary and coarser cells inside the volume. The procedure begins with splitting the volume data into octree cells iteratively until the mesh adaptation is satisfactory. The mesh adaptation can be controlled adaptively by using a feature sensitive function $F = \sum(|f^{i+1} - f^i|/\bigtriangledown f^i)$, where f is a tri-linear interpolation function inside the cell [24]. This feature sensitive error function measures the iso-contour difference between two neighboring octree levels. When the error is below a pre-defined tolerance, the octree splitting ceases. In order to obtain triangles with good quality, we restrict the neighboring level difference to be less than or equal to two. One minimizer was computed for each leaf cell to minimize a predefined Quadratic Error Function (QEF), $QEF(x) = \sum(n_i \cdot (x - p_i))^2$, where p_i and n_i are the position and normal vectors at the intersection points between the iso-surface and octree cell [6, 7].

The next stage is to analyze each sign change edge, whose one endpoint lies inside the internal volume while another lies outside. Since the adaptive octree consists of leaves with different resolution levels, each leaf may have neighbors at different levels, i.e., an edge in one leaf cell may be divided into several edges in its neighboring cells. In deciding which edge should be analyzed, we follow the DC method and always choose the minimal one. These edges are those of leaf cells that do not properly contain an edge of a neighboring leaf. In this adaptive octree, the minimal edge is either shared by four cells, and we connect the four QEF minimizers to form a quadrilateral (quad); or it is shared by three cells and we form a triangle, and we obtain a hybrid mesh consisting of quad and triangular elements. Finally we divide these quads to form a triangular mesh. One splitting diagonal is chosen to achieve better aspect ratios. By now an adaptive triangular mesh is created. Several variants of DC method have been developed as well, which introduce better topology and mesh quality [1, 9, 18, 17, 22].

Despite being adaptive and feature-preserving, one drawback of such DC method is that it fails to handle ambiguities in the original topology and it may introduce non-manifold surfaces, that is, an edge on the contour may be shared by more than two polygons. These surfaces are not only troublesome in visualization but also are problematic in further mesh processing techniques such as smoothing and parameterization. Therefore the ambiguities must be resolved.

4 Modified DC for 2D Ambiguities

Ambiguity arises when the iso-surface can be represented using different approaches and there are multiple ways to connect the minimizers. To capture the ambiguities, we need to categorize all the cells into different cases and then adjust the mesh topology accordingly. We first discuss the simple 2D ambiguity problem followed by the complex 3D problem. Here are some definitions used in the categorization.

Positive grid point: A positive grid point is a grid point whose function value is greater than or equal to the given iso-value.
Negative grid point: A negative grid point is a grid point whose function value is less than the given iso-value.
Joined grid points: If two grid points within the same cell meet the following two criteria, we call them joined grid points: (1) These two points have the same positive/negative sign; (2) there is a path between these two points, along which the interior function value between them does not change sign.
Separated grid points: If two grid points within the same cell only meet the first criterion above, we call these two points separated grid points.

4.1 Resolving 2D Ambiguities

To begin with, we now resolve the simple 2D ambiguities. Since the function values are only given at the four corners of each quadtree cell, here we choose a bi-linear interpolation, $F(\xi, \eta)$, which is written in terms of the function values at four corners F_{ij} $(i, j = 0, 1)$:

$$F(\xi, \eta) = F_{00}(1-\xi)(1-\eta) + F_{01}(1-\xi)\eta \\ + F_{10}\xi(1-\eta) + F_{11}\xi\eta. \quad (1)$$

Depending on the sign of the four grid points, we summarize all 2D possibilities into 4 categories (Cases 0, 1, 2, and 3) with the consideration of symmetry and complementary (reversing positive/negative signs). The meshes generated via the standard DC are shown in Fig. 2. DC generates a series of quads first and then splits them into triangles, therefore when we discuss about triangulation, we only consider how to generate the quads. The cell may only have one positive grid point (Case 1); or it has two neighboring positive grid points (Case 2); otherwise the cell has two positive grid points forming one diagonal: these two grid points can be either joined (Case 3-1) or separated (Case 3-2).

From Case 3-1 and Case 3-2 it is obvious that ambiguity arises when two grid points in one diagonal have the same sign. The standard DC method falls short to recognize this ambiguity and creates non-manifold meshes, as shown in Fig. 2. To discriminate these two possibilities, we find out the saddle point, where the first partial derivative for each direction is zero, and then compute the function value at the saddle point. If that value is positive (greater than or equal to the iso-value), we can tell that the positive zone spreads across the cell so that the two positive grid points are joined; otherwise the positive zone is divided and then the two positive grid points are separated.

After resolving these two ambiguous cases, there are a total of five topology cases for a 2D cell (see Fig. 2).

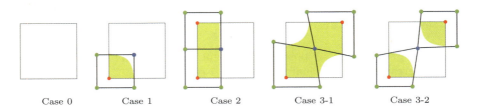

Case 0 Case 1 Case 2 Case 3-1 Case 3-2

Fig. 2. 2D topology possibilities and the meshing results via the standard DC. The cell has one or two positive grid points (red), one minimizer (blue) and up to six neighboring cell minimizers (green). The iso-surface (yellow) intersects the cell.

Simple configurations: Case 0, 1, 2 have no ambiguities. These cases can be meshed using the standard DC.

Configuration 3: There are two possible cases:
Case 3.1: Positive grid points are joined inside the cell; and the iso-surface spreads across the cell;
Case 3.2: Positive grid points are separated; and the iso-surface breaks into two pieces inside this cell.

4.2 Handling 2D Ambiguities in Triangulation

With the two ambiguities resolved, we are able to re-mesh the 2D iso-surface. For those simple cases, the traditional DC method is sufficient to represent the correct topology. For the two ambiguous cases, re-meshing is necessary. We first need to insert two new minimizers inside the cell. The minimizers are located on the intersection points of two tangent lines to the iso-surface. Next we check all the quads sharing the original minimizer, and then re-connect the quads using the schemes shown in Fig. 3. The old minimizer is then deleted since no quads connect to it anymore. Till now we obtain a hybrid mesh consisting of quads and pentagons. For a quad we pick one diagonal and divide it into two triangles. For a pentagon, we split it into three triangles.

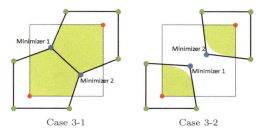

Case 3-1   Case 3-2

Fig. 3. Resolving 2D ambiguities. Each cell has positive grid points (red) and minimizers (blue); the green nodes denote the minimizers of neighboring cells.

5 Modified DC for 3D Ambiguities

Resolving 3D topology ambiguities within each cell is much more complicated. We first discuss how to resolve these ambiguities and then present the re-meshing techniques to represent the iso-surface accordingly.

5.1 Resolving 3D Ambiguities

In 3D octree cells the function values are given only at the eight grid points, similarly here we choose a tri-linear interpolation, $F(\xi, \eta, \zeta)$, which is written in terms of the function values at the eight grid points, F_{ijk} $(i, j, k = 0, 1)$:

$$F(\xi, \eta, \zeta) = F_{000}(1-\xi)(1-\eta)(1-\zeta)$$
$$+ F_{001}(1-\xi)(1-\eta)\zeta + F_{010}(1-\xi)\eta(1-\zeta)$$
$$+ F_{011}(1-\xi)\eta\zeta + F_{100}\xi(1-\eta)(1-\zeta) \quad (2)$$
$$+ F_{101}\xi(1-\eta)\zeta + F_{110}\xi\eta(1-\zeta)$$
$$+ F_{111}\xi\eta\zeta.$$

Depending on the sign of the eight grid points, there are $2^8 = 256$ possible configurations for one cubic cell. Considering all the symmetry and complementary, the 256 situations can be reduced to 14 fundamental cases, as shown in Table 1.

Some of these fundamental cases have topology ambiguities, which can be categorized into two groups: **face ambiguity** and **interior ambiguity**. Face ambiguity arises when two positive grid points occupy one cubic face diagonal, as shown in Fig. 4(a-b). The four edges of this face are all sign change edges, and we need to decide how to triangulate this local region. If the mesh is built in an inconsistent way, discrepancies such as a gap may be introduced between adjacent cells.

Meanwhile, there are additional interior ambiguities inside the cubic cell, as shown in Fig. 4(c-d). In these situations the iso-surface can either be several separated surfaces or a tunnel piece passing through the cubic cell. These interior structures generally do not cause immediate discrepancies, but they decide what the topology looks like inside the cell.

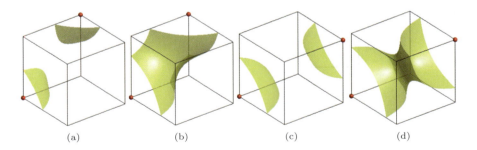

Fig. 4. Example of 3D face and interior ambiguities. The cubic cell has two positive grid points (red) and the iso-surface (yellow) intersects with it. (a-b) One face ambiguity; and (c-d) one interior ambiguity.

In order to discriminate one topology from another, we again analyze the function values of the face and interior saddle points. Similar to 2D saddle points, we find the saddle points by setting all the first partial derivatives to be 0. When analyzing face ambiguities, the saddle points are confined to stay on that very face; therefore the tri-linear interpolation degenerates to a bi-linear one, which leads to a single root. On the other hand, for interior ambiguities, the saddle points are set to be inside the cubic cell; so we need to

Table 1. 14 fundamental cases for 3D cubic cells with meshes generated using the standard DC. Each cell has a few positive grid points (red), one minimizer (blue) and several minimizers of adjacent cells (green).

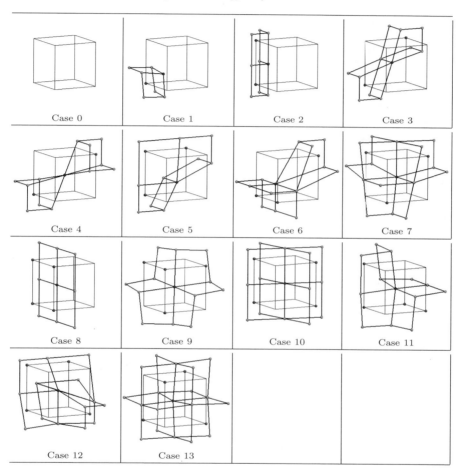

solve a quadratic equation with double roots. If the computed saddle points lie outside the specific surface or the cell, it is ignored. Till now we can compute the function values at these saddle points, and use them to categorize the topology.

After considering all face and interior ambiguities, a comprehensive set of 31 cases are summarized. Since this set is actually a detailed version of the 14 fundamental cases, therefore we use labeling notations inherited from the 14 cases. The labeling of these 31 cases denotes the topologies they belong to: the first number denotes the case number in the original 14 fundamental cases (see Table 1); the second indicates the resolution of a face ambiguity; in the end the last index clarifies an interior ambiguity. These 31 cases are described as follows (see Table 2):

Simple configurations: Case **0, 1, 2, 5, 8, 9** and **11** have neither face ambiguity nor interior ambiguity. These cases can be meshed using the standard DC method.

Configuration 3: There is only one ambiguous face. After resolving this face, two different cases exist:
Case 3-1: Positive grid points are separated;
Case 3-2: Positive grid points are joined inside the face.

Configuration 4: There is no face ambiguity but one interior ambiguity. Resolving such interior ambiguity leads to two cases:
Case 4-1-1: Positive body diagonal grid points are separated;
Case 4-1-2: Positive grid points are joined inside the cubic cell and thus form a "tunnel" structure.

Configuration 6: There is one ambiguous face; and if the positive grid points are separated on this face, additional interior ambiguity arises: the two positive grid points can be either separated or joined inside the cube. Three cases are possible:
Case 6-1-1: Positive grid points on the ambiguous face are separated and they are not joined inside the cube;
Case 6-1-2: Positive grid points on the ambiguous face are separated and they are joined inside the cube;
Case 6-2: Positive grid points are joined on the ambiguous face.

Configuration 7: There are three adjacent ambiguous faces sharing one negative grid point. If all the positive grid points on the ambiguous faces are separated, the interior ambiguity arises: the common negative grid point can be separated from other negative grid points or joined inside the cube. Considering the number of face ambiguities, five distinct cases are possible:
Case 7-1: All positive grid points are separated and they are not joined inside the cube;
Case 7-2: Positive grid points on one ambiguous face are joined while others are separated;
Case 7-3: Positive grid points on two ambiguous faces are joined while others are separated;
Case 7-4-1: All positive grid points are joined on the three ambiguous faces; the negative grid point shared by three ambiguous faces is separated from other negative grid points;
Case 7-4-2: All positive grid points are joined on the three ambiguous faces; the negative grid point shared by three ambiguous faces is joined inside with other negative grid points.

Configuration 10: For these configurations two ambiguous faces need to be handled. By reversing the positive/negative signs, we are able to ensure

Table 2. 31 cases for 3D cubic cells with meshes generated using the modified DC. Each cell has a few positive grid points (red), several minimizers (blue) and some minimizers of adjacent cells (green).

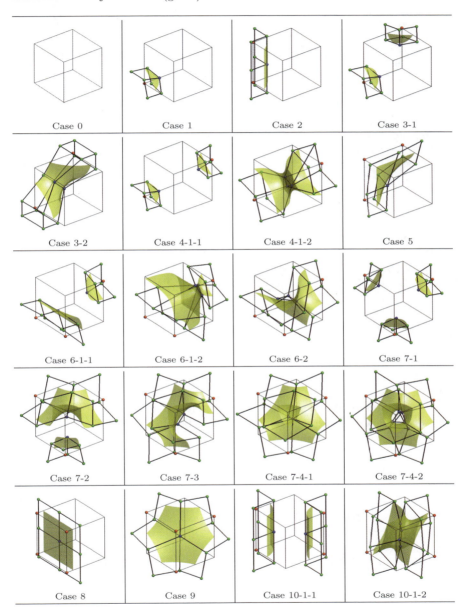

Table 2. (*continued*)

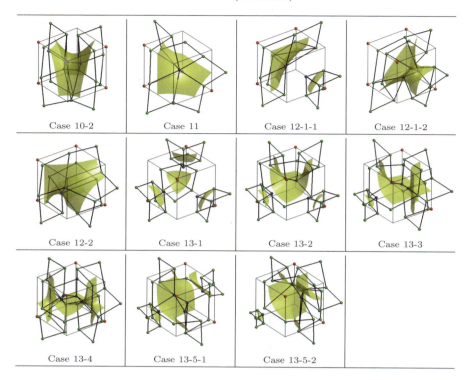

that at least one of the ambiguous faces has separated positive grid points. Particularly, we rotate the cubes so that the positive grid points on the top face are separated. When the positive grid points on the two ambiguous faces are both separated, interior ambiguities may be introduced. There are three cases belonging to Configuration 10:

Case 10-1-1: Positive grid points on the top and the bottom faces are both separated; no positives grid points on the body diagonal are joined from inside.

Case 10-1-2: Positive grid points on the top and the bottom faces are both separated; however, two grid points on the body diagonal are joined.

Case 10-2: Positive grid points on the top face are separated; however, positive grid points on the bottom are not. In this case the iso-surface forms a typical saddle shape.

Configuration 12: There are two face ambiguities in this configuration. By reversing the positive/negative signs, we are able to ensure that the positive grid points are separated on at least one ambiguity face. Interior ambiguity arises when the positive points on both ambiguous faces are separated. There are a total of four cases.

Case 12-1-1: Positive grid points are separated on both ambiguous faces; no body diagonal consisting of two positive grid points is joined from inside.
Case 12-1-2: Positive grid points are separated on both ambiguous faces; one body diagonal consisting of two positive grid points is joined inside the cube.
Case 12-2: Positive grid points are joined in one face while separated in another face. Depending on which one is joined, a mirror case is also alike.

Configuration 13: This configuration is the most complex one since its six faces are all ambiguous. By reversing the positive/negative, we can always ensure that the number of faces with separated positive grid points is greater than or equal to the number of faces with joined positive grid points. When that number is three, additional interior ambiguities may also arise. These ambiguities lead to a total of six distinct cases.
Case 13-1: Positive grid points are separated on all six ambiguous faces.
Case 13-2: Positive grid points are separated on five ambiguous faces while joined on the other one.
Case 13-3: Positive grid points are joined on four ambiguous faces while separated on the other two. Due to the nature of the tri-linear interpolation function [4], these two other faces must be adjacent to each other.

The remaining cases all have three ambiguous faces with positive grid points joined. These three faces must share one common grid point due to the nature of the tri-linear interpolation function. Depending on whether the common grid point is positive, different cases are possible. If yes, interior ambiguity arises.

Case 13-4: The common grid point of the three joined faces is negative.
Case 13-5-1: The common grid point is positive, and it is separated from all other positive grid points.
Case 13-5-2: The common grid point is positive, and it is separated from other positive ones; all negative points are joined from inside. A complementary case is also possible when we reverse all the positive/negative signs. In that case, all positive grid points are joined inside the cell while all negative ones are separated.

All of these 31 cases are shown in Table 2. For each leaf cell in the octree, we categorize it into one of the cases above. Once the classification is made, we need to modify the dual mesh and incorporate the specific topologies.

5.2 Handling 3D Ambiguities in Triangulation

It is obvious that resolving 3D ambiguities is much more complicated, not only in terms of the number of cases but also in terms of the complexity of each case. To incorporate these features, we traverse all the octree leaf cells to insert minimizers and to re-connect the triangles.

5.2.1 Adding Minimizers

To represent correct topologies, we may need multiple minimizers within one cubic cell. In order to add appropriate number of minimizers we need to analyze the intersections between the cubic cell and the iso-surface. For each cell face, there are four types of intersection according to the tri-linear interpolation in Equation (2):

Type 0: The iso-surface does not intersect the cell face; and no minimizer is needed at all.

Type 1: The iso-surface intersects the cell face only once; and one minimizer should be inserted nearby. If one face is adjacent to another one which also needs new minimizers (Type 1, Type 2), these two faces can share one common minimizer. The minimizers are located in the position where the QEF is minimized. In 2D, these minimizers are actually the intersections of the tangential lines to the isosurface.

Type 2: The iso-surface intersects the cell face twice; in this case we need at least two minimizers to approximate the iso-surface.

Type 3: The iso-surface cuts through the cell forming one "tunnel" inside; at least three minimizes are needed to approximate the interior structure.

5.2.2 Re-connecting Triangles with New Minimizers

With enough minimizers we are able to re-connect them so that the resulting triangulation reflects the correct topology. Similar to 2D cases, we first (1) find all the quads sharing the original minimizer; and then (2) decide which sign change edges these quads associate to; in the following we (3) associate the sign change edge with the appropriate minimizer(s); in the next stage we (4) re-connect the quads or pentagons using the new minimizers; and finally (5) split the quads and pentagons to form a triangular mesh.

Among these five steps the third one is the most critical. After the association between the sigh change edge and the minimizer is created, all quads associating to that sign change edge are connected to the very minimizer(s). To illustrate, we show an example depicting Case 3-2 with details, see Fig. 5.

In the beginning we have a base mesh created using the standard DC, see Fig. 5(a). To incorporate the correct topologies we take the following two steps:

Step 1 (Insert multiple minimizers): We check the intersection between each cell face and the iso-surface. Only one intersection belongs to Type 2 (back face) while all the others belong to Type 0 or 1. Therefore two new minimizers (M and P) are inserted near that face as shown in Fig. 5(b). Note that in the adjacent cell that shares the ambiguous face, similar techniques are carried out as well; therefore two new minimizers (N and Q) of the adjacent cell are also inserted.

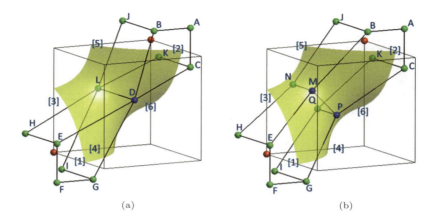

Fig. 5. The 3D re-meshing for Case 3-2. The numbers denote edge indices while the capital letters denote minimizers. The red points are positive grid points, the blue nodes are minimizers of this cell, and finally the green ones are minimizers of adjacent cells.

Step 2 (Re-connect the mesh): We traverse the quads sharing the original minimizer, node D, find their sign change edges, and then associate each of them to one minimizer. At this point we have the following six configurations:

1. Quad $\overline{DBJL}$ associates with the sign change edge [5], and it is associated with minimizer M, forming Quad $\overline{MBJN}$.
2. Quad $\overline{DLHE}$ associates with the sign change edge [3], and it is associated with minimizer M, forming Quad $\overline{MNHE}$.
3. Quad $\overline{DCKL}$ associates with the sign change edge [6], and it is associated with minimizer P, forming Quad $\overline{PCKQ}$.
4. Quad $\overline{DLIG}$ associates with the sign change edge [4], and it is associated with minimizer P, forming Quad $\overline{PQIG}$.
5. Quad $\overline{DCAB}$ associates with the sign change edge [2], and it is associated with two minimizers M and P at the same time, forming Pentagon $\overline{MPCAB}$.
6. Quad $\overline{DEFG}$ associates with the sign change edge [1]. Similar to Quad $\overline{DCAB}$, it is also associated with P and M altogether, forming Pentagon $\overline{MEFGP}$.

By now we have traversed all the quads sharing the original minimizer D and re-constructed a hybrid mesh consisting of quads and pentagons. This hybrid mesh can be easily split into triangular meshes. Following the same procedure we are able to handle any of the 31 cases and incorporate correct topologies into the triangular mesh. Fig. 6 shows the meshes of five regions before and after re-connections. These figures show the topology Case 3-1, Case 3-2, Case 4-1-2, Case 7-4-2 and Case 10-2, respectively. The green lines or the blue node denote non-manifold situations, where one edge is shared by

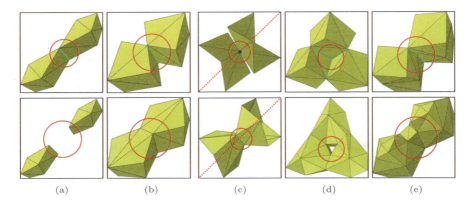

Fig. 6. Meshes of five regions before (first row) and after (second row) the reconnection. The red circles indicate the locations of the ambiguity, green lines or the blue dot indicate non-manifold situations, and red dashed lines denote the axes of a tunnel topology. (a) Case 3-1; (b) Case 3-2; (c) Case 4-1-2; (d) Case 7-4-2; and (e) Case 10-2 examples.

more than two triangles, or one vertex whose neighborhood is not like a disc. Note in (c), along the red dashed axes, an hollow tunnel topology is missing, and the our re-connection algorithm recovers it. In these five figures, it is obvious that the topology ambiguities are resolved and appropriate meshes are built.

Till now the mesh is topologically correct; however, it may have low-quality triangles that influence the convergence and stability of the numerical solutions. Therefore, we need to take an additional step to relocate these vertices as well as to improve the mesh quality.

To begin with, we need a metric to measure the quality of a triangle. Here, we choose the aspect ratio, defined as $2 * r_{in}/r_{out}$, where r_{in} is the radius of the inscribed circle, and r_{out} is the radius of the circumcircle. A perfect triangle has the aspect ratio of 1; therefore our goal is to increase the ratio toward 1. To attain this goal, a Laplacian smoothing technique is applied. Laplacian smoothing is a fast and efficient method to smooth a triangular mesh. For each mesh vertex, Laplacian smoothing relocates its position to the weighted average of the nodes connecting to it.

6 Results and Discussion

The developed algorithm is automatic and robust for meshing domains with topology ambiguities. We have applied our algorithm to three complicated datasets (Figs. 1, 7, 8). Our technique generates quality meshes with ambiguities resolved and feature preserved. The computations were based on a PC with Intel Q9600 CPU and 4GB DDR-II memories. Table 3 shows statistics

Table 3. Statistics of the generated meshes

Model	Mesh	Mesh Size (Vert ♯, Elem ♯)	Aspect Ratio (Worst, Best)	Time (s)	Ambiguity Cell #
Trabecular Bone 1 (Fig. 1)	Original	(98,694, 197,636)	(0.0006 1.0)		
	Smoothed	(98,944, 197,884)	(0.158, 1.0)	28	887
Trabecular Bone 2 (Fig. 7)	Original	(134,925, 271,224)	(0.000066, 1.0)		
	Smoothed	(135,911, 272,150)	(0.168, 1.0)	36	241
Foam (Fig. 8)	Original	(186,629, 374,420)	(0.030 1.0)		
	Smoothed	(187,046, 374,824)	(0.16, 1.0)	51	388

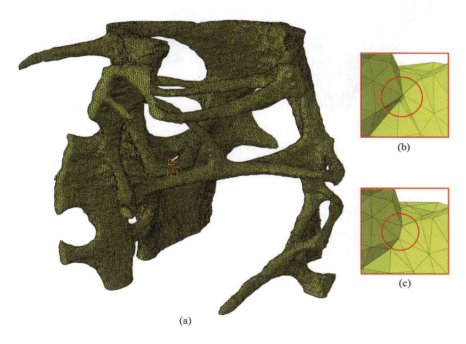

Fig. 7. (a) Triangular mesh of another trabecular bone structure. (b) and (c) show zoom-in details of a local region with topology ambiguity (see the red circle). The green line in (b) denotes one non-manifold edge, which is resolved in (c).

of the generated triangular meshes. The mesh size is increased after resolving the ambiguities, since one leaf cell may contain multiple minimizers. The mesh quality is improved greatly after smoothing. Note that the computation time includes the time for the base mesh generation, topology preservation, and the following smoothing. The smoothing process takes most of the time and has an approximately linear running time.

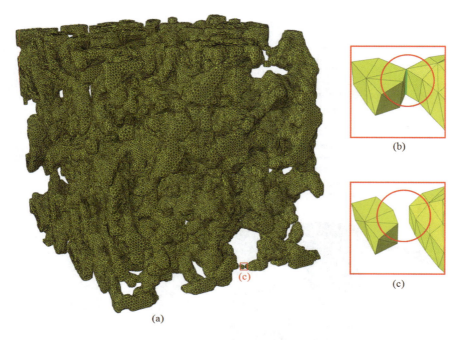

Fig. 8. (a) Triangular mesh of a foam material. (b) and (c) show zoom-in details of a local region with topology ambiguity (see the red circle). The green line in (b) denotes one non-manifold edge, which is resolved in (c).

Our technique is a combination of DC and MC. When we generate the base mesh, we analyze the sign change edges and construct a dual of the octree; when we resolve and approximate the ambiguities, we traverse all the octree leaf cells to re-connect the mesh. In this manner we have the advantages of both DC (can handle large datasets with adaptive octree; better qualities) and MC (can preserve topology ambiguities). Furthermore, to make meshes conformal across cell boundaries, the re-meshing templates should be carefully designed so that no discrepancies or gaps exist on both sides of one cell face. Meanwhile, only face ambiguities influence the meshes in other cells. As a result, if all face ambiguities are handled consistently, as what we have done in our algorithm, we will obtain conformal meshes everywhere.

7 Conclusion

We have developed an efficient and automatic method to generate triangular iso-surfaces with correct topology. We begin from the standard DC method and generate one mesh without considering topology ambiguities. To resolve these ambiguities, we first model the interior regions with a tri-linear interpolation, and then compute the face and body saddle points, and then discuss all the 31 possible configurations. After categorizing all the cells into different topology groups, we traverse these cells and re-connect the triangles according to a set of pre-defined templates. In the end we relocate the vertex positions and thus improve the mesh quality. We have already applied our approach to three complicated datasets and obtained good results. In the future we will extend our algorithm to domains with multiple components. Instead of continuous function values, these data grids have distinct component IDs. Since one cell may have up to eight components, the ambiguity resolving is much more complicated and deserves great attention.

Acknowledgement

We would like to thank W. Wang for useful discussions on topology categorization. The research was supported in part by a NSF/DoD-MRSEC seed grant.

References

1. Ashida, K., Badler, N.: Feature preserving manifold mesh from an octree. In: The 8th ACM Symposium on Solid Modelling and Applications 2003, pp. 292–297 (2003)
2. Bajaj, C., Pascucci, V., Schikore, D.: Fast isocontouring for improved interactivity. In: IEEE Symposium on Volume Visualization 1996, pp. 39–46 (1996)
3. Brodsky, D., Watson, B.: Model simplification through refinement. In: Graphics Interface, pp. 221–228 (2000)
4. Chernyaev, E.: Marching cubes 33: Construction of topologically correct isosurfaces. Technical Report CH/95-17, CERN (1995)
5. Durst, M.: Letters: Additional reference to marching cubes. Computer Graphics 22, 72–73 (1988)
6. Garland, M., Heckbert, P.: Simplifying surfaces with color and texture using quadric error metrics. In: IEEE Visualization 1988, pp. 263–269 (1988)
7. Garland, M., Shaffer, E.: A multiphase approach to efficient surface simplification. In: IEEE Visualization 2002, pp. 117–124 (2002)
8. George, P.L., Borouchaki, H.: Delaunay triangulation and meshing, application to finite elements. Hermes Science Publisher (1998)
9. Nielson, G.: Dual marching cubes. In: IEEE Visualization 2004, pp. 489–496 (2004)
10. Ju, T., Losasso, F., Schaefer, S., Warren, J.: Dual contouring of Hermite data. ACM Transactions on Graphics 21, 339–346 (2002)
11. Kanaya, T., Teshima, Y., Kobori, K., Nishio, K.: A topology-preserving polygonal simplification using vertex clustering. In: GRAPHITE, pp. 117–120 (2005)
12. Kobbelt, L., Botsch, M., Schwanecke, U., Seidel, H.: Feature sensitive surface extraction from volume data. In: SIGGRAPH 2001, pp. 57–66 (2001)
13. Lopes, A., Brodlie, K.: Improving the robustness and accuracy of the marching cubes algorithm for isosurfacing. IEEE Transactions on Visualization and Computer Graphics 9(1), 16–29 (2003)
14. Lorensen, W., Cline, H.: Marching cubes: A high resolution 3D surface construction algorithm. Computer Graphics 21, 163–169 (1987)

15. Natarajan, B.: On generating topologically correct isosurfaces from uniform samples. The Visual Computer 11, 52–62 (1994)
16. Nielson, G., Hamann: The asymptotic decider: Resolving the ambiguity in marching cubes. In: IEEE Visualization 1992, pp. 83–91 (1992)
17. Schaefer, S., Ju, T., Warren, J.: Manifold dual contouring. IEEE Transactions on Visualization and Computer Graphics 13(3), 610–619 (2007)
18. Schaefer, S., Warren, J.: Dual marching cubes: Primal contouring of dual grids. In: Computer Graphics and Applications 2004, pp. 70–76 (2004)
19. Varadhan, G., Krishnan, S., Kim, Y., Manocha, D.: Feature-sensitive subdivision and iso-surface reconstruction. In: IEEE Visualization 2003, pp. 99–106 (2003)
20. Westermann, J.R., Kobbelt, L., Ertl, T.: Real-time exploration of regular volume data by adaptive reconstruction of isosurfaces. The Visual Computer 15(2), 100–111 (1999)
21. Wilhelm, J., Van Gelder, A.: Octrees for faster isosurface generation. ACM Transactions on Graphics, 57-62 (1992)
22. Zhang, N., Hong, W., Kaufman, A.: Dual contouring with topology preserving simplification using enhanced cell representation. In: IEEE Visualization 2004, pp. 505–512 (2004)
23. Zhang, Y., Bajaj, C.: Adaptive and quality quadrilateral/hexahedral meshing from volumetric data. Computer Methods in Applied Mechanics and Engineering 195(9-12), 942–960 (2006)
24. Zhang, Y., Bajaj, C., Sohn, B.: 3D finite element meshing from imaging data. Computer Methods in Applied Mechanics and Engineering 194, 5083–5106 (2005)
25. Zhang, Y., Bajaj, C., Xu, G.: Surface smoothing and quality improvement of quadrilateral/hexahedral meshes with geometric flow. Communications in Numerical Methods in Engineering 25(1), 1–18 (2009)
26. Zhang, Y., Hughes, T., Bajaj, C.: An automatic 3D mesh generation method for domains with multiple materials. Computer Methods in Applied Mechanics and Engineering 199(5-8), 405–415 (2010)

MESH SIZE, ORIENTATION, and ANISOTROPY (Monday Afternoon)

High Quality Geometric Meshing of CAD Surfaces

Patrick Laug[1] and Houman Borouchaki[2]

[1] INRIA Paris-Rocquencourt, France
   patrick.laug@inria.fr
[2] UTT, Troyes, France
   houman.borouchaki@utt.fr

Summary. A wide range of surfaces can be defined by means of composite parametric surfaces as is the case for most CAD modelers. There are, essentially, two approaches to meshing parametric surfaces: direct and indirect. Popular direct methods include the octree-based method, the advancing-front-based method and the paving-based method working directly in the tridimensional space. The indirect approach consists in meshing the parametric domain and mapping the resulting mesh onto the surface. Using the latter approach, we propose a general "geometry accurate" mesh generation scheme using geometric isotropic or anisotropic metrics. In addition, we introduce a new methodology to control the mesh gradation for these geometric meshes in order to obtain finite element geometric meshes. Application examples are given to show the pertinence of our approach.

Keywords: parametric surface meshing, curve discretization, anisotropic meshing, mesh gradation, geometric meshes.

1 Introduction

Surface meshing is involved in many numerical fields which include the finite element method. It is a necessary step when one wants to construct the mesh of a solid domain in three dimensions. Generally, isotropic meshes are used in solid mechanics while anisotropic meshes are preferred in CFD (computational fluid dynamics) as directional fields must be captured. A wide range of surfaces can be defined by means of composite parametric surfaces. Most of the surfaces are approximated by polynomial or rational parametric patches as is the case for most CAD modelers. In this case, the indirect approach (consisting in meshing the parametric domain and mapping the resulting mesh onto the surface) is conceptually straightforward as a planar mesh is generated in the parametric domain. In this paper, we are interested in generating geometry preserving meshes called geometric meshes for finite element computation.

Despite its simplicity, the problem with the indirect approach is the generation of a mesh which conforms to the metric of the surface. Historically, people were initially interested in surface visualization using this indirect approach. In fact, they aimed to minimize the error in the polyhedral approximation of the surface indirectly in the parametric space without paying attention to the quality of the resulting mesh [1, 2, 3, 4]. The mesh in the parametric surface is usually anisotropic, due to the metric deformation from the surface to its parametric domain. Thus, for people in finite element computation, the problem is reduced to the generation of an anisotropic mesh in the parametric domain. To this end, various algorithms are proposed [5, 6, 7, 8]. In addition, one can control explicitly the accuracy of a generated element with respect to the geometry of the surface if careful attention is paid. Indeed, a mesh of a parametric patch whose element vertices belong to the surface is "geometrically" suitable if all mesh elements are close to the surface and if every mesh element is close to the tangent planes related to its vertices. A mesh satisfying these properties is called a geometric mesh. The first property allows us to bound the gap between the elements and the surface. This gap measures the largest distance between an element (any point of the element) and the surface. The second property ensures that the surface is locally of order G^1 in terms of continuity. To obtain this, the angular gap between the element and the tangent plane at its vertices must be bounded. These properties result in the definition of a mesh metric map depending of surface curvatures called geometric metrics and the goal is to generate a unit mesh (all elements are of unit size with respect to the geometric metrics).

We propose a general scheme of an indirect approach for generating isotropic and anisotropic geometric meshes of a surface constituted by a conformal assembly of parametric patches, based on the concept of metric. The different steps of the scheme are detailed and, in particular, the definition of the geometric metric at each point of the surface (internal to a patch, belonging to an interface or boundary curve, or extremity of such a curve) as well as its corresponding induced metric in parametric domains.

Isotropic or anisotropic geometric metrics can locally produce significant size variations (internal to a patch or across interface curves) and can even be discontinuous along the interface curves. The larger the rate of the mesh size variation, the worse is the shape quality of the resulting mesh. To control this size variation, various methodologies based on metric reduction have been proposed [9] in the case of a continuous isotropic metric. We introduce a novel iterative mesh gradation approach for discontinuous metrics. The approach uses a particular metric reduction procedure in order to ensure the convergence of the gradation process. In particular, we show that in the worst case the anisotropic discontinuous geometric metric map is reduced to a isotropic continuous geometric metric map for which the gradation is controlled.

In Section 2, we introduce and detail the general scheme for meshing composite parametric surfaces. The new mesh gradation control is developed in Section 3. Several application examples are provided in Section 4 to illustrate the capabilities of the proposed method. Finally, in the last section, we conclude with a few words about the prospects.

2 Methodology

A surface Σ composed of parametric patches is defined by a collection of surface patches Σ_i fitted together in a "conforming" manner (see equation (6) below) and verifying:

$$\Sigma = \bigcup_i \Sigma_i, \quad \Sigma_i = \sigma_i(\Omega_i) \tag{1}$$

where Ω_i is a domain of $\mathbb{R}^2$ (parametric domain) and σ_i is a C^1 continuous application:

$$\sigma_i : \Omega_i \subset \mathbb{R}^2 \to \Sigma_i \subset \mathbb{R}^3, \quad \begin{pmatrix} u \\ v \end{pmatrix} \mapsto \sigma_i(u,v) \in \mathbb{R}^3 \tag{2}$$

Each domain Ω_i is defined by its contour, closed and non self-intersecting, constituted by a collection of contiguous curve segments γ_{ij} in $\mathbb{R}^2$:

$$\overline{\Omega_i} = \bigcup_j \gamma_{ij}, \quad \gamma_{ij} = \omega_{ij}([a_{ij}, b_{ij}]) \tag{3}$$

with

$$\omega_{ij} : [a_{ij}, b_{ij}] \subset \mathbb{R} \to \gamma_{ij} \subset \mathbb{R}^2, \quad t \mapsto \omega_{ij}(t) \in \mathbb{R}^2 \tag{4}$$

thus verifying

$$\gamma_{ij} \cap \gamma_{ik} = \emptyset \quad \text{or} \quad e_{il} \tag{5}$$

where $\emptyset$ denotes the empty set and e_{il} a common extremity of curve segments γ_{ij} and γ_{ik}.

Surface Σ is conforming if and only if:

$$\Sigma_i \cap \Sigma_j = \emptyset \quad \text{or} \quad \bigcup_k E_{ij,k} \quad \text{or} \quad \bigcup_k \Gamma_{ij,k} \tag{6}$$

where $\exists\, l, m$ such that $E_{ij,k} = \sigma_i(e_{il}) = \sigma_j(e_{jm})$
and $\exists\, l, m$ such that $\Gamma_{ij,k} = \sigma_i(\gamma_{il}) = \sigma_j(\gamma_{jm})$.

Therefore, $\Gamma_{ij,k}$ is a boundary curve segment shared by Σ_i and Σ_j, image of two boundary curve segments γ_{il} of Ω_i and γ_{jm} of Ω_j. Thus, by considering common curve segments only once, we obtain:

$$\bigcup_i \overline{\Sigma_i} = \bigcup_j \Gamma_j \tag{7}$$

where

$$\Gamma_j \cap \Gamma_k = \emptyset \quad \text{or} \quad E_{jk} \tag{8}$$

and there exists a set of indices (i,k) such that each Γ_j equals $\sigma_i(\gamma_{ik})$.

We suppose in the following that Σ is conforming (see [10] for setting the conformity of any surface). The generation of a mesh of Σ following an indirect approach is given by the following general scheme:

1. Specification of a size map or metric map associated with points of Σ.
2. Discretization of each Γ_j.
3. Transfer of the discretization of each Γ_j onto corresponding segments γ_{ik}.
4. Mesh generation of each Ω_i from the discretization of its boundary (obtained in the previous step).
5. Mapping the mesh of each Ω_i onto Σ_i.
6. Construction of the mesh of Σ from meshes of Σ_i.

These different steps are detailed in the following (for further information, see references cited in each step):

2.1 Size Map or Metric Map

Within a classical framework, mainly two categories of size maps or metric maps can be considered. *The first category* concerns uniform meshes with a given constant size h or a given constant metric $\mathcal{M} = \frac{1}{h^2}\mathcal{I}_3$ (the size specification results in a given metric and a mesh complying with this size is a mesh whose edge length equals unity in this metric). The advantage of this kind of meshing is that it provides, in general, equilateral meshes. On the other hand, it cannot guarantee a good representation of the geometry of the domain for a given size. *The second category* concerns meshes referred to as geometric, adapted to the geometry of the patches composing the surface. To define the size or the metric at a given point of the surface, three cases are discussed hereafter: internal point, interface or boundary point and extremity point.

Internal point. An internal point P is a point belonging to the interior of a patch Σ_i. In an isotropic framework, it can be demonstrated that locally the geometric size at P must be proportional to the minimal radius of curvature $\rho_1(P)$ of patch Σ_i [11]:

$$\mathcal{M}_{iso}(\Sigma_i, P) = \frac{1}{h_1^2(P)} \mathcal{I}_3 \quad \text{with} \quad h_1(P) = \lambda_1\, \rho_1(P) \tag{9}$$

where $\lambda_1 = 2\sin\theta$, θ being the maximum angle between an element and tangent planes to the surface, or equivalently $\lambda_1 = 2\sqrt{\varepsilon(2-\varepsilon)}$, ε being the maximum relative distance between an element and the surface. In an anisotropic framework, the metric can also be deduced from the principal radii of curvature ($\rho_1(P) < \rho_2(P)$) and the principal directions of curvature (defined by two orthogonal unit vectors $\vec{v_1}(P)$ and $\vec{v_2}(P)$) of patch i [12]:

$$\mathcal{M}_{aniso}(\Sigma_i, P) = \begin{pmatrix} \vec{v_1}(P) & \vec{v_2}(P) \end{pmatrix} \begin{pmatrix} \frac{1}{h_1^2(P)} & 0 \\ 0 & \frac{1}{h_2^2(P)} \end{pmatrix} \begin{pmatrix} \vec{v_1}(P)^T \\ \vec{v_2}(P)^T \end{pmatrix} \tag{10}$$

with $h_1(P) = \lambda_1\, \rho_1(P)$ and $h_2(P) = \lambda_2\, \rho_2(P)$, where λ_1 can be defined again by $\lambda_1 = 2\sqrt{\varepsilon(2-\varepsilon)}$ and λ_2 is a smaller coefficient given by $\lambda_2 = 2\sqrt{\varepsilon\frac{\rho_1}{\rho_2}(2-\varepsilon\frac{\rho_1}{\rho_2})}$. The above anisotropic geometric metric is degenerate since the size is not defined in the direction orthogonal to the plane containing $\vec{v_1}$ and $\vec{v_2}$. In order to obtain a well-defined metric consistent with the isotropic case, we redefine the anisotropic geometric as:

$$\mathcal{M}_{aniso}(\Sigma_i, P) = \begin{pmatrix} \vec{v_1}(P) & \vec{v_2}(P) & \vec{n}(P) \end{pmatrix} \begin{pmatrix} \frac{1}{h_1^2(P)} & 0 & 0 \\ 0 & \frac{1}{h_2^2(P)} & 0 \\ 0 & 0 & \frac{1}{h_1^2(P)} \end{pmatrix} \begin{pmatrix} \vec{v_1}(P)^T \\ \vec{v_2}(P)^T \\ \vec{n}(P)^T \end{pmatrix} \tag{11}$$

where $\vec{n}(P)$ is the unit normal to the surface at the considered point. In practice, the sizes in the above metrics are bounded by specified minimal and maximal size values and thus these metrics are always well defined.

The defined geometric metrics allows us to bound by a specified threshold the angular deviation θ of each element with respect to the tangent planes at its vertices. The Hausdorff distance between each element and the surface can be expressed by these angular deviations. To bound this distance by a threshold value, it is sufficient to consider the related angular deviation and thus the corresponding geometric metric. In this case, the angular deviation θ depends on the considered vertex.

Interface or boundary point. An interface or boundary point C is a point belonging to the interior of a curve segment Γ_j. For an interface point, curve Γ_j is shared by at least two patches while for an boundary point, curve Γ_j belongs to only one patch. Let us denote by $\{\Sigma_{ij}\}$ the set of patches containing Γ_j. The geometric size at C depends on the geometric size of each Σ_{ij} and also the geometric size of curve Γ_j. If $\rho(C)$ is the radius of curvature of curve Γ_j at C, the geometric size of curve Γ_j is defined by:

$$\mathcal{M}(\Gamma_j, C) = \frac{1}{h^2(C)} \mathcal{I}_3 \quad \text{with} \quad h(C) = \lambda_1 \rho(C). \tag{12}$$

Hence, at an interface or boundary point C, several geometric metrics are defined ($\mathcal{M}_{iso}(\Sigma_{ij}, C)$ or $\mathcal{M}_{aniso}(\Sigma_{ij}, C)$ and $\mathcal{M}(\Gamma_j, C)$).

Extremity point. An extremity point E is a common extremity of a set of curves $\{\Gamma_j\}$. Each $\{\Gamma_j\}$ belongs to a set of patches $\{\Sigma_{ij}\}$. Therefore, the geometric size at E depends on the geometric size of each curve Γ_j and the geometric size of corresponding patches Σ_{ij}. Similarly, at an extremity point E, several geometric metrics ($\mathcal{M}_{iso}(\Sigma_{ij}, E)$ or $\mathcal{M}_{aniso}(\Sigma_{ij}, E)$ for all i, j such that Σ_{ij} contains a curve $\{\Gamma_j\}$ with E as extremity and $\mathcal{M}(\Gamma_j, E)$ for all j such that E is an extremity of Γ_j) are defined.

Remark: size variation. The problem with this kind of meshing (geometric meshing) is that it can produce a very important variation of the size according to the variation of curvature. The shape quality of the elements largely depends of the size variation underlying in the metric field. To remedy this, it is sufficient to modify the metric field according to the desired size variation. To control the latter, methods of size smoothing or mesh gradation control can be considered. This issue is detailed in the next section.

2.2 Discretization of Curve Segments Γ_j

The discretization of each curve segment consists in subdividing the curve by curve segments of unit length with respect to a specified isotropic metric function. For each point C of a curve, this metric length is obtained regarding the metric at the point C in the direction of the tangent to the curve. In the geometric case, as mentioned above, several metrics are defined ($\mathcal{M}_{iso}(\Sigma_{ij}, C)$ or $\mathcal{M}_{aniso}(\Sigma_{ij}, C)$) on adjacent patches, and $\mathcal{M}(\Gamma_j, C)$ on the curve). Thus the "metric length" at C is the minimum length specified by these metrics in the direction of the tangent at C to the curve. To compute the length of a curve segment with respect to a metric, a polyline approximating the curve is constructed and the length of this polyline is calculated (this length computation allows us to subdivide the curve by segments of unit length).

2.3 Inverse Mapping of the Discretization of Γ_j in Parametric Domains

The discretization of Γ_j is defined by a set of vertices ordered by their curvilinear abscissae. This discretization is mapped back to the corresponding curve segments γ_{ik} in parametric domains. The discretization of all curve segments γ in the parametric domains being well defined, the corresponding metrics in parametric domains must now be provided. These bidimensional metrics will be calculated from metrics in the tridimensional space that are defined in the following.

For an interface or boundary point C of a curve segment Γ_j belonging to a given patch Σ_{ij}, the metric $\mathcal{M}_{iso}(\Sigma_{ij}, C)$ or $\mathcal{M}_{aniso}(\Sigma_{ij}, C)$ is shrunk to fit the metric length at C in the direction of the tangent to the curve giving the new geometric metric $\overline{\mathcal{M}}_{iso}(\Sigma_{ij}, C)$ or $\overline{\mathcal{M}}_{aniso}(\Sigma_{ij}, C)$. For an extremity point E of a patch Σ_{ij}, the same procedure is applied considering each interface or boundary curve Γ_j of Σ_{ij} such that E is an extremity of Γ_j leading to different geometric metrics $\overline{\mathcal{M}}_{iso}(\Sigma_{ij}, E)$ or $\overline{\mathcal{M}}_{aniso}(\Sigma_{ij}, E)$ and we consider the new geometric metric at E the metric $\overline{\mathcal{M}}_{iso}(\Sigma_{ij}, E)$ or $\overline{\mathcal{M}}_{aniso}(\Sigma_{ij}, E)$ giving the smallest size along the tangent direction at each curve Γ_j. Thus for an extremity point E, the geometric metric with respect to a patch Σ_{ij} is such that the minimal metric length at E is satisfied.

As an illustration, Fig. 1 (left) shows an interface curve Γ shared by two patches Σ_1 and Σ_2. Using the previous notations, Γ is in fact equal to a curve Γ_j, and Σ_1 (resp. Σ_2) is equal to a patch $\Sigma_{i_1 j}$ (resp. $\Sigma_{i_2 j}$). As explained in section 2.2, the metric length at a point C belonging to the interior of Γ is the minimum length specified by the three metrics $\mathcal{M}_1 = \mathcal{M}_{aniso}(\Sigma_{i_1 j}, C)$, $\mathcal{M}_2 = \mathcal{M}_{aniso}(\Sigma_{i_2 j}, C)$ and $\mathcal{M}(\Gamma_j, C)$ in the direction of the tangent τ at C to the curve. In this example, the minimum length l_{min} is given by the latter metric. Consequently, the shrunk metrics $\overline{\mathcal{M}}_1 = \overline{\mathcal{M}}_{aniso}(\Sigma_{i_1 j}, C)$ and $\overline{\mathcal{M}}_2 = \overline{\mathcal{M}}_{aniso}(\Sigma_{i_2 j}, C)$ are represented by ellipsoids centered at C and passing through a same point of τ at a distance l_{min} of C.

On Fig. 1 (right), an extremity E is shared by two curves $\Gamma_1 = \Gamma_{j_1}$ and $\Gamma_2 = \Gamma_{j_2}$ at the boundary of patch $\Sigma = \Sigma_{ij_1} = \Sigma_{ij_2}$. The previous process gives one point on tangent τ_1 to Γ_1 and a second point on tangent τ_2 to Γ_2, and the corresponding metrics $\overline{\overline{\mathcal{M}}}_1 = \overline{\overline{\mathcal{M}}}_{aniso}(\Sigma_{ij_1}, E)$ and $\overline{\overline{\mathcal{M}}}_2 = \overline{\overline{\mathcal{M}}}_{aniso}(\Sigma_{ij_2}, E)$. In this new example, the minimal metric length is given by the second metric and thus the geometric metric at E with respect to patch Σ is $\overline{\overline{\mathcal{M}}} = \overline{\overline{\mathcal{M}}}_2$ or $\overline{\overline{\mathcal{M}}}_{aniso}(\Sigma, E) = \overline{\overline{\mathcal{M}}}_{aniso}(\Sigma_{ij_2}, E)$.

Fig. 1. Left: anisotropic metrics at an interface point C belonging to the interior of a curve segment Γ shared by two patches Σ_1 and Σ_2. Right: anisotropic metrics at a common extremity E of two curves Γ_1 and Γ_2 bounding a patch Σ.

2.4 Mesh Generation of Domains Ω_i

We use an indirect method for meshing general parametric surfaces conforming to a pre-specified metric map $\mathcal{M}_3$ (for more details, see [11]). Let Σ be such a surface parameterized by:

$$\sigma : \Omega \longrightarrow \Sigma, \qquad (u,v) \longmapsto \sigma(u,v), \qquad (13)$$

where Ω denotes the parametric domain. The Riemannian metric specification $\mathcal{M}_3$ gives the unit measure in any direction. In the geometric case this metric is defined as:

- internal point: $\mathcal{M}_{iso}(\Sigma_i)$ or $\mathcal{M}_{aniso}(\Sigma_i)$.
- interface or boundary point: $\overline{\mathcal{M}}_{iso}(\Sigma_i)$ or $\overline{\mathcal{M}}_{aniso}(\Sigma_i)$.
- extremity point: $\overline{\overline{\mathcal{M}}}_{iso}(\Sigma_i)$ or $\overline{\overline{\mathcal{M}}}_{aniso}(\Sigma_i)$.

The goal is to generate a mesh of Σ such that the edge lengths are equal to one with respect to the related Riemannian space (such meshes being referred to as "unit" meshes). Based on the intrinsic properties of the surface, namely the first fundamental form:

$$\mathcal{M}_\sigma = \begin{pmatrix} \sigma_u^T \sigma_u & \sigma_u^T \sigma_v \\ \sigma_v^T \sigma_u & \sigma_v^T \sigma_v \end{pmatrix}, \tag{14}$$

the Riemannian structure $\mathcal{M}_3$ is induced into the parametric space as follows:

$$\widetilde{\mathcal{M}}_2 = \begin{pmatrix} \sigma_u^T \\ \sigma_v^T \end{pmatrix} \mathcal{M}_3 \begin{pmatrix} \sigma_u & \sigma_v \end{pmatrix}. \tag{15}$$

The above equation is the product of three matrices respectively of order 2×3, 3×3 and 3×2, resulting in a metric of order 2×2 in the parametric domain.

Even if the metric specification $\mathcal{M}_3$ is isotropic, the induced metric in parametric space is in general anisotropic, due to the variation of the tangent plane along the surface. Finally, a unit mesh is generated completely inside the parametric space such that it conforms to the induced metric $\mathcal{M}_2$. This mesh is constructed using a combined advancing-front – Delaunay approach applied within a Riemannian context: the field points are defined after an advancing front method and are connected using a generalized Delaunay type method.

This method is efficient if the metric $\mathcal{M}_\sigma$ of the first fundamental form of the surface is well defined and its variation is bounded. If this is not the case, one can consider the metric in the vicinity of the degenerated points.

2.5 Mapping Back the Mesh of Each Ω_i onto Σ_i

The mesh of each Σ_i is constituted by vertices, images by σ_i of the vertices of the mesh of Ω_i, keeping the same connectivity. This methodology is functional if the tangent plane metric does not involve strong variations (i.e., the image of an edge of the mesh of the parametric domain is close to the straight segment joining the images of its extremities).

2.6 Construction of the Mesh of Σ from Meshes of Σ_i

The global mesh of Σ is obtained by gathering all the meshes of patches Σ_i. In this process, vertices of the discretizations of the boundary curves must not be duplicated.

3 Mesh Gradation

The metric $\mathcal{M}_3$ can locally produce important size variations, in particular in the present context of geometric mesh generation. These size variations entail a generation of elements having a poor shape quality. To remedy this, metric $\mathcal{M}_3$ can be modified while accounting for the size constraints at best and while controlling the underlying gradation, which measures the size variation in the vicinity of a vertex [9]. The general scheme of the mesh gradation methodology has several steps:

High Quality Geometric Meshing of CAD Surfaces

1. Generation of the initial geometric mesh, controlled by the geometric metrics detailed in the previous section.
2. Computation of the geometric metrics at the vertices of this mesh.
3. Modification of these metrics in order to bound the gradation by a specified threshold c_{goal}.
4. Generation of an adapted mesh controlled by the modified metrics.
5. Repetition of the three steps 2–4, once or several times.

The purpose of the step repetition is to accurately capture the surface geometry, and in practise it is applied only once. In the following, for each case of isotropic or anisotropic geometric metrics, the step 3 is detailed and the corresponding algorithm is given.

3.1 Isotropic Geometric Metrics

As indicated in section 2.3, at the end of step 2 of the general scheme, the geometric metric at each vertex of the mesh is defined by:

- If the vertex is a point P belonging to the interior of a patch Σ_i, its metric is unique and defined by $\mathcal{M}_{iso}(\Sigma_i, P)$.
- If the vertex is a point C belonging to the interior of a curve segment Γ_j, several metrics $\overline{\mathcal{M}}_{iso}(\Sigma_{ij}, C)$ are defined. However, in the case of isotropic geometric metrics, these metrics are identical for all patches $\Sigma_{i,j}$ because all these isotropic metrics give the same length in the direction of the tangent to Γ_j. This common metric is then denoted by $\overline{\mathcal{M}}_{iso}(\Sigma_{*j}, C)$.
- If the vertex is a point E being a common extremity of a set of curves $\{\Gamma_j\}$, several metrics $\overline{\mathcal{M}}_{iso}(\Sigma_{*j}, E)$ corresponding to every $\{\Gamma_j\}$ are defined. Among these metrics, there exists a metric denoted by $\overline{\mathcal{M}}_{iso}(\Sigma_{**}, E)$ which gives the smallest length in all directions. The latter metric is taken into account in the gradation control methodology.

The modification of the geometric metrics consists in locally modifying these metrics by considering the size variation on each edge of the mesh. For each edge, the modification includes two successive steps – the calculation of the shock and, if necessary, a metric update – which are detailed below.

Calculation of the shock. Let PQ be an edge, and let $\mathcal{M}(P)$ and $\mathcal{M}(Q)$ be the metrics at its extremities. If $h(P)$ and $h(Q)$ respectively represent the sizes specified by these metrics (in all directions and in particular in the direction of vector $\overrightarrow{PQ}$), let us assume without loss of generality that $h(P) \leq h(Q)$. The H-shock (or more simply the shock) $c(PQ)$ related to the edge PQ is the value:

$$c(PQ) = \left(\frac{h(Q)}{h(P)}\right)^{1/l(PQ)} \quad (16)$$

where $l(PQ)$ is the length of edge PQ in a metric interpolating the size given by the two extremity metrics $\mathcal{M}(P)$ and $\mathcal{M}(Q)$ in direction $\overrightarrow{PQ}$:

$$l(PQ) = ||\overrightarrow{PQ}|| \int_0^1 \frac{1}{h(P + t\overrightarrow{PQ})} \, dt \qquad (17)$$

Metric update. If the shock $c(PQ)$ is greater than the given threshold c_{goal}, then the size $h(Q)$ is multiplied by η, or equivalently the metric $\mathcal{M}(Q)$ is divided by η^2, where η is a size reduction factor given by:

$$\eta = \left(\frac{c_{goal}}{c(PQ)}\right)^{l(PQ)} < 1 \qquad (18)$$

General algorithm. Using the above notations, the gradation algorithm for isotropic metrics can be written in simplified pseudo-code as shown on Fig. 2. Its inputs are the mesh, the geometric metrics $\mathcal{M}$ at the mesh vertices, and the threshold c_{goal}. The outer loop runs until $c_{max} \leq c_{goal}$, where c_{max} is the maximum shock on all the edges. Consequently, in output, metrics are modified so that the gradation is bounded by the given threshold c_{goal}.

Input: mesh, $\mathcal{M}_{iso}$, c_{goal}
Repeat {
    $c_{max} = 0$
    For each edge PQ of the mesh {
        Compute $c(PQ)$, the shock on PQ
        If $(c(PQ) > c_{goal})$ update $\mathcal{M}_{iso}(Q)$
        $c_{max} = \max(c_{max}, c(PQ))$
    }
} until $(c_{max} \leq c_{goal})$
Output: $\mathcal{M}_{iso,gra} = \mathcal{M}_{iso}$

Fig. 2. Gradation algorithm in the isotropic case.

3.2 Anisotropic Geometric Metrics

In the anisotropic case, the geometric metrics at each vertex of the mesh are defined as follows:

- If the vertex is a point P belonging to the interior of a patch Σ_i, its metric is unique and defined by $\mathcal{M}_{aniso}(\Sigma_i, P)$.
- If the vertex is a point C belonging to the interior of a curve segment Γ_j, several metrics $\overline{\mathcal{M}}_{aniso}(\Sigma_{ij}, C)$ are defined. Indeed, the anisotropic metric is discontinuous at C.
- If the vertex is a point E being a common extremity of a set of curves $\{\Gamma_j\}$, several metrics $\overline{\mathcal{M}}_{aniso}(\Sigma_{ij}, E)$ corresponding to every $\{\Gamma_j\}$ are defined. In addition, for all patch Σ_{ij} containing E, the metrics $\mathcal{M}_{aniso}(\Sigma_{ij}, E)$ are also considered. Notice also that the metric is discontinuous at E.

Before we give the general algorithm for the gradation of these anisotropic metrics, several points must be clarified concerning the calculation of the shock and the metric update.

Calculation of the shock. Let PQ be an edge of a mesh of a patch Σ_i. For each extremity, for instance point P, are defined a metric $\mathcal{M}(P)$ and a direction $\vec{v}(P)$ as follows:

- If PQ is an internal edge of the mesh, $\vec{v}(P) = \vec{PQ}$ and there are three possibilities for metric $\mathcal{M}(P)$: if P belongs to the interior of Σ_i, $\mathcal{M}(P) = \mathcal{M}_{aniso}(\Sigma_i, P)$; if P belongs to the interior of a curve Γ_j, $\mathcal{M}(P) = \overline{\mathcal{M}}_{aniso}(\Sigma_{ij}, P)$; if P is an extremity, $\mathcal{M}(P) = \mathcal{M}_{aniso}(\Sigma_i, P)$ independently from any curve Γ_j.
- Otherwise, PQ belongs to the discretization of a curve Γ_j. Direction $\vec{v}(P)$ is given by the tangent to Γ_j at P (see section 2.3) and metric $\mathcal{M}(P)$ is defined by $\overline{\mathcal{M}}_{aniso}(\Sigma_{ij}, P)$.

Denoting by $h(P)$ the size specified by metric $\mathcal{M}(P)$ in direction $\vec{v}(P)$, sizes $h(P)$ and $h(Q)$ are defined at both extremities of PQ and the shock $c(PQ)$ is calculated like in the isotropic case using equations (16) and (17).

Metric update. If the shock $c(PQ)$ is greater than the given threshold c_{goal}, a metric update is necessary. The procedure detailed in section 3.1 for the isotropic case is rather straightforward: assuming that $h(P) < h(Q)$, the metric associated with Q is divided by a factor η^2. In the anisotropic case, this procedure is more complicated, as explained in the following.

Firstly, the metric map is discontinuous along interface curves and thus, as mentioned above, several metrics are associated with a given point. The metric to be updated for a point P is $\mathcal{M}(P)$ whose definition is given above, depending on edge PQ (internal or not) and on point P (internal to a patch, internal to a curve, or extremity). Careful attention must be paid if $\mathcal{M}(P) = \overline{\mathcal{M}}_{aniso}(\Sigma_{ij}, P)$, a metric defined on a curve Γ_j. In this case, the updated metric gives a new metric length in the direction of the tangent to Γ_j at P, and all the patches sharing Γ_j must be updated so that their local metrics give the same metric length at P (as in section 2.3).

A second point is that a simple homothetic reduction of a metric $\mathcal{M}(P)$ does not guarantee the convergence of the gradation process. Indeed, a reduction on one patch implies other reductions on adjacent patches, which may imply a reduction on the first patch, resulting in an endless loop. To avoid this, the key idea is to run beforehand an isotropic gradation defining an isotropic metric at each vertex. The latter is used as a lower limit for the anisotropic gradation. This methodology guarantees the convergence of the process and ensures that a smaller number of elements is generated in the anisotropic case. More precisely, let h_{lim} be the size limit given by the isotropic metric, let $h_1 < h_2$ be the sizes along the principal axes of a metric $\mathcal{M}$, and let η be the size reduction factor. To reduce metric $\mathcal{M}$ with a factor $\eta < 1$, first a homothetic reduction replaces h_1 by $h_1' = \eta h_1$ and h_2 by

$h'_2 = \eta h_2$. However, if $h'_1 < h_{lim}$, h'_1 is set to h_{lim} and h'_2 is computed so that the size in direction $\overrightarrow{PQ}$ is $h'(P) = \eta h(P)$. This procedure is illustrated on Fig. 3. Metric $\mathcal{M}$ is represented by the outer ellipse and h_{lim} is the radius of the inner circle. If η is near 1 then a homothetic reduction is made, but if η becomes smaller the metric becomes "more isotropic". The prior isotropic gradation guarantees that it is never necessary to go below the size limit h_{lim}.

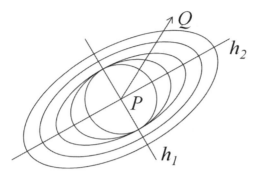

Fig. 3. Reduction of a metric complying with a lower bound h_{lim}.

Thirdly, a problem may occur during the anisotropic gradation process. If a shock $c(PQ) > c_{goal}$ is detected on an edge PQ such that $h(P) < h(Q)$, a metric update at Q may be impossible because the size limit is reached: $h_1 = h_2 = h_{lim}$. This may happen because the edges are not analyzed in the same order in the isotropic and anisotropic gradations. In this case, it is still possible to update the metric at the other extremity P; an iterative procedure finds a new reduction factor $\eta_P < 1$ such that the shock on edge PQ is less than c_{goal}, and metric $\mathcal{M}(P)$ is reduced with this factor η_P.

General algorithm. The gradation algorithm in the anisotropic case is written in simplified pseudo-code on Fig. 4 with inputs and outputs similar to the isotropic case.

4 Application Examples

The methodology is implemented in the BLSURF software package [13]. To illustrate our approach for high quality geometric meshing, two examples of CAD surfaces are presented in this section, which represent respectively a crank and a propeller. In each example, the input is an IGES file read by Open Cascade platform and the surface meshes are generated by BLSURF.

Crank. In this first example, we consider a simplified crank. Fig. 5 shows a representation of the CAD model and three isotropic geometric meshes,

```
Input: mesh, $\mathcal{M}_{aniso}$, $c_{goal}$
Run the gradation algorithm in the isotropic case, giving $\mathcal{M}_{iso,gra}$
Repeat {
    $c_{max,1} = 0$
    For each patch $\Sigma_i$ of the mesh {
        Repeat {
            $c_{max,2} = 0$
            For each edge $PQ$ of patch $\Sigma_i$ {
                Compute $c(PQ)$, the shock on $PQ$
                If $(c(PQ) > c_{goal})$ update $\mathcal{M}_{aniso}(Q)$ or else $\mathcal{M}_{aniso}(P)$
                $c_{max,2} = \max(c_{max,2}, c(PQ))$
            }
            $c_{max,1} = \max(c_{max,1}, c_{max,2})$
        } until $(c_{max,2} \leq c_{goal})$
    }
} until $(c_{max,1} \leq c_{goal})$
Adjust $\mathcal{M}_{aniso}(E)$, the metrics at the extremities
Output: $\mathcal{M}_{aniso,gra} = \mathcal{M}_{aniso}$
```

Fig. 4. Gradation algorithm in the anisotropic case.

while Fig. 6 shows four anisotropic geometric meshes. Some corresponding statistics are displayed in Table 1. The CAD surface is made up of 12 patches. For all these geometric meshes, an angle $\theta = 2$ degrees is specified, where θ is the maximum angle between each triangle and the underlying tangent planes to the surface (see section 2.1).

The first geometric mesh of Fig. 5 is isotropic, without gradation. As pointed out at the end of section 2.1, elongated triangles can be noticed because of important variations of the surface curvature. To remedy this, a gradation of 2.5 is applied on the metric field, showing an improvement of the shape quality. With a smaller threshold of 1.5, the mesh triangles become almost equilateral. The relative number of elements for these three isotropic meshes is respectively 1.000, 1.118 and 1.287.

To reduce the number of elements with the same geometric accuracy, anisotropic geometric meshes are built. The first mesh of Fig. 6, without gradation, contains less than 9.9 % of the corresponding number of triangles in the isotropic case. Again, a gradation 2.5 (resp. 1.5) has been applied to produce the second (resp. third) mesh. Finally, an even better shape quality can obtained by limiting the aspect ratio of the metrics. The fourth mesh shown is generated with a threshold of 2.5 for the metric aspect ratio. Compared with an isotropic mesh with a same gradation, the relative number of elements is respectively 0.099, 0.157, 0.240 and 0.491 (the latter having a limited anisotropy).

For the last example with 12,424 triangles, the total CPU time on a Dell Precision mobile workstation M6400 at 2.53 GHz is 0.811 seconds. This includes the input of the CAD file, the setting of the topology, the generation of the initial mesh and the two adapted meshes (cf. section 3) and the output of the mesh file. In the first adaptation, the isotropic gradation runs 5 iterations and the anisotropic gradation 2 more iterations, and in the second adaptation the number of iterations is $8 + 3$, which are executed in a negligible time.

Table 1. Crank: meshing statistics.

mesh	gradation	aspect ratio	vertices	triangles
iso	∞	∞	9826	19656
iso	2.5	∞	10988	21980
iso	1.5	∞	12644	25292
aniso	∞	∞	969	1942
aniso	2.5	∞	1726	3456
aniso	1.5	∞	3030	6064
aniso	1.5	2.5	6210	12424

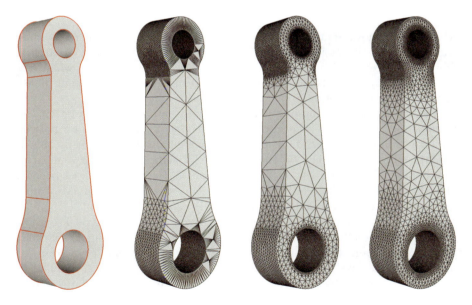

Fig. 5. Crank, from left to right: CAD model, isotropic meshes with gradation ∞, 2.5, and finally 1.5.

Propeller. In this second example, we consider a submarine propeller. Similarly to the previous section, isotropic and anisotropic meshes are shown on Fig. 7. Close up pictures are shown on 8 and meshing statistics are displayed

High Quality Geometric Meshing of CAD Surfaces

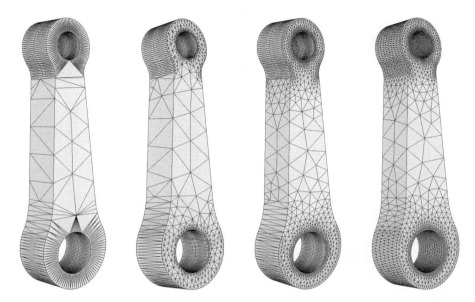

Fig. 6. Crank, from left to right: anisotropic meshes with gradation ∞, 2.5, 1.5, and finally same gradation 1.5 with aspect ratio 2.5.

in Table 2. The CAD surface is made up of 10 patches, one for each of the 7 blades and 3 for the boss. The angle specified for all these geometric meshes is now $\theta = 4$ degrees.

Fig. 7 (top left) shows an isotropic mesh of the propeller without gradation, resulting in a poor shape quality due to the varying curvatures of the curves and surfaces. This quality is improved with a gradation of 2.5 and triangles are almost equilateral with a gradation of 1.5. The relative number of elements for these three isotropic meshes is respectively 1.000, 1.006 and 1.131, showing a lesser increase than in the previous example. Anisotropic meshes with gradation ∞, 2.5, 1.5, and finally same gradation 1.5 with aspect ratio 2.5 are also generated. Compared with an isotropic mesh with a same gradation, the relative number of elements is respectively 0.402, 0.448, 0.644, 2.033. This ratio is higher than in the previous example (and can even be greater than one) because the default value of h_{min}, the minimum size of the element edges, is 100 times smaller in the anisotropic case than in the isotropic case. This minimum is reached in this example because of the high curvatures near the leading edge of each blade. Therefore, the geometric accuracy is much better in the anisotropic case, with a comparable number of elements.

For the last example with 211,745 triangles, the total time for I/O and meshing is 24.289 seconds. The first gradation runs 6 + 3 iterations and the second 16 + 3 iterations with a negligible execution time.

Table 2. Propeller: meshing statistics.

mesh	gradation	aspect ratio	vertices	triangles
iso	∞	∞	46093	92055
iso	2.5	∞	47150	92606
iso	1.5	∞	52149	104159
aniso	∞	∞	18585	37028
aniso	2.5	∞	21656	41508
aniso	1.5	∞	33600	67053
aniso	1.5	2.5	107168	211745

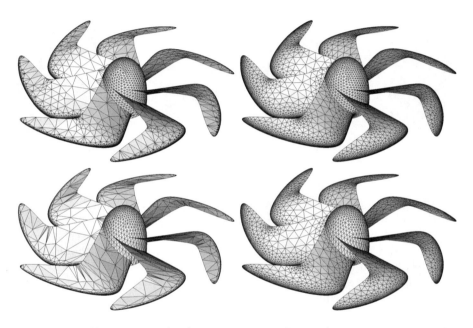

Fig. 7. Propeller: isotropic (top) and anisotropic (bottom) meshes without gradation and with gradation 1.5.

5 Conclusion

The general scheme of an indirect approach for meshing a surface constituted by a conformal assembly of parametric patches has been introduced and each step of the general scheme has been detailed. Emphasis has been placed on the geometric mesh generation, based on continuous isotropic and discontinuous anisotropic geometric metrics. In addition, a new mesh gradation control strategy for discontinuous anisotropic geometric metrics has been proposed. This strategy can be applied to control the gradation for volume meshing from a 3D continuous metric map. The proposed methodology has been applied to numerical examples, showing its efficiency.

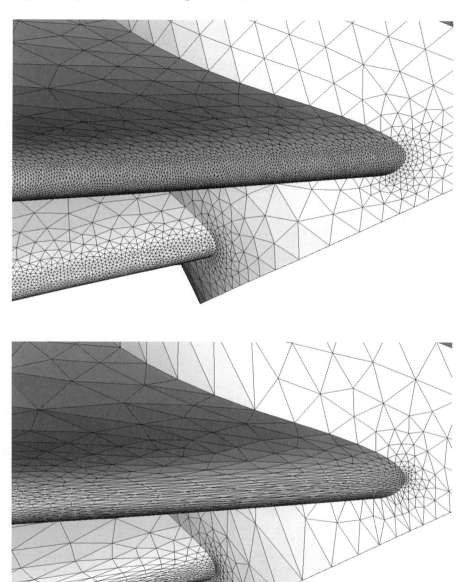

Fig. 8. Propeller: close up views of geometric meshes with gradation 1.5, either isotropic or anisotropic.

Future works include the parallelization of the geometric meshing process for large complex geometry, the parallelization of the mesh gradation strategy and the patch independent anisotropic geometric meshing.

References

1. Dolenc, A., Mäkelä, I.: Optimized triangulation of parametric surfaces. Mathematics of Surfaces IV (1990)
2. Filip, D., Magedson, R., Markot, R.: Surface algorithm using bounds on derivatives. Comput.-Aided Geom. Des. 3, 295–311 (1986)
3. Piegl, L.A., Richard, A.M.: Tessellating trimmed NURBS surfaces. Computer Aided Design 27(1), 16–26 (1995)
4. Sheng, X., Hirsch, B.E.: Triangulation of trimmed surfaces in parametric space. Computer Aided Design 24(8), 437–444 (1992)
5. Borouchaki, H., George, P.L., Hecht, F., Laug, P., Saltel, E.: Delaunay mesh generation governed by metric specifications – Part I: Algorithms. Finite Elements in Analysis and Design (FEAD) 25(1-2), 61–83 (1997)
6. Bossen, F.J., Heckbert, P.S.: A pliant method for anisotropic mesh generation. In: 5th International Meshing Roundtable 1996 Proceedings, pp. 375–390 (1996)
7. Shimada, K.: Anisotropic triangular meshing of parametric surfaces via close packing of ellipsoidal bubbles. In: 6th International Meshing Roundtable 1997 Proceedings, pp. 63–74 (1997)
8. Tristano, J.R., Owen, S.J., Canann, S.A.: Advancing front surface mesh generation in parametric space using a Riemannian surface definition. In: 7th International Meshing Roundtable 1998 Proceedings, pp. 429–445 (1998)
9. Borouchaki, H., Hecht, F., Frey, P.: Mesh gradation control. Int. J. Numer. Meth. Engng 43, 1143–1165 (1998)
10. Renaut, E.: Reconstruction de la topologie et génération de maillages de surfaces composées de carreaux paramétrés. Thèse de doctorat UTT/INRIA soutenue le 15 décembre (2009)
11. Borouchaki, H., Laug, P., George, P.L.: Parametric surface meshing using a combined advancing-front – generalized-Delaunay approach. Int. J. Numer. Meth. Engng. 49(1-2), 233–259 (2000)
12. Laug, P., Borouchaki, H.: Interpolating and meshing 3-D surface grids. Int. J. Numer. Meth. Engng 58(2), 209–225 (2003)
13. Laug, P., Borouchaki, H.: BLSURF – Mesh generator for composite parametric surfaces – User's manual. INRIA Technical Report RT-0235 (1999)

Isotropic 2D Quadrangle Meshing with Size and Orientation Control

Bertrand Pellenard[1], Pierre Alliez[1], and Jean-Marie Morvan[2,3]

[1] INRIA Sophia Antipolis - Méditerranée, France
[2] Université Lyon 1/CNRS, Institut Camille Jordan, 43 blvd du 11 Novembre 1918, F-69622 Villeurbanne - Cedex, France
 `morvan@math.univ-lyon1.fr`
[3] King Abdullah University of Science and Technology, GMSV Research Center, Bldg 1, Thuwal 23955-6900, Saudi Arabia

Summary. We propose an approach for automatically generating isotropic 2D quadrangle meshes from arbitrary domains with a fine control over sizing and orientation of the elements. At the heart of our algorithm is an optimization procedure that, from a coarse initial tiling of the 2D domain, enforces each of the desirable mesh quality criteria (size, shape, orientation, degree, regularity) one at a time, in an order designed not to undo previous enhancements. Our experiments demonstrate how well our resulting quadrangle meshes conform to a wide range of input sizing and orientation fields.

1 Introduction

Quadrangle meshes are preferred over triangle meshes in a number of applications related to computer graphics, computer aided geometric design, computational engineering and reverse engineering. However, the automatic generation of isotropic quadrangle meshes for arbitrary 2D domains is still a scientific challenge due to the variety of requirements and quality criteria sought after.

The bare minimum requirement we impose that the input domain must be tiled with only *convex* quadrangles. We also wish to generate quadrangle meshes for which, *locally*, elements i) are well shaped in the sense of being close to squares, ii) are sized in accordance to an input sizing field, iii) are oriented in accordance to an input cross field (an orientation field modulo 90 degrees) and iv) have edges aligned to the domain boundary. A more *global* requirement practitioners often desire is that meshes should predominantly be composed of regular vertices, i.e., degree-4 vertices [1] inside the domain and (in general) degree-2 vertices on the domain boundary.

While these quality requirements are widely regarded as desirable, one key meshing difficulty is that they often conflict with one another: irregular vertices are often necessary for non-trivial sizing and cross fields, but they inevitably induce shape

---
[1] A degree-k vertex has k adjacent inside cells.

distortion and hence must be avoided whenever possible; a rapidly varying sizing field naturally induces both shape and orientation distortion; similarly, a rapidly varying cross field often results in both shape and size distortion. Another key challenge comes from the fact that some of the requirements or quality criteria, although locally defined, have global constraints—e.g., the number of edges on the domain boundary must be even, and the total index of irregular vertices must obey Gauss-Bonnet theorem [13].

1.1 Previous Work

The tension between local and global criteria may explain the variety of approaches proposed so far for the automatic generation of quadrangle meshes. Including methods devised to generate quadrangle surface tilings, the rich literature on this topic contains approaches which proceed by quadrangulation [9, 3], square packing [28], advancing front [22], conversion [8], decimation [18], Morse-Smale complexes [14, 33, 17], clustering [4, 20], local and global operators [23, 19], whisker weaving [32], medial axis [24], streamlining [1, 21] and parameterization [25, 29, 7].

Among these approaches, some favor the conformance of the final mesh to an input cross field either by construction [21], or by solving for the smoothest cross field given a set of orientation constraints [7]. Conformance to an input sizing field is either derived from the triangle mesh before conversion [8], or encoded in a density function before clustering [20]. Mesh regularity is controlled either explicitly by interactively placing a small number of irregular vertices before parameterization [29], or indirectly through a smooth cross field [7]. Regularity may be improved a posteriori through, e.g., grid-preserving operators so as to generate simple base complexes [6]. Strict local conformance to both sizing and cross fields is notoriously delicate for most approaches which involve a global variational formulation, and almost none of the fine-to-coarse approaches based on decimation [18, 23] leads to meshes that conform to both sizing and cross fields.

1.2 Contribution

In this paper we present a simple and practical isotropic quadrangle meshing algorithm for arbitrary 2D domains. We place a particular emphasis on having the resulting mesh conforming to both a sizing and a cross field. Our approach differs from previous work in that its methodology can be seen as antithetical to (e.g., Delaunay) refinement algorithms that locally refine a mesh one element at a time until all quality criteria are met. Rather than continuously fixing the element that most violate *any* of the requirements as in the Delaunay refinement strategy, we instead enhance the mesh by carefully *addressing one requirement at a time, in a processing order designed not to undo previous enhancements*. Table 1 describes how each step of the algorithm improves the various desirable quality criteria for flat 2D meshes.

Table 1. Each step of the algorithm improves different quality criteria. •: criterion is partially met; ●: criterion is met; ×: criterion may not be preserved; ○: criterion is preserved.

	Size	Shape	Orientation	Degree	Regularity
Initialization (2.2)	●	•			
Relaxation (2.3)	○	●	●		
Conforming relaxation (2.4)	○	○	○	•	•
Local parameterizations (2.5)	○	○	○	•	●
Barycentric subdivision (2.6)	○	×	○	●	○
Smoothing (2.7)	○	●	○	○	○

2 Algorithm

The input of our algorithm is a closed domain, together with a sizing and a cross field. The sizing field is either user specified or automatically computed as a Lipschitz function from the local feature size estimate of the domain boundary [2]. The cross field is either specified by the user or automatically computed as the smoothest field that is tangential to the domain boundary [13]. Note that the mixed-integer approach [7] could also be used. The algorithm mainly proceeds by clustering and local parameterizations over a fine isotropic background triangle mesh $\mathcal{M}$ obtained by Delaunay refinement. We give pseudo-code of the algorithm below, while Figure 1 provides a visual depiction of its main steps.

Algorithm 1. Quadrangle meshing

Input: 2D domain, sizing field, cross field
begin
  1. Initialization (**2.2**)
  2. Relaxation (**2.3**)
  3. Conforming relaxation (**2.4**)
  4. Local parameterizations (**2.5**)
  5. Barycentric subdivision (**2.6**)
  6. Smoothing (**2.7**)
end
Output: Quadrangle mesh

2.1 Preliminaries

The main steps of the algorithm (from initialization to local parameterizations through relaxations) act on the background mesh $\mathcal{M}$. In the reminder of this paper, we will make heavy use of the following terms (see Figure 2):

- **Tile.** A tile is a union of triangles of $\mathcal{M}$ defining a subdomain homeomorphic to a disk.
- **Meta-vertex.** A meta-vertex is a vertex of $\mathcal{M}$ which is either incident to 3 or more tiles inside the domain, or incident to 2 or more tiles on the domain boundary, or incident to a single tile when the vertex is tagged as a corner boundary vertex.

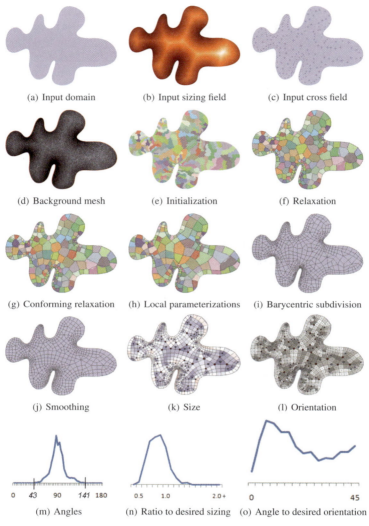

Fig. 1. Overview. The algorithm takes as input a closed domain (a), a sizing field (b) and a cross field (c). It then operates on a triangle background mesh refined according to the sizing field (d). The initialization clusters background mesh triangles (e) so that the tiling roughly meets the size and shape criteria; A relaxation (f) then improves the tiling for shape and orientation while preserving size; A conforming relaxation (g) improves the degree of the tiles and the regularity of the tiling; A series of local parameterizations (h) further improves the degrees and regularity; Barycentric subdivision (i) generates a pure quadrangle mesh; Smoothing (j) finally improves the shape of the quadrangles. We depict the conformance to the sizing field with a color ramp ranging from white to blue (resp. white to red), for elements smaller (resp. larger) than the specified sizing field (k). We depict the conformance to the cross field with a color ramp ranging from white to gray (l). Irregular vertices are outlined in red for degree excess and in blue for degree deficit. We show the distribution of angles, as well as distributions measuring conformance to sizing and orientation. 1000 quadrangles, total time: 40 s.

- **Meta-edge.** A meta-edge connects two meta-vertices through an edge path of $\mathcal{M}$ such that all edges along the path are incident to 2 tiles in the interior, or 1 tile on the boundary. A tile is thus surrounded by a cycle of meta-edges.
- **Side.** A side is a chain of meta-edges around a tile. In particular, assuming a tile is surrounded by a cycle of at least 4 meta-edges, a subset of four of its meta-vertices can be chosen to represent quadrangle corners so that the tile has 4 sides. This assignment of sides will be useful in our algorithm since we target quadrangle elements.

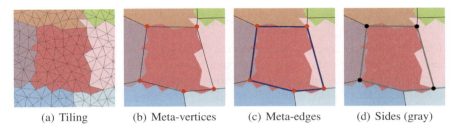

(a) Tiling   (b) Meta-vertices   (c) Meta-edges   (d) Sides (gray)

Fig. 2. Terminology defined in Section 2.1.

2.2 Initialization

We construct the background mesh $\mathcal{M}$ as a 2D constrained Delaunay triangulation that fits a polyline approximating the domain boundary. We then refine $\mathcal{M}$ through Delaunay refinement [27, 26] until all triangles are both well shaped (isotropic) and sized in accordance to the input sizing field within a small fraction (by default 0.1).

The initialization step aims at generating a tiling that conforms to the sizing field, and that roughly meets the shape criterion (see Table 1). First, we proceed by generating one tile per triangle of $\mathcal{M}$. We then recursively merge pairs of tiles using a priority queue of merging operations until the sizing field requirement is met (see Figure 1(e)). In order to favor isotropic tiles the merging operations are popped out of the queue in order of decreasing compactness, the latter being defined as the ratio between area and squared perimeter. We have experimented with another score which favors squares instead of discs but the simpler compactness score is sufficient. The idea of this step is similar to [16].

2.3 Relaxation

The first relaxation step aims at optimizing the initial tiling for the shape and orientation criteria while preserving the size (see Table 1). Optimization is carried out through a discrete clustering algorithm which operates over the background mesh so as to favor squares tiles which are both well sized and well oriented with respect to the sizing and cross fields ([31]).

Similar in spirit to [20], we consider the $L^\mathcal{R}_\infty$ metric related to a local Cartesian coordinate frame $\mathcal{R} = (\mathbf{u}, \mathbf{v})$ specified by the cross field. We compute an approximate discrete Voronoi diagram over the background mesh instead of computing the exact continuous Voronoi diagram [5], so as to deal with a simple triangle labeling procedure. The $L^\mathcal{R}_\infty$ distance between two points $p, q \in \mathbb{R}^2$ is defined as:

$$d^\mathcal{R}_\infty(p, q) = \max(|(p - q) \cdot \mathbf{u}|, |(p - q) \cdot \mathbf{v}|).$$

Using the continuous formulation the minimized energy is defined as:

$$\mathcal{G}\left(\{z_i\}_{i=1}^N, \{V_i\}_{i=1}^N\right) = \sum_{i=1}^N \int_{V_i} \rho(x) \, d^\mathcal{R}_\infty(x, z_i) \, dx,$$

where ρ is a density function defined on the domain: $\rho(x) = s(x)^{-4}$ (s denotes the sizing function to be preserved); z_i is a point generator and V_i is the Voronoi cell of z_i. Using a discrete formulation [30] now involving the background mesh $\mathcal{M}$ we consider a set of triangle generators $\{g_i\}_{i=1}^N$, and a tiling $\{V_i\}_{i=1}^N$. The energy to minimize is defined as:

$$\mathcal{H}\left(\{g_i\}_{i=1}^N, \{T_i\}_{i=1}^N\right) = \sum_{i=1}^N \left(\sum_{t_j \in T_i} \rho(c(t_j)) \text{area}(t_j) d^\mathcal{R}_\infty(c(t_j), c(\mu(T_i)))\right),$$

where $c(t)$ denotes the centroid of triangle t and μ denotes the triangle that contains the centroid of a tile. Energy $\mathcal{H}$ is iteratively minimized by alternating $L^\mathcal{R}_\infty$ discrete Voronoi partitioning and relocation of the generators to their tile centroids. Algorithm 2 provides a pseudo-code for the relaxation step, where N is the number of tiles and K is the total number of iterations.

Algorithm 2. Relaxation

Input: Initial triangle generators $\{g_i^0\}_{i=1}^N$ and corresponding tiles $\{T_i^0\}_{i=1}^N$.
begin
    while *no convergence* **do**
        Discrete partitioning $\left(\{T_i^k\}_{i=1}^N \text{ partition associated to } \{g_i^k\}_{i=1}^N\right)$
        Relocate generators to centroids $\left(\forall i, 1 \leq i \leq n, \, g_i^{k+1} = \Gamma(g_i^k)\right)$
end
Output: Optimized triangle generators $\{g_i^K\}_{i=1}^N$ and corresponding tiles $\{T_i^K\}_{i=1}^N$.

Discrete partitioning is achieved by flooding the domain one triangle at a time from their generators with a dynamic priority queue [12]. Each tile is initialized to be its triangle generator, and a priority queue is filled with (up to three) incident triangles (candidates for flooding) per generator. In order to favor square tiles, the triangles are popped out of the queue and added to tiles in increasing order of $L^\mathcal{R}_\infty$ distance to their respective triangle generator, where $\mathcal{R}$ is the local Cartesian coordinate frame specified by the cross field at the generator centroid.

Relocation is achieved through computing the (triangle) center of mass of each tile. In the continuous case the center of mass of a tile does not depend on the metric chosen to measure the object, and we observe a very similar behavior in our discrete setup. After the k-th iteration, for each tile T_i^k, we choose to find the triangle that minimizes the following energy:

$$f(t) = \sum_{t_j \in T_i^k} \rho(c(t_j)) \, \text{area}(t_j) \, \|c(t_j) - c(t)\|_2^2,$$

where $t \in T_i^k$. We denote by $\Gamma : g_i^k \mapsto g_i^{k+1}$ the operation that computes the triangle centroid of the tile T_i^k associated to g_i^k. In such discrete algorithm convergence is reached when $\forall i, 1 \leq i \leq N, \Gamma(g_i^k) = g_i^k$. As convergence for the $L_\infty^\mathcal{R}$ metric on arbitrary domains is not guaranteed, we also specify a maximum number of iterations (by default set to 50). Figure 3 depicts some iterations.

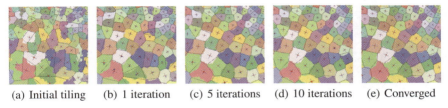

(a) Initial tiling (b) 1 iteration (c) 5 iterations (d) 10 iterations (e) Converged

Fig. 3. Discrete $L_\infty^\mathcal{R}$ iterations with a non-uniform sizing field (the cross field is shown).

As expected, the relaxation leads to a tiling of the domain with mostly square tiles which are well oriented and well sized (similar in spirit to a square packing approach [28]), even if a varying sizing field inevitably implies shape distortion. We further observe in Figure 4 that although well shaped, the tiles are generally not conforming (see the many T-junctions), hence most of them would generate degree-6 polygons and the final mesh would contain many irregular vertices (generally of degree 3). We describe next a conforming relaxation procedure which aims at generating quasi 2-conforming configurations.

2.4 Conforming Relaxation

A closer look at Figure 4 reveals that the tiling is in general already conforming in one direction. We call this configuration 1-conforming. This is explained by the fact that a general tiling of the plane with equally sized square tiles is 1-conforming (see Figure 5, left). Our goal is to further relax the tiling so as to obtain quasi 2-conforming configurations (5, middle).

An intuitive understanding of the conforming relaxation procedure can be gathered by looking at Figure 5(left) and realizing that we could shift the three square columns vertically as little as possible so as to tend toward a perfect 2-conforming configuration. In order to obtain the aforementioned shift, we propose to simply

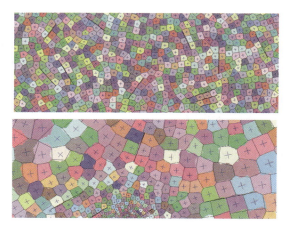

Fig. 4. After relaxation, for uniform (top) and non-uniform (bottom) sizing field. Tiles are in general conforming to one direction of the local cross field. The black lines depict some local conforming directions.

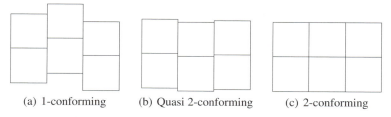

(a) 1-conforming (b) Quasi 2-conforming (c) 2-conforming

Fig. 5. Tiling with squares. Left: general configuration after relaxation. Middle: general configuration after conforming relaxation. Right: ideal configuration sought after.

shift the centroid during the relocation of a relaxation iteration. The only remaining technicality now resides in the way to compute the shift. Although simple at first glance, recovering the local 1-conforming direction (vertical or horizontal) is already non trivial, and so is finding the shift magnitude and orientation (see Figure 6). In addition, both size and cross fields vary over the domain, requiring not just shifting the tiles but also sacrificing aspect ratio to reach conforming.

Fig. 6. Shifting centroids. We depict examples of shifts with increasing ambiguity. A tile which is already quadrangle is not shifted (left). A tile with one side split into two meta-edges (middle left), the chosen quadrangle corners are depicted in black. A tile with two parallel sides split (middle). A tile with three sides split (middle right). Another ambiguous case (right).

Tiles which are already quadrangles or triangles (with 4 or 3 meta-vertices) are not shifted. While we assume that after relaxation all tiles are squares (geometry-wise), in general they are degree-6 tiles (see Figure 5(a)). We thus first need to choose which four of its meta-vertices form a square locally aligned to the cross field. We first compute the triangle centroid of the tile as during relaxation (see 2.3). Its cross (given by the input cross field) is now taken as local Cartesian coordinate frame $\mathcal{R} = (\mathbf{u}, \mathbf{v})$. Denote by $v_1, \ldots, v_p$ ($p > 4$) the meta-vertices of the tile, ordered by circulating along the tile boundary. To decide which of the meta-vertices are chosen as (ordered) corners c_1, c_2, c_3, c_4 we maximize through dynamic programming the alignment of the sides (see 2.1) with the axes of $\mathcal{R}$ through maximization of the following energy:

$$\mathcal{E} = \max_{\substack{\mathbf{a} \subset \{\mathbf{u},\mathbf{v}\} \\ \{c_1,c_2,c_3,c_4\} \subset \{v_1,\ldots,v_p\}}} \left[\min \left(|(c_2 - c_1) \cdot \mathbf{a}|, |(c_3 - c_2) \cdot \mathbf{a}^{90}|, |(c_4 - c_3) \cdot \mathbf{a}|, |(c_1 - c_4) \cdot \mathbf{a}^{90}| \right) \right],$$

where $\mathbf{a}^{90}$ denotes vector $\mathbf{a}$ (which stands for $\mathbf{u}$ or $\mathbf{v}$) rotated by 90 degrees.

In addition to assigning corners, the maximum of $\mathcal{E}$ for each side provides a local reference direction of the cross field ($\mathbf{u}$ or $\mathbf{v}$). For each meta-edge, we then compute the length of its projection on its reference direction ($\mathbf{u}$ or $\mathbf{v}$). Among all meta-edges, we then select only the ones with exactly one end meta-vertex coinciding to a corner meta-vertex, and pick the longest one, denoted by e. We then shift the centroid along a line parallel to $\mathbf{a}$, in the opposite direction to the corner meta-vertex of e. The magnitude of the shift is chosen as a fraction (0.2 in our experiments) of the length of the projection of e

(see inset). The number of iterations of conforming relaxations is a user-specified parameter (set to 20 for all examples shown). Figure 7 depicts how such conforming relaxation brings some initial 1-conforming configurations to quasi 2-conforming configurations. The latter can be fixed by a series of local parameterizations which we describe next.

2.5 Local Parameterizations

The previous conforming relaxation step favored quasi-2-conforming configurations (Figure 5(b)) that we wish to turn into 2-conforming configurations (Figure 5(c)) so as to improve both degree and regularity criteria (see Table 1) while not negatively affecting the previous efforts made by previous steps.

Figure 7 reveals that the many quasi-2-conforming configurations exhibit similar topological arrangements of the tiles. We call *butterfly* a set of four tiles incident to a short meta-edge connecting two degree-3 meta-vertices (see Figure 8). Inspired by [23] we remove many of these butterfly configurations through local parameterizations on square domains, which merge 2 degree-3 meta-vertices into one degree-4 meta-vertex.

For each butterfly, we consider the union of its four tiles A, B, C, D as a subdomain Ω_{ABCD}, and first perform a classification of its meta-vertices as depicted in

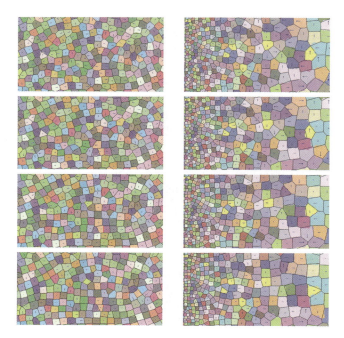

Fig. 7. Conforming relaxation with uniform sizing (left) and non-uniform sizing (right). From top to bottom: initial, 1 iteration, 5 iterations and 10 iterations. In order to bring some initial 1-conforming configurations to quasi 2-conforming configurations, centroids of two adjacent tiles are locally shifted, which can induce size distortion.

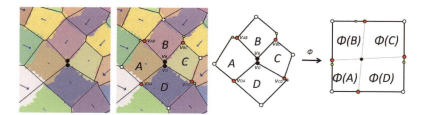

Fig. 8. Butterfly configuration. Left: a short meta-edge connects two degree-3 meta-vertices (outlined in black) which share four tiles. Middle left: among all meta-vertices of the four-tile subdomain we identify two inner meta-vertices (black), 4 interface meta-vertices separating the four tiles A, B, C, D pairwise (red), four corner meta-vertices (white) and the remaining meta-vertices (yellow). Middle right and right: the four tiles parameterized on a square. The parameterization preserves the topology of the interfaces on the boundary of the four tile subdomain. Some tiles may be smaller in parameter space so as to better fit the sizing field.

Figure 8. While the inner (black) and interface (red) meta-vertices can be easily classified from topological (adjacency) relationships between the tiles, distinguishing the four corners among the other meta-vertices (possibly many for rapidly varying sizing fields) requires incorporating a geometric criterion related to the local orientation of the tiles.

Contrary to the way we choose the corners for the conforming relaxation step, the orientation is, this time, not given by the cross field but instead depends on the relaxed four tiles (the rationale being, once again, to avoid undoing the previous enhancements). Denote by $V_{AB}, V_{BC}, V_{CD}, V_{DA}$ the four interface meta-vertices. We estimate a reference Cartesian frame by fitting two lines through principal component analysis to the segment sets $([V_{AB}, V_B], [V_{CD}, V_D])$ and $([V_{DA}, V_D], [V_{BC}, V_B])$ [11]. Among these two lines the most reliable one (i.e., the line with minimum variance in the orthogonal direction) is chosen as reference direction. To select four corners among the meta-vertices we again resort to a dynamic programming approach similar to the one used in Section 2.4.

Our goal is to parameterize Ω_{ABCD} on a square domain and to label its triangles (again with labels A,B,D,C) such that i) the chosen corner meta-vertices coincide with the corners of the square, ii) the interfaces at the boundary of Ω_{ABCD} are preserved, iii) the two inner meta-vertices are merged into a degree-4 vertex and iv) Ω_{ABCD} is partitioned with 4 tiles with a disk topology. Call ϕ the parameterization that maps Ω_{ABCD} on the square domain. We first constrain the whole boundary of $\phi(\Omega_{ABCD})$ so as to respect the chosen corners and using an arc-length parameterization in-between these corners. We then parameterize Ω_{ABCD} using the mean value coordinate approach [15] and compute, in parameter space, the intersection point $\phi(v^*)$ between the two line segments $(\phi(V_{AB})\phi(V_{CD}))$ and $(\phi(V_{BC})\phi(V_{DA}))$. The nearest vertex (of degree at least 4) from $\phi(v^*)$ is then chosen as inner vertex, which means that the issuing vertex v^* will be the center degree-4 meta-vertex. While simple at first glance once the inner and boundary vertices are decided upon, a naive triangle labeling step based on localization within quadrilaterals in parameter space can lead to improper topological partitioning. For this reason we trace four edge paths from $\phi(v^*)$ to $\phi(V_{AB}), \phi(V_{BC}), \phi(V_{CD}), \phi(V_{DA})$ in order of increasing length (shorter segments first) so as to determine proper interfaces between the triangle labels. These edge paths are constrained not to intersect except at the inner vertex and to connect no other boundary vertices than their target vertex $V_{AB}, V_{BC}, V_{CD}, V_{DA}$. Upon successful completion the triangles of Ω_{ABCD} are labeled, and this ends the butterfly removal procedure.

To avoid distorting the shape and orientation of the final mesh, we only remove butterflies whose inner meta-edge length is smaller than a fraction of the local sizing field (0.5 in our experiments). We use a dynamic priority queue to gracefully deal with butterflies in order of increasing inner meta-edge length. Figure 9 depicts how local parameterizations improve degree and regularity criteria for both uniform and non-uniform sizing field cases.

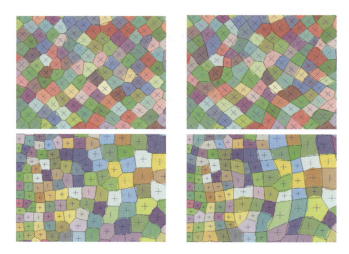

Fig. 9. Local parameterizations. Uniform sizing field (top left and top right) and non-uniform sizing field (bottom left and bottom right) before (left) and after (right) local parameterizations. The cross field is shown.

2.6 Barycentric Subdivision

After conforming relaxation there is no guarantee that all tiles are quadrangles. Experimentally we obtain an order of 75% quadrangles for uniform sizing and of 60% quadrangles for non-uniform sizing. We thus resort to barycentric subdivision (see Figure 1(i)) in order to meet the degree criterion (Table 1). The edge lengths of the quadrangles are reduced by a factor of two compared to the mesh before barycentric subdivision. This factor is taken into account during all previous steps of the algorithm.

2.7 Smoothing

Near T-junctions the previous barycentric subdivision step generates highly distorted quadrangles. We thus resort to a few iterations of Laplacian smoothing (see Figures 1(j) and 10) so as to enhance the shape criterion.

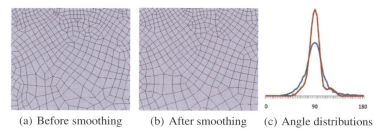

(a) Before smoothing (b) After smoothing (c) Angle distributions

Fig. 10. Smoothing. Mesh before (a) and after smoothing (b). We compare the distribution of angles before (blue) and after (red) smoothing.

3 Results

Our algorithm is implemented in C++ using the CGAL library [10]. All examples were computed on a 2.40GHz PC with a single thread. Each of our figures highlights irregular vertices in blue for degree deficit, and in red for degree excess. To depict conformance to the input cross field we use a color ramp ranging from white to gray, where white means perfect conformance and gray means 45 degree distortion. To depict conformance to the input sizing field we use a color ramp ranging from white to blue (resp. red) for tiles smaller (resp. larger) than targeted. We also depict distributions of angles, conformance to sizing, and conformance to cross field.

We ran a large number of examples of various size and complexity in order to assess our results and compare them to state-of-the-art methods. Figure 11 depicts how a final quadrangle mesh conforms to both size and orientation on a geographic map. The input sizing field is set to be the largest 1-Lipschitz function constrained to match the local feature size estimate of the input domain boundary [2]. The cross field is set to be the smoothest cross field constrained to be tangential to the input domain boundary. The distribution of angles is shown, with over 80% of angles within the interval [75 − 105]. The algorithm takes 350 seconds, with two third of the time spent on local parameterizations. The peak memory usage is 200 MBytes.

Figure 12 depicts an example with a constant cross field combined with a rapidly varying sizing field. Irregular vertices inevitably appear between dense and sparse mesh areas, and the orientation is partially distorted. Figure 13 shows a trivial domain example with a constant cross field and compares uniform vs. non-uniform sizing. Figure 14 depicts examples of uniform sizing and varying cross fields set to be smooth and tangential to the input domain boundary.

Figure 15 compares our results with [7] using a uniform sizing as this method is not primarily aimed at handling rapidly varying sizing fields. Our approach better preserves orientations near the domain boundary at the price of a larger number of irregular vertices.

We do not provide direct control over the final number of vertices of the final mesh as it depends on both the input sizing field and the number of degree-4 tiles after local parameterizations (experimentally near 70%). The efficiency of the algorithm can be improved by accelerating the relaxation step [30] and by numerical optimizations during the local parameterizations.

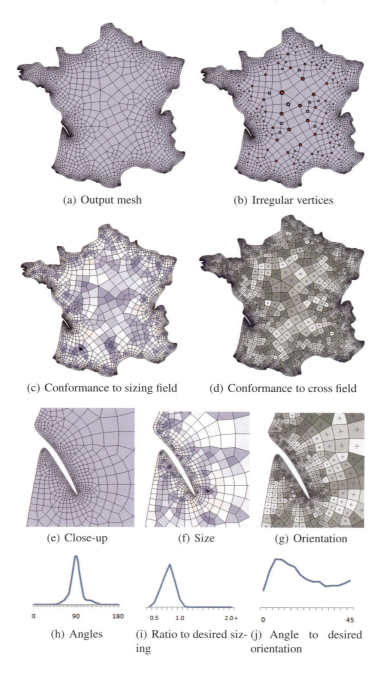

Fig. 11. France. The final mesh (a) contains 77% regular vertices (b). It conforms to the sizing field (c) as well as to the cross field (d). The close-up depicts an area where size and orientation vary rapidly. 4500 quadrangles, total time: 300 s.

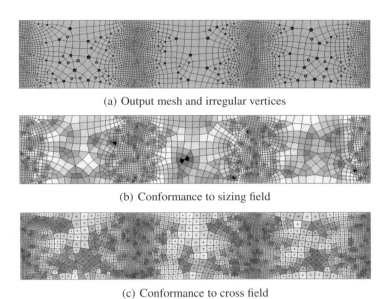

(a) Output mesh and irregular vertices

(b) Conformance to sizing field

(c) Conformance to cross field

Fig. 12. Non-uniform sizing field defined as $h(x) = 0.01\left(2 + \sin\left(6x\pi - \frac{\pi}{2}\right)\right)$, $0 < x < 1$. Total time: 480 s.

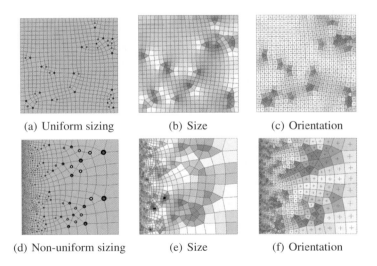

(a) Uniform sizing (b) Size (c) Orientation

(d) Non-uniform sizing (e) Size (f) Orientation

Fig. 13. Uniform vs. non-uniform sizing. The cross field is constant and axis-aligned. Total time: 30 s. and 60 s.

(a) Disc (b) Free form

Fig. 14. Non-uniform cross fields. Sizing fields are uniform. Total time: 45 s. and 450 s.

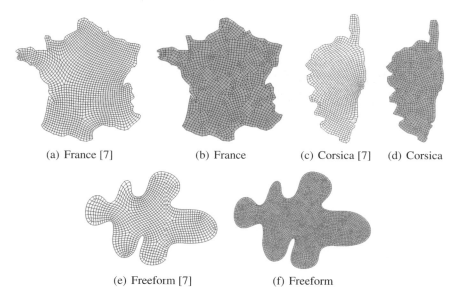

(a) France [7] (b) France (c) Corsica [7] (d) Corsica

(e) Freeform [7] (f) Freeform

Fig. 15. Comparison with Bommes et al. [7]. The cross field is set to be tangential to the input domain boundary. The meshes produced by [7] are more regular at the price of increased shearing distortion.

4 Conclusion

We proposed a principled approach for the automatic generation of quadrangle meshes from arbitrary 2D domains. The main methodology consists of enhancing a rough initial tiling by carefully addressing one meshing requirement at a time (size, shape, orientation, degree, regularity) in an order designed not to undo previous enhancements. While other similar approaches (such as [28]) only show examples on simple domains and smooth cross fields, our experiments confirm that the output quadrangles meshes conform both to the input sizing and cross fields, even on complex domains and for rapidly changing fields.

Size and orientation conformance comes at the price of a larger number of irregular vertices. For applications requiring coarse base complexes, we could potentially improve our approach by allowing the user to trade conformance to input fields for increased regularity. We also wish to improve the conforming step by resorting to a different strategy when it fails locally, and the parameterization step by making it more general for an arbitrary set of tiles clustered around a target irregular vertex.

Our approach is primarily based on relaxation and local parameterizations. Because the main steps of the algorithm only deal with labeling the triangles of a background triangle mesh, our implementation is simple and reliable, and is less prone to numerical robustness issues. Resorting to local parameterizations on trivial convex domains such as [23] provides us with scalability and robustness. For these reasons our approach could be extended to reliable quadrangle surface remeshing (including anisotropic quadrangle surface remeshing). Most of the main steps also seem to generalize to hexahedron domain tiling, except for the barycentric subdivision step. As future work we plan to alleviate this problem so as to extend the main principles behind our approach for hexahedron domain tiling.

Acknowledgments

The authors are grateful to Mathieu Desbrun for discussion and advice, and to David Bommes for experiments. This work was funded by the European Research Council (ERC Starting Grant 'Robust Geometry Processing', Grant agreement 257474), and by a French ANR Grant (GIGA ANR-09-BLAN-0331-01).

References

1. Alliez, P., Cohen-Steiner, D., Devillers, O., Lévy, B., Desbrun, M.: Anisotropic polygonal remeshing. ACM Trans. Graph. 22(3), 485–493 (2003)
2. Antani, L., Delage, C., Alliez, P.: Mesh sizing with additively weighted Voronoi diagrams. In: Proc. of the 16th Int. Meshing Roundtable, pp. 335–346 (2007)
3. Bern, M.W., Eppstein, D.: Quadrilateral meshing by circle packing. Int. J. Comput. Geometry Appl. 10(4), 347–360 (2000)
4. Boier-Martin, I.M., Rushmeier, H.E., Jin, J.: Parameterization of triangle meshes over quadrilateral domains. In: Symposium on Geometry Processing, pp. 197–208 (2004)
5. Boissonnat, J.-D., Sharir, M., Tagansky, B., Yvinec, M.: Voronoi diagrams in higher dimensions under certain polyhedral distance functions. Discrete & Computational Geometry 19(4), 485–519 (1998)
6. Bommes, D., Lempfer, T., Kobbelt, L.: Global structure optimization of quadrilateral meshes. Comput. Graph. Forum 30(2), 375–384 (2011)
7. Bommes, D., Zimmer, H., Kobbelt, L.: Mixed-integer quadrangulation. ACM Trans. Graph. 28(3) (2009)
8. Borouchaki, H., Frey, P.J.: Adaptive triangular-quadrilateral mesh generation. International Journal of Numerical Methods in Engineering 41, 915–934 (1996)
9. Bremner, D., Hurtado, F., Ramaswami, S., Sacristan, V.: Small strictly convex quadrilateral meshes of point sets. Algorithmica 38(2), 317–339 (2003)
10. CGAL, Computational Geometry Algorithms Library, http://www.cgal.org
11. Principal component analysis in CGAL, http://www.cgal.org
12. Cohen-Steiner, D., Alliez, P., Desbrun, M.: Variational shape approximation. ACM Trans. Graph. 23, 905–914 (2004)

13. Crane, K., Desbrun, M., Schröder, P.: Trivial connections on discrete surfaces. Comput. Graph. Forum 29(5), 1525–1533 (2010)
14. Dong, S., Bremer, P.-T., Garland, M., Pascucci, V., Hart, J.C.: Spectral surface quadrangulation. ACM Trans. Graph. 25(3), 1057–1066 (2006)
15. Floater, M.S.: Mean value coordinates. Comput. Aided Geom. Des. 20, 19–27 (2003)
16. Garland, M., Willmott, A.J., Heckbert, P.S.: Hierarchical face clustering on polygonal surfaces. In: SI3D, pp. 49–58 (2001)
17. Huang, J., Zhang, M., Ma, J., Liu, X., Kobbelt, L., Bao, H.: Spectral quadrangulation with orientation and alignment control. In: ACM SIGGRAPH Asia 2008 papers, SIGGRAPH Asia 2008, pp. 147:1–147:9 (2008)
18. Daniels II, J., Silva, C.T., Cohenn, E.: Localized quadrilateral coarsening. Comput. Graph. Forum 28(5), 1437–1444 (2009)
19. Lai, Y.-K., Kobbelt, L., Hu, S.-M.: An incremental approach to feature aligned quad dominant remeshing. In: Proceedings of the ACM symposium on Solid and physical modeling, SPM 2008, pp. 137–145 (2008)
20. Lévy, B., Liu, Y.: L^p centroidal voronoi tessellation and its applications. ACM Trans. Graph. 29(4) (2010)
21. Marinov, M., Kobbelt, L.: Direct anisotropic quad-dominant remeshing. In: Pacific Conference on Computer Graphics and Applications, pp. 207–216 (2004)
22. Owen, S.J., Staten, M.L., Canann, S.A., Saigal, S.: Q-morph: an indirect approach to advancing front quad meshing. International Journal for Numerical Methods in Engineering 44(9), 1317–1340 (1999)
23. Pietroni, N., Tarini, M., Cignoni, P.: Almost isometric mesh parameterization through abstract domains. IEEE Trans. Vis. Comput. Graph. 16(4), 621–635 (2010)
24. Quadros, W.R., Ramaswami, K., Prinz, F.B., Gurumoorthy, B.: Laytracks: A new approach to automated quadrilateral mesh generation using medial axis transform. In: IMR, pp. 239–250 (2000)
25. Ray, N., Li, W.C., Lévy, B., Sheffer, A., Alliez, P.: Periodic global parameterization. ACM Trans. Graph. 25(4), 1460–1485 (2006)
26. Rineau, L., Yvinec, M.: A generic software design for delaunay refinement meshing. Comput. Geom. 38(1-2), 100–110 (2007)
27. Shewchuk, J.R.: Delaunay refinement algorithms for triangular mesh generation. Comput. Geom. 22(1-3), 21–74 (2002)
28. Shimada, K., Liao, J.-H., Itoh, T.: Quadrilateral meshing with directionality control through the packing of square cells. In: IMR, pp. 61–75 (1998)
29. Tong, Y., Alliez, P., Cohen-Steiner, D., Desbrun, M.: Designing quadrangulations with discrete harmonic forms. In: Symposium on Geometry Processing, pp. 201–210 (2006)
30. Valette, S., Chassery, J.-M.: Approximated centroidal voronoi diagrams for uniform polygonal mesh coarsening. Comput. Graph. Forum 23(3), 381–390 (2004)
31. Valette, S., Chassery, J.-M., Prost, R.: Generic remeshing of 3d triangular meshes with metric-dependent discrete voronoi diagrams. IEEE Trans. Vis. Comput. Graph. 14(2), 369–381 (2008)
32. Wolfenbarger, P., Jung, J., Dohrmann, C.R., Witkowski, W.R., Panthaki, M.J., Gerstle, W.H.: A global minimization-based, automatic quadrilateral meshing algorithm. In: IMR, pp. 87–103 (1998)
33. Zhang, M., Huang, J., Liu, X., Bao, H.: A wave-based anisotropic quadrangulation method. ACM Trans. Graph. 29(4) (2010)

Anisotropic Goal-Oriented Mesh Adaptation for Time Dependent Problems

Frederic Alauzet[1], Anca Belme[2], and Alain Dervieux[2]

[1] INRIA Roquencourt, Domaine de Voluceau, 78150 Rocquencourt, France
  Frederic.Alauzet@inria.fr
[2] INRIA Sophia Antipolis, 2004 route de Lucioles, 06902 Sophia-Antipolis, France
  Anca.Belme@inria.fr, Alain.Dervieux@inria.fr

Summary. We present a new algorithm for combining an anisotropic goal-oriented error estimate with the mesh adaptation fixed point method for unsteady problems. The minimization of the error on a functional provides both the density and the anisotropy (stretching) of the optimal mesh. They are expressed in terms of state and adjoint. This method is used for specifying the mesh for a time sub-interval. A global fixed point iterates the re-evaluation of meshes and states over the whole time interval until convergence of the space-time mesh. Applications to unsteady blast-wave are presented.

Keywords: Unsteady compressible flow, goal-oriented mesh adaptation, anisotropic mesh adaptation, adjoint, metric.

Introduction

Engineering problems frequently require computational fluid dynamics (CFD) solutions with functional outputs of specified accuracy. The computational resources available for these solutions are often limited and errors in solutions and outputs are difficult to control. CFD solutions may be computed with an unnecessarily large number of mesh vertices (and associated high cost) to ensure that the outputs are computed within a required accuracy.

One of the powerful methods for increasing the accuracy and reducing the computational cost is mesh adaptation, the purpose of which is to control the accuracy of the numerical solution by changing the discretization of the computational domain according to mesh size and mesh directions constraints.

Pioneering works have shown a fertile development of Hessian-based or metric-based methods [12, 9] which rely on an ideal representation of the *interpolation error* and of the *mesh*. The "multiscale" version relies on the optimization of the $\mathbf{L}^p$ norm of the interpolation error [14]. It allows to take into account the discontinuities with higher-order convergence [18]. However, these methods are limited to the minimization of some interpolation errors for some solution fields, the "sensors", and do not take into account the PDE being solved. If for many applications, this simplifying standpoint is an advantage, there are also many applications where Hessian-based mesh adaptation is far from optimal regarding the way the degrees of freedom are distributed in the computational domain.

On the other side, *goal-oriented* mesh adaptation focuses on deriving the best mesh to observe a given output functional. Goal-oriented methods result from a series of papers dealing with *a posteriori estimates* (see e.g. [20, 6, 11, 21]). Extracting informations concerning mesh anisotropy from an *a posteriori* estimate is a difficult task. Starting from *a priori estimates*, Loseille *et al.* proposed in [17] a fully anisotropic goal-oriented mesh adaptation technique for steady problems. This latter method combines goal-oriented rationale and the application of Hessian-based analysis to truncation error.

Mesh adaptation for unsteady flows is also an active field of research and brings an attracting increase in simulation efficiency. Complexity of the algorithms is larger than for steady case: for most flows, the mesh should change during the time interval. Meshes can be moved [5], pattern-split [7, 13], locally refined [4], or globally rebuild [1]. Hessian-based methods are essentially applied with a non-moving mesh system. A mesh adaptation fixed-point method was proposed in [1]. The Hessian criteria at the different time steps of a subinterval are synthetized into a single criterion for these steps with the metric intersection [1]. A mesh-PDE solver iteration is applied on time sub-intervals. Extension to $\mathbf{L}^p$ error estimator has been proposed in [4].

The objective of this paper is the extension of goal-oriented anisotropic mesh adaptation method of [17] to the unsteady framework introduced in [1].

To this end, several methodological issues need to be addressed. First, similarly to [4], we propose a global fixed-point algorithm for solving the coupled system made, this time of three fields, the unsteady state, the unsteady adjoint state and the adapted meshes. Second, this algorithm needs to be *a priori* analyzed and its convergence rate to continuous solution needs to be optimized. Third, at the computer algorithmic level, it is also necessary to master the computational (memory and time) cost of the new system, which couples a time-forward state, a time-backward adjoint and a mesh update influenced by global statistics.

We start this paper with a formal description of the error analysis in its most general expression, then the application to unsteady compressible Euler flows is presented. In Section 3, we introduce the optimal adjoint-based metric definition and all its relative issues, then in Section 4 we present our mesh adaptation algorithm. We end this paper with numerical experiments for blast wave problems.

1 Formal Error Analysis

Let us introduce a system of PDE's in its variational formulation:

$$\text{Find } w \in \mathcal{V} \text{ such that } \forall \varphi \in \mathcal{V}, \quad (\Psi(w), \varphi) = 0 \tag{1}$$

with $\mathcal{V}$ a functional space of solutions. The associated discrete variational formulation then writes:

$$\text{Find } w_h \in \mathcal{V}_h \text{ such that } \forall \varphi_h \in \mathcal{V}_h, \quad (\Psi_h(w_h), \varphi_h) = 0 \qquad (2)$$

where $\mathcal{V}_h$ is a subspace of $\mathcal{V}$. For a solution w of state system (1), we define a *functional output* as:

$$j \in \mathbb{R} \ ; \ j = (g, w), \qquad (3)$$

where (g, w) holds for the following rather general functional output formulation:

$$(g, w) = \int_0^T \int_\Omega (g_\Omega, w) \, d\Omega \, dt + \int_\Omega (g_T, w(T)) \, d\Omega + \int_0^T \int_\Gamma (g_\Gamma, w) \, d\Gamma \, dt, \qquad (4)$$

where g_Ω, g_T, and g_Γ are assumed to be regular enough functions. We introduce the *continuous adjoint* w^*, solution of the following system:

$$w^* \in \mathcal{V}, \ \forall \psi \in \mathcal{V}, \ \left(\frac{\partial \Psi}{\partial w}(w)\psi, w^* \right) = (g, \psi) \ . \qquad (5)$$

The objective here is to estimate the following approximation error committed on the functional:

$$\delta j = j(w) - j(w_h) \ ,$$

where w and w_h are respectively solutions of (1) and (2). It is then useful to choose the test function φ_h as the discrete adjoint state, $\varphi_h = w_h^*$, which is the solution of:

$$\forall \psi_h \in \mathcal{V}_h, \ \left(\frac{\partial \Psi_h}{\partial w_h}(w_h)\psi_h, w_h^* \right) = (g, \psi_h). \qquad (6)$$

We assume that w_h^* is close to the continuous adjoint state w^*. We refer to [17] in which the following *a priori* formal estimate is finally proposed:

$$\delta j \approx ((\Psi_h - \Psi)(w), w^*) \ . \qquad (7)$$

The next section is devoted to the application of Estimator (7) to the unsteady Euler model.

2 Unsteady Euler Models

2.1 Continuous State System and Finite Volume Formulation

Continuous state system. The 3D unsteady compressible Euler equations are set in the computational space-time domain $\mathcal{Q} = \Omega \times [0, T]$, where T

is the (positive) maximal time and $\Omega \subset \mathbb{R}^3$ is the spatial domain. An essential ingredient of our discretisation and of our analysis is the elementwise linear interpolation operator. In order to use it easily, we define our working functional space as $V = \left[H^1(\Omega) \cap \mathcal{C}(\bar{\Omega})\right]^5$, that is the set of measurable functions that are continuous with square integrable gradient. We formulate the Euler model in a compact variational formulation in the functional space $\mathcal{V} = H^1\{[0,T]; V\}$ as follows:

Find $W \in \mathcal{V}$ such that $\forall \varphi \in \mathcal{V}$, $(\Psi(W), \varphi) = 0$

with $\displaystyle (\Psi(W), \varphi) = \int_\Omega \varphi(0)(W_0 - W(0)) \, d\Omega + \int_0^T \int_\Omega \varphi W_t \, d\Omega \, dt$

$\displaystyle + \int_0^T \int_\Omega \varphi \nabla \cdot \mathcal{F}(W) \, d\Omega \, dt - \int_0^T \int_\Gamma \varphi \hat{\mathcal{F}}(W).\mathbf{n} \, d\Gamma \, dt. \quad (8)$

In the above definition, W is the vector of conservative flow variables and $\mathcal{F}(W) = (\mathcal{F}_1(W), \mathcal{F}_2(W), \mathcal{F}_3(W))$ is the usual Euler fluxes given by:

$$W = \begin{pmatrix} \rho \\ \rho u_i \\ \rho e \end{pmatrix} \quad \text{and} \quad \mathcal{F}_j(W) = \begin{pmatrix} \rho u_j \\ \rho u_j u_i + p\, \delta_{ij} \\ (\rho e + p) u_j \end{pmatrix} \quad \text{with } i = \{1, 2, 3\},$$

where ρ, u_i, p and e denote respectively the fluid density, i^{th} component of the Cartesian velocity, pressure and total energy. δ_{ij} is the Kronecker delta function. Here, functions φ and W have 5 components, and therefore the product φW holds for $\sum_{k=1..5} \varphi_k W_k$. We have denoted by Γ the boundary of the computational domain Ω, $\mathbf{n}$ is the outward normal to Γ, $W(0)(\mathbf{x}) = W(\mathbf{x}, t)|_{t=0}$ for any $\mathbf{x}$ in Ω, W_0 the initial condition and the boundary flux $\hat{\mathcal{F}}$ contains the different boundary conditions.

Discrete state system. As a spatially semi-discrete model, we consider the Mixed-Element-Volume formulation [8]. As in [17], we reformulate it under the form of a finite element variational formulation, this time in the unsteady context. We assume that Ω is covered by a finite-element partition in simplicial elements denoted K. The mesh, denoted by $\mathcal{H}$ is the set of the elements. Let us introduce the following approximation space:

$V_h = \left\{ \varphi_h \in V \mid \varphi_{h|K} \text{ is affine } \forall K \in \mathcal{H} \right\}, \quad \text{and} \quad \mathcal{V}_h = H^1\{[0,T]; V_h\} \subset \mathcal{V}.$

Let Π_h be the usual $\mathcal{P}^1$ projector:

$\Pi_h : V \to V_h \quad \text{such that} \quad \Pi_h \varphi(\mathbf{x}_i) = \varphi(\mathbf{x}_i), \; \forall \, \mathbf{x}_i \text{ vertex of } \mathcal{H}.$

We extend it to time-dependent functions:

$\Pi_h : H^1\{[0,T]; V\} \to \mathcal{V}_h \quad \text{such that} \quad (\Pi_h \varphi)(t) = \Pi_h(\varphi(t)), \; \forall \, t \in [0,T].$

The weak discrete formulation writes:

$$\text{Find } W_h \in \mathcal{V}_h \text{ such that } \forall \varphi_h \in \mathcal{V}_h, \ (\Psi_h(W_h), \varphi_h) = 0,$$

with: $(\Psi_h(W_h), \varphi) = \int_\Omega \varphi(0)(\Pi_h W_h(0) - W_{0h}) \,\mathrm{d}\Omega + \int_0^T \int_\Omega \varphi \, \Pi_h W_{h,t} \,\mathrm{d}\Omega \,\mathrm{d}t$

$+ \int_0^T \int_\Omega \varphi \nabla \cdot \mathcal{F}_h(W_h) \,\mathrm{d}\Omega \,\mathrm{d}t - \int_0^T \int_\Gamma \varphi \hat{\mathcal{F}}_h(W_h) . \mathbf{n} \,\mathrm{d}\Gamma \,\mathrm{d}t + \int_0^T \int_\Omega \varphi \, D_h(W_h) \,\mathrm{d}\Omega \,\mathrm{d}t$

with $\mathcal{F}_h = \Pi_h \mathcal{F}$ and $\hat{\mathcal{F}}_h = \Pi_h \hat{\mathcal{F}}$. The D_h term accounts for the numerical diffusion. In the present study, we only need to know that for smooth fields, the D_h term is a third order term with respect to the mesh size. For shocked fields, monotonicity limiters become first-order terms.

2.2 Continuous Adjoint System and Discretization

Continuous adjoint system. We refer here to the continuous adjoint system introduced previously:

$$W^* \in \mathcal{V}, \ \forall \psi \in \mathcal{V} \ : \ \left(\frac{\partial \Psi}{\partial W}(W)\psi, W^*\right) - (g, \psi) = 0. \tag{9}$$

We recall that (g, ψ) is defined by (4). Replacing $\Psi(W)$ by its Formulation (8) and integrating by parts, we get:

$$\left(\frac{\partial \Psi}{\partial W}(W)\psi, W^*\right) = \int_\Omega (\psi(0)W^*(0) - \psi(T)W^*(T)) \,\mathrm{d}\Omega$$

$$+ \int_0^T \int_\Omega \psi \left(-W_t^* - \left(\frac{\partial \mathcal{F}}{\partial W}\right)^* \nabla W^*\right) \,\mathrm{d}\Omega \,\mathrm{d}t$$

$$+ \int_0^T \int_\Gamma \psi \left[\left(\frac{\partial \mathcal{F}}{\partial W}\right)^* W^* . \mathbf{n} + \left(\frac{\partial \hat{\mathcal{F}}}{\partial W}\right)^* W^* . \mathbf{n}\right] \mathrm{d}\Gamma \,\mathrm{d}t \tag{10}$$

The adjoint Euler equations is a system of advection equations, where the temporal integration goes backwards, *i.e.*, in the opposite direction of usual time. Thus, when solving the unsteady adjoint system, one starts at the end of the flow run and progresses back until reaching the start time.

Discrete Adjoint System. Although any consistent approximation of the continuous adjoint system could be built, we choose to build the adjoint of the discrete state defined in (9) in order to be closer to the true error from which the continuous model were derived. Consider the following semi-discrete unsteady compressible Euler model (explicit RK1 time integration):

$$\Psi_h^n(W_h^n, W_h^{n-1}) = \frac{W_h^n - W_h^{n-1}}{\delta t^n} + \Phi_h(W_h^{n-1}) = 0 \quad \text{for } n = 1, ..., N. \tag{11}$$

The time-dependent functional is discretised as follows:

$$j(W_h) = \sum_{n=1}^{N} \delta t^n j^{n-1}(W_h).$$

The problem of minimizing the error committed on the target functional $j(W_h) = (g, W_h)$, subject to the Euler system (11), can be transformed into an unconstrained problem for the following Lagrangian functional:

$$\mathcal{L} = \sum_{n=1}^{N} \delta t^n j^{n-1}(W_h) - \sum_{n=1}^{N} \delta t^n (W_h^{*,n})^T \Psi_h^n(W_h^n, W_h^{n-1})$$

where $W_h^{*,n}$ are the N vectors of the Lagrange multipliers (which are the time-dependent adjoint states). The conditions for an extremum becomes then:

$$\frac{\partial \mathcal{L}}{\partial W_h^{*,n}} = 0 \quad \text{and} \quad \frac{\partial \mathcal{L}}{\partial W_h^n} = 0, \quad \text{for } n = 1, ..., N.$$

The first condition is clearly verified from relation (11). Thus the Lagrangian multipliers $W_h^{*,n}$ must be chosen such that the second condition of extrema $\frac{\partial \mathcal{L}}{\partial W_h^n} = 0$ is verified. This gives the unsteady discrete adjoint system:

$$\begin{cases} W_h^{*,N} = 0 \\ W_h^{*,n-1} = W_h^{*,n} + \delta t^n \dfrac{\partial j^{n-1}}{\partial W_h^{n-1}} - \delta t^n (W_h^{*,n})^T \dfrac{\partial \Psi_h^{n-1}}{\partial W_h^{n-1}} \end{cases} \quad (12)$$

As the adjoint system runs in reverse time, the first expression in the adjoint system (12) is referred to as adjoint "initialization". Computing $W_h^{*,n-1}$ at time t^{n-1} requires the knowledge of state W_h^{n-1} and adjoint state $W_h^{*,n}$. Therefore, the knowledge of all states $\{W_h^{n-1}\}_{n=1,N}$ is needed to compute backward the adjoint state from time T to 0 which involves large memory storage effort.

Solve state foreward: $\Psi(W) = 0$

Solve adjoint state backward: $\Psi^*(W, W^*) = 0$

This drawback can be reduced by out-of-core storage of checkpoints (as shown in the picture below), although it implies a recomputing effort of the state W.

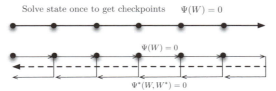

2.3 Numerical Example

The simulation of a blast in a 2D geometry representing a city is performed, see Figure 1. A blast-like initialization $W_{blast} = (10, 0, 0, 250)$ in ambient air $W_{air} = (1, 0, 0, 2.5)$ is considered in a small region of the computational domain. We perform a forward/backward computation on a uniform mesh of 22 574 vertices and 44 415 triangles. Output functional of interest j is the quadratic deviation from ambient pressure on target surface S which is a part of the higher building roof (Figure 1):

$$j(W) = \int_0^T \int_S \frac{1}{2}(p(t) - p_{air})^2 \, \mathrm{d}S \, \mathrm{d}t.$$

Figure 2 plots the density isolines of the flow at different times showing several shock waves traveling throughout the computational domain. Figure 3 depicts the associated density adjoint state progressing backward in time. The same computational time is considered for both figures.

The simulation points out the ability of the adjoint to automatically provide the sensitivity of the flow field on the functional. Indeed, at early time of the simulation (top left picture), a lot of information is captured by the adjoint, *i.e.*, non-zero adjoint values. We notice that shock waves which will directly impact are clearly detected by the adjoint, but also shocks waves

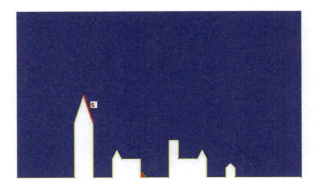

Fig. 1. Initial blast solution and location of the target surface.

reflected by the left building which will be redirected towards surface S. At the middle of the simulation, the adjoint neglects waves that are traveling in the direction opposite to S and also waves that will not impact surface S before final time T since they won't have an influence on the cost functional. While getting closer to final time T (bottom right picture), the adjoint only focuses on the last waves that will impact surface S and ignores the rest of the flow.

3 Optimal Unsteady Adjoint-Based Metric

3.1 Error Analysis Applied to Unsteady Euler Model

We replace in Estimation (7) operators Ψ and Ψ_h by their expressions given by Relations (8) and (9). In [17], it was observed that even for shocked flows, it is interesting to neglect the numerical viscosity term. We follow again this option. We also discard the error committed when imposing the initial condition. And finally, after integrating by parts, the previous error estimate leads to:

$$\delta j \approx \int_0^T \int_\Omega W^* \left(W - \Pi_h W\right)_t \mathrm{d}\Omega \, \mathrm{d}t - \int_0^T \int_\Omega \nabla W^* \left(\mathcal{F}(W) - \Pi_h \mathcal{F}(W)\right) \mathrm{d}\Omega \, \mathrm{d}t$$
$$- \int_0^T \int_\Gamma W^* \left(\bar{\mathcal{F}}(W) - \Pi_h \bar{\mathcal{F}}(W)\right). \mathbf{n} \, \mathrm{d}\Gamma \, \mathrm{d}t. \tag{13}$$

with $\bar{\mathcal{F}} = \hat{\mathcal{F}} - \mathcal{F}$. We observe that this estimate of δj is expressed in terms of interpolation errors of the Euler fluxes and of the time derivative weighted by continuous functions W^* and ∇W^*. The integrands in Error Estimation (13) contain positive and negative parts which can compensate for some particular meshes. In our strategy, we prefer to not rely on these parasitic effects and to slightly over-estimate the error. To this end, all integrands are bounded by their absolute values. Moreover, we observe that the third term introduce a dependency of the error with respect to the boundary surface mesh. In the present paper, we discard this term and refer to [17] for a discussion of the influence of it. At the end, we get:

$$(g, W_h - W) \leq \int_0^T \int_\Omega |W^*| \left|\left(W - \Pi_h W\right)_t\right| \mathrm{d}\Omega \, \mathrm{d}t + \int_0^T \int_\Omega |\nabla W^*| \left|\mathcal{F}(W) - \Pi_h \mathcal{F}(W)\right| \mathrm{d}\Omega \, \mathrm{d}t. \tag{14}$$

3.2 Continuous Mesh Model

We propose to work in the continuous mesh framework, introduced in [15, 16]. It allows us to define proper differentiable optimization, *i.e.*, to use a calculus of variations which cannot apply on the class of discrete meshes. This framework lies in the class of metric-based methods.

Unsteady Goal-Oriented Mesh Adaptation

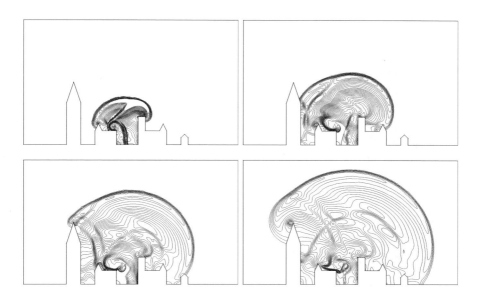

Fig. 2. 2D city blast solution state evolution. From left to right and top to bottom, snapshot of the density isolines at a-dimensional time 1.2, 2.25, 3.3 and 4.35.

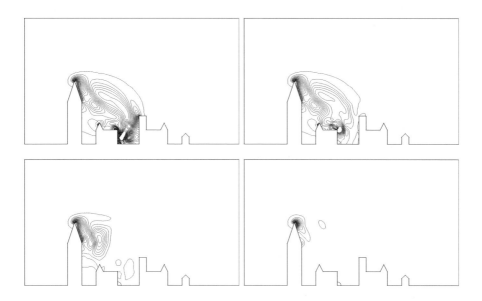

Fig. 3. 2D city blast adjoint state evolution. From left to right and top to bottom, snapshot of the density isolines at a-dimensional time 0.15, 1.2, 2.25 and 3.3.

A continuous mesh $\mathbf{M}$ of computational domain Ω is identified to a Riemannian metric field $\mathbf{M} = (\mathcal{M}(\mathbf{x}))_{\mathbf{x}\in\Omega}$. For all $\mathbf{x}$ of Ω, $\mathcal{M}(\mathbf{x})$ is a symmetric tensor having $(\lambda_i(\mathbf{x}))_{i=1,3}$ as eigenvalues along the principal directions $\mathcal{R}(\mathbf{x}) = (\mathbf{v}_i(\mathbf{x}))_{i=1,3}$. Sizes along these directions are denoted $(h_i(\mathbf{x}))_{i=1,3} = (\lambda_i^{-\frac{1}{2}}(\mathbf{x}))_{i=1,3}$ and the three anisotropic quotients r_i are defined by: $r_i = h_i^3 (h_1 h_2 h_3)^{-1}$. The node density d is equal to: $d = (h_1 h_2 h_3)^{-1} = (\lambda_1 \lambda_2 \lambda_3)^{\frac{1}{2}} = \sqrt{\det(\mathcal{M})}$. By integrating the node density, we define the complexity $\mathcal{C}$ of a continuous mesh which is the continuous counterpart of the total number of vertices:

$$\mathcal{C}(\mathbf{M}) = \int_\Omega d(\mathbf{x})\,\mathrm{d}\mathbf{x} = \int_\Omega \sqrt{\det(\mathcal{M}(\mathbf{x}))}\,\mathrm{d}\mathbf{x}.$$

Given a continuous mesh $\mathbf{M}$, we shall say, following [15, 16], that a discrete mesh $\mathcal{H}$ of the same domain Ω is a **unit mesh with respect to M**, if each tetrahedron $K \in \mathcal{H}$, defined by its list of edges $(\mathbf{e}_i)_{i=1\ldots 6}$, verifies:

$$\forall i \in [1,6], \quad \ell_\mathcal{M}(\mathbf{e}_i) \in \left[\frac{1}{\sqrt{2}}, \sqrt{2}\right] \quad \text{and} \quad Q_\mathcal{M}(K) \in [\alpha, 1] \text{ with } \alpha > 0,$$

where the length and quality in the metric are defined similarly to [15, 16].

We want to emphasize that the set of all the discrete meshes that are unit meshes with respect to a unique $\mathbf{M}$ contains an infinite number of meshes. Given a smooth function u, to each unit mesh $\mathcal{H}$ corresponds a local interpolation error $|u - \Pi u|$. In [15, 16], it is shown that all these interpolation errors are well represented by the so-called continuous interpolation error related to $\mathbf{M}$, which is expressed locally in terms of the Hessian H_u of u as follows:

$$(u - \pi_\mathcal{M} u)(\mathbf{x}, t) = \frac{1}{10}\mathrm{trace}(\mathcal{M}^{-\frac{1}{2}}(\mathbf{x})\,|H_u(\mathbf{x},t)|\,\mathcal{M}^{-\frac{1}{2}}(\mathbf{x})). \qquad (15)$$

3.3 Continuous Error Model

Working in this framework enables us to write Estimate (14) in a continuous form:

$$|(g, W_h - W)| \approx \mathbf{E}(\mathbf{M}) = \int_0^T \int_\Omega |W^*|\,|(W - \pi_\mathcal{M} W)_t|\,\mathrm{d}\Omega\,\mathrm{d}t$$
$$+ \int_0^T \int_\Omega |\nabla W^*|\,|\mathcal{F}(W) - \pi_\mathcal{M} \mathcal{F}(W)|\,\mathrm{d}\Omega\,\mathrm{d}t, \quad (16)$$

where $\mathbf{M} = (\mathcal{M}(\mathbf{x}))_{\mathbf{x}\in\Omega}$ is a continuous mesh and $\pi_\mathcal{M}$ is the continuous linear interpolate. Then, introducing the continuous interpolation error, we can write the simplified error model as follows:

$$\mathbf{E}(\mathbf{M}) = \int_0^T \int_\Omega \mathrm{trace}\left(\mathcal{M}^{-\frac{1}{2}}(\mathbf{x},t)\,\mathbf{H}(\mathbf{x},t)\,\mathcal{M}^{-\frac{1}{2}}(\mathbf{x},t)\right)\,\mathrm{d}\Omega\,\mathrm{d}t$$

with $\mathbf{H}(\mathbf{x},t) = \sum_{j=1}^{5} ([\Delta t]_j(\mathbf{x},t) + [\Delta x]_j(\mathbf{x},t) + [\Delta y]_j(\mathbf{x},t) + [\Delta z]_j(\mathbf{x},t))$, (17)

in which $[\Delta t]_j = |W_j^*| \cdot |H(W_{j,t})|$, $[\Delta x]_j = \left|\dfrac{\partial W_j^*}{\partial x}\right| \cdot |H(\mathcal{F}_1(W_j))|$,

$[\Delta y]_j = \left|\dfrac{\partial W_j^*}{\partial y}\right| \cdot |H(\mathcal{F}_2(W_j))|$, $[\Delta z]_j = \left|\dfrac{\partial W_j^*}{\partial z}\right| \cdot |H(\mathcal{F}_3(W_j))|$.

Here, W_j^* denotes the jth component of the adjoint vector W^*, $H(\mathcal{F}_i(W_j))$ the Hessian of the jth component of the vector $\mathcal{F}_i(W)$, and $H(W_{j,t})$ the Hessian of the jth component of the time derivative of W. The mesh optimization problem writes:

Find $\mathbf{M}_{opt} = \text{Argmin}_\mathbf{M} \, \mathbf{E}(\mathbf{M})$ under the constraint $\mathcal{C}_{st}(\mathbf{M}) = N_{st}$, (18)

where N_{st} is a specified total number of nodes. Since we consider an unsteady problem, the space-time (st) complexity used to compute the solution takes into account the time discretization:

$$\mathcal{C}_{st}(\mathbf{M}) = \int_0^T \tau(t)^{-1} \left(\int_\Omega d_\mathcal{M}(\mathbf{x},t)\mathrm{d}\mathbf{x}\right) \mathrm{d}t \qquad (19)$$

where $\tau(t)$ is the time step used at time t of interval $[0,T]$.

3.4 Spatial Minimization for a Fixed t

Let us assume that at time t, we seek for the optimal continuous mesh $\mathbf{M}_{go}(t)$ which minimizes the instantaneous error, *i.e.*, the spatial error for a fixed time t:

$$\tilde{\mathbf{E}}(\mathbf{M}(t)) = \int_\Omega \text{trace}\left(\mathcal{M}^{-\frac{1}{2}}(\mathbf{x},t)\,\mathbf{H}(\mathbf{x},t)\,\mathcal{M}^{-\frac{1}{2}}(\mathbf{x},t)\right)\mathrm{d}\mathbf{x}$$

under the constraint that the number of vertices is prescribed to

$$\mathcal{C}(\mathbf{M}(t)) = \int_\Omega d_{\mathcal{M}(t)}(\mathbf{x},t)\,\mathrm{d}\mathbf{x} = N(t). \qquad (20)$$

Similarly to [17], solving the optimality conditions provides the *optimal goal-oriented instantaneous continuous mesh* $\mathbf{M}_{go}(t) = (\mathcal{M}_{go}(\mathbf{x},t))_{\mathbf{x}\in\Omega}$ at time t defined by:

$$\mathcal{M}_{go}(\mathbf{x},t) = N(t)^{\frac{2}{3}} \mathcal{M}_{go,1}(\mathbf{x},t), \qquad (21)$$

where $\mathcal{M}_{go,1}$ is the optimum for $\mathcal{C}(\mathbf{M}(t)) = 1$:

$$\mathcal{M}_{go,1}(\mathbf{x},t) = \mathcal{K}(t)^{-\frac{2}{5}} (\det \mathbf{H}(\mathbf{x},t))^{-\frac{1}{5}} \mathbf{H}(\mathbf{x},t). \qquad (22)$$

with $\mathcal{K}(t) = \left(\int_\Omega (\det \mathbf{H}(\mathbf{x},t))^{\frac{1}{5}} \mathrm{d}\mathbf{x} \right)^{\frac{5}{3}}$. The corresponding optimal instantaneous error at time t writes:

$$\tilde{\mathbf{E}}(\mathbf{M}_{go}(t)) = 3\, N(t)^{-\frac{2}{3}} \left(\int_\Omega (\det \mathbf{H}(\mathbf{x},t))^{\frac{1}{5}} \mathrm{d}\mathbf{x} \right)^{\frac{5}{3}} = 3\, N(t)^{-\frac{2}{3}} \mathcal{K}(t).$$

3.5 Temporal Minimization

To complete the resolution of optimization Problem (18), we perform a temporal minimization in order to get the optimal space-time continuous mesh. In other words, we need to find the optimal time law $t \to N(t)$ for the instantaneous mesh size. Here, we only consider the simpler case where the time step τ is specified by the user as a function of time $t \to \tau(t)$. A similar analysis can be done to deal with the case of an explicit time advancing solver subject to Courant time step condition, but such an analysis is out of the scope of this proceeding.

Let us consider the case where the time step τ is specified by a function of time $t \to \tau(t)$. After the spatial optimization, the space-time error writes:

$$\mathbf{E}(\mathbf{M}_{go}) = \int_0^T \tilde{\mathbf{E}}(\mathbf{M}_{go}(t))\, \mathrm{d}t = 3 \int_0^T N(t)^{-\frac{2}{3}} \mathcal{K}(t)\, \mathrm{d}t \tag{23}$$

and we aim at minimizing it under the following space-time complexity constraint:

$$\int_0^T N(t) \tau(t)^{-1}\, \mathrm{d}t = N_{st}. \tag{24}$$

In other words, we concentrate on seeking for *the optimal distribution of $N(t)$ when the space-time total number of nodes N_{st} is prescribed*. Let us apply the one-to-one change of variables:

$$\tilde{N}(t) = N(t)\tau(t)^{-1} \quad \text{and} \quad \tilde{\mathcal{K}}(t) = \tau(t)^{-\frac{2}{3}} \mathcal{K}(t).$$

Then, our temporal optimization problem becomes:

$$\min_{\mathbf{M}} \mathbf{E}(\mathbf{M}) = \int_0^T \tilde{N}(t)^{-\frac{2}{3}} \tilde{\mathcal{K}}(t)\, \mathrm{d}t \quad \text{under constraint} \quad \int_0^T \tilde{N}(t)\, \mathrm{d}t = N_{st}.$$

The solution of this problem is given by:

$$\tilde{N}_{opt}(t)^{-\frac{5}{3}} \tilde{\mathcal{K}}(t) = const \quad \Rightarrow \quad N_{opt}(t) = C(N_{st})\, (\tau(t)\, \mathcal{K}(t))^{\frac{3}{5}}$$

Here, constant $C(N_{st})$ can be obtained by introducing the above expression in space-time complexity Constraint (24), leading to:

$$C(N_{st}) = \left(\int_0^T \tau(t)^{-\frac{2}{5}} \mathcal{K}(t)^{\frac{3}{5}}\, \mathrm{d}t \right)^{-1} N_{st},$$

which completes the description of the optimal space-time metric for a prescribed time step. Using Relation (21), the analytic expression of the optimal space-time goal-oriented metric $\mathbf{M}_{go}$ writes:

$$\mathcal{M}_{go}(\mathbf{x},t) = N_{st}^{\frac{2}{3}} \left(\int_0^T \tau(t)^{-\frac{2}{5}} \mathcal{K}(t)^{\frac{3}{5}} dt \right)^{-\frac{2}{3}} \tau(t)^{\frac{2}{5}} (\det \mathbf{H}(\mathbf{x},t))^{-\frac{1}{5}} \mathbf{H}(\mathbf{x},t). \quad (25)$$

We get the following optimal error:

$$\mathbf{E}(\mathbf{M}_{go}) = 3\, N_{st}^{-\frac{2}{3}} \left(\int_0^T \tau(t)^{-\frac{2}{5}} \mathcal{K}(t)^{\frac{3}{5}} dt \right)^{\frac{5}{3}}. \quad (26)$$

3.6 Space-Time Minimization for Time Sub-intervals

The previous analysis provides the optimal size of the adapted meshes for each time level. Hence, this analysis requires the mesh to be adapted at each flow solver time step. But, in practice this approach involves a very large number of remeshing which is CPU consuming and spoils solution accuracy due to many solution transfers from one mesh to a new one. In consequence, a new adaptive strategy has been proposed in [1, 4] where the number of remeshing is controlled (thus drastically reduced) by generating adapted meshes for several solver time steps. The idea is to split the simulation time interval into n_{adap} *sub-intervals* $[t_i, t_{i+1}]$ for $i = 1, .., n_{adap}$. Each spatial mesh $\mathbf{M}^i$ is then kept constant during each sub-interval $[t_i, t_{i+1}]$. We could consider this partition as a *time discretization of the mesh adaptation problem*.

Spatial minimization on a sub-interval. Given the continuous mesh complexity N_i for the single adapted mesh used during time sub-interval $[t_i, t_{i+1}]$, we seek for the optimal continuous mesh $\mathbf{M}^i_{go}$ solution of the following problem:

$$\min_{\mathbf{M}^i} \mathbf{E}^i(\mathbf{M}^i) = \int_\Omega \text{trace}\left((\mathcal{M}^i)^{-\frac{1}{2}}(\mathbf{x})\, \mathbf{H}^i(\mathbf{x})\, (\mathcal{M}^i)^{-\frac{1}{2}}(\mathbf{x}) \right) d\mathbf{x} \text{ such that } \mathcal{C}(\mathbf{M}^i) = N^i,$$

where hessian metric $\mathbf{H}^i$ on the sub-interval can be defined by either using an $\mathbf{L}^1$ or an $\mathbf{L}^\infty$ norm:

$$\mathbf{H}^i_{\mathbf{L}^1}(\mathbf{x}) = \int_{t_i}^{t_{i+1}} \mathbf{H}(\mathbf{x},t)\, dt \quad \text{or} \quad \mathbf{H}^i_{\mathbf{L}^\infty}(\mathbf{x}) = \Delta t_i \max_{t \in [t_i, t_{i+1}]} \mathbf{H}(\mathbf{x},t),$$

with $\Delta t_i = t_{i+1} - t_i$. Processing as previously, we get the spatial optimality condition:

$$\mathcal{M}^i_{go}(\mathbf{x}) = (N^i)^{\frac{2}{3}}\, \mathcal{M}^i_{go,1}(\mathbf{x}) \quad \text{with } \mathcal{M}^i_{go,1}(\mathbf{x}) = (\mathcal{K}^i)^{-\frac{2}{5}} (\det \mathbf{H}^i(\mathbf{x}))^{-\frac{1}{5}} \mathbf{H}^i(\mathbf{x}).$$

The corresponding optimal error $\mathbf{E}^i(\mathbf{M}_{go}^i)$ writes:

$$\mathbf{E}^i(\mathbf{M}_{go}^i) = 3\,(N^i)^{-\frac{2}{3}}\left(\int_\Omega (\det \mathbf{H}^i(\mathbf{x}))^{\frac{1}{5}}\mathrm{d}\mathbf{x}\right)^{\frac{5}{3}} = 3\,(N^i)^{-\frac{2}{3}}\,\mathcal{K}^i\,.$$

Temporal minimization for specified τ. To complete our analysis, we shall now perform a temporal minimization. After the spatial minimization, the temporal optimization problem reads:

$$\min_{\mathbf{M}} \mathbf{E}(\mathbf{M}) = 3\sum_{i=1}^{n_{adap}} (N^i)^{-\frac{2}{3}}\,\mathcal{K}^i \quad\text{such that}\quad \sum_{i=1}^{n_{adap}} N^i\left(\int_{t_i}^{t_{i+1}} \tau(t)^{-1}\mathrm{d}t\right) = N_{st}\,.$$

We set the one-to-one mapping:

$$\tilde{N}^i = N^i\left(\int_{t_i}^{t_{i+1}} \tau(t)^{-1}\mathrm{d}t\right) \quad\text{and}\quad \tilde{\mathcal{K}}^i = \mathcal{K}^i\left(\int_{t_i}^{t_{i+1}} \tau(t)^{-1}\mathrm{d}t\right)^{\frac{2}{3}},$$

then the optimization problem reduces to:

$$\min_{\mathbf{M}} \sum_{i=1}^{n_{adap}} (\tilde{N}^i)^{-\frac{2}{3}}\,\tilde{\mathcal{K}}^i \quad\text{such that}\quad \sum_{i=1}^{n_{adap}} \tilde{N}^i = N_{st}\,.$$

We deduce the optimal continuous mesh $\mathbf{M}_{go} = \{\mathbf{M}_{go}^i\}_{i=1,\ldots,n_{adap}}$ and error:

$$\mathcal{M}_{go}^i(\mathbf{x}) = N_{st}^{\frac{2}{3}}\left(\sum_{i=1}^{n_{adap}} (\mathcal{K}^i)^{\frac{3}{5}}\,\mathcal{T}^i\right)^{-\frac{2}{3}} (\mathcal{T}^i)^{-1}(\det\mathbf{H}^i(\mathbf{x}))^{-\frac{1}{5}}\mathbf{H}^i(\mathbf{x})$$

$$\mathbf{E}(\mathbf{M}) = 3\,N_{st}^{-\frac{2}{3}}\left(\sum_{i=1}^{n_{adap}} (\mathcal{K}^i)^{\frac{3}{5}}\,\mathcal{T}^i\right)^{\frac{5}{3}},$$

with $(\mathcal{K}^i)^{\frac{3}{5}} = \int_\Omega (\det\mathbf{H}^i(\mathbf{x}))^{\frac{1}{5}}\mathrm{d}\mathbf{x}$ and $\mathcal{T}^i = \left(\int_{t_i}^{t_{i+1}} \tau(t)^{-1}\mathrm{d}t\right)^{\frac{2}{5}}$.

4 From Theory to Practice

In order to remedy all the problematics relative to mesh adaptation for time-dependent simulations stated in the introduction, an innovative strategy based on a fixed-point algorithm has been initiated in [2] and fully developed in [1]. The fixed-point algorithm aims at avoiding the generation of a new mesh at each solver iteration which would imply serious degradation of the CPU time and of the solution accuracy due to the large number of mesh modifications. It is also an answer to the lag problem occurring when computing the solution at t^n and accordingly adapt the mesh at each time step. Indeed, by doing this, the mesh is always late as compared to the solution as it is not adapted for the displacement of the solution between t^n and t^{n+1}.

The basic idea consists in splitting the simulation time frame $[0, T]$ into n_{adap} adaptation sub-intervals:

$$[0, T] = [0 = t_0, t_1] \cup \ldots \cup [t_i, t_{i+1}] \cup \ldots \cup [t_{n_{adap}-1}, t_{n_{adap}}],$$

and to keep the same adapted mesh for each sub-interval. On each sub-interval, the mesh is adapted to control the solution accuracy from t^n to t^{n+1}. Consequently, the time-dependent simulation is performed with n_{adap} different adapted meshes. This can be seen as a coarse discretization of the time axis where the spatial mesh is constant for each sub-interval when the global space-time mesh is visualized, providing thus a first step in the adaptation of the whole space-time mesh.

4.1 Choice of the Goal-Oriented Metric

The optimal adapted meshes for each sub-interval are generated according to analysis of Section 3.6. In this work, the following particular choice has been made:

- the hessian metric for sub-interval i is based on a control of the temporal error in $\mathbf{L}^\infty$ norm:

$$\mathbf{H}^i_{\mathbf{L}^\infty}(\mathbf{x}) = \Delta t_i \max_{t \in [t_i, t_{i+1}]} \mathbf{H}(\mathbf{x}, t) = \Delta t_i \, \mathbf{H}^i_{\max}(\mathbf{x}),$$

- function $\tau : t \to \tau(t)$ is constant and equal to 1,
- all sub-intervals have the same time length Δt.

The optimal goal-oriented metric $\mathbf{M}_{go} = \{\mathbf{M}^i_{go}\}_{i=1,\ldots,n_{adap}}$ then simplifies to:

$$\mathcal{M}^i_{go}(\mathbf{x}) = N_{st}^{\frac{2}{3}} \left(\sum_{i=1}^{n_{adap}} (\int_\Omega (\det \mathbf{H}^i_{\max}(\mathbf{x}))^{\frac{1}{5}} \mathrm{d}\mathbf{x}) \right)^{-\frac{2}{3}} (\Delta t)^{\frac{1}{3}} \left(\det \mathbf{H}^i_{\max}(\mathbf{x}) \right)^{-\frac{1}{5}} \mathbf{H}^i_{\max}(\mathbf{x}).$$

Remark: We notice that we obtain a similar expression of the optimal metric as the one proposed in [4], but here in the goal-oriented context and by means of a space-time error minimization.

4.2 Global Fixed-Point Mesh Adaptation Algorithm

To converge the non-linear mesh adaptation problem, *i.e.*, converging the couple mesh-solution, we propose a fixed-point mesh adaptation algorithm. This is also a way to predict the solution evolution and to adapt the mesh accordingly. Nevertheless, to compute all metrics fields $\mathbf{M}^i_{go}$, we have to evaluate the global normalization term which requires the knowledge of all $\mathbf{H}^i_{\max}$. Thus, all the simulation must be performed to be able to evaluate all metrics $\mathbf{M}^i_{go}$. Similarly to [4], a global fixed point strategy covering the whole time-frame $[0, T]$, called *Global adjoint fixed-point mesh adaptation algorithm*, is considered:

```
//--- Fixed-point loop to converge the global space-time mesh adaptation
For j=1,nptfx
   //--- Solve state once to get checkpoints
   For i=1,n_adap
     • S_{0,i}^j = ConservativeSolutionTransfer(H_{i-1}^j, S_{i-1}^j, H_i^j)
     • S_i^j = SolveStateForward(S_{0,i}^j, H_i^j)
   End for
   //--- Solve state and adjoint backward and store samples
   For i=n_adap,1
     • (S*)_i^j = AdjointStateTransfer(H_{i+1}^j, (S_0*)_{i+1}^j, H_i^j)
     • {S_i^j(k), (S*)_i^j(k)}_{k=1,n_k} = SolveStateAndAdjointBackward(S_{0,i}^j, (S*)_i^j, H_i^j)
     • (H_max)_i^j = ComputeGoalOrientedHessianMetric(H_i^j, {S_i^j(k), (S*)_i^j(k)}_{k=1,n_k})
   End for
   • C^j = ComputeSpaceTimeComplexity({(H_max)_i^j}_{i=1,n_adap})
   • {M_i^j}_{i=1,n_adap} = ComputeUnsteadyGoalOrientedMetrics(C^j, {|H_max|_i^j}_{i=1,n_adap})
   • {H_i^{j+1}}_{i=1,n_adap} = GenerateAdaptedMeshes({H_i^j}_{i=1,n_adap}, {M_i^j}_{i=1,n_adap})
End for
```

5 Numerical Examples

The procedures described in this paper have been implemented in our Finite Element/Finite Volume CFD code Wolf which is thoroughly detailed in [3]. As regards the meshing part, we consider a local remeshing strategy. We use Yams [10] for the adaptation in 2D and Feflo.a [19] in 3D.

2D Blast Wave Propagation. We first apply the goal-oriented adaptive strategy to the example presented in Section 2.3. It consists in a 2D blast in a geometry representing a city. We recall that the cost function of interest j is the quadratic deviation of the pressure ambient pressure on target surface S (see Figure 1):

$$j(W) = \int_0^T \int_S \frac{1}{2}(p(t) - p_{air})^2 \mathrm{d}S \mathrm{d}t.$$

The simulation time frame is split into 30 time sub-intervals, *i.e.*, 30 different adapted meshes are used to perform the simulation. Hence, 30 checkpoints are stored for the backward computation of the unsteady adjoint.

The resulting adjoint-based anisotropic adapted meshes are shown in Figure 4. The corresponding density isolines are depicted in Figure 5. These adapted meshes indubitably illustrate that, thanks to the unsteady adjoint, the mesh adaptation only focuses on shock waves that impact the observation region and ignores other area of the flow field. Therefore, waves traveling toward the observation are accurately captured whereas the rest of the flow is poorly computed. We also observe that once waves go beyond the target surface, the mesh is no more refined even if they continue traveling throughout the computational domain. Indeed, they do not impact anymore the functional.

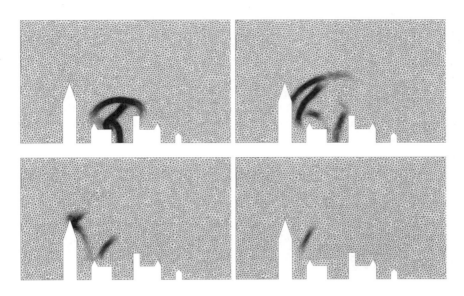

Fig. 4. 2D city blast adjoint-based adapted meshes evolution. From top to bottom and left to right, meshes corresponding to sub-intervals 8, 15, 22 and 29 at a-dimensional time 1.2, 2.25, 3.3 and 4.35.

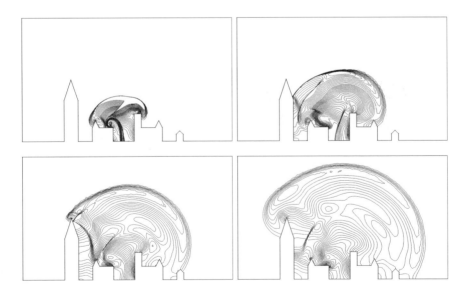

Fig. 5. 2D city blast adaptive solution state evolution. From top to bottom and left to right, density iso-lines corresponding to the end of sub-intervals 8, 15, 22 and 29 at a-dimensional time 1.2, 2.25, 3.3 and 4.35.

It is then quite interesting to compare the Hessian-based approach of [4] with our adjoint-based method. This comparison is shown in Figure 6. It demonstrates how the adjoint defines an optimal distribution of the degrees of freedom for the specific functional, while it is clear that in this context the Hessian-based approach gives a non-optimal result for the evaluation of the functional but capture accurately the whole flow.

In conclusion, if an output functional of interest is provided then the reduction of the simulation number of degrees of freedom can be even more improved by considering a goal-oriented analysis instead of an Hessian-based methodology.

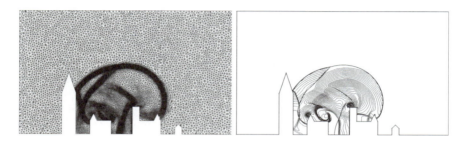

Fig. 6. Adapted mesh and corresponding density iso-lines for sub-interval 15 at a-dimensional time 2.25 obtained with the Hessian-based method of [4].

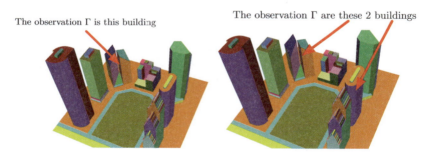

Fig. 7. 3D City test case geometry and location of target surface Γ composed of one building for simulation 1 (left) or two buildings for simulation 2 (right).

3D Blast Wave Propagation. Finally, we consider exactly the same blast test case but in a 3D city geometry. Cost function j is again the quadratic deviation from ambient pressure on target surface Γ which is composed of one building for simulation 1 or two buildings for simulation 2, see Figure 7. The simulation time frame is split into 40 time sub-intervals, *i.e.*, 40 different adapted meshes are used to perform the simulation.

The resulting adjoint-based anisotropic adapted meshes (surface and volume) for both simulations at sub-interval 10, 15 and 20 are shown in Figures 8

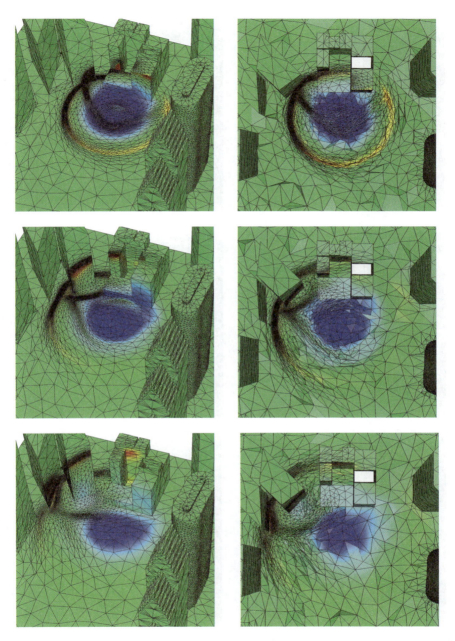

Fig. 8. 3D Blast wave propagation: simulation1. Adjoint-based anisotropic adapted surface (left) and volume (right) meshes at sub-interval 10, 15 and 20 and corresponding solution density at a-dimensioned time 5, 7.5 and 10.

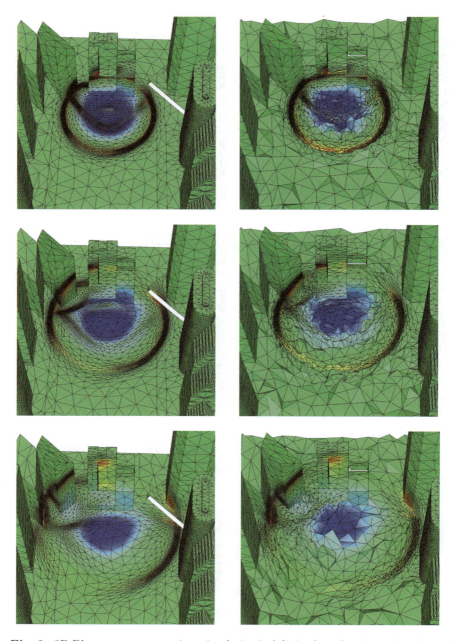

Fig. 9. 3D Blast wave propagation: simulation2. Adjoint-based anisotropic adapted surface (left) and volume (right) meshes at sub-interval 10, 15 and 20 and corresponding solution density at a-dimensioned time 5, 7.5 and 10.

and 9. It is very interesting to see that we are not restricted to just one target surface. As previously in 2D, we notice that mesh refinement only focuses on shock waves that will impact the target buildings. Other waves are neglected thus leading to a large reduction of the mesh size. To illustrate this point, we provide meshes size for several sub-intervals for simulation 2: for sub-intervals 1, 5, 10 and 20 the mesh number of vertices is respectively 1 051 805, 678 802, 233 116 and 45 500. The mesh size has been reduced by a factor 20 between the first and the twentieth sub-interval.

6 Conclusions

We have designed a new mesh adaptation algorithm which prescribes the spatial mesh of an unsteady simulation as the optimum of a goal-oriented error analysis. This method specifies both mesh density and mesh anisotropy by variational calculus. Accounting for unsteadiness is applied in a time-implicit mesh-solution coupling which needs a non-linear iteration, the fixed point. In contrast to the Hessian-based fixed-point of [1] which iterates on each sub-interval, the new iteration covers the whole time interval, including forward steps for evaluating the state and backward ones for the adjoint. This algorithm was applied to 2D and 3D blast wave Euler test cases. Numerical results demonstrate the favorable behavior expected from an adjoint-based method, which gives an automatic selection of the mesh necessary for the target output.

Several important issues remain to be addressed. Among them, the strategies for choosing the splitting in time sub-intervals and the accurate integration of time errors in the mesh adaptation process with a more general formulation of the mesh optimization problem is examined seriously in [4].

Time discretization error is not considered in this study. Solving this question is not so important for the type of calculation that are shown in this paper, but can be of paramount impact in many other cases, in particular when implicit time advancing is used. The additional effort in time-error reduction has to be integrated in the convergence analysis sketched in this paper, and the authors plan to propose this global analysis in some future.

References

1. Alauzet, F., Frey, P.J., George, P.L., Mohammadi, B.: 3D transient fixed point mesh adaptation for time-dependent problems: Application to CFD simulations. J. Comp. Phys. 222, 592–623 (2007)
2. Alauzet, F., George, P.L., Mohammadi, B., Frey, P.J., Borouchaki, H.: Transient fixed point based unstructured mesh adaptation. Int. J. Numer. Meth. Fluids 43(6-7), 729–745 (2003)

3. Alauzet, F., Loseille, A.: High order sonic boom modeling by adaptive methods. J. Comp. Phys. 229, 561–593 (2010)
4. Alauzet, F., Olivier, G.: Extension of metric-based anisotropic mesh adaptation to time-dependent problems involving moving geometries. In: 49th AIAA Aerospace Sciences Meeting and Exhibit, AIAA-2011-0896, Orlando, FL, USA (January 2011)
5. Baines, M.J.: Moving finite elements. Oxford University Press, Inc., New York (1994)
6. Becker, R., Rannacher, R.: A feed-back approach to error control in finite element methods: basic analysis and examples. East-West J. Numer. Math. 4, 237–264 (1996)
7. Berger, M., Colella, P.: Local adaptive mesh refinement for shock hydrodynamics. J. Comp. Phys. 82(1), 67–84 (1989)
8. Cournède, P.-H., Koobus, B., Dervieux, A.: Positivity statements for a Mixed-Element-Volume scheme on fixed and moving grids. European Journal of Computational Mechanics 15(7-8), 767–798 (2006)
9. Dompierre, J., Vallet, M.G., Fortin, M., Bourgault, Y., Habashi, W.G.: Anisotropic mesh adaptation: towards a solver and user independent CFD. In: AIAA 35th Aerospace Sciences Meeting and Exhibit, AIAA-1997-0861, Reno, NV, USA (January 1997)
10. Frey, P.J.: Yams, a fully automatic adaptive isotropic surface remeshing procedure. RT-0252, INRIA (November 2001)
11. Giles, M.B., Suli, E.: Adjoint methods for PDEs: a posteriori error analysis and postprocessing by duality, pp. 145–236. Cambridge University Press (2002)
12. Hecht, F., Mohammadi, B.: Mesh adaptation by metric control for multi-scale phenomena and turbulence. AIAA Paper, 97-0859 (1997)
13. Löhner, R.: Adaptive remeshing for transient problems. Comp. Methods Appl. mech. Engrg. 75, 195–214 (1989)
14. Loseille, A., Alauzet, F.: Optimal 3D highly anisotropic mesh adaptation based on the continuous mesh framework. In: Proceedings of the 18th International Meshing Roundtable, pp. 575–594. Springer, Heidelberg (2009)
15. Loseille, A., Alauzet, F.: Continuous mesh framework. Part I: well-posed continuous interpolation error. SIAM J. Numer. Anal. 49(1), 38–60 (2011)
16. Loseille, A., Alauzet, F.: Continuous mesh framework. Part II: validations and applications. SIAM J. Numer. Anal. 49(1), 61–86 (2011)
17. Loseille, A., Dervieux, A., Alauzet, F.: Fully anisotropic goal-oriented mesh adaptation for 3D steady Euler equations. J. Comp. Phys. 229, 2866–2897 (2010)
18. Loseille, A., Dervieux, A., Frey, P.J., Alauzet, F.: Achievement of global second-order mesh convergence for discontinuous flows with adapted unstructured meshes. In: 37th AIAA Fluid Dynamics Conference and Exhibit, AIAA-2007-4186, Miami, FL, USA (June 2007)

19. Loseille, A., Löhner, R.: Adaptive anisotropic simulations in aerodynamics. In: 48th AIAA Aerospace Sciences Meeting and Exhibit, AIAA-2010-169, Orlando, FL, USA (January 2010)
20. Verfürth, R.: A review of A Posteriori Error Estimation and Adaptative Mesh-Refinement techniques. Wiley Teubner Mathematics, New York (1996)
21. Wintzer, M., Nemec, M., Aftosmis, M.J.: Adjoint-based adaptive mesh refinement for sonic boom prediction. In: AIAA 26th Applied Aerodynamics Conference, AIAA-2008-6593, Honolulu, HI, USA (August 2008)

HEX MESHING (Monday Afternoon)

Automatic All-Hex Mesh Generation of Thin-Walled Solids via a Conformal Pyramid-Less Hex, Prism, and Tet Mixed Mesh[*]

Soji Yamakawa and Kenji Shimada

The Department of Mechanical Engineering Carnegie Mellon University
soji@andrew.cmu.edu,
shimada@cmu.edu

Summary. A new computational method for creating all-hex meshes of thin-walled solids is presented. The proposed method creates an all-hex mesh in a three-step process. First, a tet mesh of the target geometric domain is created, and a layer of prism elements is inserted on the boundary of the mesh. Second, a sequence of novel topological transformations and smoothing operations is applied to reduce the number of tet elements and increase the number of hex elements. The topological transformations maintain conformity of the hex, prism, and tet mixed mesh. The proposed topological transformations do not introduce pyramid elements, and therefore avoids problems imposed by pyramid elements. Third and finally, the mixed mesh is subdivided into an all-hex mesh by applying mid-point subdivision templates. Experimental results show that the proposed method creates high-quality all-hex meshes.

1 Introduction

This paper describes a new computational method for creating high-quality all-hex meshes of thin-walled solids. The proposed method creates an all-hex mesh in three steps: (1) prism-tet pillowing, (2) conformal transformation, and (3) all-hex mesh subdivision. In Step (1), the proposed method creates an all-tet mesh of the input geometric domain, and a layer of prism elements is inserted on the boundary of the tet mesh. In Step (2), a sequence of topological transformations and smoothing operations is applied so that the number of tet elements between prism elements is reduced, and some prism elements are converted to hex elements. In Step (3), the elements of the mixed mesh are subdivided into all-hex elements by applying mid-point subdivision templates as shown in Figure 1.

The main contribution of this research is in the second step. Conformal transformations reduce the number of tet elements and increase the number of hex elements while maintaining mesh conformity. There have been no previously-published topological transformations that transform a hex, prism, and tet mixed

---

[*] Provisional patent (61/520,795) has been filed on the technique described in this paper.

mesh without inserting pyramid elements. Pyramid elements are problematic in two ways: (1) many finite-element solvers reject pyramid elements and thus the mixed mesh itself becomes unusable for such solvers, and (2) unlike other three types of volumetric elements, hex, prism, and tet, a pyramid element cannot be subdivided into all-hex elements with a simple mid-point subdivision template (see Figure 1). Applying a pyramid subdivision template [1] substantially increases element count and decreases mesh quality. Problems imposed by pyramid elements are avoided in the proposed new topological transformations by not inserting such elements.

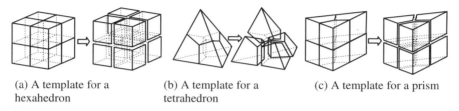

(a) A template for a hexahedron

(b) A template for a tetrahedron

(c) A template for a prism

Fig. 1. Mid-point subdivision templates

All-hex meshes tend to yield more accurate finite-element solutions than all-tet meshes, however, automatically generating all-hex meshes is much more challenging than generating all-tet meshes. The latter can be performed in a straight forward manner [2-4]. Despite two decades of research, an automated all-hex mesh generator - so-called 'push-button' all-hex mesh generator, has not yet been developed

In the effort to develop an all-hex mesh-generation scheme, mesh-conversion approach, which first creates a hex, prism, and tet mixed mesh and then converts it into an all-hex mesh, attracted much early attention since a similar approach worked well in 2D. In 2D meshing, a quad-triangle mixed mesh can be subdivided into an all-quad mesh by applying mid-point subdivision templates as shown in Figure 2.

(a) A quad-dominant to an all-quad mesh conversion

(b) Two all-quad conversion templates

Fig. 2. 2D mixed mesh to all-quadrilateral mesh conversion

Due to the algorithm's similarity, numerous attempts have been made to translate the 2D method into 3D. However, conversion from a 3D mixed mesh to an all-hex mesh poses additional, a higher level of challenges. In 2D, all elements interface over the same type of boundary, an edge, whereas in 3D, elements can interface over either a triangular or a quadrilateral face. Since a tet element consists of triangular faces alone, and a hex element of quad faces alone, it is impossible to make a conformal connection between these two types of elements.

Although numerous potential solutions have been proposed [1, 5-9], none of them can solve the tet-hex connection problem without using a complex element-subdivision template, which substantially increases the element count and creates low-quality elements.

The proposed method, to the authors' knowledge, is the first all-hex mesh-conversion approach that does not require a complex subdivision template. By avoiding a complex subdivision template, the proposed method can create a high-quality all-hex mesh with no areas of unintended non-uniform node distribution.

2 Overview of the Proposed Method

To assist the explanation and comprehension of the proposed method, the 2D analogue is presented. The 2D analogue of the proposed method creates an all-quad mesh in three steps:

Step 1: Quad-triangle pillowing (prism-tet pillowing in 3D)
Step 2: Conformal transformation
Step 3: All-quad subdivision (all-hex subdivision in 3D)

In Step 1, an all-triangular mesh (tet mesh in 3D) of the input geometric domain is created, and a layer of quad elements (prism elements in 3D) is inserted on the boundary of the triangular mesh as shown in Figure 3 (a). In Step 2, a sequence of topological transformation, detailed in Section 4, and smoothing operations are applied as shown in Figure 3 (b). One of the most important goals of this step is to reduce the number of triangular (tet in 3D) elements. In 3D, some prism elements are also converted to hex elements. In Step 3, the elements included in the mesh are subdivided into all-quad (all-hex in 3D) elements as shown in Figure 3 (c).

Step 1 of the 3D solution generates a tet mesh of the input geometric domain as shown in Figure 4 and inserts a layer of prism elements on the boundary of the tet mesh as shown in Figure 5.

Step 2 applies a sequence of topological transformations and smoothing operations resulting in a reduction of the number of tet elements and an increase in the number of hex elements while maintaining the mesh conformity. A variety of topological transformations are applied in order to achieve the two goals For example, edge-collapse transformations eliminate tet elements lying between prism elements as shown in Figure 6. These transformations in addition to other topological transformations and smoothing operations are applied until no more tet element can be removed as shown in Figure 7.

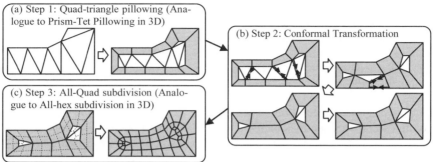

Fig. 3. 2D analogue of the proposed method

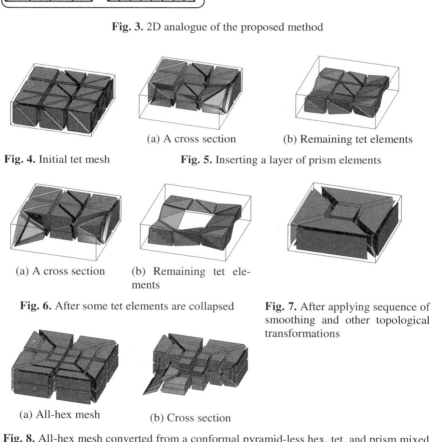

Fig. 4. Initial tet mesh

Fig. 5. Inserting a layer of prism elements

Fig. 6. After some tet elements are collapsed

Fig. 7. After applying sequence of smoothing and other topological transformations

Fig. 8. All-hex mesh converted from a conformal pyramid-less hex, tet, and prism mixed mesh

Step 3 produces an all-hex mesh by subdividing the mixed elements into all-hex elements by applying mid-point subdivision templates as shown in Figure 8. The mesh remains conformal through each step of the algorithm and no problematic pyramid element is introduced.

Step 1 can be done by a conventional tet mesh-generation scheme [2, 3, 10, 11] and a conventional boundary-layer element-generation scheme [12, 13]. Unlike the boundary-layer element generation for a CFD application, only one layer of prism elements is created in this step. For smoothing operations in Step 2, the proposed method uses combination of schemes including linear-programming method [14, 15], volume equalization [9], and angle-based smoothing method [16]. Midpoint subdivision templates used in Step 3 has long been known to exist.

The uniqueness and power of the proposed method is in the novel topological transformations used in Step 2. These transformations reduce the number of tet elements and increase the number of hex elements while maintaining mesh conformity. No existing topological transformation methods had been able to transform a hex, prism, and tet mixed mesh without introducing pyramid elements. This problem was substantially limiting the usability of a mixed meshes and made all-hex conversion very difficult. Since the proposed topological transformation schemes do not introduce pyramid elements, it effectively avoids problems imposed by the pyramid elements. Details of the topological transformations are explained in Section 4.

3 Prism Chain and Prism-Chain Pair

Many of the topological transformations used by the proposed method transform a collection of prism elements called *a prism-chain pair*. A prism-chain pair is a pair of sets of prism elements called *prism chains*. A prism chain is defined as a single prism element or a set of prism elements that are connected to at least one other member of the set by a triangular face.

Two prism chains may form a prism-chain pair when the following conditions are met: (1) the prism chains consist of same number of elements, (2) each of the elements in one prism chain has an element in the other prism chain that shares a quadrilateral face, and (3) each of the triangular faces of one prism chain has a triangular face in the other prism chain that shares an edge. A prism-chain pair has a column of quadrilateral faces, called *gluing faces*, each of which is shared by a prism element in one prism chain and a prism element in the other prism chain. A prism-chain pair also has a column of edges, each of which is shared by two triangular faces within the prism-chain pair. Such edges are called *rib edges*. Figure 9 shows an example of a prism-chain pair with two prism chains each of which consist of three prism elements.

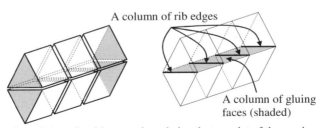

Fig. 9. A prism-chain pair with two prism chains that consist of three prisms each.

4 Topological Transformations

This section explains in detail the topological transformations used for reducing the number of tet elements and increasing the number of prism and hex elements. The proposed method uses following eight types of transformations:

- Edge Collapse
- Prism Swap
- Prism Collapse
- Prism Split
- Clamp
- Diamond Collapse
- Half- and Full-Wedge Collapse
- Prism Paring

Mesh conformity is guaranteed through the topological transformations.

Edge Collapse

An edge used only by tet elements can be collapsed, or all tet elements using the edge are deleted and two nodes of the edge are merged to one. For example, edge **AB** in Figure 10 (a) is shared by six tet elements and can be collapsed. The transformation connects the prism elements sharing node **A** and the prism elements sharing node **B** as shown in Figure 10 (b).

The main purpose of this transformation is to remove tet elements lying between prism elements. If multiple tet elements exist between opposing prism elements, the transformation needs to be applied multiple times to connect two opposing prism elements.

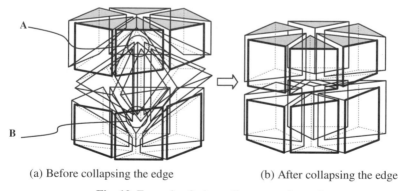

(a) Before collapsing the edge (b) After collapsing the edge

Fig. 10. Example of edge-collapse transformation

Prism Swap

The prism-swap transformation re-connects, or swaps, a column of gluing faces in a prism-chain pair. The two prism elements sharing a gluing face have a combined four edges that are not used by any of the four triangular faces of the two prism elements. Two of the four edges are connected by a gluing face, and the remaining two are not directly connected. The prism-swap transformation replaces the two prism elements with two new prism elements such that new gluing face connects the two edges that were not connected in the original configuration and disconnects the edges that where originally connected as shown in Figure 11. Shaded faces in the figure are exterior faces and are not connected to other elements.

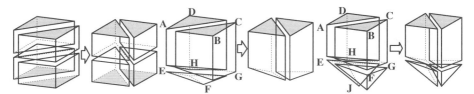

Fig. 11. Prism-swap transformation

Fig. 12. A variation of prism-swap transformation

Fig. 13. Another variation of prism-swap transformation

If a tet element shares four nodes of one end of a prism-chain pair, the tet element is removed by this transformation. In Figure 12, a sub-volume of the mesh is initially filled with a pair of prism elements **ABCEFG** and **ACDEGH** (ABC and ACD are exterior faces) and a tet element **FHGE**. The three elements can be replaced with two prism elements **ABDEFH** and **BCDFGH**.

When two tet elements sharing a triangular face are connected to one end of a prism-chain pair, the two tet elements are also replaced with two new tet elements by the prism-swap transformation such that the conformity is maintained. In Figure 13, a sub-volume of a mesh is initially filled with two prism elements **ABCEFG** and **ACDEGH** and two tet elements **FEJG** and **EHJG**. The two prism elements are replaced with two new prism elements **ABDEFH** and **BCDFGH**, and the two tet elements are replaced with two new tet elements **FEJH** and **FHJG**.

Prism Collapse

The prism-collapse transformation collapses each rib edge in the prism-chain pair into one node, and essentially deletes all the elements sharing a rib edge. A rib edge at an end of the prism-chain pair can be used by tet elements, which are deleted by the prism collapse transformation. If a rib edge is used by a non-tet element that is not included in the prism-chain pair, this transformation cannot be applied.

Prism and tet elements shown in Figure 14 can be collapsed using the prism collapse transform. Prism elements **ABDEFH**, **BCDFGH**, **EFHJKM**, **FGHKLM**, and tet elements **JKMN**, **KLMN**, can be deleted by collapsing edges **BD**, **FH**, and **KM** into one node each.

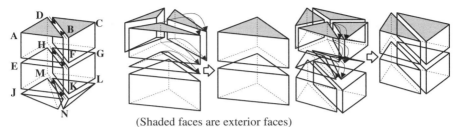

Fig. 14. A prism-chain pair and connected tet elements that can be collapsed

(Shaded faces are exterior faces)

Fig. 15. An application of the prism-collapse transformation

Fig. 16. An application of the prism-collapse transformation

This transformation is useful when three prism elements share an edge, all three prism elements are connected to a single tet element, and the other side of each prism element is exposed to the exterior of the mesh. Figure 15 shows this configuration and illustrates how the three prism elements are collapsed into one and the tet element is deleted.

This transformation is also useful when four prism elements share an edge, two of them are connected to a single tet element, and the remaining two are connected to another single tet element. This configuration, shown in Figure 16, can be collapsed from four prism and two tet elements to two prism elements.

Prism Split

The prism-split transformation splits each prism element in a prism-chain pair into two prism elements by inserting a node on each rib edge. If tet elements are using a rib edge at an end of the prism-chain pair, each of those tet elements are also split into two.

For example, consider two prism elements **ABCEFH** and **ACDHFG** that share a gluing face **AHFC** as shown in Figure 17 (a) and two tet elements **EFHJ** and **HFGJ** that are connected to the pair of the two prism elements. Node **K** is inserted on edge **AC** and node **L** is inserted on **HF**, prism element **ABCEFH** is split into two prism elements **ABKHEL** and **BCKEFL**, prism element **ACDHFG** is split into two prism elements **AKDHLG** and **CDKFGL**, tetrahedron **EFHJ** is split into two tet elements **ELHJ** and **EFLJ**, and tetrahedron **HFGJ** is split into two tet elements **HLGJ** and **FGLJ** as shown in Figure 17 (b). Node **L** can be then merged to node **J** and the four tet elements can subsequently be deleted in this case as shown in Figure 17 (c).

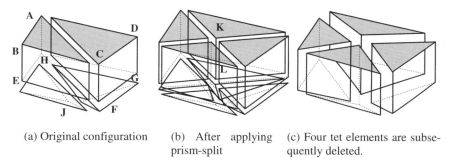

(a) Original configuration (b) After applying prism-split (c) Four tet elements are subsequently deleted.

Fig. 17. The prism-split transformation

Clamp

The clamp transformation converts three tet elements with three connected prism-chain pairs into one prism element with three chains of hex elements. This transformation reduces number of tet elements and converts pairs of prism elements into single hex elements.

The following four conditions are needed to perform the clamp transformation: (1) three tet elements share an edge and are comprised of six nodes, (2) three prism-chain pairs are connected to the three tet elements, (3) the other side of the three prism chain pairs are exposed to the exterior of the mesh, and (4) two triangular faces of the three tet element cluster that are exposed outside the three tet elements and not connected to the three prism chain pairs do not share any nodes.

Figure 18 (a) shows a configuration that satisfies the above conditions which is converted into one prism and three hex elements after using the clamp transformation. Figure 19 shows a typical pattern on which the transformation can be applied. This pattern often appears on models near the corners of thin plate features. The clamp transformation reduces the number of tet element in such geometric constructs.

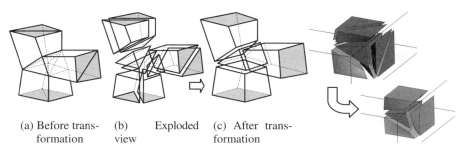

(a) Before transformation (b) Exploded view (c) After transformation

Fig. 18. An example of the clamp transformation

Fig. 19. A typical configuration that the clamp transformation can be applied

Diamond Collapse

The diamond-collapse transformation collapses and deletes three prism chains and a tet element. The transformation can be applied when an edge, E_0, is shared by three triangular and two quadrilateral faces such that the two quadrilateral faces are separated by one of the triangular faces and that the 3D elements that share edge E_0 are four prism elements plus a tet element, as shown in Figure 20 (a). Two of the three prism chains to be deleted by this transformation start from the triangular faces of the tet element sharing E_0 and continue to the boundary of the mesh. The third prism chain includes the third triangular face sharing E_0. Each prism element in the third prism chain shares a quadrilateral with a prism element in one of the first two chains. Both ends of the third prism chain must be exposed to the exterior of the boundary. If the above conditions are satisfied, edge E_0 and each edge shared by two triangular faces of the three prism chains can be collapsed to a node. The diamond-collapse transformation maintains mesh conformity.

Figure 20 illustrates an instance of the diamond-collapse transformation applied to a set of prism and tet elements resulting in only quadrilateral and triangular faces after transformation.

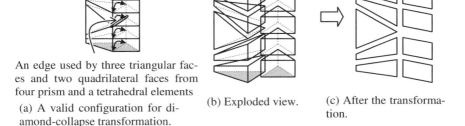

(a) A valid configuration for diamond-collapse transformation.
An edge used by three triangular faces and two quadrilateral faces from four prism and a tetrahedral elements

(b) Exploded view.

(c) After the transformation.

Fig. 20. Diamond collapse (Shaded faces are exposed to the exterior of the mesh.)

Half- and Full- Wedge Collapse

The half-wedge-collapse transformation collapses and deletes two prism chains and a tet element. When the following three conditions are met:

(1) one edge is shared by two triangular faces and a quadrilateral face from two prism and one tet elements,
(2) the other end of the two prism chains are exposed to the mesh boundary, and
(3) the two prism chains form a prism-chain pair,

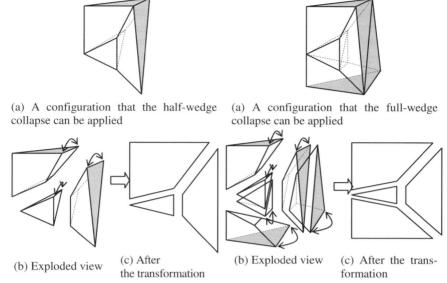

(a) A configuration that the half-wedge collapse can be applied

(a) A configuration that the full-wedge collapse can be applied

(b) Exploded view (c) After the transformation

(b) Exploded view (c) After the transformation

Fig. 21. Half-wedge collapse transformation (Shaded faces are exterior faces.)

Fig. 22. Full-wedge collapse transformation

Each of the edges in the prism-chain pair shared by two triangular faces is collapsed into one node, and the tet element and the prism elements in the prism chains are deleted as shown in Figure 21.

Figure 22 illustrates an instance of a commonly appearing variant configuration that can be addressed with the half-wedge collapse transformation. The configuration is comprised of two separate valid half-wedge collapse transformation candidates that can be optimized in a single operation such that each of the edges that would be collapsed separately can be collapsed at the same time.

Prism Paring

Prism-pairing transformation merges two prisms sharing a quadrilateral face included in a prism-chain pair as shown in Figure 23. This transformation is applicable if one of the following two conditions is satisfied: (1) both end faces of the prism-chain pair are exposed to the exterior of the mesh, or (2) the prism-chain pair is a closed prism-chain pair, i.e., none of the triangular faces is exposed to the outside of the prism-chain pair. If neither condition is satisfied this transformation will result in a non conformal mesh, and cannot be applied.

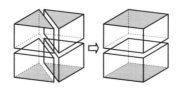

Fig. 23. A prism-chain pair with two prism chains that consist of three prisms each.

Maintaining Element Quality and Element Size

It is not trivial to estimate the resulting element quality and element size before actually applying the above-mentioned transformations. The program, therefore, checks affected elements after each transformation, and if an element violates the user-specified quality criteria, the transformation is retracted.

When a given geometric domain consists of non-thin-walled, or thick, parts, it becomes difficult to remove many of the tet elements lying between opposing prism elements without violating quality criteria. Tet elements remain unremoved in such thick parts. and hex elements subdivided from those remaining tet elements will have relatively lower quality.

5 Examples

One major advantage of the proposed method is requiring no manual volume decomposition to obtain high-quality meshes. The tet mesh of a U-shaped pipe connected to the thin plate as shown in Figure 24 demonstrates such an example. Using conventional methods, such a model must be manually decomposed into sweepable sub-domains to create a high-quality all-hex mesh. The proposed method does not require such manual decomposition. Table 1 presents metrics of the automatically generated all-hex mesh, showing that the results are of good quality.

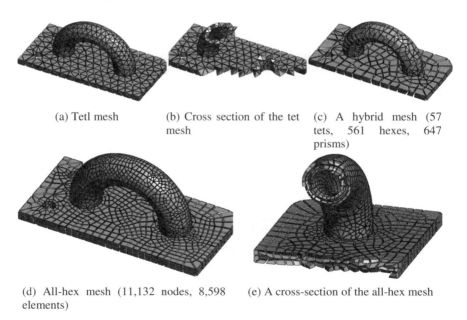

(a) Tet1 mesh

(b) Cross section of the tet mesh

(c) A hybrid mesh (57 tets, 561 hexes, 647 prisms)

(d) All-hex mesh (11,132 nodes, 8,598 elements)

(e) A cross-section of the all-hex mesh

Fig. 24. An example of non-sweepable geometry

Table 1 Statistics of the all-hex mesh of the non-sweepable geometry

Scaled Jacobian and number of elements		Quad Warpage (degree) and number of quads	
1.0 to 0.9	99	0 to 10 degree	18,277
0.9 to 0.8	575	10 to 20 degree	9,182
0.8 to 0.7	1127	20 to 30 degree	367
0.7 to 0.6	1502	30 to 40 degree	5
0.6 to 0.5	1459	Greater than 40 degree	0
0.5 to 0.4	1447		
0.4 to 0.3	1302		
0.3 to 0.2	1087		
Less than 0.2	0		
Minimum Scaled Jacobian: 0.20		Maximum Warpage Angle: 34.2 degree	

The proposed method can be applied to biological models such as that of an aorta artery as shown in Figure 25 (a). The hybrid mesh shown in Figure 25 (b) is generated from the tet mesh shown in Figure 25 (a). Subdividing the mesh yields the all-hex mesh shown in Figure 25 (c). Quality statistics of the all-hex mesh, shown in Table 2, indicate that the results are of high quality.

6 Reduced Mid-Point Subdivision

Prism elements inserted in the prism-tet pillowing step (the first step) makes a sheet of prism elements along the boundary of the tet mesh. An element included in such a sheet has a distinctive thickness direction. Each exterior face of a sheet has an opposite face, which is an interior face connected to a tet element of the mesh. Edges connecting an exterior face and the opposite face define the thickness direction. Since none of the proposed topological transformations breaks the sheet, the thickness direction remains the same through the conformal transformation step even after some prism elements in the sheet are converted to hex elements.

Table 2 Statistics of the all-hex mesh of the aorta artery

Scaled Jacobian and number of elements		Quad Warpage (degree) and number of quads	
1.0 to 0.9	38,114	0 to 10 degree	490,215
0.9 to 0.8	83,895	10 to 20 degree	128,342
0.8 to 0.7	42,740	20 to 30 degree	22,958
0.7 to 0.6	20,042	30 to 40 degree	4558
0.6 to 0.5	7,810	40 to 50 degree	920
0.5 to 0.4	3,634	50 to 60 degree	329
0.4 to 0.3	2,286	60 to 70 degree	1
0.3 to 0.2	771	Greater than 70 degree	0
0.2 to 0.1	2		
Less than 0.1	0		
Minimum Scaled Jacobian:0.199993		Maximum Warpage Angle:60.01 degree	

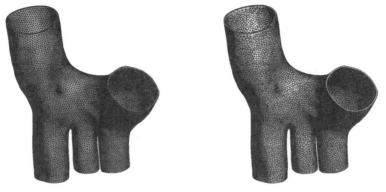

(a) Tet mesh of an aorta artery

(b) Hybrid mesh with 27 tet, 18,383 hex, and 8687 prism elements

 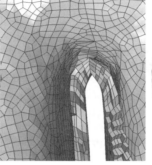 

(c) All-hex mesh with 248,971 nodes and 199,294 elements

(d) Close view of a cross section of the all-hex mesh

(e) Cross section of the all-hex mesh

Fig. 25. An aorta model

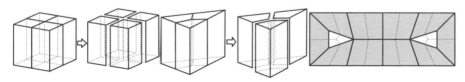

(a) Reduced mid-point subdivision template for a hex element

(b) Reduced mid-point subdivision template for a prism element

(c) A sample cross-section of an all-hex mesh with reduced mid-point subdivision

Fig. 26. Reduced mid-point subdivision

When a sheet of elements has a distinctive thickness direction, the elements do not have to be subdivided in the thickness direction in the all-hex subdivision step (the third step). For those elements, reduced mid-point subdivision templates shown in Figure 26 (a) and (b) can be applied. These subdivision templates insert a node only on the exterior face and the opposite face. No node is inserted inside the element.

The reduced mid-point subdivision templates substantially reduce the number of nodes and elements in the final all-hex mesh. However, the mesh has less freedom of the interior nodes, and the mesh-smoothing scheme may not achieve a target mesh quality due to the lack of freedom.

The decision of whether to use reduced mid-point subdivision should be made according to the requirements of the application in which the mesh is used.

7 Discussions on the Order Dependency Issue

While the result from the above topological transformations depends on the order of transformation operations, it is difficult to find the globally optimum transformation order in a practical computational time.

Nonetheless, experiments conducted in research indicate some clues of finding a good transformation order including:

(1) Locations where prism-collapse, prism-swap, and prism-split transformations can be applied are often isolated and do not depend on the order of the transformations.
(2) Diamond-collapse, wedge-collapse, and clamp transformations work more effectively after tet elements are deleted as much as possible by edge-collapse, prism-collapse, prism-swap, and prism-split transformations.

From these observations, the following steps have been tested to confirm that they work well for many cases:

1) Apply an edge-collapse transformation when it can delete a tet element lying between two nearest nodes from opposing prism elements.
2) Apply prism-collapse, prism-swap, and prism-split transformations whenever the transformation can reduce the number of tet elements.
3) Apply smoothing.
4) Repeat Steps 1 to 3 until no more tet element can be deleted.
5) Apply clamp, diamond-collapse, and wedge-collapse transformations as many times as possible.

This strategy, a greedy-search approach targeting the maximum reduction of tet elements, gives a reasonably good result in many test cases used in this research. In some cases, however, manually tweaking the order of the transformations finds a better mesh in the end. There can be a smarter automated algorithm of finding a more ideal order of transformations.

One option is to apply greedy-search approach targeting the best mesh quality rather than the maximum reduction of tet elements. In this approach, the transformation that makes the best quality elements is prioritized in Steps 1 and 2 of the above approach. The issue to be solved to implement this approach is how to compare the element quality. In a mixed-element mesh, the element quality is measured by multiple measurements such as scaled Jacobian, quad warpage, and radius-ratio. Since those measurements are highly non-linear, simply taking a weighted sum does not give a fair evaluation. In other words, this approach is a multi-objective optimization, and further research is needed to develop the good element-comparison condition for this purpose.

Another option is a trial-and-error approach. In this approach, a greedy-search approach is applied first, from which the program searches a possible improvements that could be made by applying topological transformations in a different order. Then the topological transformations applied in the first step are retracted and the mesh is brought back to the state immediately after the prism-tet pillowing step. Finally, the topological transformations are applied in the order that is modified based on the search. This trial-and-error loop can be repeated until no more improvement is gained. This approach suffers from the same issue as the quality-based greedy-search approach; it requires a good element-comparison condition to evaluate the improvement. Also, since this applies the conformal-transformation step multiple times, it takes longer time to generate a mesh. Further research is needed for developing a solution to these issues.

8 Conclusions

This paper has presented a new computational method for creating conformal pyramid-less hex, prism, and tet mixed meshes and an all-hex meshes of thin-walled solids. The proposed method creates an all-hex mesh in three steps: (1) prism-tet pillowing, (2) conformal transformation, and (3) all-hex mesh subdivision.

The significance of the new topological transformations used in Step (2) is it maintains the conformity of the hex, prism, and tet mixed mesh. Known topological transformation that modifies the topology of a mixed mesh requires pyramid elements to preserve mesh conformity. A pyramid element, however, is problematic and makes conversion from a mixed mesh to an all-hex mesh difficult. The proposed topological transformation does not introduce pyramid elements, and therefore avoids problems imposed by pyramid elements.

References

[1] Yamakawa, S., Shimada, K.: 88-Element Solution to Schneiders' Pyramid Hex-Meshing Problem. International Journal for Numerical Me-thods in Biomedical Engineering 26, 1700–1712 (2010)
[2] George, P.L.: Tet Meshing: Construction, Optimization and Adaptation. In: 8th International Meshing Roundtable, South Lake Tahoe, CA, pp. 133–141 (1999)

[3] George, P.L., Hecht, F., Saltel, E.: Automatic Mesh Generator with Specified Boundary. Computer Methods in Applied Mechanics and Engineering 92, 269–288 (1991)
[4] Yamakawa, S., Shimada, K.: Anisotropic Tetrahedral Meshing via Bubble Packing and Advancing Front. International Journal for Numerical Methods in Engineering 57, 1923–1942 (2003)
[5] Owen, S.J., Saigal, S.: H-Morph: an Indirect Approach to Advancing Front Hex Meshing. International Journal for Numerical Methods in Engineering 49, 289–312 (2000)
[6] Yamakawa, S., Gentilini, I., Shimada, K.: Subdivision Templates for Converting a Non-Conformal Hex-Dominant Mesh to a Conformal Hex-Dominant Mesh without Pyramid Elements. Engineering with Computers 27, 51–65 (2011)
[7] Mitchell, S.A.: The All-Hex Geode-Template for Conforming a Diced Tetrahedral Mesh to any Diced Hexahedral Mesh. In: 7th International Meshing Roundtable, Dearborn, MI, pp. 295–305 (1998)
[8] Yamakawa, S., Shimada, K.: HEXHOOP: Modular Template for Converting a Hex-Dominant Mesh to an ALL-Hex Mesh (2001)
[9] Yamakawa, S., Shimada, K.: HEXHOOP: Modular Templates for Converting a Hex-Dominant Mesh to an ALL-Hex Mesh. Engineering with Computers 18, 211–228 (2002)
[10] Yamakawa, S., Shaw, C., Shimada, K.: Layered Tetrahedral Meshing of Thin-Walled Solids for Plastic Injection Molding FEM. Computer-Aided Design 38, 315–326 (2006)
[11] Yamakawa, S., Shimada, K.: High Quality Triangular/Quadrilateral Mesh Generation of a 3D Polygon Model via Bubble Packing. In: 7th US National Congress on Computational Mechanics, Albuquerque, NM (2003)
[12] Garimella, R.V., Shephard, M.S.: Boundary Layer Mesh Generation for Viscous Flow Simulations. International Journal for Numerical Me-thods in Engineering 49, 193–218 (2000)
[13] Ito, Y., Nakahashi, K.: Unstructured Mesh Generation for Viscous Flow Computations. In: 11th International Meshing Roundtable, Ithaca, NY, pp. 367–378 (2002)
[14] Freitag, L.A., Plassmann, P.: Local Optimization-based Simplical Mesh Untangling and Improvement. International Journal for Numerical Methods in Engineering 49, 109–125 (2000)
[15] Freitag, L.A., Plassmann, P.E.: Local Optimization-based Untan-gling Algorithms for Quadrilateral Meshes. In: 10th International Meshing Roundtable, Newport Beach, CA, pp. 397–406 (2001)
[16] Zhou, T., Shimada, K.: An Angle-Based Approach to Two-dimensional Mesh Smoothing. In: 9th International Meshing Roundtable, New Orleans, LA, pp. 373–384 (2000)

Hexahedral Mesh Refinement Using an Error Sizing Function

Gaurab Paudel,[1] Steven J. Owen,[2] and Steven E. Benzley[1]

[1] Brigham Young University
gaurab.paudel@gmail.com, seb@byu.edu
[2] Sandia National Laboratories
sjowen@sandia.gov

Summary. The ability to effectively adapt a mesh is a very important feature of high fidelity finite element modeling. In a finite element analysis, a relatively high node density is desired in areas of the model where there are high error estimates from an initial analysis. Providing a higher node density in such areas improves the accuracy of the model and reduces the computational time compared to having a high node density over the entire model. Node densities can be determined for any model using the sizing functions based on the geometry of the model or the error estimates from a finite element analysis. Robust methods for mesh adaptation using sizing functions are available for refining triangular, tetrahedral, and quadrilateral elements. However, little work has been published for adaptively refining all hexahedral meshes using sizing functions. This paper describes a new approach to drive hexahedral refinement based upon an error sizing function and a mechanism to compare the sizes of the node after refinement.

Keywords: hexahedral, meshing, adaptation, refinement, sizing function, error estimates.

1 Introduction

Mesh adaptation based on a sizing function is not a new topic. Procedures that incorporate quadrilateral, triangular, and tetrahedral mesh adaptation that rely on error-based sizing functions are available in the literature[1, 2]. In addition, there are a few techniques that generate an initial hexahedral mesh using the geometry features of the model[3]. However, conformal hexahedral mesh refinement, relying on an error-based sizing function, is a topic of current interest.

Traditional hexahedral meshing methods have not effectively used a sizing function because of connectivity restrictions imposed by traditional generation techniques[4, 5, 6]. This paper presents a method for a conformal hexahedral mesh refinement procedure based upon a sizing function. We incorporate the hexahedral mesh refinement techniques developed by Parrish[7] with a sizing function to drive refinement. The Parrish technique uses a set of seven transition templates and incorporates a special treatment for concave regions to ensure conformal and local mesh refinement. The sizing function used is developed from computed error estimates. However, criteria such as feature size or user specifications could also be included in the sizing function. To validate the method, comparisons between the actual refined node size and the ideal target node size are presented.

2 Background

The accuracy of finite element solutions can be improved by adapting the mesh. For example, meshes can be smoothed - known as r-adaptation - to improve quality. In addition, p-adaptation, which involves increasing the degree of the basis functions of the elements in the mesh, and h-adaptation, which involves increasing the number of elements, are traditional adaptation approaches. In addition, coarsening[8] can be used to reduce the number of elements. Although coarsening, r adaptation, and p adaptation are valid methods, this paper focuses specifically on h-adaptation i.e. refining by increasing the number of elements locally, to increase accuracy.

2.1 Refinement

As shown in Figure 1, 3-refinement[7] splits and exsiting hex three times along an edge, and 2-refinement splits an exisitng hex two times along and edge. 3-refinement is simple to implement but often can over refine a region of interest. 2- refinement[9] has more constraints on it's implementation but can often provide more gradual and controlled refined regions. The work reported here is based on 3- refinement, but can be easily adapted for 2- refinement.

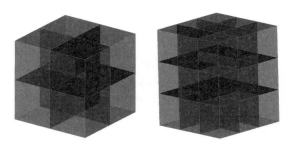

Fig. 1. 2 refinement and 3 refinement

Another common approach for adapting hexahedral meshes involves introduction of hanging nodes at edge centers as shown in Figure 2. This approach is straightforward to implement, and no transition elements to surrounding hexahedra are required. However, for this work we only consider conformal refinement techniques and do not introduce hanging nodes.

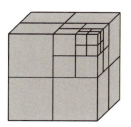 

Fig. 2. Refinement with hanging nodes compared to conformal refinement

2.2 Current Methods

We note that a mesh can be adapted before any analysis is run if an a priori knowledge of the physics and geometry of problem is known. However, a fully automatic sizing function based all-hexahedral mesh adaptation procedure, including both refinement and coarsening, based on computed error is the central motivation of this research.

There are several methods that use sizing functions to refine the nodes or to adjust the node densities at the time of the initial mesh generation. Quadros, et al.[10], and Zhang and Zhao[11] have introduced mesh refinement using a sizing function based on geometric features of the model, however, they do not discuss hexahedral mesh refinement based on the error estimates. Anderson et al.[1] developed a refining and coarsening technique that uses error estimates as the sizing function, however this method is limited to all quadrilateral elements and does not consider the hexahedral mesh. Zhang and Bajaj[12] introduce hexahedral mesh refinement using volumetric data, but do not consider converting the error estimate from the finite element analysis into a mesh size for refinement. Wada et. al[13] discuss adaptation of hexahedral meshes using local refinement and error estimates, however their method does not compare the refined size of the mesh to the target size from the error estimates.

As mentioned by Anderson[1], most of the adaptation techniques are limited to triangular and tetrahedral elements. In 2010 Kamenski[14] presented mesh adaptation using the error estimates but his method is limited to triangular elements. De Cougny and Shephard[15], discuss tetrahedral mesh adaptation but they do not consider a hexahedral technique. Kallinderis and

Vijayan[2] also discuss tetrahedral and triangular mesh refinement and coarsening but do not consider hexahedral elements. Babuska et. al[16] have presented a refinement technique based on a sizing function derived from error estimates, however their method is limited to non conformal rectangular elements with hanging nodes.

A hexahedral, when compared to a tetrahedral, mesh can provide more accurate results and, as mentioned in the introduction, is often the choice of an analyst. However, hexahedral adaptation techniques are not common. Hexahedral adaptation is a time consuming process, and requires knowledge of physics of the problem so that the generated mesh produces an acceptable error estimate from the finite element analysis. This paper presents unique and simple criteria to refine a hexahedral mesh using a sizing function and compares the refined size of the mesh with the target mesh size.

2.3 Sierra Mechanics Refinement Technique

In practice, rather than using a sizing function to drive the refinement, the error measures themselves are often utilized. For example, Sierra[17], a potential application for the work of this paper, is an advanced suite of analysis tools and provides three main approaches for driving refinement based upon an error measure. Although these techniques are currently used for tetrahedral and hanging node refinement, they could also be applicable for driving conformal hexahedral refinement in an adaptive analysis. For each of the three approaches, an error metric is computed for each element in the mesh, and the elements are ordered $M_{i=1..N}$ from minimum to maximum error. The three approaches are:

1. Percent of Elements: A threshold, a, which represents a percentage of the total number of elements N in the mesh that will be refined is specified. Starting from the element in M with the highest error and working towards the smallest error, a percent of the elements in the list are identified for refinement.
2. Percent of Max: A percentage threshold, b that represents the percentage of maximum error in the mesh that will be identified for refinement is specified. For example, if the maximum error of all elements in $M_{i=1..N}$ was 50 percent with $b = 90$ percent, then all elements with error > 5 percent would be identified for refinement.
3. Percent of Total Error: A percentage threshold, g, which represents a percent of the total error in the mesh that will be identified for refinement is specified. For example if we represent the total error of all elements in $M_{i=1..N}$ as:

$$||e||_{total} = \sum_{i=0}^{N} ||e||_i$$

Starting from the element in M with the highest error and working to the smallest error, those elements that contribute to a total error of $g.$ $||e||_{total}$ would be identified for refinement.

Sierra refinement is performed based upon one of the above approaches, followed by subsequent analysis iteration. After each iteration, elements can be again identified for refinement. This procedure continues until a convergence or error threshold has been achieved.

For the work presented in this paper, rather than using the error measure directly, a sizing function is developed from calculated error estimates. This provides the opportunity to utilize the sizing function as a general field to drive meshing or refinement. It also provides a field for which we can validate the resulting refinement operations to determine the effectiveness of the refinement algorithms at reaching the desired size.

3 Hexahedral Mesh Refinement

It is desired to have the size of a mesh be as close as possible to the sizes provided by the sizing function in order to obtain high computational accuracy in the results without significantly increasing the computation time.

3.1 Sizing Functions

Sizing functions are used as the mechanism for refining a mesh. There are several ways to generate sizing functions. Error estimates can be used to define a sizing function. Geometric characteristics, curvature, and sharp features in the model can also be used to define a sizing function [10]. Other bases for sizing functions include: the stress or strain gradients, change in the material properties, points of application of loading, and the location of boundary conditions.

For this paper, developing a sizing function based on the error estimate of an initial calculation will be used. The error estimate should be robust enough to ensure the increase in accuracy of the results, and also steer the adaptation only in the desired area of the model. The generation of the error estimate is an important area of study. For this work, error estimates are obtained from the existing finite element code. Physical phenomenon in engineering and sciences can be modeled using partial differential equations. However, complex mathematical models using the partial differential equations might not have an analytical solution. Fortunately, finite element analysis can provide an approximate solution to these complex models[18]. As these solutions are an approximation to the analytical solution, there are several sources of error.

As cited by Grastch and Bathe[18], the computation of error estimates and using it as criteria for subsequently refining the region where error estimates are high should be computationally cheaper than refining the entire model. The error estimates should be accurate enough to closely represent the unknown actual error. The goal of the error estimates is to steer the mesh adaptation. For this work, the built in error estimates produced by the simulation code, ADINA[19] are used.

3.2 Tools and Requirements

The error estimates from the finite element analysis approximate the expected error produced from the numerical model. Our refinement process is driven by a sizing function generated from error estimates. The error estimates from the finite element solvers are converted into an Exodus II[20] file, a random access, machine independent, binary file, that is used as a sizing function by the mesh generating toolkit, CUBIT[21]. The Exodus file format stores all the information about the initial mesh. It can be used for input and output of results and can also be used for post-processing of results. The Application Programming Interface (API) to create the exodus file is available in the public domain and a manual to create such file is also available. An example to create an Exodus file to drive the refinement process can be found in Reference[22].

3.3 Algorithm

Our proposed algorithm refines a hexahedral mesh locally based on the generated sizing function. The main goal of this algorithm is to complete the refinement process without the need for a user intervention. The error estimate, used to define the sizing function, determines whether a node should be considered for refinement or not. Often, after the mesh has been refined, the quality of the elements degrade. The degradation is usually a result of insertion of templates in the transition zones between the refined and non-refined regions. Hence, smoothing is performed on the elements within and near the refinement region to improve the element quality. A flowchart of the algorithm is given in Figure 3.

3.4 Algorithm Example

This section outlines the input, refinement criteria, and comparison of target and current sizes, of the algorithm. An example is used to explain the steps outlined in the algorithm. For simplicity, the algorithm example section is further divided into three sub-sections: input, refinement criteria and comparison of current size and target size.

3.5 Input

For this example, a quarter piston modeled with a load on its base plate is used. Initially, a coarse mesh, as shown in Figure 4, is generated using

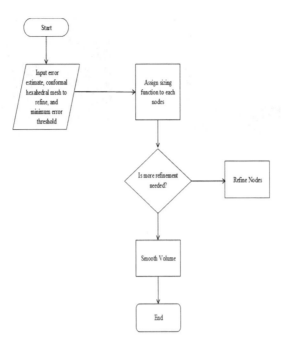

Fig. 3. Algorithm Flowchart

an all hexahedral meshing technique[21]. Next, the appropriate boundary conditions and loading, are applied. The initial mesh and boundary conditions are then exported to ADINA[19] to perform the finite element analysis. Error estimates are generated as a part of the ADINA analysis and the band plot of the error estimate is shown in Figure 4. The error estimate from the finite element analysis is then written in Exodus II, a binary file format, and is used to compute the sizing function to drive the refinement process.

3.6 Refinement Criteria

After each node is assigned a scalar error estimate, it is compared with the minimum specified threshold allowable error, g, for the particular problem. Normally the value for g and the error estimates are scalar values between 0 and 100, representing a percent error. Nodes with error estimates greater than the allowable user defined error are identified for further refinement. Nodes with error estimates lower than g are not refined but may be subsequently smoothed to improve the mesh quality. In this section the specific criteria is presented, based on the sizing function computed from error estimate, to identify the nodes needed for refinement. The terms basic to this technique are defined below.

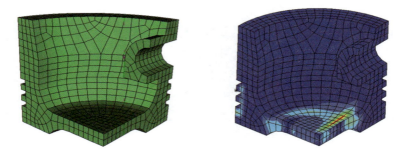

Fig. 4. Initial coarse mesh and Band plot of error estimate

The size of a node, h_a, is computed using Equation 1. h_a is the average of the lengths of edges attached to the node.

$$h_a = \frac{1}{n}\sum_{i=1}^{n} l_i \qquad (1)$$

where: $\qquad h_a =$ size of a node

$l_i =$ length of ith edge

$n =$ number of edges attached to the node

The relationship between error and mesh size can be approximated, for elasticity and heat problems, from the Poisson heat equation[23] as:

$$|e| = Ch^2 \qquad (2)$$

where: $\qquad C =$ a constant

$h =$ the element edge length

If the error and the element edge length for a node in the mesh are known, then C_n for that node can be computed as:

$$C_n = \frac{e_n}{h_a^2} \qquad (3)$$

where:

C_n a constant for the node

e_n error estimate at the node

$h_a =$ size of a node as defined in Equation (1)

The user provides a threshold error measure for the entire mesh, $|e| = g$ at which below g is acceptable. For this work, the target size is computed from the error measure but it can also be computed from the geometric feature or size provided by the user. Then, a target size from the above equation at the node is computed as:

$$t_s = \sqrt{\frac{g}{C_n}} \qquad (4)$$

where $\qquad t_s =$ target node size

$g =$ minimum allowable error provided by user

$C_n =$ a constant computed from Equation 3

The size ratio, S_r, is defined in Equation 5 as the ratio of the target size of the node, t_s, determined using the error estimates, to the actual size of the node, h_a.

$$S_r = \frac{h_a}{t_s} \qquad (5)$$

where: $\qquad S_r =$ node size ratio

$h_a =$ actual node size as defined in Equation (1)

$t_s =$ desired node size determined from the error measure

The terms defined in Equations 1 through 5 are used to identify the nodes that require refinement. It is assumed that the nodes with a size ratio of less than or equal to one are acceptable and need no refinement. Similarly, if the size ratio at the node approaches 3, then it indicates that at least one refinement operation should be performed. The value of 3 is assigned because 3-refinement is used as the mechanism for refinement. It is assumed that when the refinement is performed the size of a node will decrease by a factor of three and hence a size ratio of one is obtained after the refinement is performed. Since each split operation will reduce the local element size by a factor of 3, as the size ratio approaches 9, the node will be marked for two split operations. Likewise, a size ratio that approaches 27 will be marked for three splits. For this work, using an average of powers of three for the thresholds seemed to provide acceptable results. Table 1 summarizes the approach.

If the size ratio is below the allowable threshold, g, then no refinement is performed on the node. It is assumed that it is perfect size or it is over refined. The nodes with the size ratio between 2.0 and 6 are identified for the first level of refinement. These nodes are identified for the first level of refinement because they have size ratio near 3 and less than 9. Hence, after the refinement their size ratio should come close to 1. Similarly, the nodes with size ratio

Table 1. Size ratio range and number split operation

Size Ratio Range	Number of refinement split operations
0 - 2	0
2 - 6	1
6 - 18	2
> 18	3

between 6 and 18 are identified for the second level of refinement based on the fact that these nodes have size ratio near 9 and less than 27. Hence, when they are refined the new size ratio should be close to 1. The nodes with size ratios greater than 18 are identified for three levels of refinement. This criteria for refinement is continued until less than 10 percent of total nodes meet the size ratio criteria. This 10 percent is chosen to ensure that computation time is not wasted performing a refinement that will not gain a significant level of accuracy in the finite element solution. This criteria serves as the exit criteria for the refinement process. Figure 5 shows the refined mesh of the quarter piston with load on its base.

3.7 Comparison of Current and Target Size

One of the means to determine if the algorithm is performing adequately to achieve the desired accuracy in the results is to compare the refined size of node to the target size of the node. The node size ratio, S_r, is used as the criteria to compare the efficiency of the algorithm. It is assumed that the algorithm should perform such that the size ratio of all the nodes should be less than or equal to 1.0. If a coarsening algorithm were to be implemented, size ratios less than 1 would be minimized and all the size ratios would be close to

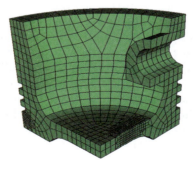

Fig. 5. Refined mesh of quarter piston

1.0. It is recognized that an exact match everywhere where $S_r = 1$ is impossible, however the approach presented here can be validated by statistically examining how close final mesh matches the intended sizing function.

Based on Equation 5, a size ratio for the node before the refinement and after the refinement is computed. Theoretically, there should not be any change in the volume of elements with nodes having size ratios below 1.0 and the volume of elements with nodes having size ratios greater 3 should be reduced significantly after the refinement. Also, there should be less than 10 percent of the volume that fall in the size ratio greater than 2.0. The plot in Figure 6, shows that for this problem, there is not much change in the volume for size ratios below 1.0 and in additon, there was not much initial volume with size ratio greater than 3. Although there is some volume greater than 2.0, this condition prevails because the refinement is deemed complete if there are less than 10 percent of nodes that require refinement. In Figure 7, the change in the percentage of volume falling in the particular size ratio before and after refinement is shown. Note that that most of the change is around size ratio 1.0 and there is negative change in volume for size ratio greater than 2.0. This shows that most of the nodes have size ratio 1.0 after the refinement and refinement is taking place in the node with size ratio greater than 2.0.

4 Example

This section gives a complete example of all hexahedral sizing based refinement. Shown are the initial coarse mesh, the band plot of the stress error provided by the finite element analysis, the refined mesh, a histogram showing the results of refinement, and the analysis using the refined mesh. All initial meshes were generated with CUBIT[21] and the finite element analysis was performed using ADINA[19].

Since, when a node is refined and three nodes are created, it is difficult to get size ratio exactly 1.0. Also, addition of templates in the transition zones tends to over refine the mesh. The goal is therefore is to get most of the nodes with size ratio close to 1.0 and 2.0.

In this example, a gear rotating about its axis is modeled. A torque is applied at the center of the gear and three teeth are constrained. Figure 8 shows the initial mesh and the band plot of error estimates. For this example the minimum threshold error used is 8 percent.

Figure 9 shows the refined mesh. Note that refinement is implemented around the constrained teeth where there is a high error estimate. Figure 10 is a plot of the volume before and after the refinement vs. size ratio. Notice that most of the volume is below a size ratio of 2.0. Also note that there is little change in volume for size ratios less than 1.0. In Figure 11, the change in the percentage of volume falling in the particular size ratio before and after refinement is shown. Notice that most of the change is around a size ratio of 1.0 and there is negative change in volume for size ratio greater than 2.0. This shows that most of the nodes have a size ratio of 1.0 after the refinement

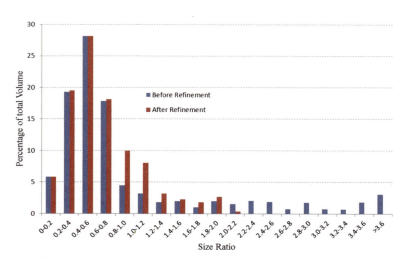

Fig. 6. Plot of size ratio and percentage of total volume before and after refinement

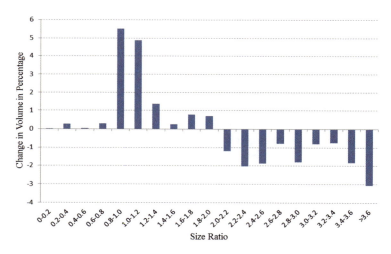

Fig. 7. Plot of size ratio and percentage of total volume before and after refinement

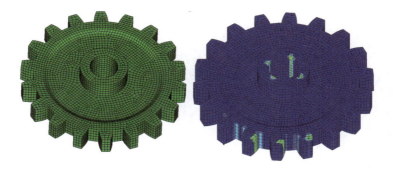

Fig. 8. Initial Mesh and Error band plot of gear example

and that refinement is taking place in the volume where the size ratio greater than 2.0. Also, it should be considered that when the coarsening algorithm is implemented the volumes should come close to the size ratio 1.0, the ideal element size ratio. Figure 12 shows the error estimate of an analysis on the refined mesh.

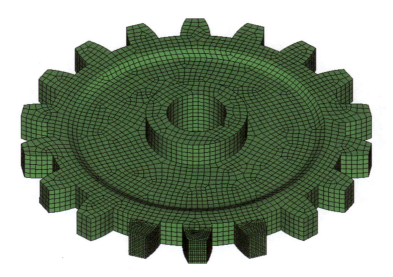

Fig. 9. Refined Mesh of gear example

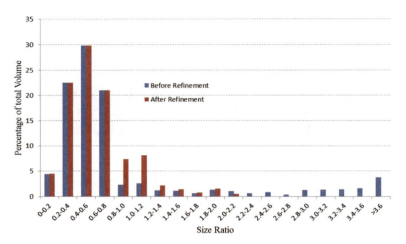

Fig. 10. Plot of percentage of total volume and size ratio before and after refinement for gear model

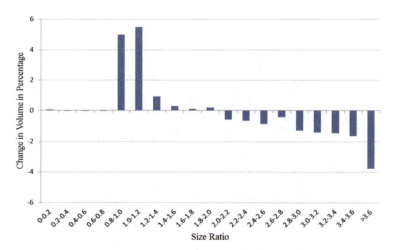

Fig. 11. Plot of size ratio and change in volume in percentage

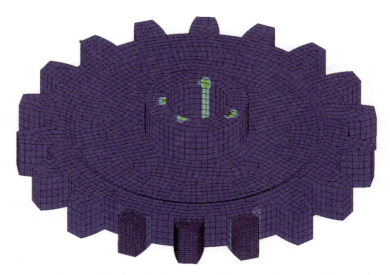

Fig. 12. Band plot of error on the refined gear mesh

5 Conclusions and Future Work

This paper presents a sizing function algorithm that selects the nodes in all hexahedral meshes for refinement and then generates a refined all conformal hexahedral mesh for subsequent finite element analysis. The sizing function is developed from error estimates from an intial analysis of the problem. Only elements in the volumes indicated size changes are modified. As a result, the locally refined mesh is able to capture the physics of the problem more accurately, with a minimum increase in the computation time thus providing high efficiency and accruacy in the finite element solution. The results from the examples shown in this paper are promising but improvements can be done on this technique to work more efficiently.

This work provides a platform for the total adaptation of hexahedral elements. Only refinement was considered here, but coarsening[8] could be included. Including coarsening would provide a method to fully adaptive hexahedral meshes.

Currently, this method uses 3-refinement which often over refines the mesh and does not provide good gradation between refined and coarsened volumes. When this technique is used with the 2-refinement technique developed by Edgel et al. [9] it should provide more gradation in the refinement.

In this work, sizing function is developed using computed error estimates. However, other criteria such as feature size or user specified field function could be included in the sizing function.

References

1. Anderson, B.D., Benzley, S.E., Owen, S.J: Automated All-Quadrilateral Mesh Adaptation: In Proceedings of 18th International Meshing Roundtable pp. 557-574 (2009)
2. Kallinderis, Y., Vijayan, P.: Adaptive Refinement-Coarsening Scheme for Three-Dimensional Unstructured Meshes. American Institute of Aeronautics and Astronautics Journal 31, 1440–1447 (1993)
3. Zhang, H., Zhao, G.: Adaptive hexahedral mesh generation based on local domain curvature and thickness using a modified grid-based method. Finite Elements in Analysis and Design 43(9), 691–704 (2007)
4. Knupp, P.M.: Next Genration Sweep Tool: A Method For Generating All-Hex Meshes on Two-And-One-Half Dimensional Geometries. In: Proceedings of 7th International Meshing Roundtable, pp. 505-513 (1998)
5. Scott, M.A., Earp, M., Benzley, S.E.: Adaptive Sweeping Techniques. In: Proceedings of 14th International Meshing Roundtable, pp. 417–432 (2005)
6. Cook, W.A., Oaks, W.R.: Mapping Methods for Generating Three-Dimensional Meshes. Computers in Mechanical Engineering 1, 67–72 (1983)
7. Parrish, M., Borden, M., Staten, M., Benzley, S.E.: A Selective Approach to Conformal Refinement of Unstructured Hexahedral Finite Element Meshes. In: Proceedings of 16th International Meshing Roundtable, pp. 251–268 (2007)
8. Woodbury, A.C., Shepherd, J.F., Staten, M.L., Benzley, S.E.: Localized Coarsening of Conforming All-Hexahedral Meshes. In: Proceedings of 17th International Meshing Roundtable, pp. 603–619 (2008)
9. Edgel, J.: An Adaptive Grid-Based All Hexahedral Meshing based on 2- Refinement. M.S. Thesis, Brigham Young University (2010)
10. Quadros, W.R., Vyas, V., Brewer, M., Owen, S., Shimada, K.: A computational framework for generating sizing function in assembly meshing. In: Proceedings of 14th International Meshing Roundtable, pp. 55–72 (2005)
11. Zhang, H., Zhao, G.: Adaptive hexahedral mesh generation based on local domain curvature and thickness using a modified grid-based method. Finite Elements in Analysis and Design 43(9), 691–704 (2007)
12. Zhang, Y., Bajaj, C.: Adaptive and Quality Quadrilateral/Hexahedral Meshing From Volumetric Data. In: Proceedings of 13th International Meshing Roundtable, pp. 365–376 (2004)
13. Wada, Y., Okuda, H.: Effective adaptation technique for hexahedral mesh: Concurrency and Computation. Practice and Experience 14(6-7), 451–463 (2002)
14. Kamenski, L.: A Study on Using Hierarchical Basis Error: I. In: Proceedings of 19th International Meshing Rountable, pp. 297–314 (2010)
15. Cougny, H.L.D., Shephard, M.L.: Parallel Refinement and Coarsening of Tetrahedral Meshes. International Journal for Numerical Methods in Engineering 99(46), 1101–1125 (1999)
16. Babuska, I., Hugger, J., Strouboulis, T., Koops, K., Gangara, S.K.: The asymptotically optimal meshsize function for bi-p degree interpolation over rectangular element. Journal of Computational and Applied Mathematics 90, 185–221 (1998)
17. Stewart, J.R., Edwards, H.C.: A framework approach for developing parallel adaptive multiphysics applications. Finite Elements in Analysis and Design 40(12), 1599–1617 (2004)

18. Grtsch, T., Bathe, K.J.: A posteriori error estimation techniques in practical finite element analysis. Computer and Structures 83, 235–265 (2005)
19. ADINA AUI 8.5.2, ADINA R and D, Inc. (2008), http://www.adina.com
20. Schoof, L.A., Yarberry, V.R.: Exodus II:Finite Element Data Model. Sandia National Laboratory, Albuquerque, NM, SAND92-2137 (1995)
21. Cubit 12.1 user documentation. Research report, Sandia National Laboratories, Albuquerque, N.M. (2010)
22. Paudel, G.: Hexahedral Mesh Refinement using an Error Sizing Function. MS Thesis, Brigham Young University, Provo, UT (2011)
23. Malvern, L.E.: Introduction to the Mechanics of a Continuous Medium. Prentice-Hall, New Jersey (1969)

Parallel Hex Meshing from Volume Fractions

Steven J. Owen, Matthew L. Staten, and Marguerite C. Sorensen

Sandia National Laboratories*, Albuquerque, New Mexico, U.S.A.
sjowen@sandia.gov, mlstate@sandia.gov, mcsoren@sandia.gov

Summary. In this work, we introduce a new method for generating Lagrangian computational meshes from Eulerian-based data. We focus specifically on shock physics problems that are relevant to Eulerian-based codes that generate volume fraction data on a Cartesian grid. A step-by-step procedure for generating an all-hexahedral mesh is presented. We focus specifically on the challenges of developing a parallel implementation using the message passing interface (MPI) to ensure a continuous, conformal and good quality hex mesh.

Keywords: grid-based, overlay grid, hexahedral mesh generation, parallel meshing, non-manifold.

1 Introduction

Computational simulation must often be performed on domains where materials are represented as scalar quantities or volume fractions at cell centers of a Cartesian or octree-based grid. Common examples include bio-medical, geotechnical or shock physics calculations where interface boundaries are represented only as discrete statistical approximations. Sandia Lab's, CTH code, is an example of an application that utilizes an Eulerian grid as its computational domain. The results of a CTH calculation are represented as volume fractions in the individual cells of the domain. In practice, this is represented as a 3-dimensional array of scalar values ranging from 0.0 to 1.0, where 1.0 represents material that completely fills the volume of the cell, and 0.0 represents the absence of material. Values that fall between represent the percentage of material, by volume, that is filling the volume of the cell.

We wish to provide a capability for the results from an Eulerian-based code to be used as input to a Langrangian, or finite element based code. To accomplish this, the scalar volume fraction data array must be interpreted

---

* Sandia is a multiprogram laboratory operated by Sandia Corporation, a Lockheed Martin Company for the United States Department of Energy's National Nuclear Security Administration under contract DE-AC04-94AL85000.

and converted into a boundary aligned hexahedral mesh that is of sufficient quality to be used in a finite element calculation.

In this work we introduce new approaches to solving the all-hex meshing problem from volume fraction data that specifically address the problem in the context of distributed memory parallel processing. We also introduce improved methods applicable for both serial and parallel processing. For example a new primal-contouring approach is introduced for defining the material domains. We describe a step-by-step procedure that includes new methods for node smoothing, resolving non-manifold conditions as well as defining geometry for parallel subdomains.

The development of general-purpose unstructured hexahedral mesh generation procedures for an arbitrary domain have been a major challenge for the research community. A wide variety of techniques and strategies have been proposed for this problem. It is convenient to classify these methods into two categories: *geometry-first* and *mesh-first*. In the former case, a topology and geometry foundation is used upon which a set of nodes and elements is developed. Historically significant methods such as plastering [1], whisker weaving [2] and the more recent unconstrained plastering [3] can be considered *geometry-first* methods. These methods begin with a well defined boundary representation and progressively build a mesh. Most of these methods define some form of advancing front procedure that requires resolution of an interior void and have the advantage of conforming to a prescribed boundary mesh where resulting element quality is normally high at the boundary. Although work in the area is on-going, the ability to generalize these techniques for a comprehensive set of B-Rep configurations has proven a major challenge and has yet to prove successful for a broad range of models.

In contrast, the mesh-first methods start with a base mesh configuration. Procedures are then employed to extract a topology and geometry from the base mesh. These methods include grid-overlay or octree methods. In most cases these methods employ a Cartesian or octree refined grid as the base mesh. Because a complete mesh is used as a starting point, the interior mesh quality is high, however the boundary mesh produced cannot be controlled as easily as in geometry-first approaches. As a result the mesh may suffer from reduced quality at the boundary and can be highly sensitive to model orientation. In spite of some of the weaknesses of grid-overlay methods, they have proven effective in a variety of applications, especially those with minimal topology or feature capture requirements. In particular, bio-medical models [4] [5] [6], metal forming applications [7] [8], and viscous flow [9] methods. The method we describe in this work also utilizes a mesh-first approach.

As one of the first to propose an automatic overlay-grid method, Schneiders [7] developed techniques for refining the grid to better capture geometry. He utilized template-based refinement operations, later extended by Ito [6] and H. Zhang [10] to adapt the grid so that geometric features such as curvature, proximity and local mesh size could be incorporated. For our implementation, we do not require sharp feature recovery or adaptively refined hexes. Instead, we focus on generation of a high resolution homogeneous hex mesh from a *flat* grid of volume fraction data.

Y. Zhang [4] [5] and Yin [11] independently propose an approach known as the *dual contouring* method that discovers and builds features into the model as the procedure progresses. The dual contouring method for generating a hexahedral mesh described by Y. Zhang [4] begins by computing intersections of the geometry with edges in the grid. Intersection locations are used to approximate normal and tangent information for the geometry. One point per intersected grid cell is then computed using a minimization procedure that is based upon Hermite approximations from the tangents computed at the grid edges. The base mesh in this case is defined as the dual of the Cartesian grid, using the cell centroids and interpolated node locations at the boundary. The proposed method, although similar in many respects, in contrast uses the grid itself as the basis for the FEA hex mesh rather than the dual of the grid to compute the intersection and normal information. As a result, nodes of the *primal* grid are used in the final mesh rather than generating the mesh from the *dual* entities. This new *primal contouring* approach that we introduce is important for our parallel implementation which avoids splitting individual cells in the domain across multiple processors and also avoids additional interpolation of cell centered data to the nodes of the grid.

2 Algorithm

The following is a brief outline of the procedure used for generating a hexahedral mesh from volume fraction data in parallel.

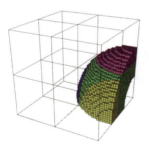
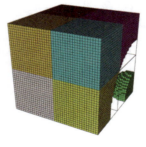
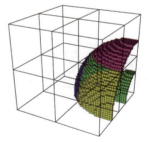

Fig. 1. 8 domains defined showing cells with volume fraction = 1.0

Fig. 2. Cells with volume fraction = 0.0

Fig. 3. Cells with volume fraction between 0.0 and 1.0

It is assumed that the application that provides volume fraction data, also provides a parallel decomposition of an axis aligned global Cartesian grid. Figure 1 shows an example of an eight processor decomposition containing volume fraction data spread across subdomains. The number of parallel processors used is the same as the number of rectilinear subdomains, where each

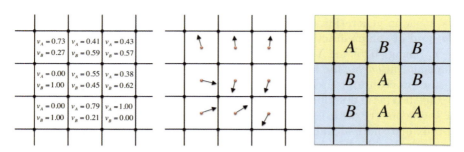

Fig. 4. Establish parallel Cartesian grid

Fig. 5. Estimate gradients at cell centers

Fig. 6. Assign materials to cells

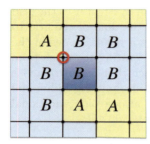

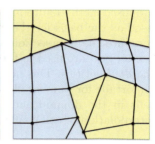

Fig. 7. Resolve non-manifold cases

Fig. 8. Compute virtual edge crossings

Fig. 9. Move grid points to iso-surface

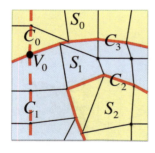

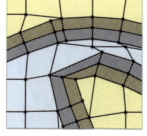

Fig. 10. Create geometry definition

Fig. 11. Insert hex buffer layer

Fig. 12. Smooth

processor assumes the task of generating the complete hex mesh for its portion of the global grid. Also illustrated in figures 1 to 3, each cell of the global Cartesian grid has been assigned at least one scalar volume fraction value, assumed to be at the cell-center of each cell of the grid. Multiple materials may also be represented, where for each cell in the grid, exactly N_{mat} scalar values, v_n are provided, where N_{mat} is the number of materials represented in the model and for each cell in the grid, $\sum_{n=0}^{n<N_{mat}} v_n \leq 1.0$. For each domain in the global Cartesian grid, the following represents the procedure performed

on each processor and is illustrated in figures 4 to 12 and further outlined in the following 9 steps:

Procedure for Generating A Parallel Hex Mesh from Volume Fractions

1. Establish Parallel Cartesian Grid: A light-weight grid data structure with ghosted cells is established to store and work on the data. (Fig. 4)
2. Estimate Gradients at Cell Centers: Based upon the cell-centered data field, gradient vectors are approximated. (Fig. 5)
3. Assign Materials: A dominant material is identified for each cell in the Cartesian grid. (Fig. 6)
4. Resolve Non-Manifold Cases: Cells are added or deleted from the base set of hexes for each material to resolve cases that would result in non-manifold connections in the final mesh. (Fig. 7)
5. Compute Virtual Edge Crossings: Identify virtual edges (connecting adjacent cell centers) that have endpoints bounding the iso-value. (Fig. 8)
6. Move Grid Points to Iso-Surface: Interpolating the virtual edge crossing data, virtual cell centers (grid nodes) are projected to an interpolated iso-surface. (Fig. 9)
7. Create Geometry Definition: A geometry description comprised of volumes, surfaces, curves and vertices is established. (Fig. 10)
8. Insert Hex Buffer Layer: A layer of hexes is inserted at iso-surface boundary and hex elements generated with geometry associativity. (Fig. 11)
9. Smooth: Smoothing procedures are employed for curves, surfaces and volume mesh entities to improve mesh quality. (Fig. 12)

Sections 2.1 to 2.9 provide a more in-depth description of each of the steps of this procedure.

2.1 Establish Parallel Cartesian Grid

For our application, the size of the axis-aligned Cartesian grid for a processor is determined by the Eulerian shock hydro code. This is convenient, since the same domain distribution used in the physics code, can be used in the Lagrangian mesh generation procedure. We can define the Cartesian grid on processor rank p as $\Omega_M^p = \{M_i^r | r = 0, 1, 2, 3\}$ where for example M^0 is a node of the grid, M^1 is an edge, and so forth. The location of grid nodes and size of cells of Ω_M^p is established by defining three independent arrays in each coordinate direction: $X_\Omega = \{x_0, x_1, x_2, \cdots, x_{nx+1}\}, Y_\Omega = \{y_0, y_1, y_2, \cdots, y_{ny+1}\}, Z_\Omega = \{z_0, z_1, z_2, \cdots, z_{nz+1}\}$, where nx, ny and nz are the number of cells in the grid in coordinate directions x, y and z respectively. Subsequent algorithms described here, utilize the entities $M_i^r | r = 0, 1, 2, 3$, however for our purposes, a lightweight representation of Ω_M^p is established, implicitly defining M_i^r only as needed.

For parallel efficiency, we also establish two layers of ghost cells on Ω_M^p. The global domain, Ω_M^G is itself described as a 3D Cartesian grid containing

N_p subdomains $\Omega_M^p | 0 \le p < N_p$ and $N_p = N_I \cdot N_J \cdot N_K$ where N_I, N_J, N_K are the number of subdomains in each Cartesian direction. For the general case of a processor subdomain on the interior of Ω_M^G, up to 26 neighboring subdomains may be present. Communication is established from processor $p_{I,J,K}$ to all of its neighboring processors $p_{I\pm,J\pm,K\pm}$ and their associated Cartesian grids, $\Omega_M^{p\pm}$. Figure 13 illustrates the communication for a simple 2D configuration. Once ghosted cells have been communicated, $X_\Omega, Y_\Omega, Z_\Omega$ and nx, ny, nz are augmented appropriately for Ω_M^p.

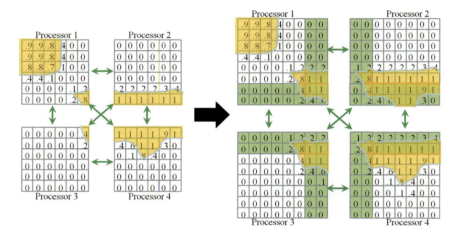

Fig. 13. Two layers of ghosting for each domain is established

2.2 Estimate Gradients at Cell Centers

Our objective is to generate hexes so that interfaces between materials are captured. To do so, we can begin by approximating the gradient at cell centers for each material defined on the domain. For each cell, exactly N_{mat} scalar values $v_n = v_0, v_1, ... v_{N_{mat}-1}$ are provided. We can represent the gradient for material n at cell center (i, j, k) as $\nabla v_n(i, j, k) = \left(\frac{\partial v_n}{\partial x}, \frac{\partial v_n}{\partial y}, \frac{\partial v_n}{\partial z} \right)$. For each cell $M_{i,j,k}^3$ in the grid, the differences of 26 neighboring values Δv_n, and cell center coordinate locations $(\Delta x, \Delta y, \Delta z)$ at $M_{i\pm,j\pm,k\pm}^3$ can be used to approximate the gradient. Solving for $\left(\frac{\partial v_n}{\partial x}, \frac{\partial v_n}{\partial y}, \frac{\partial v_n}{\partial z} \right)$ in equation (1) produces the least squares approximation to the gradient for material n.

$$\begin{bmatrix} \sum \Delta x_i^2 & \sum \Delta x_i \Delta y_i & \sum \Delta x_i \Delta z_i \\ \sum \Delta x_i \Delta y_i & \sum \Delta y_i^2 & \sum \Delta y_i \Delta z_i \\ \sum \Delta x_i \Delta z_i & \sum \Delta y_i \Delta z_i & \sum \Delta z_i^2 \end{bmatrix} \left\{ \begin{array}{c} \frac{\partial v}{\partial x} \\ \frac{\partial v}{\partial y} \\ \frac{\partial v}{\partial z} \end{array} \right\} = \begin{bmatrix} \sum \Delta x_i \Delta (v_n)_i \\ \sum \Delta y_i \Delta (v_n)_i \\ \sum \Delta z_i \Delta (v_n)_i \end{bmatrix} \quad (1)$$

The gradients in much of the grid will be undefined where Δv_n is small or zero. Since the interfaces we seek are defined only where $|\Delta v_n| \gg 0$, we can normally ignore cases where the gradient is undefined.

Note also that for cells at the boundary of the grid, fewer than 26 adjacent cells are available to for the summations in equation (1) to compute ∇v_n. Where neighboring processors are present, this will result in inconsistent results for gradients at the processor boundaries computed on the outer layer of ghost cells. If not resolved, this may effect the smoothness of the grid across processors and whether the nodes conform at all. To avoid this condition, once the gradients are computed, communication is established with neighboring processors, $\Omega_M^{p\pm}$ and gradients in the outer layer of ghost cells on each processor are sent and received via MPI.

2.3 Assign Materials

In this step we must assign each cell $M^3_{i,j,k}$ in the grid to one of the N_{mat} materials in the model. Illustrated in figure 6, this is done simply by identifying material n in $M^3_{i,j,k}$ with the greatest volume fraction v_n. In many cases, the void space, or absence of any material is required to be meshed. In this case, we keep track of the void as a separate material where the volume fraction is defined as $v_{void} = 1.0 - \sum v_n$. Separate lists of cells for each material are then maintained.

2.4 Resolve Non-manifold Cases

Figure 6 shows a simple 2D case where materials A and B meet at a non-manifold point. This configuration results in an invalid finite element mesh and must be resolved prior to generating the mesh. Figure 7 shows a simple resolution of the condition by reclassifying the assigned material in one of the cells from material A to B. For 3D, seven unique cases have been identified, as shown in figure 14, where a non-manifold condition may exist at a node. Algorithm 1 illustrates how 3D non-manifold resolution is accomplished by temporarily modifying the volume fraction v_n by a small value, ε until all non-manifold conditions have been resolved. To avoid oscillation, the value for ε_k incrementally increases using a prime-like progression of floating point values. The function *resolve_non_manifold_at_node()* in algorithm 1 identifies which of the seven unique 3D non-manifold conditions exist at a node

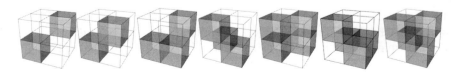

Fig. 14. Seven unique cases for non-manifold conditions in 3D at a node

and enumerates hexes to be added or subtracted for the current material. In practice, algorithm 1 normally converges within 2 to 3 iterations.

$\varepsilon_k = \{0.05, 0.07, 0.11, 0.13, 0.17, ...\}; k = 0;$
while *non-manifold condition exists* **do**
    foreach *material* $n = 0, 1, ... N_{mat-1}$ **do**
        //initialize lists of hexes to add and subtract for material n
        $L_{add}(M^3) = \varnothing, L_{sub}(M^3) = \varnothing;$
        foreach *non-manifold* $(M_i^0) \in \Omega_M^p$ **do**
            $M_I^3 = M_j^3 | j = 0, 1, ...7 \in M_i^0;$
            //add to lists L_{add} and L_{sub} from hexes M_I^3
            $resolve\_non\_manifold\_at\_node(M_I^3, L_{add}, L_{sub});$
        end
        $v_n(M_i^3) =$ volume fraction of material n for $M_i^3;$
        foreach $M_i^3 \in L_{add}$ **do** $v_n(M_i^3) = v_n(M_i^3) + \varepsilon_k$;
        foreach $M_i^3 \in L_{sub}$ **do** $v_n(M_i^3) = v_n(M_i^3) - \varepsilon_{k+1}$;
    end
    reclassify material assignment for M_i^3 in $\Omega_M^p;$
    communicate material assignment of ghost cells to $\Omega_M^{p\pm};$
    $k = k + 1;$
end

Algorithm 1. Algorithm for resolving non-manifold conditions in Ω_M^p

Note also that a parallel communication step is required following each iteration of this procedure. Because identification of non-manifold conditions depends upon checking the status of all surrounding cells of M_i^0, the non-manifold state of grid nodes at the boundary can only be established through interprocessor communication.

2.5 Compute Virtual Edge Crossings

The next stage in defining the material interfaces for the hexes is to compute *virtual edge crossings*. We would like to compute all locations on Ω_M^p where the iso-value $s = 0.5$ crosses one of the edges of the grid. We will use these locations in the next section to help move the grid nodes to the interpolated iso-surface. We choose $v_n = 0.5$ as the most likely volume fraction value where the interface surface will exist. For convenience, rather than directly computing the crossing locations on the edges, M_i^1 of the grid, we define the *virtual edges* of the grid as the segments connecting the midpoints of adjacent cells. We can uniquely identify a virtual edge in the grid from a face M_i^2. The midpoint of its two adjacent cells, $M_{j=0,1}^3$ can be defined as $P_{j=0,1} = center(M_{j=0,1}^3)$. Similarly, the volume fraction for material n in each adjacent cell to face M_i^2 can be defined as $(v_n)_{j=0,1}$. Equation (2) can then be used to compute a location $P_{cross}(M_i^2)$ for any virtual edge where its two associated cells are assigned to different materials, $material(M_0^3) \neq material(M_1^3)$.

$$P_{cross} = P_0 + \frac{s - (v_n)_0}{(v_n)_1 - (v_n)_0}(P_1 - P_0) \quad (2)$$

Also required is the normal or gradient of material n at the location of P_{cross} which can be interpolated similarly. The gradient at cells $M_{j=0,1}^3$ can be described as $(\nabla v_n)_{j=0,1}$, (see section 2.2).

$$N_{cross} = (\nabla v_n)_0 + \frac{s - (v_n)_0}{(v_n)_1 - (v_n)_0}[(\nabla v_n)_1 - (\nabla v_n)_0] \quad (3)$$

Since P_{cross} and N_{cross} values may be used frequently, they are precomputed for each material and associated with its respective virtual edge M_i^2 in Ω_M^p.

2.6 Move Grid Points to Iso-Surface

With information computed up to this stage of the procedure, we can now compute locations for nodes in the grid that will form the interface between materials. A node M_i^0 in Ω_M^p is defined as *movable* if at least 2 unique materials are identified from its eight surrounding cells $M_{j=0,1,...7}^3$. One common material assigned to all $M_{j=0,1,...7}^3$, indicates an interior node.

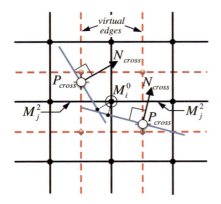
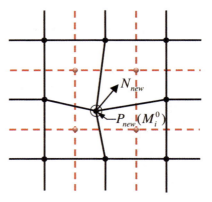

Fig. 15. Variables used to move node M_i^0 to an interpolated iso-surface

Fig. 16. Node M_i^0 after having been moved to the iso-surface

Figures 15 and 16 illustrate the procedure for computing the new location, P_{new} for M_i^0 that has been identified as at a material interface. The 12 grid faces in Ω_M^p that contain node M_i^0 can be defined as $M_{j=0,1,...11}^2$. These faces also uniquely define a virtual cell who's 12 virtual edges are defined by $M_{j=0,1,...11}^2$ with M_i^0 as the centroid. Using the values computed for P_{cross}

and N_{cross} for each M_j^2 (see section 2.5) we can define tangent planes at the material interface that cross the virtual edges. Equation (4) computes P_{new} by taking an average of the projection of M_i^0 onto each of the tangent planes.

$$P_{new} = \frac{1}{nc} \sum_{i=0}^{i<nc} P_0 - (N_{cross})_i \cdot (P_0 - (P_{cross})_i) \times (N_{cross})_i \qquad (4)$$

$$(N_{new})_n = \left| \sum_{i=0}^{i<nc_n} (N_{cross})_i \right| \qquad (5)$$

where nc is the number of virtual edge crossings where a value for P_{cross} has been computed on the virtual cell surrounding M_i^0 and P_0 is the initial location of M_i^0. We also compute $(N_{new})_n$ as the normalized average of $(N_{cross})_i$. This provides the local surface normal information needed for the subsequent buffer layer insertion and smoothing operations. The subscript n in equation (5) indicates that a separate normal is computed with respect to each material using only $(N_{cross})_i$ vectors that originate from material n.

Note that we have not distinguished between materials for the computation of P_{new}. As a result, all materials that have virtual edge crossings defined on M_i^0's virtual cell, will contribute to the new location. Where multiple materials meet at M_i^0, this has the effect of averaging the contribution from all materials resulting in a reasonable approximation of the surfaces at that point.

2.7 Create Geometry Definition

Having captured the material interfaces in Ω_M^p, we turn our attention to improving the mesh so that hexes are of sufficient quality for FE analysis. Prior to doing this, however, we have found it useful to generate a boundary representation or *B-Rep* of the hex structure that captures the material interfaces and domain boundaries. This will prove valuable in the next stages when we add a buffer layer of hexes, smooth the elements as well as assisting in encapsulation and transfer of data. The B-Rep on processor p, which we will define as Ω_G^p consists of entities $\{G_i^r | r = 0, 1, 2, 3\}$ where G^0 is a vertex of the B-Rep, G^1 is a curve, G^2 a surface and G^3, a volume. To generate the B-Rep, we must find groups of grid entities, M_i^r of dimension r that will be assigned to, or *owned* by corresponding B-Rep entities, G_i^r of the same dimension. Figure 17 shows an example of the resulting B-Rep on one processor subdomain of the model shown in figures 1 to 3. Note that geometry entities are generated at the processor interfaces. To better facilitate smoothing, one layer of ghost cells is also used in the definition of the geometry. This ensures that two layers of overlapping hexes are established between domains Ω_G^p and $\Omega_G^{p\pm}$ facilitating the subsequent Jacobi smoothing procedure.

Starting with volumes, G_i^3 are established by gathering contiguous sets of cells $M_{i,j,k}^3$ with the same material assignment. Surfaces G_i^2 are then built by

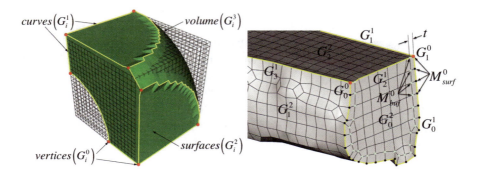

Fig. 17. A B-Rep, Ω_G^p, is built from the entities of the Cartesian grid Ω_M^p

Fig. 18. Buffer layer hexes are generated to be continuous across processor boundaries

skinning or traversing the faces M_i^2 at the boundary of each volume, including the boundaries at processor subdomains. Grid faces, M_i^2 at a subdomain boundary can be distinguished from those on the interior of the grid which will facilitate generation of curves, G_i^1 from grid edges M_i^1 along their interface. Curves can also be generated along the edges where at least three materials meet. Finally, vertices G_i^0 are established where more than two surfaces share a common grid node M_i^0.

2.8 Insert Hex Buffer Layer

The first stage in improving element quality is to insert a layer of hexes at all material interfaces. To accomplish this we identify all nodes on the material interfaces, M_{surf}^0. These are the same nodes for which we have computed P_{new} and N_{new} in equations 4 and 5. For each M_{surf}^0, one offset buffer layer node M_{buf}^0 must be generated for each material adjacent M_{surf}^0. The location of M_{surf}^0 can be defined as:

$$P(M_{buf}^0)_n = P(M_{surf}^0) + tN(M_{surf}^0)_n \tag{6}$$

where t is the thickness of the buffer layer and $N(M_{surf}^0)_n$ is the normal for material n computed in equation 5. We chose the thickness of the buffer layer to be $\frac{1}{4}$ the diagonal distance of the grid cell.

Because the buffer layer must be continuous across subdomain boundaries, the geometry ownership described in section 2.7 can be used to aid in defining its placement. Figure 18 shows an example where a volume G^3 intersects the subdomain boundary. In this case, three surfaces, $G_{0,1,2}^2$ enclose the volume, however, only one surface, G_1^2, contains nodes M_{surf}^0. Instead, surfaces G_0^2 and G_2^2 are used only to *cap* the volume where it intersects the boundary.

For our purposes, we designate G_1^2 an *interior* surface, and G_0^2 and G_2^2 as *capping* surfaces. Using the same criteria, in this example we designate the curves $G_{0,1,3}^1$ as *interior* and G_2^1 as *capping*.

With this information, we visit each capping entity $(G_i^{1,2})_{cap}$ and insert buffer edges or quads at its boundaries. For example, for curve G_2^1 in figure 18, two buffer mesh edges M_{buf}^1 are inserted into G_2^1 at vertices G_0^0 and G_1^0. In a similar manner, we insert buffer quad elements M_{buf}^2 in surface G_0^2 only adjacent to its interior curve G_0^1. Finally, the buffer hex elements M_{buf}^3 are constructed adjacent the quads on interior surface G_1^2.

2.9 Smooth

We now turn our attention to improving the element quality by smoothing. We require that the subdomain boundaries not impose artificial constraints on the quality and location of the nodes. Furthermore we require parallel-serial consistency, or in other words, the same mesh must be generated regardless of the number of processors or the location of subdomain boundaries. To accomplish this we establish communication between Ω_M^p and neighbors, $\Omega_M^{p\pm}$.

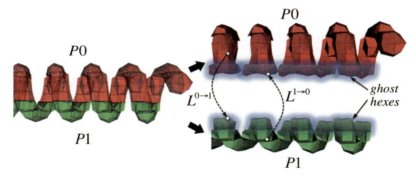

Fig. 19. Node locations on ghost hexes are are communicated following each smoothing iteration

Following each iteration of smoothing, to ensure a continuous mesh with identical node locations computed at subdomain boundaries, a communication step must be performed. To do so, we require all nodes on processor p that are within its ghost layer, receive updated nodal coordinates from its neighbors $p\pm$. To facilitate this, we first identify lists $L(M_i^0)^{p \to p\pm}$ of all nodes M_i^0 in Ω_M^p that are part of the two ghost layers at its boundary, and that have been previously assigned to a geometry entity $G^{0,1,2,3}$. This includes any nodes generated to form the buffer layer from entities in the ghost layers. Each list, $L(M_i^0)^{p \to p\pm}$, will contain nodes that are ghosted from a unique neighboring processor, $p\pm$ to the host processor p. The locations of the nodes in $L(M_i^0)^{p \to p\pm}$ on processor p, are then sent to neighboring processors $p\pm$. Neighbor processors, $p\pm$ then determine the corresponding node in $\Omega_M^{p\pm}$ for each location sent to it from processor p and form corresponding lists $L(M_i^0)^{p\pm \to p}$. These lists comprise all nodes which must be sent to neighboring processors following each iteration of smoothing. Corresponding nodes M_i^0 on $\Omega_M^{p\pm}$ from locations received from Ω_M^p can be identified using a spatial tree search since locations should be identical on both processors. The lists $L(M_i^0)^{p\pm \to p}$ must be set up prior to any smoothing, since node locations will change once smoothing has begun. Figure 19 illustrates a simple two processor example where node locations in each of the lists $L^{0 \to 1}$ and $L^{1 \to 0}$ are communicated.

Node smoothing consists of several iterations of successively smoothing nodes M_j^0 on geometric entities starting with curves $G_i^1(M_j^0)$, surfaces $G_i^2(M_j^0)$, and then volumes $G_i^3(M_j^0)$. Rather than traditional Gauss-Seidel smoothing that relies on a order-dependent incremental update of node locations, we use a Jacobi approach that uses the initial locations of the nodes at the start of the iteration.

For most cases we can neglect curve smoothing, as most interior and capping curves reside in ghost regions which are updated by the parallel communication discussed previously. For curves at the absolute domain boundaries a simple one-dimensional Laplacian smooth is performed on a piecewise quadratic approximation of the nodes.

For interior surfaces, since an explicit representation of the surface is not available, we use a quadric approximation of the nodes M_{surf}^0 [13] that is centered at the node to be smoothed $P_k(x_k, y_k, z_k)$.

$$Q_k(x,y) = z_k + a_{k2}(x - x_k) + a_{k3}(y - y_k) + a_{k4}(x - x_k)^2 \\ + a_{k5}(x - x_k)(y - y_k) + a_{k6}(y - y_k)^2 \quad (7)$$

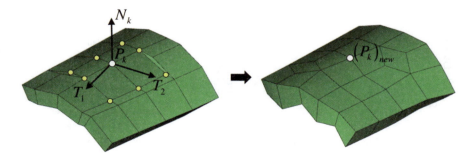

Fig. 20. Quadric approximation of surface from surrounding nodes at P_k is performed

We first transform nodes attached by edges to P_k to a local coordinate system centered at P_k with orientation defined by (N_k, T_1, T_2) as shown in figure 20, where N_k is the surface normal at P_k and (T_1, T_2) are orthogonal tangent vectors. Coefficients $a_{k2,k3,...,k6}$ for equation (7) can then be computed by solving the linear system:

$$\begin{bmatrix} \sum w_i x^2 & \sum w_i xy & \sum w_i x^3 & \sum w_i x^2 y & \sum w_i x^2 y^2 \\ \sum w_i xy & \sum w_i y^2 & \sum w_i x^2 y & \sum w_i xy^2 & \sum w_i xy^3 \\ \sum w_i x^3 & \sum w_i x^2 y & \sum w_i x^4 & \sum w_i x^3 y & \sum w_i x^2 y^2 \\ \sum w_i x^2 y & \sum w_i xy^2 & \sum w_i x^3 y & \sum w_i x^2 y^2 & \sum w_i xy^3 \\ \sum w_i xy^2 & \sum w_i y^3 & \sum w_i x^2 y^2 & \sum w_i x^2 y^2 & \sum w_i y^4 \end{bmatrix} \begin{Bmatrix} a_{k2} \\ a_{k3} \\ a_{k4} \\ a_{k5} \\ a_{k6} \end{Bmatrix} = \begin{Bmatrix} \sum w_i xz \\ \sum w_i yz \\ \sum w_i x^2 z \\ \sum w_i xyz \\ \sum w_i y^2 z \end{Bmatrix} \quad (8)$$

where $x = x_i - x_k$, $y = y_i - y_k$, $z = z_i - z_k$ and w_i is an inverse distance weight. A Laplacian smoothing operation can then be performed on node P_k to get a smoothed location P'_k in the local coordinate system. The point P'_k is then projected to the quadric surface, also in the local coordinate system using:

$$(P_k)_{local} = \begin{Bmatrix} (P'_k - P_k) \cdot T_1 \\ (P'_k - P_k) \cdot T_2 \\ a_{k2} x_k + a_{k3} y_k + a_{k4} x_k^2 + a_{k5} x_k y_k + a_{k6} y_k^2 \end{Bmatrix} \quad (9)$$

Finally, the new location $(P_k)_{new}$ in the original coordinate system is computed as:

$$(P_k)_{new} = P_k + (P_k)_{local}^T \begin{Bmatrix} T_1 \\ T_2 \\ N_k \end{Bmatrix} \quad (10)$$

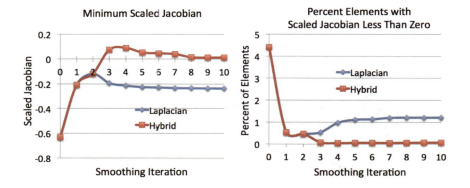

Fig. 21. Smoothing

For volumes, we also use a Jacobi Laplacian smoothing method to smooth interior nodes $G_i^3(M_j^0)$. In practice we have found that Laplacian smoothing alone is not sufficient to generate acceptable quality. Instead, after an initial two iterations of Laplacian smoothing, an optimization-based approach using the Mesquite toolkit [12] is used for subsequent iterations. This requires both untangling and mesh optimization which the ShapeImprovement tool in Mesquite is able to provide. Figure 21 shows typical results over 10 iterations of Jacobi smoothing on a model with approximately 500,000 elements. The left graph compares minimum Scaled Jacobian using just Laplacian smoothing vs. using combined Laplacian and the Mesquite ShapeImprovement smoother after the second iteration. On the right, results from the same model are shown, except that the percent of elements with Scaled Jacobian less than zero is displayed. These results show that two iterations of Laplacian smoothing, combined with two iterations of ShapeImprovement optimization is sufficient to drive the elements to a computable range.

In practice, Laplacian smoothing is much faster than optimization-based smoothing and is able to make enormous improvements with very small cost. For this reason, we do not start with the Mesquite ShapeImprovement optimization. Also, to improve efficiency, we limit application of the ShapeImprovement procedure to only those patches of elements where the scaled Jacobian mesh quality falls below a threshold of 0.2. These small patches of elements must be consistently identified on all processors, therefore, an additional communication step is required to communicate element patches smoothed in ghost regions.

3 Examples

We show several examples to illustrate the proposed capability. The first example shown in figures 22 and 23 illustrates several geometric primitives, or *shapes diatoms*, that have been converted to volume fractions on a Cartesian grid and then hex meshed. Scalability results, shown in figure 22 are still

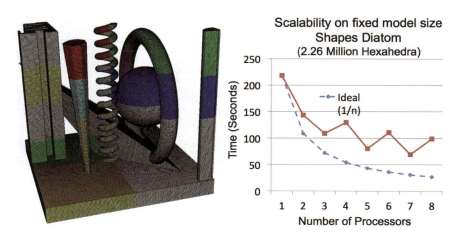

Fig. 22. Hex mesh constructed on eight processors and its associated timing data

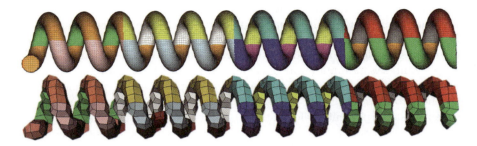

Fig. 23. Detail of helix in shapes diatom model in fig. 22 at two different resolutions

preliminary and illustrate up to an eight processor decomposition generating approximately 2.26 million elements. Timing results include I/O beginning with initial volume fraction data with one file per processor and resulting in a Lagrangian mesh file for each processor that contain conforming node locations at the subdomain interfaces. The oscillating pattern observed in the timing results is most likely an artifact of the domain decomposition of odd vs even configurations. Similar timing results were also observed in subsequent examples. The two helix images shown in figure 23 are an enlarged view of one of the objects in figure 22 modeled at two different resolutions where color represent different processors. Note the smoothness of the surfaces at higher resolution even across processor boundaries.

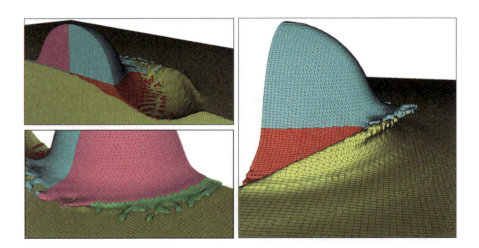

Fig. 24. Hex mesh of ball impact on plate at one selected time step.

In figure 24 a series of results at time step intervals have been computed with CTH of a sphere impacting a plate. The results have been exported as volume fraction data and processed at one selected time step using the proposed procedure. In figure 25 we illustrate a hex mesh at two time steps of a simulated pipe bomb explosion that was computed with CTH.

Fig. 25. Close-up view of hex meshes generated at two different time steps of a simulated pipe bomb explosion

Fig. 26. Hex mesh of simulated grain microstructure with 15 different materials.

The final example, shown in figure 26 is a hex mesh generated from volume fraction data, also computed from CTH, representing a simulated microstructure of a material having columnar type grain structure with isolated porosity. This model contains 15 materials and 1.6 million elements generated on eight processors.

4 Conclusion

This work introduces a step-by-step procedure for generating a hexahedral mesh from volume fraction data defined on a Cartesian grid. Contributions include improved methods applicable for both serial and parallel processing including a new primal-contouring approach for defining multiple material domains, new methods for node smoothing, resolving non-manifold conditions as well as defining geometry for parallel subdomains. We recognize that we are still in the research phases of this project, and that there are many areas left to explore. We would anticipate that the results of such research would move this technology towards a scalable tool that can be robustly used for coupling Eulerian and Lagrangian codes. We do however offer that these methods improve on existing techniques proposed in the literature particularly as they apply to parallel mesh generation using overlay grid methods.

References

1. Blacker, T.D., Meyers, R.J.: Seams and Wedges in Plastering:A 3D Hexahedral Mesh Generation Algorithm. Engineering with Computers 2(9), 83–93 (1993)
2. Tautges, T.J., Blacker, T.D., Mitchell, S.A.: The Whisker Weaving Algorithm: A Connectivity-Based Method for Constructing All-Hexahedral Finite Element Meshes. International Journal for Numerical Methods in Engineering 39, 3327–3349 (1996)
3. Staten, M.L., Kerr, R.A., Owen, S.J., Blacker, T.D.: Unconstrained Paving and Plastering: Progress Update, In: Proceedings, 15th International Meshing Roundtable 469–486 (2006)
4. Zhang, Y., Bajaj, C.L.: Adaptive and Quality Quadrilateral/Hexahedral Meshing from Volumetric Data. Computer Methods in Applied Mechanics and Engineering 195, 942–960 (2006)
5. Zhang, Y., Hughes, T.J.R., Bajaj, C.L.: Automatic 3D Mesh Generation for a Domain with Multiple Materials. In: Proceedings of the 16th International Meshing Roundtable, pp. 367–386 (2007)
6. Ito, Y., Shih, A.M., Soni, B.K.: Octree-based reasonable-quality hexahedral mesh generation using a new set of refinement templates. International Journal for Numerical Methods in Engineering 77(13), 1809–1833 (2009)
7. Schneiders, R., Schindler, F., Weiler, F.: Octree-based Generation of Hexahedral Element Meshes. In: Proceedings of the 5th International Meshing Roundtable, pp. 205–216 (1996)
8. Kwak, D.Y., Im, Y.T.: Remeshing for metal forming simulations - Part II: Three-dimensional hexahedral mesh generation. International Journal for Numerical Methods in Engineering 53, 2501–2528 (2002)
9. Tchon, K.F., Hirsch, C., Schneiders, R.: Octree-based Hexahedral Mesh Generation for Viscous Flow Simulations. American Institute of Aeronautics and Astronautics A97-3247O 781–789 (1997)
10. Zhang, H., Zhao, G.: Adaptive hexahedral mesh generation based on local domain curvature and thickness using a modified grid-based method. Finite Elements in Analysis and Design 43, 691–704 (2007)
11. Yin, J., Teodosiu: Constrained mesh optimization on boundary. Engineering with Computers 24, 231–240 (2008)
12. Brewer, M., Freitag-Diachin, L., Knupp, P., Leurent, T., Melander, D.: The Mesquite Mesh Quality Improvement Toolkit. In: Proceedings, 12th International Meshing Roundtable, pp. 239–250 (2003)
13. Jones, N.L.: Solid Modeling of Earth Masses for Applications in Geotechnical Engineering. Dissertation, University of Texas, Austin (1990)

Volumetric Decomposition via Medial Object and Pen-Based User Interface for Hexahedral Mesh Generation

Jean Hsiang-Chun Lu[1], Inho Song[1], William Roshan Quadros[2], and Kenji Shimada[1]

[1] Carnegie Mellon University, Pittsburgh, PA
   hsiangcl@andrew.cmu.edu, {songphd,shimada}@cmu.edu
[2] Sandia National Laboratories*, Albuquerque, NM
   wrquadr@sandia.gov

Summary. This paper describes an approach that combines the volumetric decomposition suggestions and a pen-based user interface (UI) to assist in the geometry decomposition process for hexahedral mesh generation. To generate the suggestions for decomposition, a 3D medial object (MO) is first used to recognize and group sweepable regions. Second, each sweepable region of the original model is visualized using different colors. Third, the ideal cutting regions to create cutting surfaces are highlighted. Based on the visual suggestions, users then create cutting surfaces with the pen-based UI. The models are then decomposed into sweepable sub-volumes following the MO based suggestions. The pen-based UI offers three types of tools to create cutting surfaces: (1) Freeform based tool, (2) B-REP based tool, and (3) MO based tool. The pen-based UI also selects a suitable type of tool automatically based on users input. The proposed approach has been tested on industrial CAD models and hex meshing results are presented.

Keywords: 3D medial object, pen-based user interface, hexahedral mesh generation, geometric decomposition, sweeping.

1 Introduction

Since hex meshing is difficult for complex shapes, volumetric decomposition is routinely used for meshing industrial parts. Compared to other volumetric mesh types, a hexahedral element yields more accurate solution [1] and hence hex meshing is the only option in many simulations. Unfortunately, existing

---

* Sandia is a multiprogram laboratory operated by Sandia Corporation, a Lockheed Martin Company for the United States Department of Energy's National Nuclear Security Administration under contract DE-AC04-94AL85000.

all hex meshing methods are not able to handle general shapes [2,3,4,5,6,7] or end up generating poor quality elements [8,9]. Performing volumentric decomposition reduces the complex shape into hex meshable sub-domains. High quality hex meshes can then be obtained by meshing simpler sub-domains.

Over the last decade, although volumetric decomposition has been extensively researched, fully automatic decomposition is still a challenging problem. With fully automatic approach, non-meshable sub-volumes still remain and require further manual decomposition. Also, fully automatic approach usually does not work on general solids. They work on only a special class of models such as many-to-many sweepable models. Unfortunately, manual decomposition capabilities in the current packages such as ABAQUS [10] and CUBIT [11] are highly labor-intensive and time consuming. Research has focused on providing better user interface (UI) to increase user productivity in the manual decomposition process [12, 13]; however, this work still requires user expertise and domain knowledge to determine sweepable regions and ideal cutting surfaces.

The medial object (MO) of a 3D shape is a skeleton representation that contains rich geometric information as shown in Fig. 1 and is of reduced dimension. The concept of medial object was originally proposed by Blum [14], which was called medial axis transformation (MAT). MAT combines medial axis and the radius of the inscribed circle, and has been applied to mesh generation [15].

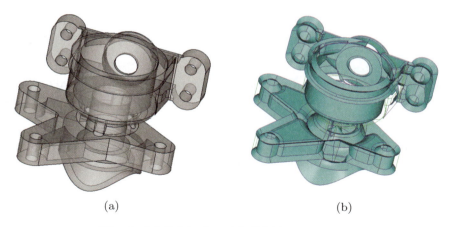

Fig. 1. (a) Original model. (b) Medial object.

In this paper, we propose an intelligent and user friendly pen-based UI that provides decomposition suggestions and automates the selection of different components to make decomposition semi-automatic. The main three steps of our approach are as follows: (1) the MO segmentation and grouping algorithm detects the sweepable sub-regions and visualize them with different colors. (2) The user use the pen-based UI and three types of cutting surface creating tools to decompose the given model. (3) The model can be all hex meshed after conducting imprinting and merging.

One of the primary contributions of this paper is to automatically detecting sweepable sub-regions via MO. Our approach uses the 3D MO to detect sweepable sub-regions and ideal cutting regions. The sweepable sub-regions are visualized on the original geometry with different colors, and the ideal cutting regions are highlighted using a virtual volume. These MO-based decomposition suggestions are combined with the pen-based UI to make manual volumetric decomposition more precise and user friendly. Our MO-based algorithm to detect sweepable sub-regions has been tested on industrial models.

The paper is organized as follows: Section 2 discusses the related work. Section 3 covers the basic characteristics of MO and MO terminologies. Section 4 gives an overview of the automatic detection of sweepable region via MO. Section 5 presents the details of the algorithm to detect sweepable regions. Section 6 describes the three types of cutting-surface creation tools in our pen-based UI. Section 7 demonstrates the decomposition and meshing results on industrial models.

2 Related Work

2.1 Automatic Decomposition

Lu et al. [16] described an automatic approach that uses local geometric information and feature recognition algorithms to subdivide meshable regions. Their approach left the remaining pieces for further manual operations. White et al. [17] automatically decomposed multi-sweep volumes into many-to-one sweepable volumes. Shin et al. [18, 19] describe swept volume decomposition method that works for geometries that can be decomposed into linearly swept volumes. These automatic decomposition methods only work for specific types of geometries. The decomposition of general solid is not always possible, and there are usually unmeshable sub-volumes remaining.

Medial axis transformation have been used to guide the decomposition of volumes into meshable primitives [8,9,20]. Li et al. [21] use the midpoint subdivision and integer programming to create simplified meshable sub-regions. Though these methods work for general solids, poor quality elements are generated.

2.2 Manual Decomposition

CUBIT offers an Immersive Topology Environment for Meshing (ITEM) [12]. ITEM determines possible cutting positions and displays suggested cutting surfaces for the user. However, the resulting cuts do not always produce meshable sub-volumes. It is because of the failure in recognizing complex features, and recognizing sweepable sub-regions. Lu et al. [13] proposed a pen-based UI that beautifies users freehand strokes to create precise cutting surfaces. This work requires user expertise and domain knowledge to determine the cutting position because it does not recognize sweepable regions.

2.3 MO for Mesh Generation

In the past, MO has been used in mesh generation as MO reduces the 3D meshing problem into 2D meshing on medial surface. Donaghy et al. [22] reduced model dimension for further analysis by using MO to classify several topology features such as end region of beams, concave and convex corners. Sampl [23] first meshed the medial faces, and then extruded the mesh on both side of the medial until the mesh intersects the boundary of the object. Chong et al. [24] used medial surface to reduce the complex original model to recognize features by identifying edge types,and by using classified medial information to treat different geometry combinations. In all these above approaches the goal is to generate a mesh than to decompose the domain into sweepable regions. The aim of this paper is to use MO to generate sweepable sub-domains so that high quality hex meshes can be generated using sweeping hex meshing algorithm.

3 The Medial Object

The Medial Axis Transform (MAT) – one of the most popular and mathematically well studied skeleton was initially defined by Blum [14] for biological shapes. In planar domain, MAT is defined as the locus of the centre of the maximal ball as it rolls inside an object, along with the associated radius function. The 3D equivalent is the locus of the center of maximal sphere and its associated radius (see Fig. 2(a)).

3.1 MO Terminology

- MO curve: a curve that connects two MO vertices.
- MO face: a surface bounded by MO curves.
- MO: a set of connected MO faces.
- MO patch: a set of 2-manifold MO faces connected by MO curves.
- MO group: a set of non-manifold MO patches that can be generated by extruding a set of connected 1D MO curves.
- MO segmentation: the process of generating MO patches from a MO.
- MO grouping: the process of generating MO groups from MO patches.

- Sweepable sub-volume: a sub-domain of the original volume that is sweepable, i.e., a 2.5D region of the original volume. A sweepable sub-volume consists of multiple source and target surfaces. Both a MO patch and a MO group will have a corresponding sweepable sub-volume.
- Valence: a medial curve is incident on one or more medial faces. The number of medial faces that are incident on a medial face determines the valence of a medial curve. If a medial curve is shared by two medial faces, the valence is two (CN-2); if it is shared by three medial surfaces, the valence is three (CN-3); and so on.
- Defining entity: the original model's geometric entities that the maximal sphere touches. These entities define the MO and hence are refered as defining entities As seen in Fig. 2(b), the defining entities of the medial face are two boundary surfaces on the model. The medial edge shown in Fig. 2(c) has three defining entities such as two surfaces and one edge on the original model.
- Trimmed MO: the MO without the medial faces/curves that touch the model boundary as seen in Fig. 3.

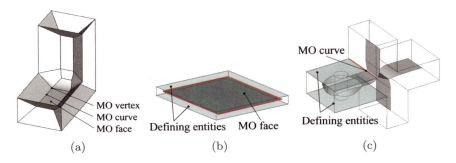

Fig. 2. (a) The MO face, MO curve and MO vertex of a L-shaped block. (b) The defining entities of a medial face. (c) The defining entities of a medial edge.

4 Automatic Detection of Sweepable Regions via MO

Instead of using traditional B-REP, we use MO skeleton representation in detecting sweepable regions. MO reduces the model dimension from 3D to 2D and makes the detection of sweepable regions easier. It is difficult to detect sweepable regions using B-REP based feature recognition techniques [25].

The two key concepts used in our approach are as follows: (1) A MO patch which is 2-manifold has a corresponding 3D sub-volume which is sweepable in 3D, and (2) a MO group which is non-manifold and is extrudable from 1D MO curves has a corresponding 3D sub-volume which is sweepable in 3D.

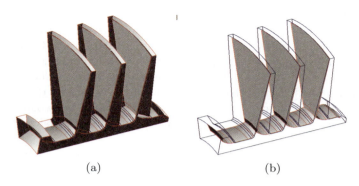

Fig. 3. (a) Original MO with the original models. (b) Trimmed MO.

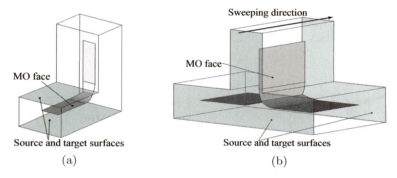

Fig. 4. (a) A MO face defined by its top and bottom surfaces, which are the sweep source and target surfaces of the region. (b) The volume can be made sweepable if two end surfaces are assigned as the source and target surfaces of a sweeping operation.

The following sections describe the detail of our algorithm that recognizes sweepable regions from MO. The process is shown in Fig. 5. Instead of using the original MO, the trimmed MO is used in our segmentation and grouping algorithm. Trimmed MO has also been used by Quadros et al. [26] as they are suitable for engineering purposes.

5 Details of the MO Segmentation and Grouping Algorithm

This section describes the three steps to generate the decomposition suggestions: (1) detecting sweepable regions via medial object (MO), (2) followed by visualizing different sweepable regions on the original model, and then (3) highlighting the ideal cutting regions if required.

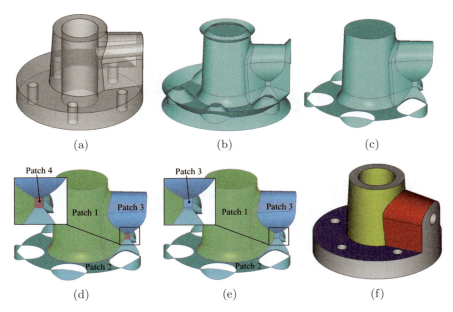

Fig. 5. (a) Original model (b) MO. (c) Trimmed MO. (d) Four 2-manifold patches. (e) Non-manifold patch obtain by uniting patches 3 and 4. (f) Different colors are assigned to each sweepable sub-volume on the original model.

5.1 Generating MO 2-manifold Patches via Segmentation

We first look at the definition of the medial face: a medial face is defined by the locus of the center of an inscribed sphere. In other words, two of the defining entities of a medial face must be the boundary surfaces of the original model. Therefore, a medial face represents a sub-volume that can be made sweepable if two defining surfaces are set as its source and target surfaces of a sweeping algorithm. As shown in Fig. 6(a), the source and target surfaces are labeled as "A" and "B".

When another sweepable sub-volume (source and target surfaces are labeled as "C" and "D" in Fig. 6(b)) is united with the adjacent sub-volume, two medial faces are connected together through a CN-2 curve. This united volume is sweepable with source surfaces A, C and target surfaces B, D. Therefore, if medial faces are connected through a CN-2 curve, their defining entities should be grouped together to represent source and target surfaces of a united sweepable sub-volume. The MO patch made by many medial faces connected through CN-2 is always 2-manifold. The 2-manifold segmentation stops at a CN-3 curve to keep the corresponding sub-volume sweepable.

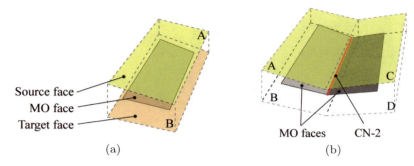

Fig. 6. A MO face represents a sweepable volume. The two defining entities are assigned as sweep source and target. (b) Two MO faces connected through a CN-2 curves represent two sweepable volumes are united. Surfaces A, C are the sweep source and surface B, D are the target.

Fig. 7 demonstrates our breadth-first-traversal MO segmentation algorithm on a MO. At the initial state (Fig. 7 (a)), we start from the largest MO face (L). In state 2, L is assigned to a group (yellow) and its neighbor MO faces are detected (Fig. 7 (b)). In state 3, the neighbor face across a CN-2 joins the current yellow patch (Fig. 7 (c)); other neighbor faces across a CN-3 remain unvisited (painted in white). The neighbors of the newly added faces are detected. In state 4, one more neighbor faces across a CN-2 joins the yellow patch (Fig. 7 (d)); the other two neighbors across a CN-3 remain unvisited. In state 5, the yellow patch has no more neighbors across CN-2 curves to visit (Fig. 7 (e)). The yellow patch is now complete. Then we start over again from the largest unvisited MO face (painted in green). One neighbor face across a CN-2 is detected and joins the green group. In state 6 (Fig. 7 (f)), the green group has no more neighbors across CN-2 s to visit. The green patch is now complete. Search for the largest unvisited MO face (painted in grey) and repeat the grouping procedure. In state 7 (Fig. 7 (g)), search for the largest unvisited MO face (paint in light blue). This MO face has no unvisited neighbor faces, and forms a stand alone patch. In the final state, only one unvisited face remains (painted in dark blue). This face forms another patch. Fig. 7(h) shows segmentation result for this trimmed MO. The algorithm is shown in Algorithm 1.

5.2 Generating MO Groups

MO faces that share the same CN-3 curve can be visualized by extruding their end curves along the CN-3 curve. The 3D sub-volumes that correspond to MO faces are also sweepable. As shown in Fig. 8, the MO is split into different sweepable patches by the CN-3 curve based on the 2-manifold segmentation as explained in previous section. Patches 1 and 2 are both sweepable along their respective sweeping directions defined by their defining entities However,

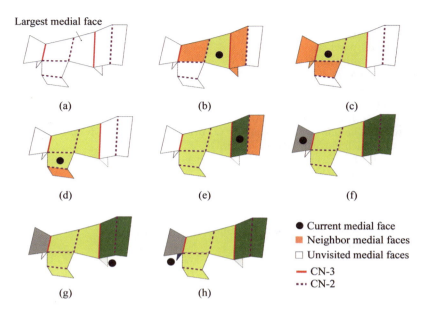

Fig. 7. A detailed MO segmentation example on a trimmed MO.

sub-volume corresponding to patches 1 and 2 is already sweepable without any decomposition if surfaces A and B are used as source and target, and swept along the CN-3 curve. Therefore, we unite these two patches together, and reduce the total number of sub-volume. If the MO patches that share the same CN-3 curve also share the same end entities as the sweep source and target, these MO patches will be grouped to represent a sweepable sub-volume. The advantage is that it reduces the number of sub-volumes and it assists in obtaining conformal mesh between the sub-volumes.

5.3 Highlight Ideal Cutting Regions

The model shown in Fig. 9(a) contains two sweepable groups but does not have any existing model entities to create a cutting surface. This happens when the defining surfaces of each sweepable MO group are overlapping. Our method detects the medial face that connects different MO groups and uses it to create a virtual volume. The overlapping portion of the virtual volume with the original models indicates the ideal cutting region, and the intersecting entities are highlighted. The user can then use the pen-based UI (Fig. 9(c)) to create a cutting surface at the highlighted region (Fig. 9(b)) using the freeform strokes. Conducting a decomposition operation in that region creates two sub-volumes that are sweepable.

Algorithm 1. MO segmentation algorithm

Input: The original model and its trimmed MO list $TrimmedMOFaceList$;
Output: MO patches;
 1: **while** $TrimmedMOFaceList$ is not empty **do**
 2:    Search for largest MO face (L) from $TrimmedMOFaceList$
 3:    Append L to patch S_i
 4:    Append L to $CurrentMOFaceList$
 5:    Remove L from $TrimmedMOFaceList$
 6:    **while** The number of $CurrentMOFaceList$'s neighbor MO faces $! = 0$ **do**
 7:      **for** each current MO face C **do**
 8:        **if** the valence of C's child MO edge $== 2$ **then**
 9:          **for** each child MO edge **do**
10:            Search for parent MO faces (P)
11:            Append P to S_i
12:            Append P to $CurrentMOFaceList$
13:            Remove P from $TrimmedMOFaceList$
14:          **end for**
15:        **end if**
16:        Remove C from $CurrentMOFaceList$
17:      **end for**
18:      Remove L from $CurrentMOFaceList$
19:    **end while**
20:    Remove P from $CurrentMOFaceList$
21:    $i++$
22: **end while**

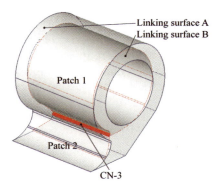

Fig. 8. A CN-3 curve splits two sweepable patches. Patches 1 and 2 share the same end MO curves and are sweepable by making linking surface A and B as source and target surfaces, repectively.

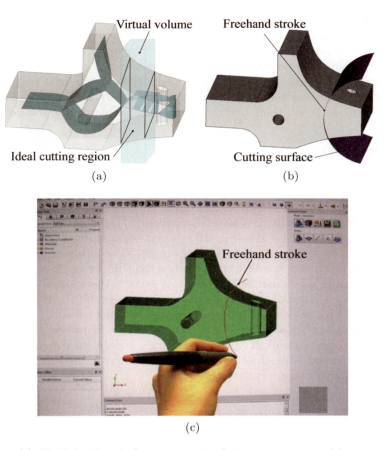

Fig. 9. (a) Highlight the ideal cutting region between two sweepable regions. (b) A cutting surface is created in the highlighted region with the pen-based UI in (c). (c) A user uses the pen-based UI to draw a freehand stroke.

6 Pen-Based UI for Cutting Surface Creation Tools

This section describes the pen-based UI that enables the user to specify precise cutting surfaces with freehand strokes using one of the three tools: (1) Freeform based, (2) B-REP (boundary representation) based, or (3) MO based cutting tool. The pen-based UI intelligently selects the proper tool based on users input.

6.1 Freeform Based Cutting Tool

The pen-based UI takes user's freehand strokes as input, smoothes, beautifies, and snaps the strokes to create more precise cutting surfaces [13]. The freehand strokes are beautified to common geometries such as lines, arcs, circles, and splines. Based on their position, strokes are also aligned to existing geometric features of the model surface. The supported alignment types include offset, overlap, perpendicular, and concentric. The mesh quality around the cutting area is improved by the alignment. Then the stroke is swept along the surface normal or the view direction to create a freeform cutting surface. This cutting surface is then used to decompose the volume into sub-volumes.

6.2 B-REP Based Cutting Tool

The B-REP based tool is able to automate the decomposition operation based on the entities selected by the user. The current implementation includes, (1) creating a cutting surface by extending a selected surface, (2) creating a cutting surface by revolving a selected surface, (3) creating a cutting surface by revolving a selected curve (Fig. 10(c)), and (4) creating a cutting surface by using a loop of bounding curves (Fig. 10(d)). The automatic selection of one of the four cutting operations is based on the number and type of selected geometric entities. If a user selects only one surface, the selected surface will be extended and the volume will be decomposed (Fig. 10(a)). If two entities are selected with the second one periodic, the first selected entity is assigned as a profile, and revolved about the axis defined by second selected entity as seen in Fig. 10(b) and (c). If one or more curves are selected, we first check if it form(s) a closed loop. If so, then the selected curve is used as a bounding curve to create a cutting surface (Fig. 10(d)).

6.3 MO Based Cutting Tool

The MO based tool uses MO to generate cutting surfaces. Fig. 11(a) shows part of the trimmed MO and a CN-3 curve. First resample the CN-3 curve and obtains the tangent point set, where the maximal spheres touch the model surfaces (Fig. 11(b), (c)). Next connect those tangent points to create polylines (Fig. 11 (d)).

Similar to the freeform based cutting tool that snaps the freehand stroke to existing entities [13], the MO based tool snaps the freehand stroke to these polylines. Then the user lofts (Fig. 11 (e)) the snapped strokes to create a cutting surface.

7 Results and Discussion

We have tested our MO assisted pen-based UI for volumetric decomposition against industrial models. These models are first imported in neutral STEP format into CUBIT. We then perform CAD cleanup operations such

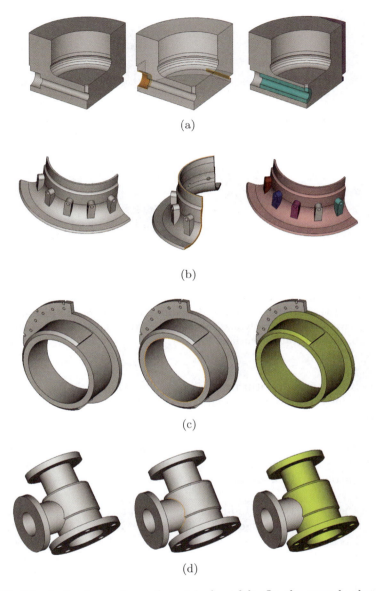

Fig. 10. The first column shows the original models. On the second column, the selected are highlighted in orange color. The last column shows the decomposition result using B-REP tool. (a) Two selected surfaces are extended. (b) A surface is revolved about the axis defined by another selected curve. (c) A curve is revolved about the axis defined by another selected curve. (d) Many selected curves from a close loop, and a cutting surface is lofted to decompose the model.

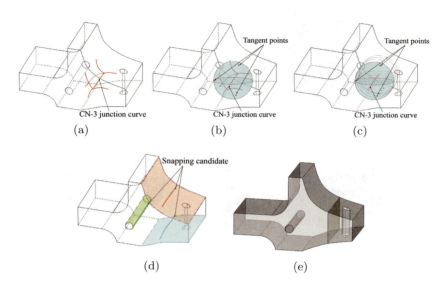

Fig. 11. (a) Part of the trimmed MO and the CN-3 junction curve. (b) The tangent points touch the model boundary. (c) The tangent points of the resampled CN-3 junction curve. (d) Create polylines by connecting tangent points. (e) Loft the two polylines to create a cutting surface.

as CAD Repair and defeaturing if needed. The volumetric decomposition is then performed using our proposed pen-based UI using MO. The volumetric decomposition step is very critical as it reduces the complex input model into simpler sub-volumes that are meshable. In order to obtain conformal mesh, we perform imprinting and merge operations. Finally, the traditional sweeping algorithms are called on the imprinted sub-volumes. Very high quality hex meshes can be obtained using the decomposition followed by the sweeping approach.

Fig. 12 shows a typical industrial model that cannot be meshed using the traditional sweeping algorithms. Trimmed MO shown in Fig. 12(b) is generated as the end branches of MAT are not useful for engineering purposes. The trimmed MO is then used to detect sweepable regions as shown in Fig. 12(c). The same figure also shows a hex mesh on the different sub-volumes using different colors. In this particular example, the model is sub-divided into ten sweepable sub-volumes. It took nine cutting surfaces to decompose this model. All the cutting surfaces are generated using one common gesture with the pen-based UI. The average mesh quality obtained on this model is 0.9404 as shown in Fig. 12(d). We are able to achieve such a high quality hex mesh because our approach decomposes the complex domain into simpler sub-volumes that can be meshed using the sweeping algorithm. Generally, the

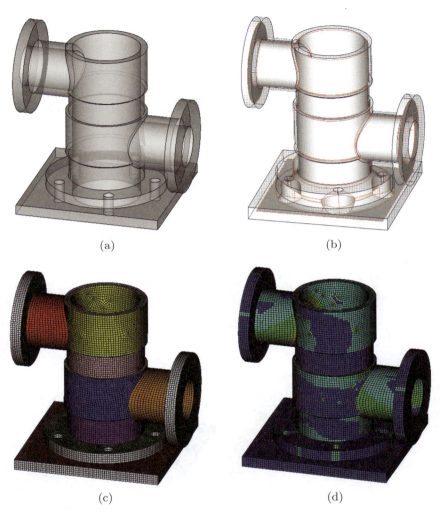

Fig. 12. (a) Original model. (b) MO with the original model boundary (black). (c) The all hex mesh on MO-based decomposition suggestion. (d) The mesh quality (Scaled Jacobian).

sweeping algorithm leads to a better mesh quality than other hex meshing algorithms.

Fig. 13(a) shows another complex model that cannot be meshed using existing sweeping algorithms. Fig. 13(b) shows the trimmed MO that contains many non-manifold regions. Fig. 13(c) shows the sub-volumes obtained using MO-based visual suggestion and the pen-based UI. The model is sub-divided into sixteen sweepable sub-volumes. The sweepable regions are color coded using different colors to provide visual aid to the users. Pen-based UI that provides three types of cutting tool is used to decompose the model based on the MO grouping suggestions. The system automatically selects the correct decomposition command based on the type and the number of geometric entities selected. Proposed approach uses an internal geometric kernel to perform the low level decomposition operations. Also, the imprint and merge operations are performed using the geometric kernel. The final conformal hex mesh shown in Fig. 13(d) contains 93653 elements with an average quality of 0.9345 (Scaled Jacobian).

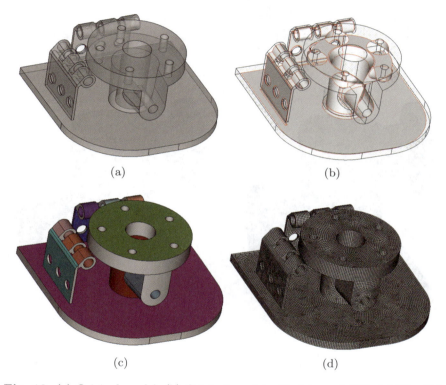

Fig. 13. (a) Original model. (b) Original model with trimmed MO. (c) The MO segmentation and grouping result. Sixteen groups are painted in different colors. (d) The all hex mesh output.

8 Conclusion

The paper presents an intelligent and user friendly pen-based UI that provides MO-based decomposition suggestions and automates the selection of difference components to make volumetric decomposition semi-automatic. One of the primary contributions of this paper is the MO segmentation and grouping algorithm that detects sweepable regions and provides visual aid for ideal cutting region. Another contribution is the pen-based UI that creates precise cutting surfaces using freeform strokes, existing B-Rep entities, and MO. In the future, cutting surface creation via MO and selection of cutting tools will be further automated. The proposed approach has been tested on industrial models by generating high quality hex meshes using sweeping algorithms.

Acknowledgement. The authors would like to thank Dr. Geoffrey Butlin, Mr. Henry Bucklow, Mr. Robin Fairey, Mr. Mark Gammon and Mr. Mike Field for assisting with the medial related work in CADFIX. The authors would also like to thank Dr. Soji Yamakawa for his valuable advise; Mr. Ved Vyas for his expertise with CUBIT; and Mr. Karthik Srinivasan for collecting MAT related articles.

References

1. Yamakawa, S., Gentilini, I., Shimada, K.: Subdivision templates for converting a non-conformal hex-dominant mesh to a conformal hex-dominant mesh without pyramid elements. Engineering with Computers 27, 51–65 (2011)
2. Blacker, T.D., Meyers, R.J.: Seams and wedges in plastering: A 3D hexahedral mesh generation algorithm. Engineering with Computers 9, 83–93 (1993)
3. Hariya, M., Nishigaki, I., Kataoka, I., Hiro, Y.: Automatic hexahedral mesh generation with feature line extraction. In: Proceeding of the 15th International Meshing Roundtable, pp. 453–468 (2006)
4. Marechal, L.: A new approach to octree-based hexahedral meshing. In: Proceeding of the 16th International Meshing Roundtable, pp. 209–221 (2001)
5. Yamakawa, S., Shimada, K.: HEXHOOP: modular templates for converting a hex-dominant mesh to an all-hex mesh. Engineering with Computers 18, 211–228 (2002)
6. White, D.R., Tautges, T.J.: Automatic scheme selection for toolkit hex meshing. International Journal for Numerical Methods in Engineering 49(1-2), 127–144 (2000)
7. Quadros, W.R., Shimada, K.: Hex-layer: Layered all-hex mesh generation on thin section solids via chordal surface transformation. In: Proceeding of the 11th International Meshing Roundtable, pp. 169–182 (2002)
8. Price, M.A., Armstrong, C.G., Sabin, M.A.: Hexahedral mesh generation by medial surface subdivision: part I. solids with convex edges. International Journal for Numerical Methods in Engineering 38(19), 3335–3359 (1995)
9. Price, M.A., Armstrong, C.G.: Hexahedral mesh generation by medial surface subdivision: part II. solids with flat and concave edges. International Journal for Numerical Methods in Engineering 40(1), 111–136 (1997)
10. Simulia Corp., Abaqus product description, version 6.9, http://www.simulia.com/products/abaqus_fae.html
11. Sandia National Laboratories, Cubit 12.1 on-line user's manual: web cutting, http://cubit.sandia.gov/help--version12.1/cubithelp.htm

12. Owen, S.J., Clark, B., Melander, D.J., Brewer, M.L., Shepherd, J., Merkley, K.G., Ernst, C., Morris, R.: An immersive topology environment for meshing. In: Proceeding of the 16th International Meshing Roundtable, pp. 553–577 (2007)
13. Lu, J.H.-C., Song, I.H., Quadros, W.R., Shimada, K.: Pen-based user interface for geometric decomposition for hexahedral mesh generation. In: Proceedings of the 19th International Meshing Roundtable, pp. 263–278 (2010)
14. Blum, H.: A transformation for extracting new descriptors of shape. In: Models for the Perception of Speech and Visual Form, pp. 362–380 (1967)
15. Quadros, W.: Extraction and applications of skeletons in finite element mesh generation. In: Proceeding of the 7th International Conference on Engineering Computational Technology (2010)
16. Lu, Y., Gadh, R., Tautges, T.J.: Feature based hex meshing methodology: feature recognition and volume decomposition. Computer-Aided Design 33(3), 221–232 (2001)
17. White, D.R., Saigal, S., Owen, S.J.: Ccsweep: automatic decomposition of multi-sweep volumes. Engineering with Computers 20, 222–236 (2004)
18. Shih, B.-Y., Sakurai, H.: Automated hexahedral mesh generation by swept volume decomposition and recomposition. In: Proceeding of the 5th International Meshing Roundtable, pp. 273–280 (1996)
19. Shih, B.-Y., Sakurai, H.: Shape recognition and shape-specific meshing for generating all hexahedral meshes. In: Proceeding of the 6th International Meshing Roundtable, pp. 197–209 (1997)
20. Ang, P.Y., Armstrong, C.G.: Adaptive curvature-sensitive meshing of the medial axis. In: Proceeding of the 10th International Meshing Roundtable, pp. 155–165 (2001)
21. Li, T.S., McKeag, R.M., Armstrong, C.G.: Hexahedral meshing using midpoint subdivision and integer programming. Computer Methods in Applied Mechanics and Engineering 124(1-2), 171–193 (1995)
22. Donaghy, R.J., Armstrong, C.G., Price, M.A.: Dimensional reduction of surface models for analysis. Engineering with Computers 16, 24–35 (2000)
23. Sampl, P.: Semi-structured mesh generation based on medial axis. In: Proceeding of the 9th International Meshing Roundtable, pp. 21–32 (2000)
24. Chong, C.S., Kumar, A.S., Lee, K.H.: Automatic solid decomposition and reduction for non-manifold geometric model generation. Computer-Aided Design 36(13), 1357–1369 (2004)
25. Sun, R., Gao, S., Zhao, W.: An approach to B-rep model simplification based on region suppression. Computers and Graphics 34(5), 556–564 (2010)
26. Quadros, W.R., Shimada, K., Owen, S.J.: Skeleton-based computational method for the generation of a 3D finite element mesh sizing function. Engineering with Computers 20, 249–264 (2004)

TET, HYBRID, and POLYHEDRAL MESHING

(Tuesday Morning Parallel Session)

Automatic Decomposition and Efficient Semi-structured Meshing of Complex Solids

Jonathan E. Makem, Cecil G. Armstrong, and Trevor T. Robinson

School of Mechanical and Aerospace Engineering, Queen's University, Belfast BT9 5AH, N. Ireland
j.makem@qub.ac.uk

Summary. In this paper, a novel approach to automatically sub-divide a complex geometry and apply an efficient mesh is presented. Following the identification and removal of thin-sheet regions from an arbitrary solid using the thick/thin decomposition approach developed by Robinson *et al* [1], the technique here employs shape metrics generated using local sizing measures to identify long, slender regions within the thick body. A series of algorithms automatically partition the thick region into a non-manifold assembly of long, slender and complex subregions. A structured anisotropic mesh is applied to the long/slender bodies and the remaining complex bodies are filled with unstructured isotropic tetrahedra. The resulting semi-structured mesh possesses significantly fewer degrees of freedom than the equivalent unstructured mesh, validating the effectiveness of the approach.

Keywords: Automatic decomposition, geometric reasoning, efficient meshing, metric field, anisotropic meshing.

1 Introduction

A major advantage of applying unstructured tetrahedral meshes to complex geometries for structural problems is the robustness of the current automatic tet mesh generators. One disadvantage is their limited capability for generating appropriate anisotropic stretched meshes. For CFD problems, meshes with very large aspect ratios can be routinely generated and adaptively refined [2]. However in structural problems a mesh density of one or two elements through the thickness of a thin sheet or across the cross-section of a long/slender region is often sufficient. This means that local, mesh-based approaches to anisotropic refinement are difficult. Alternatively, by generating a "mixed" or "hybrid" mesh consisting of structured meshes on specific regions of the model, combined with unstructured meshes on more complex areas, it is possible to achieve a dramatic reduction in degrees of freedom and thereby improve the efficiency of the analysis process. Consequently, this requires a process to partition the model into sub-regions where the different meshing strategies may be applied. This may be achieved by identifying easily mappable regions and sweepable volumes such as thin-sheets or long, slender sections. Structured anisotropic pentahedral or hexahedral meshes may be applied to these regions, and the remaining complex volumes in the model may be filled with unstructured tetrahedral elements to produce an efficient, semi-structured, mesh of the model.

Geometric properties of solids such as the medial object have been used in the past to decompose complex geometries into thick and thin sub-regions [1]. A structured thin-sheet mesh could be applied to the thin-sheet regions and merged with isotropic unstructured elements in the adjoining thick regions. Even though a significant reduction in degrees of freedom was achieved using this approach, it was less than expected as the long, slender areas of the thick-region consumed a lot of nodes when meshed with tetrahedra. However, if these long, slender regions could be identified, instead of being tet meshed they could be filled with a structured, swept mesh comprising anisotropic elements, thereby reducing the mesh density even further.

The objective of this research is to develop an automatic approach that uses an *a priori* knowledge of shape properties to identify long, slender regions in complex solids for the application of an efficient semi-structured mesh. Local sizing measures such as edge length and curvature, face width and curvature and local 3D thickness are employed to generate metric tensor fields which identify meshable sub-regions within complex volumes. Intelligent routines interrogate the geometry and automatically partition the body, isolating the long/slender regions from residual complex regions. Appropriate meshing strategies are applied to the respective sub-regions and the metric fields are then used to grade the mesh along the length of each slender region ensuring a smooth transition with isotropic elements in residual complex areas. The effectiveness of the approach is demonstrated on a complex model with a substantial reduction in degrees of freedom achieved.

The remainder of this paper is organized as follows: Section 2 reviews related work on geometric reasoning for meshing applications; Section 3 explains how the metric fields are generated using local sizing measures; Section 4 describes the approach for identifying the long/slender regions and partitioning the body into a non-manifold assembly of meshable subregions; Section 5 details the results for a complex model and makes a comparison between the semi-structured mesh and the equivalent unstructured tetrahedral mesh; Section 6 presents conclusions and future work.

2 Previous Work

Thakur *et al* [3] gives a comprehensive summary of the state of the art in CAD model simplification techniques for meshing. Research of particular note to the work in this paper is the prototype thick-thin decomposition process developed by Robinson *et al* [1] provided a capability for locating thin-sheet regions in complex solid geometries. This functionality is now available in the commercial CAE tool CADfix, by Transcendata [4]. The approach uses the 3D medial object (MO) [5] of the solid to determine the local thickness followed by a 2D MO computation on the mid-surfaces. By comparing the diameter of the 2D MO (which is an approximation of the lateral dimensions of the object) to the thickness of the 3D MO, it is possible to determine if the region is thick or thin. If the diameter of the inscribed circle on the 2D MO is large in comparison to the thickness, the region will be a thin sheet. Luo et al. [6] addressed the very similar problem of finding thin sections for the generation of prismatic p–version finite element meshes, using an octree-based approach to identify medial surface points and local thin sections.

Although efficiently meshing the thin-sheet regions significantly reduces the degrees of freedom in the model, it was noticed that there was still a lot of degrees of freedom consumed when long/slender regions such as flanges were filled with an unstructured tetrahedral mesh.

Price *et al* [7, 8] proposed a logical approach to firstly partition a complex 3D object into meshable regions and secondly apply a structured mesh to these regions. The medial surface was used to sub-divide the solid into hex-meshable subregions known as primitives. After all the subregions have been formed, each primitive is meshed using a midpoint subdivision technique [9]. Mesh compatibility between the primitives can be controlled by an integer programming technique described by Tam *et al* [10]. However, the approach has its limitations as it relies on a robust 3D medial object computation. Moreover, a comprehensive treatment of all possible shape features has not been developed.

Tchon *et al* [11] used an *a priori* knowledge of model geometry to generate a Riemannian metric based on local curvature and thickness. An isotropically refined octree grid is used as a support medium to generate the shape metrics. The anisotropic sizing information provided by the metrics is then used to refine the mesh. Some impressive results have been achieved with this approach with a dramatic reduction in element count. However, the technique relies on an initial mesh and the quality of the elements generated after the refinement process is questionable as their shape is not perfectly cubical. Other authors [12, 13] have reported similar geometry-based methods for mesh adaption and refinement. However, all of them relate to the relocation or refinement of an initial mesh. The focus here has been to use similar shape measures to identify regions of the model for structured meshing, rather than local mesh.

Zhao *et al* [14, 15] proposed a technique for the adaptive generation of an initial mesh based on the geometric features of a solid model. A special refinement field based on curvature and thickness was constructed to control mesh size and density distribution. The boundary of the adapted indentation mesh produced was matched to the boundary of the solid model using a threading method. Although the resultant mesh on a complex model possessed fewer degrees of freedom than a structured swept mesh applied to a manually sub-divided version, the quality of the mesh produced was not satisfactory as a significant number of singularities were evident.

White *et al* [16] developed an interesting algorithm for automatically decomposing multi-sweep volumes into volumes that can be swept meshed. This is achieved by discretizing the linking faces of volumes with quad elements which provide a layering system to traverse the volume vertically. Target faces are pushed through the volume onto opposing source faces providing an imprint that governs the decomposition of the solid. The newly generated volumes are all sweep meshable with a single target face. However, the major limitation with the approach is that it can only be applied to sweepable geometries and cannot be used to decompose more complex models. Consequently, this identifies a requirement for a more robust technique for the automatic decomposition of complex models that will successfully partition any dumb solid geometry into an assortment of sub-regions, including sweepable volumes, where various structured meshing strategies may be applied.

3 Geometric Reasoning Using Local Sizing Measures

The approaches and methodology reported in the forthcoming sections have been implemented in the CADfix software package using an API based on the Python programming language. Assuming that the thin-sheet regions have been identified and partitioned out using the aforementioned thick/thin decomposition tool developed by Robinson *et al* [1], decomposition of the remaining thick region into long/slender and chunky bodies may be achieved using a shape metric.

3.1 Metric Classification

The goal is to generate within the CAD model ellipsoids representing the shapes of different regions. The principal axes of an ellipsoid may be used to define the target element size in three dimensions [17].

Table 1. Ellipsoid classification

Ellipsoid Type	Diagram	Criteria
Long/ Slender		$V_3 \gg V_1$ $V_3 \gg V_2$ $V_1 \approx V_2$
Thin-sheet		$V_3 \gg V_1$ $V_2 \gg V_1$ $V_2 \approx V_3$
Isotropic		$V_1 \approx V_2 \approx V_3$

If an ellipsoid has one principal direction that is much greater than the other two, this identifies a slender region where the mesh can be extruded in the direction of the largest principal axis. Conversely, areas of the model where each axis is similar in magnitude and the ellipsoid is spherical identify regions where the target element size will be similar in all directions, requiring an unstructured isotropic tetrahedral mesh. The three main types of ellipsoids are displayed in table 1. Note that regions represented by thin-sheet ellipsoids will have been removed by the thick/thin tool prior to implementing the approach presented here.

Within the procedures described in this paper, an ellipsoid is generated on every edge with its centre at the mid-point of the edge. The edge length and curvature are used to determine the length of the ellipsoid in the edge direction. In terms of long/slender ellipsoids, this sizing measure is normally employed to gauge the size of the largest principal axis which provides the sweep direction for the extruded mesh. The other sizing measures of face width and curvature and local 3D thickness are employed to determine the extent of the remaining two principal axes. If a face is planar, the length of an axis may be calculated using the 2D medial axis on the face or an approximation to it, whereas the axis length on non-planar surfaces may be determined using a curvature based searching algorithm for a pre-defined sag value, d. For other scenarios such as concave edges and corners ray casting is used to assess axis length by giving an approximation of the local thickness.

3.2 Metric Sizing Using Edge Length and Curvature

If the edge is straight, then half the edge length is used for the extent of the ellipsoid axis by default, as shown in figure 1. However, if an edge is curved, it is necessary to employ a curvature sensitive approach to check the size of the ellipsoid axes along curved edges and surfaces. Along an edge, the change in tangent vector, T_i, quantified by the angle α defined in equation 1, is used to determine the size of the first ellipsoid axis, V_i, as shown in figure 2. When α reaches a tolerance value the ellipsoid axis vector V_i can be determined.

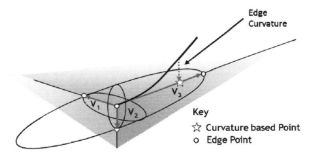

Fig. 1. 1D metric generated using edge length and curvature

$$\alpha = \cos^{-1}\left(\frac{T_i \cdot T_{i+1}}{|T_i||T_{i+1}|}\right) \quad (1)$$

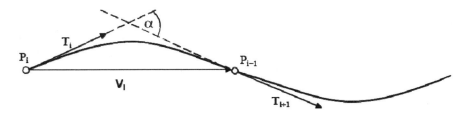

Fig. 2. Metric sizing using edge curvature

3.3 Metric Sizing Using Face Width and Surface Curvature

If the surfaces adjoining an edge are curved, a similar strategy is applied to assess the size of the second and third axes. The surface curvature is used to determine the size of the ellipsoid axis based on a sag value, d. For a pre-defined d value, the size of the vector orthogonal to the tangent vector at the centre point on the edge can be calculated, as shown in figure 3.

In cases where the surface is planar, the 2D medial object is used to determine the local face width and the length of the ellipsoid axis in this direction.

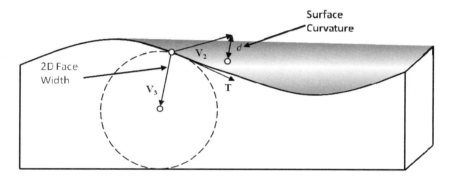

Fig. 3. Sizing of ellipsoid axis using surface curvature and face width.

On a parametric surface, as shown in figure 4, for a given tangent vector, **T** and normal vector **N** at point P on a given edge, the sag value, d at point Q on the surface is equivalent to the projection of **PQ** onto **N**, as described in equation 2.

$$d = \mathbf{PQ} \cdot \mathbf{N} \qquad (2)$$

d can also be described as the perpendicular distance from Q to the tangent plane at P. In parametric terms, P and Q can be represented by the points $x(u, v)$ and $x(u + du, v + dv)$. Consequently, using Taylor's theorem [18], equation 3 can be modified to give:

$$d = \frac{1}{2} d^2 \mathbf{x} \cdot \mathbf{N} + O(du^2 + dv^2) \qquad (3)$$

Thus, for a pre-defined sag value, d, equation can be used to determine the location of point Q and the extent of the ellipsoid axis, **V**.

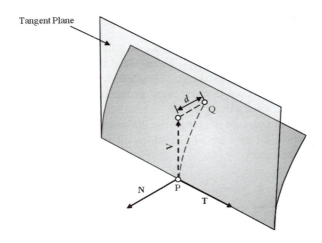

Fig. 4. Metric sizing for a given sag value on a parametric surface.

3.4 Metric Sizing Using Local Thickness

Local 3D thickness is used as a sizing measure to gauge the extent of the ellipsoid axis for scenarios like concave and convex edges. For ellipsoids generated on concave or convex edges ray casting is used to assess the thickness of the body in that region, as shown in figure 5. An axis length based on the thickness of the body is then compared to potential axes generated using other sizing measures and the shortest axis is always selected for the ellipsoid in that direction.

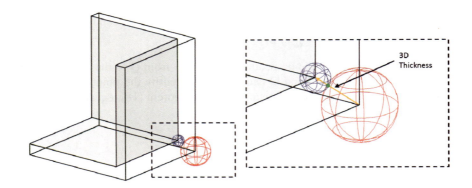

Fig. 5. Metric sizing by assessing local 3D thickness.

4 Automatic Partitioning of Complex Geometry

After a thick/thin decomposition has been performed on a complex geometry, as shown in figure 6, a search is then performed on the thick body (in red) for slender regions by generating ellipsoids at the mid-point of every edge.

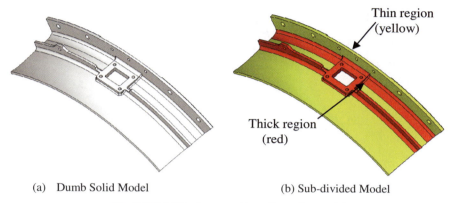

(a) Dumb Solid Model (b) Sub-divided Model

Fig. 6. Thick/thin decomposition of complex model

4.1 Identification of Long/Slender Regions

The aspect ratio of an ellipsoid compares the length of its longest axis to the length of its other axes. Critical aspect ratio (CAR) is a user specified criteria that is used to determine above which ratio it is considered an ellipsoid axis is much longer than the other two. For a pre-defined critical aspect ratio (CAR), all long/slender ellipsoids are identified using the algorithm detailed in figure 7. Where D represents the dimension of the ellipsoid (for long/slender ellipsoids $D = 1$), V_1, V_2 and V_3 are vectors defining the three principal axes of the ellipsoid, A is the vector with the greatest magnitude and B defines a set of vectors excluding the vector with the greatest magnitude. All of these long/slender ellipsoids for the thick region on the model shown in figure 6 are shown in figure 8.

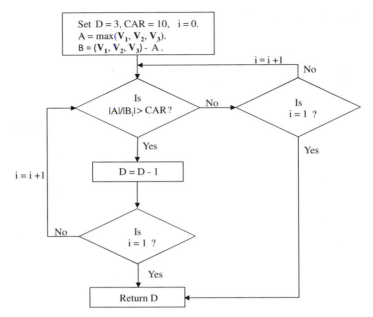

Fig. 7. Selection process for 1D ellipsoids using a critical aspect ratio.

Fig. 8. Long/slender ellipsoids on the thick region of complex model.

After all the long/slender ellipsoids are generated, a "closed-loop" searching algorithm is then initiated that uses the ellipsoids to look for closed loops of surfaces which each bound a slender section. For any given edge on which a long/slender ellipsoid is located, the algorithm searches for an adjoining surface that has another long/slender ellipsoid on one of its bounding edges. If such an ellipsoid exists,

Automatic Decomposition and Efficient Semi-structured Meshing of Complex Solids 209

this establishes a direction in which to continue the search, in either a clockwise or anti-clockwise fashion. The search is repeated for the other surface and if successful the process continues until a surface is located that contains an edge with the ellipsoid that was used for the initial iteration. At this point, a long/slender region comprising a closed loop of surfaces is identified and these surfaces are then removed from the search. A new edge is selected and the process is repeated until all closed loop surface groups have been identified in the model. Table 2 describes the closed-loop searching algorithm for long/slender extruded regions. The results of the closed-loop search are shown in figure 9. The algorithm successfully identifies five slender regions, where each region is highlighted in a different colour. Note that the bottom-left arm of the model is composed of two separate long/slender regions. The different surfaces identifying those regions are in different colours on the left and right image.

Fig. 9. Long/slender regions identified on complex model

Table 2. Closed-loop searching algorithm.

Stage	Diagram
1. Select an ellipsoid.	
2. Locate an ellipsoid on the edge of an adjoining surface.	
3. Continue the search in the search direction.	
4. Stop when a closed loop is formed	

4.2 Automatic Partitioning into Meshable Sub-regions

After the groups of surfaces that define each slender region have been identified, the next stage of the process involves generating meshable slender bodies by partitioning the surface groups using cutting planes. Open ended slender regions, which don't require any cutting planes, can be identified because the 'capping' surface shares an edge with all of the surfaces bounding the sweepable region. Cutting planes are generated using the edge tangent and surface normal at an offset (calculated as a fraction of the shortest edge length) away from the long/slender-complex region interface, as shown in figure 10. The cutting plane(s) for each edge are grouped into a set and given an ID based on the surface group.

After associating each surface group with a cutting plane, the body is automatically partitioned into a non-manifold assembly of long/slender and residual complex bodies in CADfix, using series of cutting commands. The cutting plane ID's are then used to identify if a body is chunky or slender. Any body that lies between the cutting planes in a set is slender. The result of the automatic partitioning process is shown in figure 11.

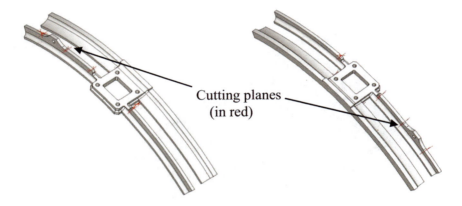

Fig. 10. Automatic generation of cutting planes

Finally, by combining the partitioned thick region with the thin region generated from the thick/thin decomposition it's possible to achieve the complete decomposition of the model into long/slender, thin-sheet and residual complex sub-regions, as shown in figure 12.

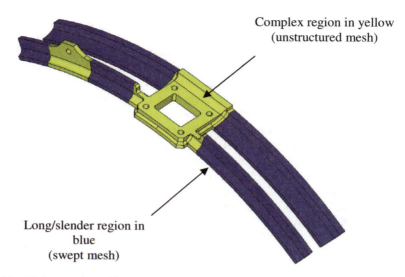

Fig. 11. Automatic partitioning into long/slender and complex sub-regions

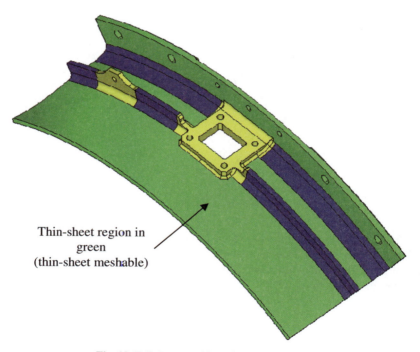

Fig. 12. Full decomposition of complex model

5 Meshing

After the thick region has been sub-divided a semi-structured mesh can be applied using various meshing algorithms. An extruded mesh is applied to the 1D regions using a 1D sweeping algorithm. The 3D complex volumes are filled with an unstructured isotropic tetrahedral mesh. Some refinements to ensure better matching of element size between 3D unstructured and 1D sweep meshes are planned. A 48% reduction in degrees of freedom is achieved with the semi-structured mesh (15,773 nodes) of the non thin-sheet region, shown in figure 13 compared to the unstructured tetrahedral mesh (30,108 nodes) of the component shown in figure 14.

Fig. 13. Unstructured tetrahedral mesh of thick region

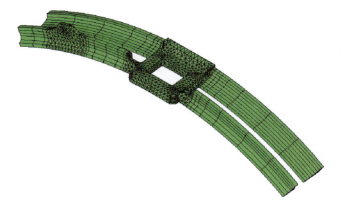

Fig. 14. Semi-structured hybrid mesh of thick region

The ellipsoids are not only used to partition the model but also grade the mesh in the long/slender regions. Each ellipsoid on the edge of a 1D sweepable region is used to define a "source field" on the model in CADfix. These source fields control the growth of the mesh and ensure a smooth transition between the 1D and 3D regions. In any case, the effectiveness of the mesh adaption technique is clearly demonstrated, generating an efficient mesh with a substantial reduction in degrees of freedom.

6 Conclusions

An automatic approach for sub-dividing a general dumb solid into meshable sub-regions has been developed. The approach relies on a robust thick/thin decomposition of the model. Provided the thick bodies can be identified, they can be partitioned into slender (1D) and chunky (3D) regions. Consequently, by applying a different meshing strategy to each region in the assembly of thin-sheet (2D), long/slender (1D) and chunky (3D) bodies, a semi-structured mesh can be generated, producing significantly fewer degrees of freedom and thereby enhancing the efficiency of the numerical analysis process.

Novel techniques for long/slender region identification have been employed that use shape metrics based on local sizing measures such as edge length, surface curvature, face width and local 3D thickness. After building an association between the shape metrics and the model geometry, a "closed loop" searching algorithm successfully locates closed loops or groups of surfaces that form slender regions within the model. Additional algorithms analyse these surface groups and generate cutting planes in strategic locations that are then used to automatically partition the model to produce a fully sub-divided, non-manifold assembly of meshable bodies. Future work will focus will be on testing the approach on assemblies of industrial complexity, quantifying the accuracy of the highly stretched meshing produced and the reduction in degrees of freedom attained.

Acknowledgements

The research leading to these results has received funding from the Eurpean Community's Seventh Framework Programme (FP7/2007-2013) under grant agreement no. 234344 (www.cresendo-fp7.eu). The authors would like to acknowledge Transcendata for their support throughout the course of the work and industrial partners for their support in providing complex models.

References

1. Robinson, T.T., Armstrong, C.G., Fairey, R.: Automated mixed dimensional modelling from 2D and 3D CAD models. Finite elements in analysis and design 47, 151–165 (2011)

2. Loseille, A., Dervieux, A., Alauzet, F.: Fully anisotropic goal-oriented mesh adaptation for 3D steady Euler equations. Journal of Computational Physics 229, 2866–2897 (2010)
3. Thakur, A., Banerjee, A.G., Gupta, S.K.: A survey of CAD model simplification techniques for physics-based simulation applications. Computer-Aided Design 41, 65–80 (2009)
4. CADfix, Transcedata,
   http://www.transcendata.com/products/cadfix/
   (accessed July 4, 2011)
5. Blum, H.: A transformation for extracting new descriptors of shape. In: Models for the Perception of Speech and Visual Form, pp. 362–380. MIT Press, Cambridge (1967)
6. Luo, X.-J., Shephard, M.S., Yin, L.-Z., O'Bara, R.M., Nastasi, R., Beall, M.W.: Construction of near optimal meshes for 3D curved domains with thin sections and singularities for p-version method. Engineering with Computers 26, 215–229 (2010)
7. Price, M.A., Armstrong, C.G.: Hexahedral mesh generation by medial surface subdivision: part I. International Journal for Numerical Methods in Engineering 38, 3335–3359 (1995)
8. Price, M.A., Armstrong, C.G.: Hexahedral mesh generation by medial surface subdivision: part II. Solids with flat and concave edges. International Journal for Numerical Methods in Engineering 40, 111–136 (1997)
9. Li, T.S., McKeag, R.M., Armstrong, C.G.: Hexahedral meshing using midpoint subdivision and integer programming. Computer Methods in Applied Mechanics and Engineering 124, 171–193 (1995)
10. Tam, T.H.K., Armstrong, C.G.: Finite element mesh control by integer programming. International Journal for Numerical Methods in Engineering 36, 2581–2605 (1993)
11. Tchon, K., Khachan, M., Guibault, F., Camarero, R.: Three-dimensional anisotropic geometric metrics based on local domain curvature and thickness. Computer-Aided Design 37, 173–187 (2005)
12. Shimanda, K., Mori, N., Kondo, T.: Automated mesh generation for sheet metal forming simulation. Int. J. Vehicle Des. 21, 278–291 (1999)
13. Frey, P.J. About surface meshing. In: Ninth International Meshing Roundatble, Sandia National Laboratories, New Mexico, pp 123 - 126
14. Zhao, G., Zhang, H.: Adaptive hexahedral mesh generation based on local domain curvature and thickness using a modified grid-based method. Journal of Finite Elements in Analysis and Design 43, 691–704 (2007)
15. Zhao, G., Zhang, H., Cheng, L.: Geometry-adaptive generation algorithm and boundary match method for initial hexahedral element mesh. Journal of Engineering with Computers 24, 321–339 (2008)
16. White, D.R., Saigal, S., Owen, S.J.: CCSweep: automatic decomposition of multi-sweep volumes. Journal of Engineering with Computers 20, 222–236 (2004)
17. Frey, P.J., George, P.J.: Mesh Generation, p. 337. Wiley (2008)
18. Lipschutz, M.M.: Schaums Outline of Theory and Problems of Differential Geometry, p. 176. McGraw-Hill (1969)

The Cutting Pattern Problem for Tetrahedral Mesh Generation

Xiaotian Yin[1], Wei Han[1], Xianfeng Gu[2], and Shing-Tung Yau[1]

[1] Mathematics Department, Harvard University, MA, U.S.A.
{xyin,weihan,yau}@math.harvard.edu
[2] Computer Science Department, Stony Brook University, NY, U.S.A.
gu@cs.sunysb.edu

Summary. In this work we study the following cutting pattern problem. Given a triangulated surface (i.e. a two-dimensional simplicial complex), assign each triangle with a triple of ± 1, one integer per edge, such that the assignment is both complete (i.e. every triangle has integers of both signs) and consistent (i.e. every edge shared by two triangles has opposite signs in these triangles). We show that this problem is the major challenge in converting a volumetric mesh consisting of prisms into a mesh consisting of tetrahedra, where each prism is cut into three tetrahedra. In this paper we provide a complete solution to this problem for topological disks under various boundary conditions ranging from very restricted one to the most flexible one. For each type of boundary conditions, we provide efficient algorithms to compute valid assignments if there is any, or report the obstructions otherwise. For all the proposed algorithms, the convergence is validated and the complexity is analyzed.

Keywords: cutting pattern, graph labeling, tetrahedral mesh, prism.

1 Introduction

1.1 Motivation

Volumetric meshes are widely used in various areas, such as geometric modeling, computer aided design and physical simulation. Tetrahedral mesh is one of the most commonly used representations, because it is a simplicial complex that allows many topological algorithms and finite element solvers to be applied. There are many methods to generate tetrahedral meshes, such as Advancing Front techniques [1, 2], Octree methods [3] and Voronoi Delaunay based methods [4, 5, 6].

In this work we look at a specific problem: converting a *prismal mesh* to a tetrahedral mesh without introducing new vertices. A prismal mesh consists of a set of prisms, which are volumetric elements bounded by two triangles from top and bottom and three quadrilaterals from sides. A typical example is a 3D space-time slab, which is generated by extruding a 2D triangular mesh in temporal direction. It has been shown in [7] that such a prismal mesh needs to be split into a tetrahedral mesh to adapt existing unstructured tetrahedral solvers. It has also been shown in [8, 9] that such a conversion is necessary in computer graphics, especially when one wants to apply efficient algorithms for volume rendering and iso-contouring that exist for purely tetrahedral meshes.

The conversion from a prismal mesh to a tetrahedral mesh has been addressed in a lot of work, such as [8, 9, 10, 11, 12]. However, none of these methods allows the user to control the triangulation on the boundary of the output volumetric mesh. In another word, they only solve the free boundary version of this problem. In this work, we aim at tackling this problem in a more general setting that includes both free boundary and fixed boundary cases, where the latter incurs more challenges than the former.

In order to do this conversion without introducing new vertices, an intuitive way is to cut each prism into three tetrahedra using two planes passing through certain corners of the prism (figure 1b-d). The options of cutting a prism can be encoded as a 3-tuple of ± 1, where $+1$ (or -1) means the corresponding wall is sliced along the diagonal (figure 1a left) or anti-diagonal (figure 1a right). Therefore, the conversion of the whole volumetric mesh can be reduced to assigning the underlying triangular surface mesh with a set of 3-tuples of ± 1.

This ± 1 assigning task is challenging in multiple aspects.

- In a local consideration, not every combination of three ± 1 represents a valid cutting, see figure 1(e) for a counterexample. A valid cutting should have both $+1$ and -1 in each triangle.
- In a global consideration, the cutting must be consistent; i.e. for any pair of prisms adjacent to each other by a quadrangular wall, their cutting lines must meet on that wall. It means the cutting options in the bottom triangles should have opposite signs across any common edge shared by two triangles.

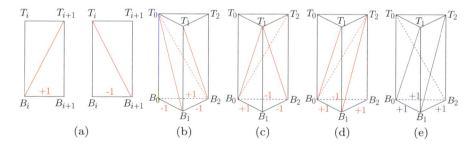

Fig. 1. Options of cutting a prism into tetrahedra are encoded by assigning $+1$ or -1 to each edge of the bottom triangle $B_0 B_1 B_2$ (a). A valid cutting pattern consists of both signs (b, c and d), while an invalid one consists of only one sign (e).

- From the user's point of view, the user should be able to control the cutting options on the boundary surface of the volumetric mesh, and the internal cutting should be subject to the boundary values.

Considering all these requirements, we convert this mesh conversion problem to an equivalent 2D graph problem in the next section.

1.2 Problem Statement

The major problem that we need to solve is as follows.

Definition 1 (The Cutting Pattern Problem). *Given a triangular mesh D, assign each triangle with a 3-tuple of ± 1, one integer per edge, such that:*
- *Every triangle has both $+1$ and -1 in its 3-tuple;*
- *Every edge shared by two adjacent triangles is assigned with opposite values in these triangles;*

In practice, the assignments on boundary edges may or may not be prescribed. If they are prescribed, there are various ways to set the boundary values. In this work, we consider the following types of boundary conditions, varying from the most restricted one to the most flexible one.

Definition 2 (Boundary Conditions). *The cutting pattern problem can be subject to the following types of boundary conditions:*

1. *Restricted boundary: Every boundary edge of D has a prescribed ± 1 assignment. In particular,*
   - *For any triangle with only one edge exposed on the boundary of D, the prescribed assignment for this boundary edge can take an arbitrary value of ± 1;*
   - *For any triangle with at least two edges exposed on the boundary of D, two of such boundary edges must have opposite prescribed assignments.*
2. *General boundary: Every boundary edge of D has a prescribed assignment that can take an arbitrary value of ± 1;*
3. *Free boundary: None of the boundary edges has a prescribed assignment.*

These boundary conditions reflect different levels of user control in the original problem of triangulating a prismal mesh, that is, whether and how one wants to control the boundary triangulation of the output tetrahedral mesh. And as discussed later, different boundary conditions result in different solvability of the problem and different ways to find a solution if there is any. If the boundary triangulation is prescribed without any special constraint, it corresponds to the general boundary condition and the problem may or may not have a solution. If the boundary triangulation is prescribed and satisfies the restricted boundary condition, the problem can always be solved using our algorithm. If the user does not want to specify the boundary triangulation, our algorithm can automatically specify the boundary and is guaranteed to find a triangulation of the whole volume.

1.3 Contributions

The cutting pattern problem is the major challenge of converting a prismal mesh to a tetrahedral mesh. If the former can be solved, so is the latter. Meanwhile, the cutting pattern problem itself is an interesting graph problem that can be categorized as graph labeling (see [13] for a survey). To our best knowledge, however, this specific problem has not been addressed in the literature.

In this paper we make efforts to tackle this problem for topological disks with a single boundary, and propose a complete set of solutions under all kinds of boundary conditions. In particular:

1. For restricted boundary conditions, we show that solution always exists and can be found using an efficient algorithm proposed in the paper. (section 2)
2. For general boundary conditions, we show that solution exists if there is no cutting pattern obstruction (definition 6). We also propose an efficient algorithm that either reports such an obstruction if there is any or outputs a valid solution otherwise. (section 3)
3. For free boundary conditions, we show that the problem can be always turned into a restricted one, therefore solution always exists and can be found using an algorithm modified from the restricted one. (section 4)

For the algorithm proposed for each case, the convergence is validated and the complexity is analyzed.

1.4 Definitions and Notations

Here we introduce the concepts and notations used in later elaborations.

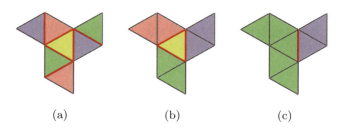

Fig. 2. Separating edges (a), breaking edges (b) and a minimal set of breaking edges (c).

In a triangular mesh D consisting of a set of triangles $F = \{f_1, f_2, \cdots, f_n\}$, the *edge-valence* (or *valence* in short) of a triangle $f_i \in F$ is the number of triangles adjacent to f_i across some edges. It should be a non-negative integer that is at most 3. A triangle is called a *dangling triangle* if its valence is 1. A dangling triangle has exactly two edges exposed on the boundary of D. A triangle is called a *singular triangle* if its valence is 0. In a singular triangle, all three edges are on the boundary of D.

For an edge e shared by two triangles $f_1, f_2 \in D$, we say it is of (i,j)-*type* if f_1 and f_2 have valence i and j respectively ($1 \leq i \leq j \leq 3$). An edge e is called a *separating edge* if it is shared by two triangles and the end vertices of e are both on the boundary of D (figure 2a). An edge e is called a *breaking edge* if it is a separating edge and at least one of its adjacent triangles has valence 3 (figure 2b and 2c).

A *sub-component* C is a sub-mesh of D that is *edge-connected* (or *connected* in short), meaning that every triangle in C is adjacent to at least one other triangle in C across a common edge.

Given an edge e in triangle f, its *assignment* (denoted as $a(e, f)$) is an integer 0 (*unsolved*) or ± 1 (*solved*). An edge e is *completely solved* (or *solved* for short) if and only if it is solved in every triangle enclosing e. An triangle f is *completely solved* (or *solved* for short) if and only if all three edges of f have been solved. A mesh D (or sub-component C) is *completely solved* (or *solved* for short) if and only if all its triangles are solved. Obviously, a valid cutting pattern over a given mesh D is a set of ± 1 assigned to every edge in every triangle such that both completeness and consistency are satisfied.

Given a mesh D (or sub-component C), its *boundary assignment* is the set of assignments of its boundary edges. A boundary assignment is *safe* if and only if every dangling or singular triangle f (if there is any) has two boundary edges with opposite non-zero assignments $+1$ and -1 (figure 3a). It is *moderately dangerous* if and only if it is not safe and at least two boundary edges have opposite non-zero assignments (figure 3b). It is *extremely dangerous* if and only if it is not safe and all the non-zero assignments on boundary edges equal to a single value $a \in \{+1, -1\}$ (figure 3c), which is called the *forbidding value* of this boundary assignment. The last two cases are both called *dangerous* boundary assignments.

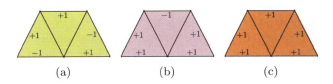

Fig. 3. Sub-components with different types of boundary assignments: (a) safe, (b) moderately dangerous, (c) extremely dangerous.

2 Solution for Restricted Boundary Conditions

For the cutting pattern problem under restricted boundary conditions, we propose the following algorithm to look for a valid cutting pattern. We will show that this algorithm always ends up with a valid cutting pattern and therefore gives a constructive proof for the existence of solutions under restricted boundary conditions.

Given a topological disk represented as a triangular mesh D, the goal is to compute a valid cutting pattern for D, i.e. to assign every edge in every triangle with an integer $b \in \{+1, -1\}$. Upon input, all the boundary edges have prescribed assignments ± 1 while all the other edges are initialized with assignment 0. The input mesh D is processed iteratively. In each iteration, the unsolved part of D is partitioned into a set of sub-components that have special types, and each sub-component is either completely solved or partially solved. After this, the remaining unsolved triangles will be brought into another iteration. Repeat this process until the whole mesh is solved.

Algorithm 2.1 (Cutting Pattern Algorithm - I)

- *Input: A triangular mesh D with a restricted boundary condition.*
- *Output: A valid cutting pattern for D.*
- *Procedures:*
   1. *Initialize all the internal edges of D with assignment 0 (i.e. unsolved).*
   2. *Repeat the following procedures on D until no unsolved triangle left.*
      a) *Partition D into a minimal set of sub-components of basic types and solve the newly exposed boundary edges (section 2.1).*
      b) *Solve each sub-component completely or partially according to its type (2.2 and 2.3).*
      c) *If there is no unsolved triangle, exit; otherwise, set D to be the unsolved part and go back to step 2a.*

2.1 Constructing Sub-components

In step 2a we partition mesh D into a set of sub-components $\{C_i \mid 1 \leq i \leq k\}$ that are *triangle-disjoint* (i.e. there is no triangle $f \in S_i \cap C_j$, $i \neq j$) and *covering* (i.e. $D = \cup_{i=1}^{k} C_i$). Figure 4 shows several examples of mesh partitioning. Furthermore, each sub-component should have one of the following basic types:

Definition 3 (Classification of Sub-components)

- *Aggregated sub-component (figure 6): a sub-component where every triangle has edge-valence at least 2;*
- *Linear sub-component (figure 5): a sub-component consisting of a sequence of triangles, where two triangles at the ends have edge-valence 1 (i.e. dangling triangles) and all the others have edge-valence 2;*

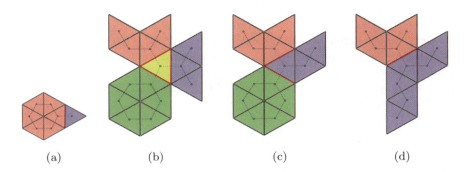

Fig. 4. Partition a mesh into a minimal set of sub-components. (a) to (d) show several different meshes and their partitions.

- *Singular sub-component: a sub-component consisting of only one triangle with edge-valence 0 (i.e. a singular triangle).*

One can get such a partition by slicing D along all the breaking edges. However, such a partition may have unnecessary small pieces (figure 2b). What we need in the algorithm is a *minimal set of sub-components* of basic types, or equivalently a set of *maximal sub-components*, meaning that the union of any two or more sub-components in this set is not of any basic type (figure 2c). Therefore, we need to find *a minimal set of breaking edges*, meaning that a smallest set of breaking edges so that the resulting sub-components are all maximal. This can be done by addressing all the breaking edges in D and then exclude unnecessary ones by checking other edges in two enclosing triangles.

The major reason of choosing a minimal set rather than the full set of breaking edges is to simplify the algorithm and the analysis. For example, as we will show later, this will result in simple cases to be considered when solving a breaking edge, and therefore simplify the proof of convergence of the algorithm.

Once the minimal set of breaking edges (denoted as E_b) is addressed, we partition the given mesh into a set of sub-components, and every edge $e \in E_b$ will be exposed on the boundary of two sub-components and needs to be solved here. In order to solve edges in E_b, we first classify them based on their positions in the mesh and the types of their adjacent sub-components.

Definition 4 (Classification of Breaking Edges). *A breaking edge e shared by sub-components C_1 and C_2 can only have one of the following types:*

- *A2A: if both C_1 and C_2 are aggregated;*
- *A2Lm (or Lm2A): if C_1 is aggregated and C_2 is linear, and e belongs to a triangle in the middle of C_2;*
- *A2Le (or Le2A): if C_1 is aggregated and C_2 is linear, and e belongs to a triangle at one end of C_2;*

- *A2S (or S2A)*: if C_1 is aggregated and C_2 is singular;
- *Lm2Lm*: if both C_1 and C_2 are linear, and e belongs to a middle triangle in C_1 and another middle triangle in C_2;
- *Lm2Le (or Le2Lm)*: if both C_1 and C_2 are linear, and e belongs to a middle triangle in C_1 and an end triangle in C_2;
- *Lm2S (or S2Lm)*: if C_1 is linear and C_2 is singular, and e belongs to a middle triangle in C_1;

Note that these are not all the combinations of two sub-components with basic types. However, other combinations, including Le2Le, Le2S and S2S, are impossible in our algorithm. For example, a Le2Le combination means that two linear sub-components connect to each other at one end and can thus form a larger linear sub-component; therefore it breaks the rule of maximal sub-components and is not a valid combination. The combination of Le2S and S2S can be excluded in a similar way.

Once the breaking edges are classified, they can be solved according to their types using the following procedure.

Procedure 2.1 (Solving A Breaking Edge). *Given a breaking edge e shared by two triangles f_1 and f_2 in sub-components C_1 and C_2 respectively, do the following:*

- *If e is A2A (i.e. between aggregated C_1 and aggregated C_2): Assign e with $+1$ on C_1 side and -1 on C_2 side.*
- *If e is A2Lm or Lm2Lm: similar to the A2A case.*
- *If e is A2Le (i.e. between aggregated C_1 and linear C_2): We will first solve e on C_2 side and then on C_1 side. On C_2 side, e is a boundary edge in an end triangle f_2. Let e' be the other boundary edge in f_2;*
  - *If e' is already solved in f_2, suppose its assignment is $a'_2 \in \{+1, -1\}$, then assign e in f_2 with $a_2 = -a'_2$.*
  - *If e' is not solved in f_2 yet, then assign e with an arbitrary value $a_2 \in \{+1, -1\}$ (so that e' can be solved later with assignment $a'_2 = -a_2$).*

  Once e receives assignment $a_2 \in \{+1, -1\}$ in $f_2 \in C_2$, we assign e with $a_1 = -a_2$ in $f_1 \in C_1$.
- *If e is Lm2Le: similar to the A2Le case.*
- *If e is A2S (i.e. between aggregated C_1 and singular C_2): We will first solve e on C_2 side and then on C_1 side. On C_2 side, there is only one triangle $f_2 \in C_2$, and e is one of the three edges in f_2, let e' and e'' be the other two. Then within f_2,*
  - *If at least one other edge (say e') is already solved in f_2, suppose its assignment is $a'_2 \in \{+1, -1\}$, then assign e with $a_2 = -a'_2$ in f_2.*
  - *If neither e' nor e'' is solved in f_2, then assign e with an arbitrary value $a_2 \in \{+1, -1\}$ (so that e' and e'' can be solved later with assignment $a_2' = -a_2$ and $a_2'' = -a_2$).*

  Once e receives assignment $a_2 \in \{+1, -1\}$ in $f_2 \in C_2$, we assign e with $a_1 = -a_2$ in $f_1 \in C_1$.
- *If e is Lm2S: similar to the A2S case.*

This procedure guarantees that every breaking edge in E_b receives opposite assignments on two sides, and every resulting sub-component has a safe boundary assignment. After this process, every sub-component C has a solved boundary that satisfies the restricted boundary condition and is ready to be solved for the inner edges based on the type of C. For a singular sub-component, all the edges are on the boundary and are already solved, no further process is needed. For linear and aggregated sub-components, their inner edges need extra efforts to solve (see section 2.2 and 2.3 respectively).

2.2 Solving Linear Sub-components

A linear sub-component C consists of a sequence of triangles,

$$f_0, f_1, \cdots, f_n$$

where f_0 and f_n are end triangles (with valence 1) that each has two edges $e^B_{i,1}$ and $e^B_{i,2}$ on the boundary of C ($i = 0, n$), while any other f_i is a middle triangle (with valence 2) that has only one edge e^B_i on the boundary of C ($1 \leq i \leq n-1$). These boundary edges are already solved. Meanwhile, each pair of consecutive triangles f_{i-1} and f_i ($1 \leq i \leq n$) share a common internal edge e^I_i; these internal edges need to be solved here.

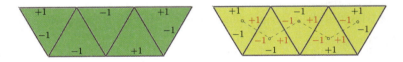

Fig. 5. A linear sub-component can be completed solved via a stabbing line (dotted).

In order to solve the internal edges, we stab the sequence of triangles with a line from f_0 all the way to f_n (figure 5). Let's take an arbitrary value $b \in \{+1, -1\}$ as an initial assignment. When the line penetrates an internal edge e^I_i, we assign b and $-b$ to its entrance side (in f_{i-1}) and exit side (in f_i) respectively. In the end, all the internal edges are completely solved and the whole sub-component is thus solved.

2.3 Solving Aggregated Sub-components

Since every sub-component in our algorithm is guaranteed to be a topological disk, every aggregated sub-component C must have only one boundary, and there must be a layer of most outside triangles (the *frontier*) that can be distinguished from the rest (the *interior*) in the following way (see figure 6).

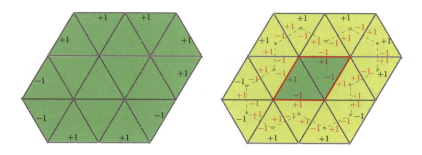

Fig. 6. An aggregated sub-component is only solved in the frontier (in yellow); the interior (in green) is left to later iterations.

Consider the dual graph C^*. Any triangular face $f_i \in C$ corresponds to a vertex $v_i^* \in C^*$, edge $e_j \in C$ to edge $e_j^* \in C^*$, vertex $v_k \in C$ to face $f_k^* \in C^*$. Note that C is a topological disk and every face $f_i \in C$ has at most one boundary edge. Accordingly, C^* is also a topological disk and its boundary is a loop connecting n vertices sequentially: $v_1^*, v_2^*, \cdots, v_n^*, v_1^*$. This boundary loop of C^* corresponds to a looping sequence of n triangular faces in C:

$$f_1, f_2, \cdots, f_n, (f_1)$$

These triangles are called the *frontier* of C, denoted as C^F, and the remaining triangles the *interior*, denoted as C^I. In this step we will solve C^F completely.

Edges in C^F can be grouped into three sets E^B, E^I and E^O, where E^B consists of edges on the boundary loop of C, E^O consists of edges on the border between C^F and C^I, and E^I consists of edges e_i^I shared by consecutive triangles f_{i-1} and f_i in C^F. Edges in E^B have already been solved, while the other two sets need to be solved here.

To solve E^I, we use a stabbing process similar to that for linear sub-components. We take arbitrary $b \in \{+1, -1\}$ as the initial assignment, stab the sequence of triangles in C^F with a circle; when penetrating an edge $e_i^I \in E^I$, assign it with b on the f_{i-1} side and $-b$ on the f_i side.

To solve E^O, we first consider the C^I side, where E^O serves as the boundary of C^I. Check every face $f \in C^I$ that contains at least one edge in E^O. If f contains only one such edge, we assign it with $+1$. If f contains at least two such edges, we assign $+1$ to one of them and -1 to the rest. Once an edge $e \in E^O$ is solved on the C^I side with assignment b, we assign it on the C^F side with an opposite value $-b$.

Once all the edges in $E^I \cup E^O$ are solved, the whole frontier C^F is completely solved, while C^I is only solved on its boundary and the remainder will be recursively solved later.

2.4 Validation and Analysis

In this part we discuss the convergence and complexity of algorithm 2.1. From the above discussion about the algorithm, several invariants of the algorithm can be induced.

Definition 5 (Invariants of Algorithm 2.1)

- *Completeness: whenever a triangle is completely solved, its 3-tuple of edge assignments should consist of both $+1$ and -1;*
- *Consistency: whenever an internal edge (shared by two triangles) is completely solved, it should have opposite assignments in two enclosing triangles;*
- *Safeness: the unsolved part (a sub-mesh) should have a safe boundary assignment at the beginning of step 2a, and every sub-component should also have a safe boundary assignment at the end of step 2a.*

We can prove that all these invariants hold in the algorithm. To do this, we need to guarantee that none of these invariants can be violated in our algorithm.

Completeness and consistency can only be violated when an edge or a face is solved. In algorithm 2.1 there are three procedures that need to be checked. The first is in step 2a, where breaking edges between adjacent sub-components are solved (section 2.1). The second and third are both in step 2b, where linear and aggregated sub-components are solved (section 2.2 and 2.3). These violations are guaranteed not to happen by Lemma 1, 2 and 3 respectively.

Safeness can only be violated when edges are newly exposed to the boundary of a sub-mesh or a sub-component. In algorithm 2.1 there are two procedures that need to be checked. The first is in step 2a, where breaking edges between adjacent sub-components are solved (section 2.1). The second is in step 2b, where aggregated sub-components are solved in the frontier (2.3). These violations are guaranteed not to happen by Lemma 4 and 5 respectively.

Lemma 1. *The completeness and consistency invariants hold when solving a breaking edge using procedure 2.1.*

Proof. In procedure 2.1, we solve all the breaking edges along which the unsolved sub-mesh is partitioned into basic type sub-components. Given such a breaking edge e shared by $f_1 \in C_1$ and $f_2 \in C_2$,

- If e is A2A: e is assigned with $+1$ and -1 in $f_1 \in C_1$ and $f_2 \in C_2$ respectively, therefore consistency holds on e.
- If e is A2Lm or Lm2Lm: similar to the A2A case.
- If e is A2Le: On C_2 side, f_2 is an end triangle that has two boundary edges, e and e'. In f_2, e always receives an assignment opposite to that of e', therefore completeness holds for f_2. On C_1 side, e in f_1 always receives an assignment opposite to that of e in f_2, therefore consistency holds for e.

- If e is Lm2Le: similar to the A2Le case.
- If e is A2S: On C_2 side, f_2 is a singular triangle that all three edges are on the boundary. In f_2, e always receives an assignment opposite to that of another edge $e' \in f_2$, therefore completeness holds for f_2. On C_1 side, e in f_1 always receives an assignment opposite to that of e in f_2, therefore consistency holds for e.
- If e is Lm2S: similar to the A2S case.

Lemma 2. *The completeness and consistency invariants hold when solving a linear sub-component.*

Proof. In section 2.2 we use a stabbing procedure to solve all the internal edges $\{e_1^I, e_2^I, \cdots, e_n^I\}$ in a linear sub-component C and thus solve all the triangles $\{f_0, f_1, \cdots, f_n\}$.

Every internal edge $e_i^I = f_{i-1} \cap f_i$ $(1 \leqslant i \leqslant n)$ receives assignment b and $-b$ in f_{i-1} and f_i respectively, where b is an initial assignment arbitrarily chosen from $\{+1, -1\}$. Therefore consistency holds for every newly solved e_i^I.

Every middle triangle f_i $(1 \leqslant i \leqslant n-1)$ has two internal edges e_i^I and e_{i+1}^I and receives opposite assignments on them during the stabbing. In addition, due to the safeness invariant, every end triangle (f_0 and f_n) already has opposite assignments on two boundary edges before the stabbing. Therefore completeness holds for every triangle in C.

Lemma 3. *The completeness and consistency invariants hold when solving an aggregated sub-component.*

Proof. In section 2.3 we solve the internal edges E^I of the frontier layer C^F and the edges E^O between the frontier C^F and the interior C^I, thus all the faces $\{f_1, f_2, \cdots, f_n\}$ in C^F are solved.

E^I are solved by a stabbing procedure, where every $e_i^I = f_{i-1} \cap f_i$ receives assignment b and $-b$ in f_{i-1} and f_i respectively, where b is an initial assignment arbitrarily chosen from $\{+1, -1\}$. Therefore consistency holds for every edge in E^I.

E^O are first solved on C^I side and then on C^F side, and every edge in E^O receives opposite assignments on two sides. Therefore consistency holds for every edge in E^O as well.

Every triangle f_i in C^F has two internal edges e_i^I and e_{i+1}^I and receives opposite assignments on them during the stabbing. Therefore completeness holds for every triangle solved in this process.

Lemma 4. *The safeness invariant holds when solving a breaking edge using procedure 2.1.*

Proof. In procedure 2.1, we partition the unsolved sub-mesh along a set of breaking edges. These edges are exposed on the boundary of resulting sub-components. Given such a breaking edge e shared by $f_1 \in C_1$ and $f_2 \in C_2$, we need to check whether their assignments violate the safeness of the boundary of C_1 and C_2.

- If e is A2A: since C_1 aggregated, f_1 is not singular or dangling, therefore any assignment will be safe; i.e. safeness holds for C_1. The same argument holds for C_2 as well.
- If e is A2Lm or Lm2Lm: similar to the A2A case.
- If e is A2Le: On C_1 side, f_1 is not singular or dangling, therefore any assignment will be safe. Therefore safeness holds for C_1. On C_2 side, f_2 is an end (dangling) triangle containing two boundary edges e and e', they receive opposite assignments. By definition this does not violate the safeness on the boundary of C_2.
- If e is Lm2Le: similar to the A2Le case.
- If e is A2S: On C_1 side, f_1 is not singular or dangling, therefore any assignment will be safe. Therefore safeness holds for C_1. On C_2 side, f_2 is a singular triangle containing three boundary edges, two of them receive opposite assignments. By definition this does not violate the safeness on the boundary of C_2.
- If e is Lm2S: similar to the A2S case.

Lemma 5. *The safeness invariant holds when solving an aggregated sub-component.*

Proof. In section 2.3 we completely solve the frontier C^F of a sub-component C and expose the edges in E^O as the boundary of the interior C^I. Those new boundary edges E^O are solved on C^I side. Recall that for face $f \in C^I$ that has at least two boundary edges (in E^O), this f receives assignments of both signs. By definition this gives a safe boundary assignment for C^I, therefore safeness holds for C^I.

Based on the above lemmas, we have the following conclusion about the convergence of the algorithm:

Theorem 1. *Algorithm 2.1 always terminates (due to the safeness invariant); at termination, all the triangles are solved and the resulting assignments represent a valid cutting pattern (due to the completeness and consistency invariants).*

Now consider the time complexity of this algorithm. Let $|V|$, $|E|$ and $|F|$ be the number of vertices, edges and triangles in the input mesh D. The cost of the algorithm lies in the following aspects.

- Addressing breaking edges to partition the unsolved part of D. Since a breaking edge is always connecting boundary vertices, we only need to check the boundary vertices of C; and to avoid unnecessary breaking edges, we only need to check its two adjacent triangles. Therefore the cost in each iteration is bounded by the number of boundary vertices of the unsolved part of D in that iteration. Since every vertex in the original mesh D appears at most once on the boundary of the unsolved part, the total cost for partitioning is bounded by $O(|V|)$.

- Solving edges. This happens when we solve a linear sub-component completely or solve an aggregated sub-component in its frontier. From an overall view, every edge in the original mesh D is solved only once; therefore this part is bounded by $O(|E|)$.
- Addressing the frontier of aggregated sub-components. From an overall view, every triangle in D will appear in the frontier of an aggregated sub-component for at most once; therefore this part is bounded by $O(|F|)$.

In summary, the time complexity of algorithm 2.1 is $O(|V| + |E| + |F|)$, which is linear.

3 Solution for General Boundary Conditions

A general boundary condition is different to a restricted one in that the boundary assignment for the input mesh is allowed to be dangerous, and it degenerates to the latter if the boundary assignment is safe. Inspired by this fact, we design an algorithm that tries to resolve all the danger and tries to turn this general problem into a restricted one. If the danger cannot be removed from some part of the mesh, the algorithm will report this part as an obstruction and terminate. Otherwise, it will continue to run as the restricted algorithm.

Algorithm 3.1 (Cutting Pattern Algorithm - II)

- *Input: A triangular mesh D with a general boundary condition.*
- *Output: A valid cutting pattern for D or an unsolvable sub-mesh.*
- *Procedures:*
    1. *Check the boundary assignment. If it is safe, jump to step 4; otherwise, do the following.*
    2. *Partition mesh D into a minimal set of basic type sub-components, try to solve the breaking edges between sub-components. If any obstruction is detected, report it and exit. (section 3.1)*
    3. *Solve moderately dangerous sub-components. (section 3.2)*
    4. *Run algorithm 2.1 on the unsolved sub-mesh from the previous step.*

In this algorithm there are two steps where we need to resolve danger, step 2 (section 3.1) and step 3 (section 3.2).

3.1 Resolving Danger via Partition

The first place that we can resolve danger in algorithm 3.1 is step 2. In this step, we first partition the input mesh D into a minimal set of basic type sub-components (as in section 2.1) along a set of breaking edges (which will be initialized with assignment 0), then try to solve these breaking edges using the following procedure.

Procedure 3.1 (Resolving Danger via Partition). *Given a minimal set of basic type sub-components, solve the breaking edges between sub-components as follows.*

1. Repeat the following until all the extremely dangerous sub-components (if there is any) have been transformed into moderately dangerous or safe ones:
   a) Address all the extremely dangerous sub-components, put them in a set $\mathfrak{C}$;
   b) For every extremely dangerous sub-component $C \in \mathfrak{C}$, check each adjacent sub-component C'_i as the following, where we use e_i^B to denote the boundary edge between C and C'_i.
      - If there is an adjacent C'_i that is safe, then assign e_i^B with $a_i = -b$ in C (thus C becomes moderately dangerous or safe) and $a'_i = b$ in C'_i (thus C'_i remains safe). Remove C from $\mathfrak{C}$.
      - Otherwise, if there is an adjacent C'_i that is moderately dangerous, then assign e_i^B with $a_i = -b$ in C (thus C becomes moderately dangerous or safe) and $a'_i = b$ in C'_i (thus C'_i remains moderately dangerous or becomes safe). Remove C from $\mathfrak{C}$.
      - Otherwise, if there is an adjacent C'_i that is extremely dangerous with opposite forbidding value $-b$, then assign e_i^B with $a_i = -b$ in C (thus C becomes moderately dangerous or safe) and $a'_i = b$ in C'_i (thus C'_i becomes moderately dangerous or safe). Remove C from $\mathfrak{C}$.
      - Otherwise, all the adjacent sub-components must be extremely dangerous and have the same forbidding value b; Then keep C in $\mathfrak{C}$ to wait for a second chance at a later point when some adjacent sub-component is transformed into a less dangerous one.
   c) Check the set $\mathfrak{C}$:
      - If $\mathfrak{C}$ is empty, jump to step 2 of this procedure.
      - Otherwise, if $\mathfrak{C}$ contains less sub-components than it does before this iteration, go back to step 1a and start another iteration.
      - Otherwise, no one in $\mathfrak{C}$ can be resolved, report $\mathfrak{C}$ and exit the procedure.
2. For every sub-component C that is moderately dangerous, do the following:
   a) For every new boundary edge e_i^B in C that has assignment 0 (i.e. still unsolved):
      - If the corresponding adjacent sub-component C'_i is moderately dangerous, then assign e_i^B with an arbitrary value $a_i \in \{+1, -1\}$ in C (thus C remains moderately dangerous or becomes safe) and $a'_i = -a_i$ in C'_i (thus C'_i remains moderately dangerous or becomes safe).
      - Otherwise, the corresponding adjacent sub-component C'_i must be safe, then assign e_i^B with arbitrary value $a_i \in \{+1, -1\}$ in C (thus

C remains moderately dangerous or becomes safe) and $a'_i = -a_i$ in C'_i (thus C'_i remains safe).

As one can see from the procedure, solving breaking edges will make a dangerous boundary assignment less or equally dangerous but not more dangerous. For example, an extremely dangerous one can be transformed to another extremely dangerous one, a moderately dangerous one, or even a safe one; A moderately dangerous one can be transformed to another moderately dangerous one or a safe one, but not to an extremely dangerous one; A safe one can only remain safe and cannot be transformed to a dangerous one.

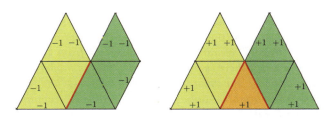

Fig. 7. Two examples of cutting pattern obstruction (definition 6).

At the end of step 1c in procedure 3.1, we detect a set of sub-components with extremely dangerous boundary assignments that cannot be transformed to less dangerous ones (figure 7). This set of unsolvable sub-components is an obstruction that keeps the algorithm from finding a valid solution.

Definition 6 (Cutting Pattern Obstruction). *A cutting pattern obstruction in an input mesh D is a maximal edge-connected component $D' \subseteq D$ consisting of a set of sub-components $\mathbb{C} = \{C_i \mid 1 \leq i \leq m\}$, such that:*

- *Every pair of sub-components C_i and C_j are triangle-disjoint to each other.*
- *Every C_i is a maximal basic type sub-component of basic types. Or equivalently, the set $\mathbb{C}$ is a minimal set of basic type sub-components.*
- *All of the sub-components in $\mathbb{C}$ have extremely dangerous boundary assignments and share the same forbidding value $a \in \{+1, -1\}$.*

3.2 Solving Moderately Dangerous Sub-components

After the previous step, if no obstruction is detected, we will reach step 3 in algorithm 3.1, which is a second place where we can resolve danger. At this point, all the extreme dangerous sub-components have been transformed, only moderately dangerous and safe ones are left. In this step, we will solve all the moderately dangerous sub-components.

A moderately dangerous sub-component, by definition, must be linear. Recall that a linear sub-component with safe boundary conditions can be completely solved using a stabbing procedure (section 2.2). As we will show in below, moderately dangerous sub-components can also be completely solved using a different stabbing procedure.

Given a moderately dangerous sub-component C, it contains a sequence of triangles $[f_0, \cdots, f_n]$. And by definition there must be two boundary edges $e_{i_1}^B$ and $e_{i_2}^B$ in two different triangles f_{i_1} and f_{i_2} ($i_1 < i_2$) that have opposite assignments, $a\left(e_{i_1}^B, f_{i_1}\right) = -a\left(e_{i_2}^B, f_{i_2}\right) = b \in \{+1, -1\}$. Taking f_{i_1} and f_{i_2} as two intermediate stops, we can partition C into three segments (in general) that can be stabbed separately (see figure 8).

- $C_{01} = [f_0, \cdots, f_{i_1}]$: stab this segment with a line from f_0 to f_{i_1} using initial assignment $-a(e_{0,1}^B, f_0)$ (i.e. opposite to the assignment for one of the boundary edges in f_0).
- $C_{12} = [f_{i_1}, \cdots, f_{i_2}]$: stab this segment with a line from f_{i_1} to f_{i_2} using initial assignment $-b$ (i.e. opposite to the assignment for boundary edge $e_{i_1}^B$ in f_{i_1}).
- $C_{2n} = [f_{i_2}, \cdots, f_n]$: stab this segment with a line from f_n to f_{i_2} using initial assignment $-a(e_{n,1}^B, f_n)$ (i.e. opposite to the assignment for one of the boundary edges in f_n).

Note that if there are more than one pair of $(e_{i_1}^B, e_{i_2}^B)$ with opposite signs, one can choose an arbitrary pair and there is no impact on the convergence and correctness of the above algorithm. As another note, if $e_{i_1}^B$ (or $e_{i_2}^B$) appears in an end triangle of C, segment C_{01} (or C_{02}) would degenerate and therefore only two segments left, the algorithm can also work. With all these considerations, after this step there is no dangerous sub-components left. All the reminders are safe ones and can be solved using algorithm 2.1.

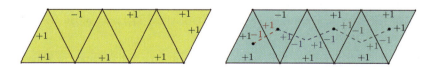

Fig. 8. A moderately dangerous sub-component must be linear; it can be completely solved via a stabbing line with three segments (separated by solid dots).

3.3 Validation and Analysis

In this part we discuss the convergence and complexity of algorithm 3.1. From the above discussion, this algorithm differs from the restricted algorithm 2.1 by several extra preprocessing steps 1, 2 and 3. As we will see in Lemma 6, this part of the algorithm always terminates.

Lemma 6. *The preprocess part (step 1, 2 and 3) of algorithm 3.1 always terminates; at termination, it either reports a cutting pattern obstruction, or outputs an unsolved sub-mesh with a safe boundary assignment.*

Proof. We check each preprocess step separately.

- In step 1: we check whether the boundary assignment is safe. If it is safe, the given mesh is left to step 4; otherwise, goes to step 2.
- In step 2: from the discussion of procedure 3.1 in section 3.1, if there is any cutting pattern obstruction in the input mesh D, it will be reported from step 1c in procedure 3.1, and the whole algorithm terminates. Otherwise, all the extremely dangerous sub-components are downgraded to moderately dangerous ones or safe ones (step 1), and then all the moderately dangerous sub-components are transformed to either moderately dangerous ones or safe ones (step 2). After this process, we have a set of basic type sub-components with completed boundary assignments, either moderately dangerous or safe.
- In step 3: based on the discussion in section 3.2, all the moderately dangerous sub-components are completely solved using a stabbing procedure, which guarantees to produce a valid set of assignments on those linear sub-components. After this process, the unsolved sub-mesh consists of a set of basic type sub-components with safe boundary conditions.

After step 3, if no obstruction is reported, the algorithm will get into step 4 with an unsolved sub-mesh with a safe boundary assignment, which has been proved in Theorem 1 to be solvable using algorithm 2.1. Therefore we can conclude:

Theorem 2. *Algorithm 3.1 always terminates; at termination, it either reports a cutting pattern obstruction or generate a valid cutting pattern for the original mesh.*

Now consider the time complexity of the algorithm. The cost of the algorithm lies in the following aspects.

- Step 1. This step only involves the boundary edges and is therefore bounded by $O(|E|)$.
- Step 2. The major cost lies in procedure 3.1. Step 1 in this procedure checks every extremely dangerous sub-component (for at most twice) and its adjacent sub-components, the cost is bounded by the total number of breaking edges in D. Similarly, step 2 in this procedure checks every moderately dangerous sub-component and its adjacent sub-components, the cost is also bounded by the total number of breaking edges in D. The total cost of this part is therefore bounded by $O(|E|)$.
- Step 3. Solving a moderately dangerous sub-component has the same complexity as that for solving a linear sub-component. Therefore this part is bounded by $O(|E|)$.

- Step 4. This step runs algorithm 2.1, which is bounded by $O(|V|+|E|+|F|)$ according to section 2.4.

In summary, the time complexity of algorithm 3.1 is $O(|V| + |E| + |F|)$, which is linear.

4 Solution for Free Boundary Conditions

Under a free boundary condition, the boundary edges of the input mesh D will not be prescribed with ± 1. Instead, these values can be set by the algorithm. It turns out that for a topological disk D, we can always generate a safe boundary assignment for D, so that the problem is converted to a restricted one (section 2). The algorithm pipeline is outlined in below.

Algorithm 4.1 (Cutting Pattern Algorithm - III)
- *Input: A triangular mesh D with a free boundary condition.*
- *Output: A valid cutting pattern for D.*
- *Procedures:*
  1. Generate a safe boundary assignment for D.
  2. Run algorithm 2.1 on D with restricted boundary condition.

Step 1 can be easily implemented. Check every triangle $f \in D$ that contains at least one boundary edge. If f contains only one boundary edge, assign it with $+1$. If f contains at least two such edges, assign $+1$ to one of them and -1 to the rest.

Obviously step 1 is guaranteed to terminate in time of $O(|F|)$. Therefore the convergence and complexity of this algorithm follows the results in section 2.4 for algorithm 2.1.

5 Conclusion and Discussion

In this paper we study the cutting pattern problem, a 2D graph labeling problem that is the major challenge of converting a prismal mesh to a tetrahedral mesh. We solve this problem for topological disks with a single boundary under three different boundary conditions. For restricted and free boundary conditions, we propose algorithms that is guaranteed to find a valid cutting pattern, therefore prove that solution always exists under these conditions. For general boundary conditions, we show that solutions exist if there is no cutting pattern obstruction; an algorithm is proposed to detect such obstructions if there is any or find a valid cutting pattern otherwise.

Note that for general boundary conditions, we only show that the absence of obstructions is a sufficient condition for the existence of solutions; but we are inclined to believe it is also necessary, and this will be an interesting topic for future exploration. Another interesting topic would be generalizing this work to other input meshes with more complicated topologies.

Acknowledgement

This work was partially supported by ONR N000140910228, NSF CCF-1081424, NSF Nets 1016829, NIH R01EB7530 and DARPA 22196330-42574-A. The first two authors were also supported by Harvard University Mathematics Department Funds. The authors would like to acknowledge and thank the anonymous reviewers for their insightful and constructive comments.

References

1. Löhner, R., Parikh, P.: Generation of three-dimensional unstructured grids by the advancing-front method. International Journal for Numerical Methods in Fluids 8(10), 1135–1149 (1988)
2. Möller, P., Hansbo, P.: On advancing front mesh generation in three dimensions. International Journal for Numerical Methods in Engineering 38(21), 3551–3569 (1995)
3. Shephard, M.S., Georges, M.K.: Automatic three-dimensional mesh generation by the finite octree technique. International Journal for Numerical Methods in Engineering 32(4), 709–749 (1991)
4. Weatherill, N.P., Hassan, O.: Efficient three-dimensional delaunay triangulation with automatic point creation and imposed boundary constraints. International Journal for Numerical Methods in Engineering 37(12), 2005–2039 (1994)
5. Edelsbrunner, H.: Geometry and topology for mesh generation. Cambridge University Press (2001)
6. Dey, T.K.: Delaunay mesh generation of three dimensional domains. Technical report, Ohio State University (2007)
7. Froncioni, A.M., Labbé, P., Garon, A., Camarero, R.: Interpolation-free spacetime remeshing for the burgers equation. Communications in Numerical Methods in Engineering 13(11), 875–884 (1997)
8. Max, N., Becker, B., Crawfis, R.: Flow volumes for interactive vector field visualization. In: Proceedings of the 4th IEEE Conference on Visualization, pp. 19–24 (1993)
9. Albertelli, G., Crawfis. R.A.: Efficient subdivision of finite-element datasets into consistent tetrahedra. In: Proceedings of the 8th IEEE Conference on Visualization, pp. 213–219 (1997)
10. Pirzadeh, S.: Advancing-layers method for generation of unstructured viscous grids. In: The 11th AIAA Applied Aerodynamics Conference, pp. 420–434 (1993)
11. Löehner, R.: Matching semi-structured and unstructured grids for navier-stokes calculations. In: the 11th AIAA Computational Fluid Dynamics Conference, pp. 555–564 (1993)
12. Dompierre, J., Labbé, P., Vallet, M.-G., Camarero, R.: How to subdivide pyramids, prisms, and hexahedra into tetrahedra. In: Proceedings of the 8th International Meshing Roundtable, pp. 195–204 (1999)
13. Gallian, J.A.: A dynamic survey of graph labeling. The Electronic Journal of Combinatorics 17 (2010)

Parametrization of Generalized Primal-Dual Triangulations

Pooran Memari[1,2], Patrick Mullen[1], and Mathieu Desbrun[1]

[1] California Institute of Technology
[2] CNRS - LTCI, Télécom ParisTech

Summary. Motivated by practical numerical issues in a number of modeling and simulation problems, we introduce the notion of a *compatible dual complex* to a primal triangulation, such that a simplicial mesh and its compatible dual complex (made out of convex cells) form what we call a *primal-dual triangulation*. Using algebraic and computational geometry results, we show that compatible dual complexes exist only for a particular type of triangulation known as weakly regular. We also demonstrate that the entire space of primal-dual triangulations, which extends the well known (weighted) Delaunay/Voronoi duality, has a convenient, geometric parametrization. We finally discuss how this parametrization may play an important role in discrete optimization problems such as optimal mesh generation, as it allows us to easily explore the space of primal-dual structures along with some important subspaces.

1 Introduction

Mesh generation traditionally aims at tiling a bounded spatial domain with simplices (triangles in 2D, tetrahedra in 3D) so that any two of these simplices are either disjoint or sharing a lower dimensional face. The resulting triangulation provides a discretization of space through both its primal (simplicial) elements *and* its dual (cell) elements. Both types of element are crucial to a variety of numerical techniques, finite element (FE) and finite volume (FV) methods being arguably the most widely used in computational science. To ensure numerical accuracy and efficiency, specific requirements on the size and shape of the primal (typically for FE) or the dual elements (typically for FV) in the mesh are often sought after.

Primal/Dual Pairs. A growing trend in numerical simulation is the simultaneous use of primal *and* dual meshes: Petrov-Galerkin finite-element/finite-volume methods (FE/FVM, [BP83, McC89, PMH07]) and exterior calculus based methods [Bos98, DKT07, GP10] use the ability to store quantities on both primal and dual elements to enforce (co)homological relationships in, e.g., Hodge theory. The choice of the dual, defined by the location of the dual vertices, is however not specified a priori. A very common dual to a triangulation in $\mathbb{R}^d$ is the cell complex which uses the circumcenters of each d-simplex as dual vertices. If the initial triangulation is Delaunay (i.e., satisfying the empty circumsphere property [Ede87]), this dual is simply the Voronoi diagram of the primal vertices, and its nice properties of non-self-intersection, convexity, and orthogonality of the primal and dual elements have led to its use in countless papers in computational sciences. The barycentric dual, for which barycenters are used instead of circumcenters, is used for certain finite-volume computations, but it fails to satisfy both the orthogonality and (more importantly) convexity properties on general triangulations.

Towards Generalized Primal/Dual Meshes. While the Delaunay-Voronoi duality [PS85, Ede87] is one of the cornerstones of meshing methods and, as such, has been extensively used in diverse fields, more general dualities are often desired. In biology for example, Voronoi cells (along with their dual triangulations) were initially identified as closely approximating a variety of monolayer cells and epithelial cells in tissue [OBSC00]; but computational biologists are now seeking generalizations of Voronoi diagrams to parameterize a larger set of convex polyhedral tilings [Mjo06] to model the development of early animal tissues and shoot meristems. The Weighted-Delaunay/Laguerre duality was also advocated recently in [MMdGD11] to help provide lower error bounds on computations. Building on Schlegel diagrams and a number of results in algebraic and computational geometry, we present an even more general primal-dual pairs of complexes that we denote as *primal-dual triangulations*.

Contributions. While Delaunay/Voronoi or Weighted-Delaunay/Laguerre duality assumes orthogonality between primal and dual complexes, we relax this requirement in our work: we investigate a general notion of dual complex to a triangulation which we call a *compatible dual complex*, where *the dual complex is only assumed to be a union of convex straight-edge polytopes*, with an adjacency graph matching the triangulation's adjacency graph— but where k-simplices of the triangulation are not necessarily orthogonal to their associated $(n-k)$-cells of the dual complex. While we will show that *any* two-dimensional triangulation admits a compatible dual complex, this property is no longer true in dimension three and above. We introduce a proper characterization of primal-dual triangulations, which results in a simple parametrization of the whole space of primal-dual triangulations. Finally, we discuss potential applications of our contributions, for instance in mesh optimization and clustering, as our parametrization allows us to easily explore a space of primal-dual structures much larger than the space of orthogonal primal-dual structures.

2 Preliminaries

Before delving into our contributions, we start by reviewing important notions related to triangulations in different fields (such as combinatorial, computational, and algebraic geometry) that we build upon and extend in subsequent sections.

2.1 Complex, Subdivision, and Triangulation

A **cell complex** in $\mathbb{R}^d$ is a set K of convex polyhedra (called cells) satisfying two conditions:

1. Every face of a cell in K is also a cell in K, and
2. If C and C' are cells in K, their intersection is either empty or a common face of both.

A **simplicial complex** is a cell complex whose cells are all simplices. The **body** $|K|$ of a complex K is the union of all its cells. When a subset P of $\mathbb{R}^d$ is the body of a complex K, then K is said to be a **subdivision** of P; if, in addition, K is a simplicial complex, then K is said to be a **triangulation** of P. For a set X of points in $\mathbb{R}^d$, a triangulation of X is a simplicial complex K for which each vertex of K is in X.

Note that in the definition of a triangulation of X, we do not require all the points of X to be used as vertices; a point $\mathbf{x_i} \in X$ is called *hidden* if it is not used in the triangulation. A triangulation of X with no hidden points is called a *full* triangulation of X.

2.2 Triangulations in $\mathbb{R}^d$ through Lifting in $\mathbb{R}^{d+1}$

Let $\mathbf{X} = \{\mathbf{x_1}, \ldots, \mathbf{x_n}\}$ be a set of points in $\mathbb{R}^d$. A simple way of constructing a triangulation of $\mathbf{X}$ is through the following *lifting procedure*: take an arbitrary function $L : \mathbf{X} \longrightarrow \mathbb{R}$ called the *lifting function*; consider the points $(\mathbf{x_i}, L(\mathbf{x_i})) \in \mathbb{R}^{d+1}$, i.e., the points of $\mathbf{X}$ *lifted* onto the graph of L; in the space $\mathbb{R}^{d+1}$, consider Conv(L) the convex hull of vertical rays $\{(x_i, l)| \ l \geq L(\mathbf{x_i}), l \in \mathbb{R}, x_i \in X\}$; the bounded faces of Conv(L), i.e. faces which do not contain vertical half lines, form the **lower envelope** of the lifting L. If the function L is generic (see [GKZ94] Chap. 7), the orthogonal projection (onto the first d coordinates) of the lower envelope of L produces a triangulation of $\mathbf{X}$.

It is clear that the above lifting procedure may produce triangulations for which not all points of $\mathbf{X}$ are vertices. A triangulation of a set $\mathbf{X}$ of points obtained through lifting is full (i.e., has no hidden points) if and only if all the points $(\mathbf{x_i}, L(\mathbf{x_i}))$ lie on the **lower envelope of** L (or, in other words, if function L can be extended, through linear interpolation in the triangles, to a **convex** piecewise-linear function).

2.3 Regular Triangulations and Subdivisions

Let $\mathbf{X}$ be a finite set of points in $\mathbb{R}^d$. A triangulation obtained by orthogonally projecting the lower envelope of a lifting of $\mathbf{X}$ in $\mathbb{R}^{d+1}$ onto the first d coordinates is called a *regular triangulation* ([Zie95], Definition 5.3).

More generally, a subdivision T of a polytope $P \subset \mathbb{R}^d$ is regular if it arises from a polytope $Q \subset \mathbb{R}^{d+1}$ in the following way:

- The polytope P is the image $\pi(Q) = P$ of the polytope Q via the projection that deletes the last coordinate.
- The complex T is the projection under π of all the lower faces of Q. We call F a lower face of Q if for every $\mathbf{x} \in F$ and a real number $\lambda > 0$, $\mathbf{x} - \lambda \mathbf{e}_{d+1} \notin Q$. Informally, they are the faces that you can see from P if you "*look up*" at Q.

Regular triangulations have appeared in different mathematical contexts and are known under different names, such as weighted Delaunay triangulations (see next section) or coherent triangulations [GKZ94]. While every point set admits regular triangulations (see [DLRS], Proposition 2.2.4.), not every triangulation is regular; Figure 4 (left) illustrates a classical non-regular triangulation.

2.4 Schlegel Diagrams

Schlegel diagrams are a related mathematical notion based on a construction very similar to the lifting procedure defined above. They have been proven an important tool for studying combinatorial and topological properties of polytopes, as well as for visualizing four-dimensional polytopes [Ban90].

A Schlegel diagram is the (*perspective*) projection of a polytope from $\mathbb{R}^{d+1}$ to $\mathbb{R}^d$ through a point p outside of the polytope, above the center of a facet f. All vertices and edges of the polytope are projected onto the hyperplane of that facet in the following way: For any vertex of the polytope, the line from p through the vertex meets the hyperplane of f at the image of the vertex. If two vertices are connected by an edge in the polytope, then the image of the edge is the segment joining the images of the two vertices.

If the polytope is convex, there exists a point near the facet f which maps all other facets inside f, so no edges need to cross in the projection. In other words, the Schlegel diagram of a convex polytope is the *projection* of the polytope's skeleton on one of its faces (the nodes corresponding to the vertices which do not belong to that face must lie inside the face). In this case the resulting entity is a polytopal subdivision of the facet in $\mathbb{R}^d$ that is combinatorially equivalent to the original polytope, see Figure 1.

Intuitively, by sending the base point of a Schlegel diagram at infinity we obtain a regular subdivision (see [Zie95] Prop. 5.9, or [Tho06]). We will use this equivalence between regular subdivisions and Schlegel diagrams (illustrated in Figure 2) later in this paper.

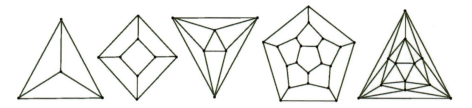

Fig. 1. The Schlegel diagrams of the five polyhedra (tetrahedron, cube, octahedron, dodecahedron and icosahedron).

Fig. 2. By sending the base point of a Schlegel diagram at infinity we obtain a regular subdivision (after [Sch03]).

Proposition 1 ([Zie95]). *If T is a Schlegel diagram, T is a regular subdivision of $|T|$.*

2.5 Weighted Delaunay Triangulations

A special choice for the lifting function produces the well-known and widely-used Delaunay triangulation (see [Raj94, Mus97] for their properties, and [PS85] for numerous applications). Indeed, let X be a set of points in $\mathbb{R}^d$. Consider the *lifting* of the points in X onto the surface of the paraboloid $h(\mathbf{x}) = \|\mathbf{x}\|^2$ in $\mathbb{R}^{d+1}$; i.e., each $\mathbf{x_i} = (a_1, \ldots, a_d) \in X$ gets mapped to $(\mathbf{x_i}, h_i) \in \mathbb{R}^{d+1}$ with $h_i = \|\mathbf{x_i}\|^2 = a_1^2 + \cdots + a_d^2$. Then the orthogonal projection of the lower envelope of this lifting produces a (full) triangulation coinciding with the **Delaunay triangulation** of $\mathbf{X}$.

A regular triangulation can now be seen as a generalization of the Delaunay triangulation as follows. We first define a **weighted point set** as a set $(\mathbf{X}, W) = (\mathbf{x_1}, w_1), \ldots, (\mathbf{x_n}, w_n)$, where $\mathbf{X}$ is a set of points in $\mathbb{R}^d$, and $\{w_i\}_{i \in [1,\ldots,n]}$ are real numbers called weights. The **weighted Delaunay triangulation** of $(\mathbf{X}, W)$ is then the triangulation of $\mathbf{X}$ obtained by projecting the lower envelope of the points $(\mathbf{x_i}, \|\mathbf{x_i}\|^2 - w_i) \in \mathbb{R}^{d+1}$. Note that a weighted Delaunay triangulation can now have hidden points.

Notice also that given a lifting function L and its values $l_i = L(\mathbf{x_i})$ at the points of X, one can always define weights to be the difference between the paraboloid and the function L, $w_i = \|\mathbf{x_i}\|^2 - l_i$. We conclude that the notions of regular triangulations and weighted Delaunay triangulations are *equivalent*.

2.6 Generalized Voronoi Diagrams (Power Diagrams)

Delaunay triangulations (resp., weighted Delaunay triangulations) can also be obtained (or defined) from their dual *Voronoi diagrams* (resp., *power diagrams*). Let $(\mathbf{X}, W) = \{(\mathbf{x_i}, w_i)\}_{i \in I}$ be a weighted point set in $\mathbb{R}^d$. The power of a point $\mathbf{x} \in \mathbb{R}^d$ with respect to a weighted point $(\mathbf{x_i}, w_i)$ (sometimes referred to as the Laguerre distance) is defined as $d^2(\mathbf{x}, \mathbf{x_i}) - w_i$, where $d(.,.)$ stands for the Euclidean distance. Using this power definition, to each x_i we associate its weighted Voronoi region $V(\mathbf{x_i}, w_i) = \{\mathbf{x} \in \mathbb{R}^d|\ d^2(\mathbf{x}, \mathbf{x_i}) - w_i \leq d^2(\mathbf{x}, \mathbf{x_j}) - w_j, \forall j\}$. The power diagram of $(\mathbf{X}, W)$ is the cell complex whose cells are the weighted Voronoi regions.

Note that when the weights are all equal, the power diagram coincides with the Euclidean Voronoi diagram of $\mathbf{X}$. Power diagrams are well known to be dual to weighted Delaunay triangulations, as we review next.

2.7 Power Diagram vs. Weighted Delaunay Triangulation

The **dual** of the power diagram of $(\mathbf{X}, W)$ is the weighted Delaunay triangulation of $(\mathbf{X}, W)$. This triangulation contains a k-simplex with vertices $\mathbf{x_{a_0}}, \mathbf{x_{a_1}}, \ldots, \mathbf{x_{a_k}}$ in $\mathbf{X}$ if and only if $V(x_{a_0}, w_{a_0}) \cap V(x_{a_1}, w_{a_1}) \cap \cdots \cap V(x_{a_k}, w_{a_k}) \neq \emptyset, \forall k \geq 0$. While many other generalization of Voronoi diagrams exist, they do not form straight-edge and convex polytopes, and are thus not relevant here.

Geometric Property. We finally review an interesting geometric property of Voronoi and power diagrams, as it will become useful later on. Consider the affine functions $\sigma_i(x) = -2\mathbf{x_i} \cdot \mathbf{x} + \|\mathbf{x_i}\|^2$ for $i = 1, \ldots, n$: their graphs are obviously hyperplanes of $\mathbb{R}^{d+1}$ that are tangent to the paraboloid $h(\mathbf{x}) = \|\mathbf{x}\|^2$ at point $\mathbf{x_i}$. Let us call these hyperplanes H_i, and let H_i^+ denote the half-space lying above H_i. The *minimization diagram* of the σ_i is obtained by projecting the lower envelop of $H_1^+ \cap \cdots \cap H_n^+$ orthogonally onto $\mathbb{R}^d$. However, for any $\mathbf{x}$, $\operatorname{argmin}_i \|\mathbf{x} - \mathbf{x_i}\|^2 = \operatorname{argmin}_i(-2\mathbf{x_i} \cdot \mathbf{x} + \|\mathbf{x_i}\|^2)$. Thus, one concludes that the Euclidean Voronoi diagram of $\mathbf{X}$ is the orthogonal projection of the skeleton of the lower envelop

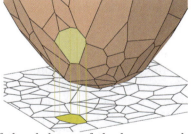

of $H_1^+ \cap \cdots \cap H_n^+$—see inset and [ES86]. More generally, the power diagram of $(\mathbf{X}, W)$ is the orthogonal projection of the skeleton of the lower envelop of the intersection of half-spaces lying above hyperplanes $(-2\mathbf{x_i} \cdot \mathbf{x} + \|\mathbf{x_i}\|^2 - w_i)$: such a hyperplane is indeed the tangent plane to the paraboloid $h(\mathbf{x}) = \|\mathbf{x}\|^2$ at point $\mathbf{x_i}$ translated down by w_i. This geometric fact can be interpreted as a hyperplane assignment to each (weighted) Voronoi region, and it will be used in the forthcoming proof of our Theorem 1.

3 Compatible Dual Complexes of Triangulations

We now show that the notion of mesh duality can be extended so that the dual complex is defined geometrically, and independently from the triangulation—while the combinatorical compatibility between the triangulation and its dual is maintained.

Definition 1 (Simple Cell Complex). *A cell complex K in $\mathbb{R}^d$ is called **simple** if every vertex of K is incident to $d+1$ edges. K is called labeled if every d-dimensional cell of K is assigned a unique label; in this case, we write $K = \{C_1, \ldots, C_n\}$, where n is the number of d-dimensional cells of K, and C_i is the i-th d-dimensional cell.*

Definition 2 (Compatible Dual Complex). *Let T be a triangulation of a set $\mathbf{X} = \{\mathbf{x_1}, \ldots, \mathbf{x_n}\}$ of points in $\mathbb{R}^d$, and $K = \{C_{i_1}, \ldots, C_{i_n}\}$ be a labeled simple cell complex, i.e. there is a one-to-one correspondence between x_p and C_{i_p}. K is called a **compatible dual complex** of T if, for every pair of points $\mathbf{x_p}$ and $\mathbf{x_q}$ that are connected in T, C_{i_p} and C_{i_q} share a face.*

This compatibility between K and T is purely combinatorial, i.e., it simply states that the connectivity between points induced by K coincides with the one induced by T. Notice that the cell C_{i_p} associated to the point $\mathbf{x_p}$, does not necessarily contain $\mathbf{x_p}$ in its interior. Moreover, the edge $[\mathbf{x_p}, \mathbf{x_q}]$ and its *dual* $C_{i_p} \cap C_{i_q}$ are not necessarily *orthogonal* to each other, unlike most conventional geometric dual structures. Consequently, we can generalize the notion of mesh duality through the following definition:

Definition 3 (Primal-Dual Triangulation (PDT)). *A pair (T, K) is said to form a d-dimensional **primal-dual triangulation** if T is a triangulation in $\mathbb{R}^d$ and K is a compatible dual complex of T. If every edge $[\mathbf{x_p}, \mathbf{x_q}]$ and its dual $C_{i_p} \cap C_{i_q}$ are orthogonal to each other, the pair (T, K) is said to form an **orthogonal primal-dual triangulation**.*

3.1 Characterization of Primal-Dual Triangulations

An immediate question is whether any triangulation can be part of a PDT. We first characterize the triangulations that admit a compatible dual complex through the following two definitions:

Definition 4 (Combinatorial Equivalence). *Two triangulations T and T' are combinatorially equivalent if there exists a labeling which associates to each point $\mathbf{x_i}$ in T a point $\mathbf{x'_i}$ in T' so that the connectivity between $\mathbf{x_i}$'s induced by T matches the connectivity between the $\mathbf{x'_i}$'s induced by T'.*

Definition 5 (Combinatorially Regular Triangulations (CRT)). *A triangulation T of a d-dimensional point set X is called a **combinatorially regular triangulation** if there exists a d-dimensional point set X' admitting a regular triangulation T' such that T and T' are combinatorially equivalent.*

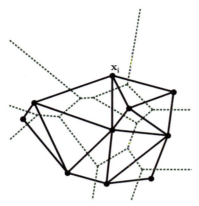
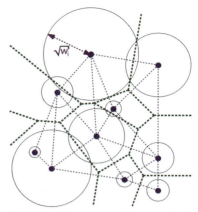

(a) A triangulation T (solid line) and a compatible (non-orthogonal) dual complex K (dashed line).

(b) K is the power diagram dual to a regular triangulation T' (thin dashed line).

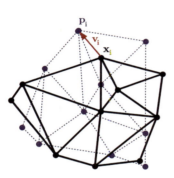
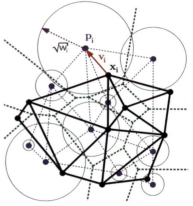

(c) The triangulation T is combinatorially equivalent to the regular triangulation T'.

(d) PDT (T, K) is parameterized through $\mathbf{x}_i$ (points), $\mathbf{v}_i$ (displacements), and w_i (weights).

Fig. 3. Primal-Dual Triangulation, with its primal triangulation, dual complex, and combinatorially equivalent regular triangulation separately displayed for clarity.

Remark: these CRT triangulations have been introduced in [Lee91] under the name of *weakly* regular triangulations, since a displacement of their vertices suffices to make them regular. Figure 4 (after [Lee91]) shows an example of a combinatorially regular triangulation which is not, itself, regular.

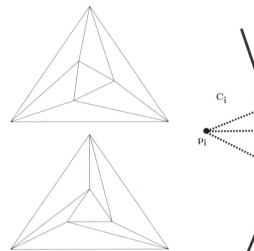 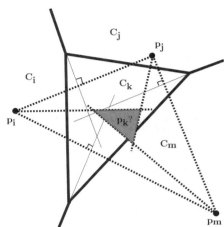

(a) A regular triangulation (top), once deformed (bottom), becomes a combinatorially regular triangulation which is *not*, itself, regular.

(b) A cell complex (solid line) that does not admit a primal orthogonal triangulation (after [Aur87a]).

Fig. 4. Classical Examples, showing the existence of non-regular triangulations (a) and of dual complexes without associated primal triangulations (b).

Existence of PDTs in 2D. The 2D case is rather simple, due to this result:

Proposition 2 ([Lee91]). *Any 2-dimensional triangulation is combinatorially regular.*

Proof. This is a straightforward corollary of a classical theorem of Steinitz [Ste22] which implies that for every complex in the plane whose edge graph is three-connected, there exists a convex 3-polyhedron with isomorphic boundary. Since any 2-dimensional triangulation T is trivially edge three-connected, there exists a convex 3-polyhedron P with isomorphic boundary. Therefore, the orthogonal projection of the boundary of P onto the plane is a regular triangulation which is isomorphic to T. □

Therefore, every 2D triangulation T can be part of a PDT pair (T, K).

Existence of PDTs in Higher Dimensions. In higher dimensions (three and above), however, the situation is rather different: the set of regular triangulations does not contain all possible triangulations between points as stated in the next proposition.

Proposition 3 ([Grü03]). *For $d \geq 3$, there exist d-D triangulations which are not combinatorially regular.*

This is equivalent to the fact that there are simplicial diagrams that are not combinatorially equivalent to any Schlegel diagram. Chronologically, the first example was found by Grünbaum in 1965 (see [Grü03] pages 219-224): he presented a simplicial 3-diagram with 7 vertices which is not combinatorially equivalent to any Schlegel diagram of 4-polytopes. Before this result, there had been several attempts to prove that any simplicial diagram is combinatorially equivalent to a Schlegel diagram (also called polytopal or polytopical)[1], see also [CH80]. The simplest non-combinatorially regular examples are the Brucker sphere and the Barnette sphere (short proofs are available in [Ewa96]Sec III.4).

3.2 PDT=CRT

We now show that combinatorially regular triangulations are *the only ones that admit compatible dual complexes*. The proof revolves around a theorem due to Aurenhammer:

Every simple cell complex in $\mathbb{R}^d$, $d \geq 3$, is a Schlegel diagram.[2]

This theorem was proved in [Aur87a] through an iterative construction which is valid in any dimension $d \geq 3$. We use this theorem to prove the following theorem which surprisingly implies that in higher dimensions there are triangulations that do *not* admit a dual complex:

Theorem 1 (PDT Characterization). *A d-dimensional triangulation T admits a compatible (not necessarily orthogonal) dual complex if and only if T is combinatorially regular.*

Proof. Let T be a triangulation combinatorially equivalent to a regular triangulation T'. The conventional dual of T', a power diagram, is thus a compatible dual of T as well.

Now suppose that T is a triangulation with a compatible dual complex K. According to Aurenhammer's theorem, K is a Schlegel diagram, *i.e.*, to each cell C_j of K we can assign a hyperplane H_j in $\mathbb{R}^{d+1}$ so that the lower envelop of these hyperplanes projects orthogonally onto K (see also [Sha93]). On the other hand, the tangent planes of the $(d+1)$-dimensional paraboloid *span* the whole space of non vertical hyperplane directions in $\mathbb{R}^{d+1}$. Therefore for each j, there exists a point $\mathbf{p_j} \in \mathbb{R}^d$ whose tangent plane to the paraboloid in $\mathbb{R}^{d+1}$ is parallel to H_j. In that case, one can easily conclude that K is the power diagram of $(\mathbf{p_j}, w_j), 1 \leq j \leq n$, where the weight w_j depends on the distance between the two parallel hyperplanes mentioned

---

[1] These terms are commonly used to study the combinatorics of polytopes and diagrams, often independently from their embedding. Therefore, the reader should be aware of some ambiguities in the definition of these notions in the literature, between being a Schlegel diagram or being combinatorially equivalent to it.

[2] Note that Aurenhammer employs the equivalent name of polytopical diagram.

before. Now consider the weighted Delaunay triangulation T' of weighted points $(\mathbf{p_j}, w_j), 1 \leq j \leq n$. Since K is a compatible dual complex for both T and T', we conclude that T and T' are combinatorially equivalent. Since T' is regular, T is a combinatorially regular triangulation. □

Existence of PDT for Given Dual Complexes. We can also discuss whether a dual complex can be part of a PDT pair to further characterize the space of primal-dual pairs we will be parameterizing. In dimension $d \geq 3$, Aurenhammer's theorem directly implies that every simple cell complex is dual to a regular triangulation. But this same theorem is not true in 2D, and Figure 4 (right) shows a counter example presented by Aurenhammer [Aur87a]—proving that not every 2D simple cell complex is the dual of a regular triangulation; thus PDTs do not capture all 2D convex cell complexes.

4 Parameterizing Primal-Dual Triangulations

We have established that primal-dual triangulations cover all dual complexes in $d \geq 3$ but only those which come from a regular triangulation for $d = 2$; they also cover all 2D triangulations, but only triangulations which admit a dual in $d \geq 3$. We now focus on parameterizing the whole space of primal-dual triangulations with n points in $\mathbb{R}^d$ by simply adding parameters at the points. We then explore a geometric interpretation of this intrinsic parametrization as well as its properties.

The proof of Theorem 1 leads us very naturally to a parametrization of all the triangulations that admit a compatible dual complex:

Definition 6. *A **parameterized primal-dual triangulation** is a primal-dual triangulation parameterized by a set of triplets $(\mathbf{x_i}, w_i, \mathbf{v_i})$, where $\mathbf{x_i}$ is the* position *in $\mathbb{R}^d$ of the ith node, w_i is a real number called the* weight *of x_i, and $\mathbf{v_i}$ is a d-dimensional vector called the* displacement vector *of $\mathbf{x_i}$. The triangulation associated with the triplets $(\mathbf{x_i}, w_i, \mathbf{v_i})$ is defined such that its dual complex K is the* power diagram *of weighted points $(\mathbf{p_i}, w_i)$, where $\mathbf{p_i} = \mathbf{x_i} + \mathbf{v_i}$.*

The dual complex K can be seen as the generalized Voronoi diagram of the $\mathbf{x_i}$'s for the distance $d(\mathbf{x}, \mathbf{x_i}) = \|\mathbf{x} - \mathbf{x_i} - \mathbf{v_i}\|^2 - w_i$. When the vectors $\mathbf{v_i}$ are all null, the parameterized primal-dual triangulation T is regular, thus perpendicular to its dual K, and the pair (T, K) forms an orthogonal primal-dual triangulation. This proves that weighted Delaunay triangulations are sufficient to parameterize the set of all orthogonal primal-dual triangulations (see also [Gli05]). The displacement vectors extend the type of triangulations and duals we can parameterize.

Geometric Interpretation. This parametrization can be seen as a very natural extension of the geometric interpretation mentioned in Section 2.6. Indeed, PDTs in $\mathbb{R}^d$ are simply *defined by assigning one hyperplane H_i in $\mathbb{R}^{d+1}$ per* *point* x_i: the orthogonal projection of the lower envelop of these hyperplanes $\{H_i\}_{i=1,\ldots,n}$ will form the dual complex of the PDT, inducing the primal triangulation. Weights and displacements vectors serve as a means to encode the choice of hyperplanes.

We now provide some geometric properties of the primal-dual triangulation parameterized by a set of triplets $(\mathbf{x_i}, w_i, \mathbf{v_i})$.

Lemma 1 (Characterization of PDT triplets). *Let (T, K) be a PDT parameterized by $(\mathbf{x_i}, w_i, \mathbf{v_i})$. The set of triplets $\{(\mathbf{x'_i}, w'_i, \mathbf{v'_i})\}$ parameterizes the same primal-dual triangulation (T, K) if and only if $\mathbf{x'_i} = \mathbf{x_i}$ and there exist a constant $\alpha \in \mathbb{R}$, a positive constant $\beta \in \mathbb{R}^+$, and a vector $\mathbf{v}$ such that $w'_i = w_i - 2\mathbf{v} \cdot (\mathbf{x_i} + \mathbf{v_i}) + \alpha$ and $\mathbf{v'_i} = \beta(\mathbf{x_i} + \mathbf{v_i}) - \mathbf{x_i} + \mathbf{v}$.*

Proof. Let us denote by R and R' the weighted Delaunay triangulations of weighted points $(\mathbf{x_i} + \mathbf{v_i}, w_i)$ and $(\mathbf{x_i} + \mathbf{v'_i}, w'_i)$ respectively. K is then the power diagram of both weighted point sets $(\mathbf{x_i} + \mathbf{v_i}, w_i)$ and $(\mathbf{x_i} + \mathbf{v'_i}, w'_i)$, and is dual to both R and R'. However, two regular triangulations have the same power diagram only if they are homothetic (see, for instance, [Aur87b] for a proof). Hence, there exist a positive constant $\beta > 0$, and a vector $\mathbf{v}$ such that $\mathbf{x_i} + \mathbf{v'_i} = \beta(\mathbf{x_i} + \mathbf{v_i}) + \mathbf{v}$. On the other hand, it is easy to show that the power diagram of the set of weighted points $(\mathbf{p_i}, w_i)$, translated by a vector $-\mathbf{v}$, coincides with the power diagram of $(\mathbf{p_i}, w_i - 2\mathbf{v} \cdot \mathbf{p_i} + \alpha)$. Also, such a translated power diagram can be changed back to the original diagram by simply adding $\mathbf{v}$ to each displacement vector $\mathbf{v_i}$. This implies that $(\mathbf{x_i}, w_i - 2\mathbf{v} \cdot (\mathbf{x_i} + \mathbf{v_i}) + \alpha, (\beta-1)\mathbf{x_i} + \beta\mathbf{v_i} + \mathbf{v})$ and $(\mathbf{x_i}, w_i, \mathbf{v_i})$ parameterize the same PDT. $\square$

This lemma characterizes the classes of *equivalent* triplets parameterizing the same PDT. Using this characterization, we now provide a set of constraints for the parameters of the triplets which allow us to avoid redundancy between equivalent triplets and define an efficient parametrization for the space of primal-dual triangulations:

Theorem 2 (PDT Parametrization). *There is a bijection between all primal-dual triangulations in $\mathbb{R}^d$ and sets of triplets $(\mathbf{x_i}, w_i, \mathbf{v_i})$, $1 \leq i \leq n$, where $\mathbf{x_i}, \mathbf{v_i} \in \mathbb{R}^d$, $w_i \in \mathbb{R}$ with $\sum_i w_i = 0$, $\sum_i \mathbf{v_i} = 0$, and $\sum_i \|\mathbf{x_i} + \mathbf{v_i}\|^2 = \sum_i \|\mathbf{x_i}\|^2$.*

Proof. The proof is provided in two parts: i) For any triplet $(\mathbf{x_i}, w_i, \mathbf{v_i})$, there exists an equivalent triplet $(\mathbf{x'_i}, w'_i, \mathbf{v'_i})$ which fulfills the conditions of the theorem. Indeed, the choice of:

$$\beta = \sqrt{\frac{n \sum_i \|\mathbf{x_i}\|^2 - \|\sum_i \mathbf{x_i}\|^2}{n \sum_i \|\mathbf{x_i} + \mathbf{v_i}\|^2 - \|\sum_i (\mathbf{x_i} + \mathbf{v_i})\|^2}},$$

$$\mathbf{v} = \frac{1-\beta}{n} \sum_i \mathbf{x_i} - \frac{\beta}{n} \sum_i \mathbf{v_i}, \quad \alpha = \frac{2}{n} \mathbf{v} \cdot \sum_i (\mathbf{x_i} + \mathbf{v_i}) - \frac{\sum_i w_i}{n}$$

in the characterization of Lemma 1 gives the desired triplet. ii) Suppose that both (x_i, w_i, v_i) and $(\mathbf{x_i}, w_i - 2\mathbf{v} \cdot \mathbf{x_i} - 2\mathbf{v}.\mathbf{v_i} + \alpha, (\beta - 1)\mathbf{x_i} + \beta\mathbf{v_i} + \mathbf{v})$ fulfill the conditions, for some constants α and $\beta > 0$, and a constant vector $\mathbf{v}$. A direct computation implies $\alpha = 0$, $\mathbf{v} = 0$ and $\beta = 1$. Therefore for each class of equivalent triplets, there is a single triplet which verifies the conditions of the parametrization. □

Remark: using this parametrization, the particular case of Delaunay / Voronoi PDT of a set of points $\{\mathbf{x_i}\}_{i=1..n}$ is naturally parameterized by triplets $(\mathbf{x_i}, 0, 0)$. Similarly, the weighted Delaunay / Power PDT is parameterized by triplets $(\mathbf{x_i}, w_i, 0)$. Note also that the condition $\sum_i w_i = 0$ may be replaced by $\min_i w_i = 0$, by simply subtracting the minimum weight from all the weights of triplets. This new condition implies that all the weights are positive, which may be useful in some applications.

5 Conclusions

In this paper, we introduced the notion of compatible dual complex for a given triangulation in $\mathbb{R}^d$, and discussed the conditions under which an arbitrary triangulation admits a compatible, possibly non-orthogonal dual complex (and vice-versa). Note that our only assumption on the dual is that it is made out of *convex polytopes*, thus reducing the space of possible primal-dual pairs to a computationally-convenient subset for which basis functions and positive barycentric coordinates are easily defined. We also pointed out a link to a previously-introduced notion of weakly regular triangulation by Lee in the nineties, and that there are triangulations that do *not* admit a dual complex. We derived a natural parametrization of all non-orthogonal primal-dual structures in $\mathbb{R}^d$ by means of plane assignments in $\mathbb{R}^{d+1}$.

Besides the theoretical interest of these new primal-dual structures, we anticipate numerous applications. We believe that our results can benefit mesh optimization algorithms as we provide a particularly convenient way to explore a large space of primal-dual structures. We recently provided a first step in this direction by designing pairs of primal-dual structures that optimize accuracy bounds on differential operators using our parametrization [MMdGD11], thus extending variational approaches designed to improve *either* primal (Optimal Delaunay Triangulations [ACSYD]) *or* dual (Centroidal Voronoi Tessellations [LWL+09]) structures. To some extent, our approach can even help to deal with situations where the primal triangulation is given and cannot safely be altered: for instance, moving vertices and/or changing the connectivity of a triangle mesh in $\mathbb{R}^3$ is potentially harmful, as it affects the surface shape. Still, the ability to optimize weights to drive the selection of the dual mesh is very useful. We can easily optimize primal-dual triangulations (meshes) by minimizing a functional (energy) with respect to weights. The connectivity is kept intact, regardless of the weights—only the position and shape of the compatible dual is optimized. Our 2D and 3D experiments [MMdGD11] show that only optimizing the weights is particularly simple and beneficial on a number of meshes. Fig. 5 depicts a triangle mesh of a hand and its intrinsic dual before and after weight optimization, showing a drastic reduction in the number of negative dual edges—thus providing a practical alternative to the use of intrinsic Delaunay meshes advocated in [FSBS06].

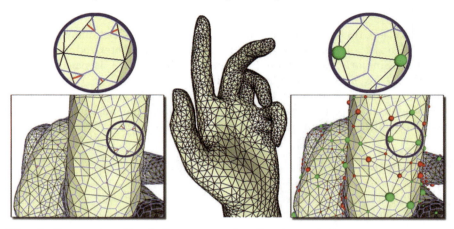

Fig. 5. Improving Dual Structure of a Surface Mesh: For a given triangulation (center) there are several triangles whose circumcenter is far outside the triangle (left, lines in red). By optimizing only the weights the new dual vertices are better placed inside the unchanged triangles (right) while keeping primal/dual orthogonality.

As another illustrative example, Fig. 6 shows that even an optimized Delaunay triangulation (ODT mesh) with exceptionally high-quality tetrahedra [TWAD09] can be made significantly better centered (i.e., with dual vertices closer to the inside of their associated primal simplex) using a simple weight optimization. Note also that in this example the number of tetrahedra with a dual vertex outside of the primal tet dropped from 17041 on the ODT mesh to 5489 on the optimized mesh—a two third reduction of *outcentered* tetrahedra. While all these results only explored the orthogonal primal-dual triangulations, we expect that better results would be obtained with our parametrization if we relax the orthogonality constraint in favor of arbitrary convex dual cells. The use of vertex weights to ensure boundary and feature protection as proposed in [CDL08] would be interesting as well in this context. In addition to applications in mesh optimization, as we mentioned in the introduction, modeling (as in computational biology) that uses *convex space tilings* could directly use our parametrization of PDTs. Clustering techniques based on k-means may also benefit from parameterizing clusters by more than just centers, as weights and vectors add more flexibility to the segmentation of input data.

Future Work. Enforcing proper embedding of a PDT can be crucial in some applications. While sufficient conditions on the weights $\{w_i\}$ and displacements

Fig. 6. Improving Dual Structures of 3D Meshes: The dual of a high-quality ODT mesh of the Bimba con Nastrino (left, cross-section; 195K tets, 36K vertices) can be optimized by improving minimal dual edge length and self-centeredness [MMdGD11] (middle; weights are displayed according to sign (red/green) and magnitude (radius)). When we single out the tetrahedra with a distance between weighted circumcenter and barycenter greater than 0.5% of the bounding box, one can see the optimized mesh (right) is significantly better than the original ODT (right, inset), even if the primal triangulations are exactly matching.

$\{\mathbf{v_i}\}$ are easy to derive (one can, for instance, limit each $\mathbf{v_i}$ to stay within the ball centered on $\mathbf{x_i}$ and of radius $\sqrt{w_i}$), it could be beneficial to have less constraining conditions. Other necessary and sufficient conditions to enforce, for instance, that primal vertices are placed within their associated dual cells (or vice-versa) could be also useful. Finally, we wish to study possible links between primal-dual triangulations and an algebraic-geometric construction due to Gelfand, Kapranov, and Zelevinsky [GKZ94], where each (embedded) triangulation of a set of n points $\mathbf{x_1}, \ldots, \mathbf{x_n}$ in $\mathbb{R}^d$ is associated with a point in $\mathbb{R}^n$. The i-th coordinate of this new point is the total volume of all simplices incident to x_i in the triangulation, and the convex hull of these new points is called the secondary polytope of the point set. This hull has dimension $n - d - 1$, and interestingly, its vertices correspond exactly to the regular triangulations of the point set. This construction helps translate geometric questions into combinatorial questions about polyhedral fans. Using a similar construction for combinatorially regular triangulations could extend some existing results for regular triangulations—in particular, algorithmic and enumerative questions. For instance, the decision of whether a given triangulation is regular is easily reduced to linear programming; however, determining whether a given triangulation is combinatorially regular seems to be very hard in general: a result by Richter-Gebert ([RG96] corollary 10.4.1) states that already in dimension 4 there are infinitely many minor-minimal non-combinatorially regular diagrams. In other words, combinatorial types of d-diagrams with n vertices may not be characterized by excluding a finite set of forbidden minors. We plan to investigate a weaker variant of this problem restricted to a subspace of combinatorially regular triangulations by focusing on rational embedding of diagrams that are easier to study for enumeration purposes. This natural restriction of the space of diagrams links our problem to the topic of tropical geometry, whose objects are rational polyhedral complexes in $\mathbb{R}^d$.

Acknowledgment. The authors wish to thank Fernando de Goes for his help on this project. Partial funding was generously provided by the National Science Foundation through grants CCF-1011944, CCF-0811373, and CMMI-0757106.

References

[ACSYD] Alliez, P., Cohen-Steiner, D., Yvinec, M., Desbrun, M.: Variational tetrahedral meshing. In: ACM SIGGRAPH 2005 Courses (2005)

[Aur87a] Aurenhammer, F.: A criterion for the affine equivalence of cell complexes in $\mathbb{R}^d$ and convex polyhedra in $\mathbb{R}^d + 1$. Discrete and Computational Geometry 2(1), 49–64 (1987)

[Aur87b] Aurenhammer, F.: Recognising polytopical cell complexes and constructing projection polyhedra. Journal of Symbolic Computation 3(3), 249–255 (1987)

[Ban90] Banchoff, T.F.: Beyond the Third Dimension: Geometry, Computer Graphics, and Higher Dimensions. W.H. Freeman (1990)

[Bos98] Bossavit, A.: Computational Electromagnetism. Academic Press, Boston (1998)

[BP83] Baligaa, B.R., Patankarb, S.V.: A Control Volume Finite-Element Method For Two-Dimensional Fluid Flow And Heat Transfer. Numerical Heat Transfer 6, 245–261 (1983)

[CDL08] Cheng, S.-W., Dey, T.K., Levine, J.: Theory of a practical Delaunay meshing algorithm for a large class of domains. In: Bhattacharya, B., et al. (eds.) Algorithms, Architecture and Information Systems Security, World Scientific Review, vol. 3, pp. 17–41 (2008)

[CH80] Connelly, R., Henderson, D.W.: A convex 3-complex not simplicially isomorphic to a strictly convex complex. In: Mathematical Proceedings of the Cambridge Philosophical Society, vol. 88, pp. 299–306. Cambridge Univ. Press (1980)

[DKT07] Desbrun, M., Kanso, E., Tong, Y.: Discrete differential forms for computational modeling. In: Bobenko, A., Schröder, P. (eds.) Discrete Differential Geometry. Springer, Heidelberg (2007)

[DLRS] De Loera, J.A., Rambau, J., Santos, F.: Triangulations: Applications, Structures and Algorithms. In: Algorithms and Computation in Mathematics. Springer, Heidelberg (to appear)

[Ede87] Edelsbrunner, H.: Algorithms in Combinatorial Geometry. Springer, Heidelberg (1987)

[ES86] Edelsbrunner, H., Seidel, R.: Voronoi diagrams and arrangements. Discrete Comput. Geom. 1, 25–44 (1986)

[Ewa96] Ewald, G.: Combinatorial convexity and algebraic geometry. Springer, Heidelberg (1996)

[FSBS06] Fisher, M., Springborn, B., Bobenko, A.I., Schröder, P.: An algorithm for the construction of intrinsic delaunay triangulations with applications to digital geometry processing. In: ACM SIGGRAPH Courses, pp. 69–74 (2006)

[GKZ94]   Gelfand, I.M., Kapranov, M.M., Zelevinsky, A.V.: Discriminants, resultants, and multidimensional determinants. Springer, Heidelberg (1994)
[Gli05]   Glickenstein, D.: Geometric triangulations and discrete Laplacians on manifolds. Arxiv preprint math/0508188 (2005)
[GP10]   Grady, L.J., Polimeni, J.R.: Discrete Calculus: Applied Analysis on Graphs for Computational Science. Springer, Heidelberg (2010)
[Grü03]   Grünbaum, B.: Convex polytopes. Springer, Heidelberg (2003)
[Lee91]   Lee, C.W.: Regular triangulations of convex polytopes. In: Gritzmann, P., Sturmfels, B. (eds.) Applied Geometry and Discrete Mathematics–The Victor Klee Festschrift. DIMACS Series in Discrete Mathematics and Theoretical Computer Science, Amer. Math. Soc., vol. 4, pp. 443–456 (1991)
[LWL+09]   Liu, Y., Wang, W., Lévy, B., Sun, F., Yan, D.M., Lu, L., Yang, C.: On centroidal voronoi tessellation energy smoothness and fast computation. ACM Transactions on Graphics (TOG) 28(4), 1–17 (2009)
[McC89]   McCormick, S.F.: Multilevel Adaptive Methods for Partial Differential Equations. SIAM (1989)
[Mjo06]   Mjolsness, E.: The Growth and Development of Some Recent Plant Models: A Viewpoint. Journal of Plant Growth Regulation 25(4), 270–277 (2006)
[MMdGD11]   Mullen, P., Memari, P., de Goes, F., Desbrun, M.: HOT: Hodge-Optimized Triangulations. ACM Trans. Graph. 30(4), 103:1–103:12 (2011)
[Mus97]   Musin, O.R.: Properties of the Delaunay triangulation. In: Proceedings of the Thirteenth Annual Symposium on Computational Geometry, p. 426. ACM (1997)
[OBSC00]   Okabe, A., Boots, B., Sugihara, K., Chiu, S.N.: Spatial tessellations: Concepts and applications of Voronoi diagrams. In: Probability and Statistics, 2nd edn. Wiley (2000)
[PMH07]   Paluszny, A., Matthäi, S., Hohmeyer, M.: Hybrid finite element finite volume discretization of complex geologic structures and a new simulation workflow demonstrated on fractured rocks. Geofluids 7, 186–208 (2007)
[PS85]   Preparata, F.P., Shamos, M.I.: Computational Geometry: An Introduction. Springer, Heidelberg (1985)
[Raj94]   Rajan, V.T.: Optimality of the Delaunay triangulation in $\mathbb{R}^d$. Discrete and Computational Geometry 12(1), 189–202 (1994)
[RG96]   Richter-Gebert, J.: Realization spaces of polytopes. Citeseer (1996)
[Sch03]   Schwartz, A.: Constructions of cubical polytopes. PhD diss., TU Berlin (2003)
[Sha93]   Shah, N.R.: A parallel algorithm for constructing projection polyhedra. Information Processing Letters 48(3), 113–119 (1993)
[Ste22]   Steinitz, E.: Polyeder und raumeinteilungen. Encyclopädie der mathematischen Wissenschaften 3(9), 1–139 (1922)
[Tho06]   Thomas, R.R.: Lectures in geometric combinatorics. Amer. Mathematical Society (2006)
[TWAD09]   Tournois, J., Wormser, C., Alliez, P., Desbrun, M.: Interleaving delaunay refinement and optimization for practical isotropic tetrahedron mesh generation. ACM Trans. Graph. 28, 75:1–75:9 (2009)
[Zie95]   Ziegler, G.M.: Lectures on polytopes. Springer, Heidelberg (1995)

Geometrical Validity of Curvilinear Finite Elements

Amaury Johnen[1], Jean-François Remacle[2], and Christophe Geuzaine[1]

[1] Université de Liège, Department of Electrical Engineering and Computer Science, Montefiore Institute B28, Grande Traverse 10, 4000 Liège, Belgium
{ajohnen,cgeuzaine}@ulg.ac.be
[2] Institute of Mechanics, Materials and Civil Engineering, Université catholique de Louvain, Avenue Georges-Lemaître 4, 1348 Louvain-la-Neuve, Belgium
jean-francois.remacle@uclouvain.be

Summary. In this paper, we describe a way to compute accurate bounds on Jacobians of curvilinear finite elements of all kinds. Our condition enables to guarantee that an element is geometrically valid, i.e., that its Jacobian is strictly positive everywhere in its reference domain. It also provides an efficient way to measure the distortion of curvilinear elements. The key feature of the method is to expand the Jacobian using a polynomial basis, built using Bézier functions, that has both properties of boundedness and positivity. Numerical results show the sharpness of our estimates.

Keywords: Finite element method, high-order methods, mesh generation, Bézier functions.

1 Introduction

There is a growing consensus in the Finite Element community that higher-order discretization methods will replace at some point the solvers of today, at least for part of their applications. These high-order methods require a good accuracy of the geometrical discretization to be accurate—in other words, such methods will critically depend on the availability of high-quality curvilinear meshes.

The usual way of building such curvilinear meshes is to first generate a straight sided mesh. Then, mesh entities that are classified on the curved boundaries of the domain are curved accordingly [1, 2, 3]. Some internal mesh entities may be curved as well. If we assume that the straight sided mesh is composed of well shaped elements, curving elements introduces a "shape distortion" that should be controlled so that the final curvilinear mesh is also composed of well shaped elements. The optimization of the shape distortion is a computationally expensive operation, especially when applied globally over the full mesh. It is thus crucial to be able to get fast and accurate bounds on the distortion in order to 1) evaluate the quality of the elements during the optimization process; and 2) reduce the sets of elements to be optimized, so that the optimization can be applied locally, i.e., only where it is necessary.

In this paper we present a method to analyze curvilinear meshes in terms of their elementary Jacobians. The method does not deal with the actual generation/optimization of the high order mesh. Instead, it provides an efficient way to guarantee that each curvilinear element is geometrically valid, i.e., that its Jacobian is strictly positive everywhere in its reference domain. It also provides a way to measure the distortion of the curvilinear element. The key feature of the method is to adaptively expand the elementary Jacobians in a polynomial basis that has both properties of boundedness and positivity. Bézier functions are used to generate these bases in a recursive manner. The proposed method can be either used to check the validity and the distortion of an existing curvilinear mesh, or embedded in the curvilinear mesh generation procedure to assess the validity and the quality of the elements on the fly. The algorithm described in this paper has been implemented in the open source mesh generator Gmsh [4], where it is used in both ways.

2 Curvilinear Meshes, Distortion and Jacobian Bounds

Let us consider a mesh that consists of a set of straight-sided elements of order p. Each element is defined geometrically through its nodes $\mathbf{x}_i$, $i = 1, \ldots, N_p$ and a set of Lagrange shape functions $\mathcal{L}_i^{(p)}(\boldsymbol{\xi})$, $i = 1, \ldots, N_p$. The Lagrange shape functions (of order p) are based on the nodes $\mathbf{x}_i$ and allow to map a reference unit element onto the real one:

$$\mathbf{x}(\boldsymbol{\xi}) = \sum_{i=1}^{N_p} \mathcal{L}_i^{(p)}(\boldsymbol{\xi}) \, \mathbf{x}_i. \tag{1}$$

The mapping $\mathbf{x}(\boldsymbol{\xi})$ should be bijective, which means that it should admit an inverse. This implies that the Jacobian $\det \mathbf{x}_{,\boldsymbol{\xi}}$ has to be strictly positive. In all what follows we will always assume that the straight-sided mesh is composed of well-shaped elements, so that the positivity of $\det \mathbf{x}_{,\boldsymbol{\xi}}$ is guaranteed. This standard setting is presented on Figure 1 for the quadratic triangle.

Let us now consider a curved element obtained after application of the curvilinear meshing procedure, i.e., after moving some or all of the nodes of the straight-sided element. The nodes of the deformed element are called $\mathbf{X}_i$, $i = 1 \ldots N_p$, and we have

$$\mathbf{X}(\boldsymbol{\xi}) = \sum_{i=1}^{N_p} \mathcal{L}_i^{(p)}(\boldsymbol{\xi}) \, \mathbf{X}_i. \tag{2}$$

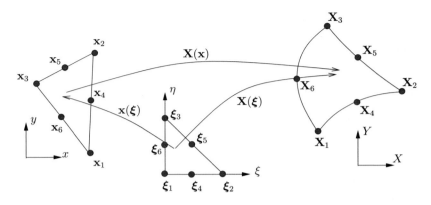

Fig. 1. Reference unit triangle in local coordinates $\boldsymbol{\xi} = (\xi, \eta)$ and the mappings $\mathbf{x}(\boldsymbol{\xi})$, $\mathbf{X}(\boldsymbol{\xi})$ and $\mathbf{X}(\mathbf{x})$.

Again, the deformed element is valid if the Jacobian $J(\boldsymbol{\xi}) := \det \mathbf{X}_{,\boldsymbol{\xi}}$ is strictly positive everywhere over the $\boldsymbol{\xi}$ reference domain. The Jacobian J, however, is not constant over the reference domain, and computing $J_{\min} := \min_{\boldsymbol{\xi}} J(\boldsymbol{\xi})$ is necessary to ensure positivity.

The approach that is commonly used is to sample the Jacobian on a very large number of points. Such a technique is however both expensive and not fully robust since we only get a necessary condition. In this paper we follow a different approach: because the Jacobian J is a polynomial in $\boldsymbol{\xi}$, J can be interpolated exactly as a linear combination of specific polynomial basis function over the element. We would then like to obtain provable bounds on $J_{\min}$ by using the properties of these basis functions.

In addition to guaranteeing the geometrical validity of the curvilinear element, we are also interested in quantifying the distortion of the curvilinear element, i.e., the deformation induced by the curving. To this end, let us consider the transformation $\mathbf{X}(\mathbf{x})$ that maps straight sided elements onto curvilinear elements. It is possible to write this determinant in terms of the $\boldsymbol{\xi}$ coordinates as:

$$\det \mathbf{X}_{,\mathbf{x}} = \frac{\det \mathbf{X}_{,\boldsymbol{\xi}}}{\det \mathbf{x}_{,\boldsymbol{\xi}}} = \frac{J(\boldsymbol{\xi})}{\det \mathbf{x}_{,\boldsymbol{\xi}}}. \tag{3}$$

We call $\mathbf{X}(\mathbf{x})$ the distortion mapping and its determinant $\delta(\boldsymbol{\xi}) := \det \mathbf{X}_{,\mathbf{x}}$ the distorsion. The distorsion δ should be as close to $\delta = 1$ as possible in order not to degrade the quality of the straight sided element. Elements that have negative distorsions are of course invalid but elements that have distorsions $\delta \ll 1$ or $\delta \gg 1$ lead to some alteration of the conditioning of the finite element problem. In order to guarantee a reasonable distortion it is thus necessary to find a reliable bound on $J_{\min}$ and $J_{\max} := \max_{\boldsymbol{\xi}} J(\boldsymbol{\xi})$ over the whole element.

Note that many different quality measures can be defined based on the Jacobian J. For example, one could look at the Jacobian divided by its average over the element instead of looking at the distortion. In any case, we see that obtaining bounds on $J_{\min}$ and $J_{\max}$ is still the main underlying challenge.

3 Jacobian Bounds for Second Order Planar Triangles

We start our analysis with the particular case of second order triangles for which a direct computation of $J_{\min}$ is relatively easy. The determinant $J(\boldsymbol{\xi}) = J(\xi, \eta)$ for a planar triangle at order p is a polynomial in ξ and η of order at most $2(p-1)$. For quadratic planar triangles, $J(\xi, \eta)$ is therefore quadratic at most in ξ and η.

The geometry of the six-node quadratic triangle is shown in Figure 1. Inspection reveals two types of nodes: corners (1, 2 and 3) and midside nodes (4, 5 and 6). If J_i is defined as $J(\xi, \eta)$ evaluated at node i, it is possible to write the Jacobian exactly as a finite element expansion whose coefficients are the Jacobian at the nodes:

$$J(\xi,\eta) = J_1 \underbrace{(1-\xi-\eta)(1-2\xi-2\eta)}_{\mathcal{L}_1^{(2)}(\xi,\eta)} + J_2 \underbrace{\xi(2\xi-1)}_{\mathcal{L}_2^{(2)}(\xi,\eta)} + J_3 \underbrace{\eta(2\eta-1)}_{\mathcal{L}_3^{(2)}(\xi,\eta)} +$$
$$J_4 \underbrace{4(1-\xi-\eta)\xi}_{\mathcal{L}_4^{(2)}(\xi,\eta)} + J_5 \underbrace{4\xi\eta}_{\mathcal{L}_5^{(2)}(\xi,\eta)} + J_6 \underbrace{4(1-\xi-\eta)\eta}_{\mathcal{L}_6^{(2)}(\xi,\eta)}. \qquad (4)$$

In equation (4), the functions $\mathcal{L}_i^{(2)}(\xi, \eta)$ are the equidistant quadratic Lagrange shape functions that are commonly used in the finite element community [5].

We first show how to compute the exact minimal Jacobian $J_{\min}$. Then we examine different bounds that can be provided on $J_{\min}$ by exploiting the properties of the basis used in the Jacobian expansion.

3.1 Exact Computation of $J_{\min}$

From equation (4), the stationnary point of J can be computed by solving

$$\frac{\partial J}{\partial \xi} = \frac{\partial J}{\partial \eta} = 0, \qquad (5)$$

which leads to the following linear system of two equations and two unknowns ξ_{sta} and η_{sta}:

$$\begin{bmatrix} 4(J_1+J_2-2J_4) & 4(J_1-J_4+J_5-J_6) \\ 4(J_1-J_4+J_5-J_6) & 4(J_1+J_3-2J_6) \end{bmatrix} \begin{pmatrix} \xi_{sta} \\ \eta_{sta} \end{pmatrix} = \begin{pmatrix} -(-3J_1-J_2+4J_4) \\ -(-3J_1-J_3+4J_6) \end{pmatrix}. \qquad (6)$$

Algorithm 1 allows to compute the minimal Jacobian over one quadratic planar element exactly. If the minimum of the function is outside of the element,

it computes the minimum on its border assuming a function $\text{MINQ}(a,b,c)$ that computes
$$\text{MINQ}(a,b,c) = \min_{x \in [0,1]} a\, x^2 + b\, x + c. \tag{7}$$

Algorithm 1. Exact computation of J_{min} over a quadratic triangle

---

1 compute nodal Jacobians J_i, $i = 1, \ldots, 6$;
2 compute ξ_{sta}, η_{sta} as in equation (6);
3 **if** $\eta_{sta} > 0$ **and** $\xi_{sta} > 0$ **and** $1 - \xi_{sta} - \eta_{sta} > 0$ **then**
4 $\quad \big|\quad J_{\min} = \min(J(\xi_{sta}, \eta_{sta}), J_1, J_2, J_3)$;
5 **else**
6 $\quad \big|\quad m_1 = \text{MINQ}(2(J_1 + J_2 - 2J_4), -3J_1 - J_2 + 4J_4, J_1)$;
7 $\quad \big|\quad m_2 = \text{MINQ}(2(J_1 + J_3 - 2J_6), -3J_1 - J_3 + 4J_6, J_1)$;
8 $\quad \big|\quad m_3 = \text{MINQ}(2(J_2 + J_3 - 2J_5), -3J_2 - J_3 + 4J_5, J_2)$;
9 $\quad \big|\quad J_{\min} = \min(m_1, m_2, m_3)$;
10 **return** $J_{\min}$;

Although Algorithm 1 is quite simple, applying similar techniques for higher order elements would become extremely expensive computationally. Instead of trying to evaluate $J_{\min}$ directly, we should try to compute (the sharpest possible) bounds in a computationally efficient manner.

3.2 The Principle for Computing Bounds on $J_{\min}$

It is obvious that a necessary condition for having $J(\xi,\eta) > 0$ everywhere is that $J_i > 0$, $i = 1, \ldots, 6$. Yet, this condition is not sufficient. The expression (4) does not give more information because the quadratic Lagrange shape functions $\mathcal{L}_i^{(2)}(\xi,\eta)$ change sign on the reference triangle. What polynomial basis should we chose to obtain usable bounds?

The first idea is to expand (4) into monomials, which gives:

$$J(\xi,\eta) = J_1 + (-3J_1 - J_2 + 4J_4)\xi + (-3J_1 - J_3 + 4J_6)\eta + \\ 4(J_1 - J_4 + J_5 - J_6)\xi\eta + 2(J_1 + J_2 - 2J_4)\xi^2 + 2(J_1 + J_3 - 2J_6)\eta^2. \tag{8}$$

Every monomial being positive on the reference triangle, we have now a set of sufficient conditions that can be written as

$$4J_4 \geq 3J_1 + J_2, \quad 4J_6 \geq 3J_1 + J_3, \quad J_1 + J_5 \geq J_4 + J_6, \quad J_1 + J_2 \geq 2J_4, \quad J_1 + J_3 \geq 2J_6.$$

However these constraints do not provide a usable bound on $J_{\min}$ and break the symmetry of the expression with respect to a rotation of corner nodes.

A second idea is to expand (4) in terms of the second order hierarchical basis functions $\psi_i(\xi,\eta)$, $i = 1, \ldots, 6$, which are also positive on the triangle [6]:

$$J(\xi,\eta) = J_1 \underbrace{(1-\xi-\eta)}_{\psi_1(\xi,\eta)} + J_2 \underbrace{\xi}_{\psi_2(\xi,\eta)} + J_3 \underbrace{\eta}_{\psi_3(\xi,\eta)} + (4J_4 - 2J_1 - 2J_2)\underbrace{(1-\xi-\eta)\xi}_{\psi_4(\xi,\eta)} +$$
$$(4J_5 - 2J_3 - 2J_2)\underbrace{\xi\eta}_{\psi_5(\xi,\eta)} + (4J_6 - 2J_1 - 2J_3)\underbrace{(1-\xi-\eta)\eta}_{\psi_6(\xi,\eta)}. \quad (9)$$

This last expression has the right symmetry, and leads to the following validity conditions:

$$J_1 \geq 0, \ J_2 \geq 0, \ J_3 \geq 0, \ 4J_4 \geq 2J_1 + 2J_2, \ 4J_5 \geq 2J_2 + 2J_3, \ 4J_6 \geq 2J_3 + 2J_1. \quad (10)$$

Writing $J(\xi,\eta) := \sum_{i=1}^{6} \psi_i(\xi,\eta) K_i$, we have

$$\min_{\xi,\eta} J(\xi,\eta) = \min_{\xi,\eta}\left(\sum_i \psi_i(\xi,\eta) K_i\right) \geq \min_{\xi,\eta}\left(\sum_i \psi_i(\xi,\eta)\right)\min_i K_i = \min_i K_i,$$

because $\sum_i \psi_i = 1 + \xi + \eta - \xi^2 - \eta^2 - \xi\eta$ has its minimum on the corner vertices (where its value is equal to 1). And since K_i, $i = 1,\ldots,3$ are values of the Jacobian, they form an upper bound on it. Thus, expansion (9) leads to the following estimate for the minimum of the Jacobian over the triangle:

$$J_{\min} \geq \min\{J_1, J_2, J_3, 4J_4 - 2J_1 - 2J_2, 4J_5 - 2J_2 - 2J_3, 4J_6 - 2J_3 - 2J_1\}$$
$$\leq \min\{J_1, J_2, J_3\}. \quad (11)$$

It is easy to see that the estimate is however of very poor quality: for an element that has a constant and positive J, (11) simply tells us that $J_{\min} \geq 0$.

In order to find a sharper estimate, instead of the hierarchical quadratic functions $\psi_i(\xi,\eta)$, we can use the quadratic triangular Bézier functions $\mathcal{B}_i^{(2)}(\xi,\eta)$ [7]:

$$J(\xi,\eta) = J_1\underbrace{(1-\xi-\eta)^2}_{\mathcal{B}_1^{(2)}(\xi,\eta)} + J_2\underbrace{\xi^2}_{\mathcal{B}_2^{(2)}(\xi,\eta)} + J_3\underbrace{\eta^2}_{\mathcal{B}_3^{(2)}(\xi,\eta)} +$$
$$\left(2J_4 - \frac{1}{2}(J_2+J_1)\right)\underbrace{2\xi(1-\xi-\eta)}_{\mathcal{B}_4^{(2)}(\xi,\eta)} + \left(2J_5 - \frac{1}{2}(J_3+J_2)\right)\underbrace{2\xi\eta}_{\mathcal{B}_5^{(2)}(\xi,\eta)} +$$
$$\left(2J_6 - \frac{1}{2}(J_1+J_3)\right)\underbrace{2\eta(1-\xi-\eta)}_{\mathcal{B}_6^{(2)}(\xi,\eta)}. \quad (12)$$

Since $\sum_{i=1}^{6}\mathcal{B}_i^{(2)}(\xi,\eta) = 1$, we obtain the following estimate

$$J_{\min} \geq \min\left\{J_1, J_2, J_3, 2J_4 - \frac{J_1+J_2}{2}, 2J_5 - \frac{J_2+J_3}{2}, 2J_6 - \frac{J_3+J_1}{2}\right\}$$
$$\leq \min\{J_1, J_2, J_3\}. \quad (13)$$

One can show that this estimate is always better than the one using the hierarchical basis. It provides two conditions on the geometrical validity of the triangle: a *sufficient* condition (if $\min\{J_1, J_2, J_3, 2J_4 - \frac{J_1+J_2}{2}, 2J_5 - \frac{J_2+J_3}{2}, 2J_6 - \frac{J_3+J_1}{2}\} > 0$, the element is valid) and a *necessary* condition (if $\min\{J_1, J_2, J_3\} < 0$, the element is invalid). However, these two conditions are sometimes insufficient to determine the validity of the element, as the bound (13) is often not sharp enough (having $\min\{2J_4 - \frac{J_1+J_2}{2}, 2J_5 - \frac{J_2+J_3}{2}, 2J_6 - \frac{J_3+J_1}{2}\} < 0$ does not imply that the element is invalid).

A sharp necessary and sufficient condition on the geometrical validity of an element can be achieved in a general way by refining the Bézier estimate adaptively so as to achieve any prescribed tolerance—and thus provide bounds as sharp as necessary for a given application.

4 Adaptive Jacobian Bounds for Arbitrary Curvilinear Finite Elements

In order to explain the adaptive bound computation let us first focus on the one-dimensional case, for "line" finite elements. Since Bézier functions can be generated for all types of common elements (triangles, quadrangles, tetrehedra, hexahedra and prisms), the generalization to 2D and 3D elements will be straightforward.

4.1 The One-Dimensional Case

In 1D the Bézier functions are the Bernstein polynomials:

$$\mathcal{B}_k^{(n)}(\xi) = \binom{n}{k}(1-\xi)^{n-k}\xi^k \qquad (\xi \in [0,1] \; ; \; k = 0, ..., n) \qquad (14)$$

where $\binom{n}{k} = \frac{n!}{n!(n-k)!}$ is the binomial coefficient. The Bézier interpolation requires $n+1$ control values b_i. We have

$$J(\xi) = \sum_{k=0}^{N_n} \mathcal{B}_k^{(n)}(\xi)\, b_k. \qquad (15)$$

Bernstein-Bézier functions have the nice following properties : (i) they form a partition of unity which means that $\sum_{k=0}^{n} \mathcal{B}_k^{(n)}(\xi) = 1$ for all $\xi \in [0,1]$ and (ii) they are positive which means that $\mathcal{B}_k^{(n)}(\xi) \geq 0$ for all $\xi \in [0,1]$. This leads to the well known property of Bézier interpolations:

$$\min_{\xi \in [0,1]} J(\xi) \geq b_{\min} = \min_i b_i \quad \text{and} \quad \max_{\xi \in [0,1]} J(\xi) \leq b_{\max} = \max_i b_i. \qquad (16)$$

Moreover, they always present control values that are values of the interpolated function. Let assume they are ordered at the K_f first indices, we have

$$\min_{\xi \in [0,1]} J(\xi) \leq \min_{i < K_f} b_i \quad \text{and} \quad \max_{\xi \in [0,1]} J(\xi) \geq \max_{i < K_f} b_i. \tag{17}$$

Since Lagrangian and Bézier functions span the same function space, computation of the Bézier values b_i from the nodal values J_i (and convertly) is done by a transformation matrix. The tranformation matrix $\boldsymbol{T}_{\mathcal{B} \to \mathcal{L}}^{(n)}$, which computes nodal values from control values, is created by evaluating Bézier functions at sampling points:

$$\boldsymbol{T}_{\mathcal{B} \to \mathcal{L}}^{(n)} = \begin{bmatrix} \mathcal{B}_0^{(n)}(\xi_0) & \dots & \mathcal{B}_n^{(n)}(\xi_0) \\ \mathcal{B}_0^{(n)}(\xi_1) & \dots & \mathcal{B}_n^{(n)}(\xi_1) \\ \vdots & \ddots & \vdots \\ \mathcal{B}_0^{(n)}(\xi_n) & \dots & \mathcal{B}_n^{(n)}(\xi_n) \end{bmatrix}.$$

The inverse transformation is $\boldsymbol{T}_{\mathcal{L} \to \mathcal{B}}^{(n)} = \boldsymbol{T}_{\mathcal{B} \to \mathcal{L}}^{(n)}{}^{-1}$ and from the expression of the interpolation of the Jacobian (15), we can write

$$\begin{aligned} J &= \boldsymbol{T}_{\mathcal{B} \to \mathcal{L}}^{(n)} B \\ B &= \boldsymbol{T}_{\mathcal{L} \to \mathcal{B}}^{(n)} J, \end{aligned} \tag{18}$$

where B and J are the vectors containing respectively the b_i's and the J_i's.

4.2 Adaptive Subdivision

Let assume that the domain is divided into Q parts and that the qth of them interpolates the Jacobian on $[a, b]$ ($a < b$). Then, the variable ξ varies in this interval while the new one varies from 0 to 1 and the new interpolation must verify

$$J^{[q]}(\xi^{[q]}) = \sum_{k=0}^{N_n} \mathcal{B}_k^{(n)}(\xi^{[q]}) \, b_k^{[q]} = \sum_{k=0}^{N_n} \mathcal{B}_k^{(n)}(\xi(\xi^{[q]})) \, b_k \qquad (\xi^{[q]} \in [0,1]), \tag{19}$$

with $\xi(\xi^{[q]}) = a + (b - a)\, \xi^{[q]}$. Considering the nodes $\xi_k^{[q]}$ such that $\xi_k^{[q]} = \xi_k$ ($k = 0, \dots, n$) (i.e., such that they are ordered like the sampling points), the expression (19) reads

$$\boldsymbol{T}_{\mathcal{B} \to \mathcal{L}}^{(n)} B^{[q]} = \begin{bmatrix} \mathcal{B}_0^{(n)}(a + (b-a)\,\xi_0) & \dots & \mathcal{B}_n^{(n)}(a + (b-a)\,\xi_0) \\ \mathcal{B}_0^{(n)}(a + (b-a)\,\xi_1) & \dots & \mathcal{B}_n^{(n)}(a + (b-a)\,\xi_1) \\ \vdots & \ddots & \vdots \\ \mathcal{B}_0^{(n)}(a + (b-a)\,\xi_n) & \dots & \mathcal{B}_n^{(n)}(a + (b-a)\,\xi_n) \end{bmatrix} B = \boldsymbol{T}_{\mathcal{B} \to \mathcal{L}}^{(n)\,[q]} B,$$

where $B^{[q]}$ is the vector containing control values of the related subdomain. This implies that

$$B^{[q]} = \left[T^{(n)}_{\mathcal{L} \to \mathcal{B}} \, T^{(n)}_{\mathcal{B} \to \mathcal{L}}^{[q]} \right] B = M^{[q]} B. \tag{20}$$

Each set of new control values bounds the Jacobian on its own subdomain and we have:

$$b'_{\min} = \min_{i,q} b_i^{[q]} \leq J_{\min} \leq \min_{i<K_f,q} b_i^{[q]} \tag{21}$$

and

$$\max_{i<K_f,q} b_i^{[q]} \leq J_{\max} \leq b'_{\max} = \max_{i,q} b_i^{[q]}. \tag{22}$$

If an estimate is not sufficiently sharp, we can thus simply subdivide the appropriate parts of the element. This leads to a simple adaptive algorithm, exemplified in Figure 2. In this particular case the original estimate (16)-(17) is not sharp enough ($J_{\min} \in [-2.5, 1]$). After one subdivision, the Jacobian is proved to be positive on the second subdomain. The first subdomain is thus subdivided once more, which proves the validity. In practice, as will be seen in Section 5, a few levels of refinement lead to the desired accuracy. The convergence of the subdivision can be proven to be quadratic [8, 9].

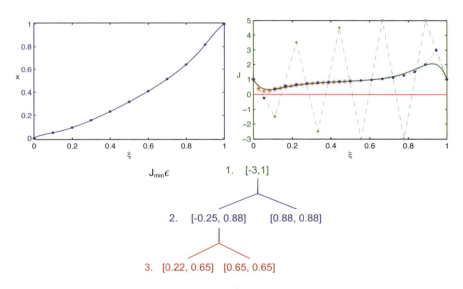

Fig. 2. Top left: One-dimensional element mapping $x(\xi)$. Top right: Exact Jacobian $J(\xi)$ (solid green), control values on the original control points (dashed green) and two adaptive subdivisions (blue and red). Bottom: Estimates of $J_{\min}$ at each step in the adaptive subdivision process.

4.3 Extension to Higher Dimensions

The extension of the method to higher dimensions is straightforward, provided that Bézier functions can be generated and that a subdivision scheme is available: Jacobians J are polynomials of ξ, η in 2D and of ξ, η, ζ in 3D.

For high order triangles, the Bézier triangular polynomials are defined as

$$\mathcal{T}_{i,j}^{(p)}(\xi, \eta) = \binom{p}{i}\binom{p-i}{j} \xi^i \eta^j (1-\xi-\eta)^{p-i-j} \qquad (i+j \leq p).$$

It is possible to interpolate any polynomial function of order at most p on the unit triangle $\xi > 0, \eta > 0, \xi + \eta < 1$ as an expansion into Bézier triangular polynomials. Recalling that, for a triangle at order p, its Jacobian $J(\xi, \eta)$ is a polynomial in ξ and η at order at most $n = 2(p-1)$, we can write

$$J(\xi, \eta) = \sum_{i+j \leq n} b_{ij} \mathcal{T}_{i,j}^{(n)}(\xi, \eta).$$

It is indeed possible to compute J in terms of Lagrange polynomials

$$J(\xi, \eta) = \sum_i J_i \mathcal{L}_i^{(n)}(\xi, \eta)$$

where the J_i are the Jacobians calculated at Lagrange points. It is then easy to find a transformation matrix $T_{\mathcal{LB}}^n$ such that

$$B = T_{\mathcal{LB}}^n J,$$

where B and J are the vectors containing respectively the control values of the Jacobian b_{ij} and the J_i's. As an example, for quadratic triangles we obtain

$$T_{\mathcal{LB}}^2 = \begin{pmatrix} 1 & 0 & 0 & 0 & 0 & 0 \\ 0 & 1 & 0 & 0 & 0 & 0 \\ 0 & 0 & 1 & 0 & 0 & 0 \\ -1/2 & -1/2 & 0 & 2 & 0 & 0 \\ 0 & -1/2 & -1/2 & 0 & 2 & 0 \\ -1/2 & 0 & -1/2 & 0 & 0 & 2 \end{pmatrix}, \qquad (23)$$

which directly provides the estimate (13).

Other element shapes can be treated similarly. For quadrangles, tetrahedra, prisms and hexahedra, the Bézier are functions respectively:

$$\mathcal{Q}_{i,j}^{(p)}(\xi, \eta) = \mathcal{B}_i^{(p)}(\xi) \, \mathcal{B}_j^{(p)}(\eta) \qquad (i \leq p, \, j \leq p),$$

$$\mathcal{T}_{i,j,k}^{(p)}(\xi, \eta, \zeta) = \binom{p}{i}\binom{p-i}{j}\binom{p-i-j}{k} \xi^i \eta^j \zeta^k (1-\xi-\eta-\zeta)^{p-i-j-k}$$

$$(i+j+k \leq p),$$

$$\mathcal{P}_{i,j,k}^{(p)}(\xi,\eta,\zeta) = \mathcal{T}_{i,j}^{(p)}(\xi,\eta)\,\mathcal{B}_k^{(p)}(\zeta) \qquad (i+j \leq p,\, k \leq p)$$

and

$$\mathcal{H}_{i,j,k}^{(p)}(\xi,\eta,\zeta) = \mathcal{B}_i^{(p)}(\xi)\,\mathcal{B}_j^{(p)}(\eta)\,\mathcal{B}_k^{(p)}(\zeta) \qquad (i \leq p,\, j \leq p,\, k \leq p).$$

Matrices of change of coordinates can then be computed inline for every polynomial order, and bounds of Jacobians computed accordingly. In all cases the subdivision scheme works exactly in the same way as for lines.

4.4 Implementation

As mentioned in Section 2, the Jacobian bounds can be used to either make the distinction between valid and invalid elements with respect to a condition on $J_{\min}$, or to measure the quality of the elements by systematically computing $J_{\min}$ and $J_{\max}$ with a defined precision.

In both cases the same operations are executed on each element. First, the Jacobian is sampled on a determined number of points N_s, equal to the dimension of the Jacobian space, and so to the number of Bézier functions. Second, Bézier values are computed. Then adaptive subdivision is executed if necessary. Algorithm 2 shows in pseudo-code the algorithm used to determine whether the Jacobian of the element is everywhere positive or not.

Algorithm 2. Check if an element is valid or invalid

Input: a pointer to an element.
Output: *true* if the element is valid, *false* if the element is invalid

1. set sampling points P_i, $i = 1, \ldots, N_s$;
2. compute Jacobians J_i at points P_i;
3. **for** $i = 1$ **to** N_s **do**
4.   **if** $J_i <= 0$ **then return** *false*;

5. compute Bézier coefficients b_i, $i = 1, \ldots, N_s$ using (18);
6. $i = 1$;
7. **while** $i \leq N_s$ **and** $b_i > 0$ **do**
8. $i = i + 1$;

9. **if** $i > N_s$ **then return** *true*;
10. call algorithm 3 with b_i as arguments and return its output;

Algorithm 3 can be further improved by optimzing the loop on line 5, by first selecting q for which we have the best chance to have a negative Jacobian (line 4, algo 3). However, in practice, this improvement is not significant since the only case for which we can save calculation is for invalid elements—and the proportion of them which require subdivision in order to be detected is usually small. Note that we may also want to find, for example, all the

Algorithm 3. Compute the control values of the subdivisions

Input: Bézier coefficients b_i, $i = 1, \ldots, N_s$
Output: *true* if the Jacobian on the domain is everywhere positive, *false* if not

1 compute new Bézier coefficients $b_i^{[q]}$, $q = 1, \ldots, Q$ as in equation (20);
2 **for** $q = 1$ **to** Q **do**
3   **for** $i = 1$ **to** K_f **do**
4   **if** $b_i^{[q]} <= 0$ **then return** *false*;

5 **for** $q = 1$ **to** Q **do**
6     $i = 1$;
7   **while** $i \leq N_s$ **and** $b_i^{[q]} > 0$ **do**
8         $i = i + 1$;
9   **if** $i \leq N_s$ **then**
10   call algorithm 3 with $b_i^{[q]}$ as arguments and store *output*;
11   **if** *output* = *false* **then return** *false*;

12 **return** *true*;

---

elements for which the Jacobian is somewhere smaller than 20% of its average. We then just have to compute this average and replace the related lines (4 and 7 for algorithm 2).

Another possible improvement is to relax the condition of rejection. We could accept elements for which all control values are positive but reject an element as soon as we find a Jacobian value smaller than a defined percent of the average Jacobian. The computational gain can be significant, since elements that were classified as good and which needed a lot of subdivisions (and have a Jacobian close to zero) will be instead rapidly be detected as invalid.

More interestingly, the computation of sampled Jacobians and the computation of Bézier control values in algorithm 2 can easily be executed for a whole groups of elements at the same time. This allows to use efficient BLAS 3 (matrix-matrix product) functions, which significantly speeds up the computations.

The algorithm using the BLAS3 approach is implemented in the open source mesh generator Gmsh [4] as the `AnalyseCurvedMesh` plugin, and was used for all the tests presented in the next section.

5 Numerical Results

We start by comparing the new adaptive computation of Jacobian bounds with the brute-force sampling of the Jacobian for the detection of invalid high-order triangles.

The points at which we sample the Jacobian for the brute-force method are taken as the nodes of an element of order k. We started the test for $k=1$ and we incremented k until the brute-force approach detected all the invalid elements. We still executed the algorithm 10 times (while incrementing k) so as to plot the change in the number of invalid element detected. In order to make the comparison as fair as possible, we have implemented the brute-force computation as efficiently as possible, i.e., for k ($> n$) sufficiently large we sample the Jacobian on the points computed for an element at order n (the order of the Jacobian) and then compute the desired Jacobian values by a matrix-vector product, just like in our own adaptive method.

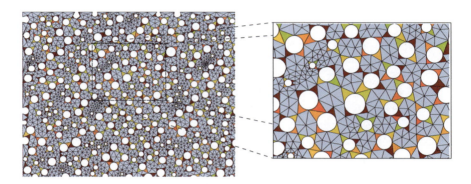

Fig. 3. Two-dimensional mesh with sixth order triangles; 23.6% of the elements are curved. The straight element are in blue and the invalid are in dark red. The other are colored in function of the distortion. They are green if they are nearly straight ($J_{\min}/J_{max} \simeq 1$) and rather light red if really distorted ($J_{\min}/J_{max} \simeq 0$).

We consider the two-dimensional microstructure with circular holes depicted in Figure 3, meshed with 331,050 sixth-order triangles. In this mesh 78,180 triangles are curved, and 45,275 are invalid. The new algorithm successfully detects all the 45,275 invalid elements in 6.194s. Some elements needed as much as 8 levels of subdivisions in order to be classified: see Table 1. The brute-force approach required 666 sample points per triangle in order to detect all the invalid elements, and took 4 times longer. But far worse, increasing the number of sampling points beyond 666 can actually lead to a decreased accuracy of the prcediction, as shown in Figure 4.

Let us now examine the use of the adaptive Jacobian bounds in the curvilinear meshing algorithms as implemented in Gmsh. We consider the mesh of a rather coarse version of the world ocean. In our CAD model, shorelines are described using cubic B-splines: for example, Europe and Asia are discretized by only one B-spline with about 3,500 control points. The description of this kind of meshing procedure is described in [10]. The quadratic triangular mesh is generated as follows. We first generate a straight sided mesh

Table 1. Number of elements detected as valid or invalid at each stage of the adaptive algorithm; 5 % of the curved elements had to be subdivided adaptively.

	Valid curved elements	Invalid curved elements
First stage	29303	44967
1 subdivision	2436	-
2 subdivisions	1119	299
3 subdivisions	23	-
4 subdivisions	10	4
5 subdivisions	9	2
6 subdivisions	5	-
7 subdivisions	-	2
8 subdivisions	-	1

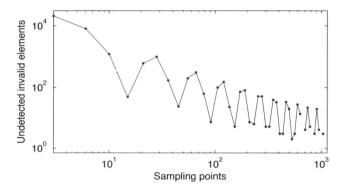

Fig. 4. Number of undetected invalid elements using brute-force sampling of the Jacobian. The three data points not displayed correspond to the correct result, i.e., when no invalid triangle is left undetected.

(see Figure 5/(a)). Then, every mesh edge that is classified on a model edge is curved by snapping its center vertex on the model edge. High order nodes are then inserted in the middle of every edge that is classified on a model face (see Figure 5/(b)). This simple procedure does not guarantee that the final mesh is valid. In our case, 175 elements are invalid. Then, a global elasticity analogy is applied to the quadratic mesh that enables to reduce the number of invalid elements to 70 (see Figure 5/(c)). Then local optimizations are perfomed to remove all invalid elements (see Figure 5/(d)). The final curvilinear mesh contains about 30% of curved elements. During the meshing process, the adaptive Jacobian bound computation allowed to detected all invalid elements (the worst distorsion that was observed was $\delta = -4.49702$). After optimization, the final mesh is composed of elements that have a distortion $\delta > 0.1$.

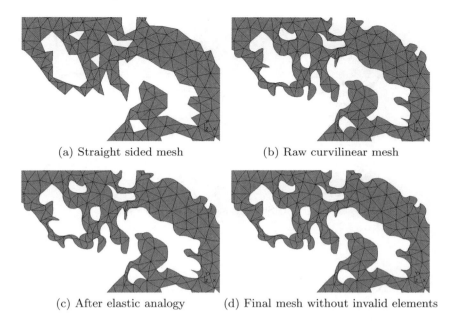

Fig. 5. The four stages of the curvilinear mesh procedure for the world ocean, meshed with second order triangles.

Finally, the same procedure is applied to the meshing of the STEP model of a rotor. After generating a first order mesh (Figure 6(a)) and snapping the high-order vertices on the geometrical model (Figure 6(b)), the adaptive Jacobian bound computation allowed to pinpoint all the invalid elements. The final, locally optimized mesh is displayed in Figure 6(c).

6 Conclusion

In this paper we presented a way to compute accurate bounds on Jacobians of curvilinear finite elements, based on the efficient expansion of these Jacobians in terms of Bézier functions. The proposed algorithm can either be used to determine the validity or invalidity of curved elements, or provide an efficient way to measure their distortion. Triangles, quadrangles, tetraheda, prisms and hexahedra can be analyzed using the same algorithm, which is available in the open source mesh generator Gmsh. Numerical tests show that the method is robust, and a user-defined error tolerance permits to adjust the accuracy vs. computational time ratio.

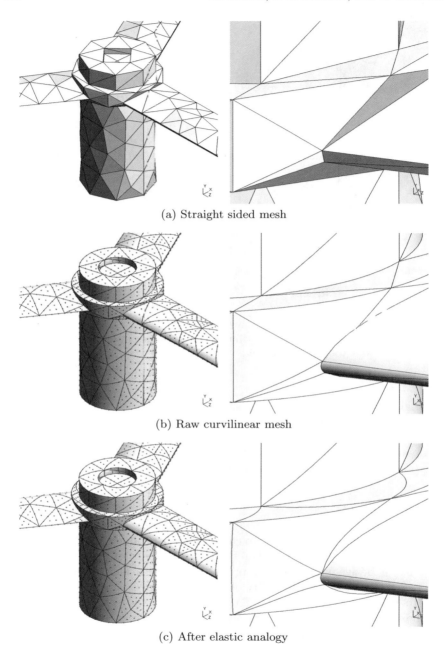

(a) Straight sided mesh

(b) Raw curvilinear mesh

(c) After elastic analogy

Fig. 6. Curvilinear mesh of a rotor using fourth-order curved triangles.

Acknowledgement

This research project was funded in part by the Walloon Region under WIST 3 grant 1017074 (DOMHEX).

Authors gratefully thank E. Bechet from the University of Liège and K. Hillewaert from Cenaero for insightful discussions about Bézier functions and curvilinear mesh generation. Authors also thank V. D. Nguyen for providing the microstructure geometry used in Figure 3.

References

1. Dey, S., O'Bara, R.M., Shephard, M.S.: Curvilinear mesh generation in 3D. Computer Aided Geom. Design 33, 199–209 (2001)
2. Shephard, M.S., Flaherty, J.E., Jansen, K.E., Li, X., Luo, X., Chevaugeon, N., Remacle, J.-F., Beall, M.W., O'Bara, R.M.: Adaptive mesh generation for curved domains. Applied Numerical Mathematics 52, 251–271 (2005)
3. Sherwin, S.J., Peiró, J.: Mesh generation in curvilinear domains using highorder elements. International Journal for Numerical Methods in Engineering 53, 207–223 (2002)
4. Geuzaine, C., Remacle, J.-F.: Gmsh: a three-dimensional finite element mesh generator with built-in pre- and post-processing facilities. International Journal for Numerical Methods in Engineering 79(11), 1309–1331 (2009)
5. Hughes, T.: The Finite Element Method. Dover (2003)
6. Babuška, I., Szabó, B., Actis, R.L.: Hierarchic models for laminated composites. International Journal for Numerical Methods in Engineering 33, 503–535 (1992)
7. Farin, G.E.: Curves and surfaces for CAGD: a practicle guide. Morgan-Kaufmann (2002)
8. Lane, J.M., Riesenfeld, R.F.: A theoretical development for the computer generation and display of piecewise polynomial surfaces. IEEE Transactions on Pattern Analysis and Machine Intelligence 2(1), 35–46 (1980)
9. Cohen, E., Schumacker, L.L.: Rates of convergence of control polygons. Computer Aided Geometric Design 2, 229–235 (1985)
10. Lambrechts, J., Comblen, R., Legat, V., Geuzaine, C., Remacle, J.-F.: Multiscale mesh generation on the sphere. Ocean Dynamics 58, 461–473 (2008)

Uniform Random Voronoi Meshes

Mohamed S. Ebeida and Scott A. Mitchell

Sandia National Laboratories, P.O. Box 5800, Albuquerque, NM 87185-1318
msebeid@sandia.gov

Summary. We generate Voronoi meshes over three dimensional domains with prescribed boundaries. Voronoi cells are clipped at one-sided domain boundaries. The seeds of Voronoi cells are generated by maximal Poisson-disk sampling. In contrast to centroidal Voronoi tessellations, our seed locations are unbiased. The exception is some bias near concave features of the boundary to ensure well-shaped cells. The method is extensible to generating Voronoi cells that agree on both sides of two-sided internal boundaries.

Maximal uniform sampling leads naturally to bounds on the aspect ratio and dihedral angles of the cells. Small cell edges are removed by collapsing them; some facets become slightly non-planar.

Voronoi meshes are preferred to tetrahedral or hexahedral meshes for some Lagrangian fracture simulations. We may generate an ensemble of random Voronoi meshes. Point location variability models some of the material strength variability observed in physical experiments. The ensemble of simulation results defines a spectrum of possible experimental results.

1 Introduction

1.1 Mesh Terminology

A Voronoi mesh has different local structure than the more familiar tetrahedral and hexahedral meshes. However, a triangulation, a hex mesh, a Voronoi diagram, and a watertight input domain are all examples of *simplicial complexes*. Space is subdivided into geometric and discrete *faces*: faces intersect at a subface or not at all. We call two-dimensional faces *facets*. In a simplicial complex faces are not necessarily simplices, the convex hull of $d+1$ vertices.

For simplicity of exposition, we assume that all faces are geometrically flat. However, the implementation does not require this and the method is extensible to solid modeling engines with curved domain boundaries. If the domain boundary is faceted and boundary edges are marked as lying on a curved facet, we may recover some of the curvature.

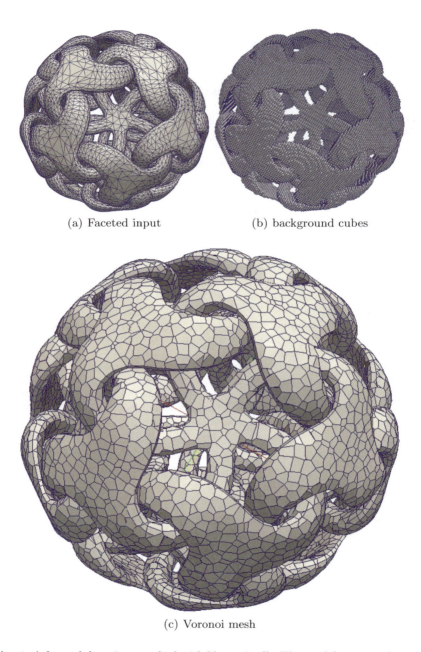

Fig. 1. A faceted domain remeshed with Voronoi cells. The model is topmod-test.stl [20], based on the dodecahedron. There are curved surfaces with narrow regions.

1.2 Maximal Poisson-disk Sampling (MPS)

Maximal Poisson-disk sampling (MPS) selects random points $\{x_i\} = X$, from a domain, $\mathcal{D}$. There is an exclusion/inclusion radius r: *empty disk* means no two sample points are closer than r to one another; and *maximal* means samples are generated until every location is within r of a sample. $\mathcal{D}_i$ is the subregion of $\mathcal{D}$ outside the r-disks of the first i samples. For a *bias-free* (a.k.a. unbiased) sampling procedure, the probability P of selecting a point from a disk-free subregion Ω is proportional to Ω's area.

$$\text{Bias-free: } \forall \Omega \subset \mathcal{D}_{i-1} : P(x_i \in \Omega) = \frac{\text{Area}(\Omega)}{\text{Area}(\mathcal{D}_{i-1})} \tag{1a}$$

$$\text{Empty disk: } \forall x_i, x_j \in X, i \neq j : ||x_i - x_j|| \geq r \tag{1b}$$

$$\text{Maximal: } \forall p \in \mathcal{D}, \exists x_i \in X : ||p - x_i|| < r \tag{1c}$$

A maximal r-disk sample (1b) (1c) is equivalent to a maximal sample of non-overlapping $r/2$-disks, known as a random close packing. Sphere packings appear frequently in nature: e.g. sand, atoms in a liquid, trees in a forest. Processes generating packings include random sequential adsorption, the hard-core Gibbs process, and the Matérn second process. Algorithmically, by successively generating points and rejecting those violating (1b) it is easy to get a near-maximal sample if run-time is unimportant. In recent years the community has developed unbiased MPS algorithms with near linear performance [10, 9, 13]. There are variations based on advancing fronts that have biased point locations, violating (1a), but may be more efficient [30].

1.3 Voronoi Diagrams

A point *seed* x_i defines a Voronoi cell, V, the subset of the domain that is closer to that seed than any other seed [11]. The cell equation is related [25] to the maximal sampling condition (1c).

$$V_i = \{p\} \in \mathcal{D} : \forall j, ||p - x_i|| \leq ||p - x_j|| \tag{2}$$

For point sets, a dual of the Voronoi diagram is a Delaunay triangulation.[1] Voronoi meshes differ from the more familiar unstructured primal meshes. Primal elements are simplices — or perhaps squares or hexahedra, a.k.a. cuboids — with a fixed number of subfaces with a particular structure. Vertices may be in an arbitrary number of elements. (The maximum degree is related to the minimum angle.) For Voronoi meshes the situation is reversed by dimension. Vertices have nominally fixed degree: e.g. three edges in two-dimensions, barring extra cocircularity. But cells have arbitrary subfaces, and relationships between subfaces are variable, position dependent. Traversing an element may involve walking dynamic datastructures. For performing analysis, developing shape functions for Voronoi elements is non-trivial [2].

---
[1] Georgy Voronoy being the doctoral advisor of Boris Delone.

Many codes are available for computing Voronoi tessellations, and related structures, for point sets. Fortune [11] provides a very fast algorithm for 2d, and Qhull [6] up to 4d. (Qhull is used in the current version of Matlab.) Some codes extend to arbitrary-dimensions, weighted points, handle degeneracies, and use a variety of algorithms [1].

Voronoi meshing algorithms add seeds at selected locations for cell quality, and cells conform to the domain boundary. Codes are less common and less universal because the desired seed locations are application specific. Many methods are based on Centroidal Voronoi Tessellations (CVTs), where each seed lies at the center of mass of its cell. This is usually achieved through iterative adjustment of seed location. See Du et al. [7] for a survey of CVTs. While it is possible to generate well shaped cells using CVTs, the geometric regularity of seed locations that arises is particularly undesirable for our fracture simulations. Clipped Voronoi diagrams truncate cells at the domain boundary. This is efficient if some background mesh of the domain is already available [32] and can answer point-location queries. (In contrast we achieve efficiency by exploiting the locality arising from our dense uniform sampling.)

Voronoi diagrams for 2d curved surfaces embedded in 3d have numerous applications, especially remeshing surfaces for a reduced numbers of points or improved quality. Since they are hard to compute, the restricted Voronoi diagram [31] approximates geodesic distance by straightline distance.

1.4 Simplicial Meshes

Methods for generating primal meshes are more developed because they have more applications. Triangular and tetrahedral meshes based on the Delaunay principle are ubiquitous in finite element methods. A common method is Delaunay refinement [5, 24, 28, 29]. Fu [12] provides a complete remeshing pipeline using a CVT. Some primal meshing codes will generate a Voronoi diagram of the mesh points [27], but this is not the same as selecting Steiner points to generate Voronoi cells meeting specific quality requirements.

Disk packings have been used to generate primal meshes, because of the relationship between empty Delaunay circles and disks maximally covering the domain. In prior work [8] we generated constrained Delaunay triangular meshes of two-dimensional non-convex domains using maximal Poisson-disk sampling. Miller et al. [17] used maximal disk packings to generate tetrahedral meshes with bounded radius-edge condition. Shimada and Gossard uses a form of disk packing ("Bubble Meshes") for 2d and 3d domains [25] and curved surfaces [26]; point locations are iteratively adjusted with a force network, and points are added and deleted.

1.5 Application Needs

Some applications, notably fracture mechanics, prefer Voronoi meshes. (A.k.a. "Voronoi froths" because of their similar geometry to soap bubbles.) See Bishop [2] for an overview of fully Lagrangian fracture simulations over Voronoi froths, including element formulation and cell movement.

Fracture simulations over structured grids and CVTs produce unrealistic cracks. For MPS Voronoi froths, the orientation of edges with respect to the coordinate system is uniformly random, and has other desirable statistical properties [4]; cracks initiate and propagate more realistically [3]. Families of random meshes are desired because they represent the random variation in material strengths, leading to variations in crack locations and propagation directions. The ensemble of simulation results predicts a range of possible experimental outcomes. For other physics simulations, a mesh ensemble can also be used to detect dependence on mesh artifacts.

Voronoi tessellations model polycrystalline structures well. Infinite domains are common: no internal boundaries and periodic boundary conditions. Each cell is a grain, a region with a particular crystal orientation. Polycrystal simulations often model fracture, the material being weak at grain boundaries. However, grains are typically divided into many primal finite elements, which models finer scale phenomena than our target fracture simulations. Seed locations model crystal initiation sites — CVT is often used — and so are fundamental to the domain and not a free choice as in larger scale fracture mechanics. However, both applications have in common the problem of obtaining good quality cells by removing small cell features [22].

Fracture domains are often rectangular blocks with internal boundaries; non-convex domains with non-trivial holes are also common. The mesh must be *constrained* to contain boundary features: each feature must be represented by a well-defined submesh, an exact subset of mesh elements. To achieve this we clip Voronoi cells by boundary facets. No part of the domain boundary is strictly-interior to a cell. Seeds are placed on concave boundary edges to ensure that a Voronoi element is visible to its seed, and is star-shaped.

1.6 Cell Quality

Quality metrics for simplicial meshes are well developed [16]. Paoletti considers mesh smoothing to optimize the interpolation error for convex polyhedral cells [21]. For Voronoi meshes, typical metrics include the *aspect ratio* of cells, the ratio of the radii of the smallest containing sphere to largest contained sphere. MPS leads to absolute bounds on both radii, and consequently a bound on the aspect ratio. MPS also leads to natural bounds on dihedral angles, similar to the angle-bounds on triangles from a Delaunay triangulation of a maximal sampling; see Section 3. There is a weak relationship between Voronoi cell quality and dual Delaunay tetrahedra quality. Except where cells are clipped by the domain boundary, Voronoi vertices are the centers of circumspheres of Delaunay tetrahedra. A cell contains all the centers of all the spheres of the tetrahedra sharing a common vertex. Hence an upper bound on the radius of Delaunay spheres bounds the outsphere radius of Voronoi cells. The Voronoi insphere radius is related to the shortest Delaunay edge, which in turn is related to the smallest Delaunay circumsphere and smallest angle. However, it appears impossible to convert tetrahedral quality directly into Voronoi cell quality, because each Voronoi cell depends on multiple tetrahedra.

For fracture over Voronoi meshes, as in many simplicial mesh applications, the timestep is determined by the shortest edge length. Since there is no natural bound on cell edge length, short edges are collapsed to increase the timestep. In 3d meshes, collapsing edges causes facets to be non-planar. (A facet may also be non-planar because it lies on a curved domain boundary.) The non-planarity reduces the simulation accuracy, by impeding facture formation. Thus collapsing short edges that are too long causes problems, and these two motivations are in competition. Additional measures may be important: e.g. electromagnetic simulations desire facets of uniform area.

1.7 Summary of Contribution

We are able to construct good quality Voronoi meshes for non-convex domains in three dimensions, suitable for fracture simulations.

The maximal sphere-packing approaches to primal meshing [8, 17] differ from the current work mainly in how the domain boundary is handled. Primal approaches actively sample the boundary of the domain in a hierarchy by dimension: sample domain vertices, then edges, facets, and finally the interior of the domain. This is done to avoid a sample point (simplex vertex) arbitrarily close to the boundary, leading to unbounded simplex angles. Boundary sampling is not needed as much to achieve good quality Voronoi cells. We presample **large-angle** domain edges. Non-manifold internal boundaries must also be sampled. In contrast, primal meshing algorithms often take no special care around concave features, and small-angle features are more problematic. For bounded aspect ratio, we may do some preprocessing around small-angle features as well. Small edges are collapsed.

The mesh we create is the clipped Voronoi diagram of the seeds: cells are truncated by the domain boundary. This is necessary because unclipped cells may be infinite and have vertices outside the domain. The dual of the Voronoi mesh are tetrahedra, except where small edges have been collapsed. However, except where we seeded reflex edges, these tetrahedra do not have vertices on the boundary of the domain: they are inside the domain. Thus our Voronoi mesh is not the dual of a body-fitted tetrahedral mesh. The Voronoi submesh on the domain boundary is not an ordinary 2d Voronoi diagram. (However, relative-interior 2-cells may be a **weighted** Voronoi diagram of the projection of nearby seeds to the domain boundary.) All of the cells are convex, except concave-edge cells may be merely star-shaped, visible to the seed on the edge. These may be further subdivided into convex cells for fracture simulations.

Our algorithm also differs from the literature in how the Voronoi diagram is constructed. Our MPS algorithm [10] provides a background grid of cubes — see Figure 1.1 — and links between a sample seed and the cube that geometrically contains it. This provides locality information, so determining which seeds are adjacent is not a time consuming step. Iteratively intersecting Voronoi cells with perpendicular-bisector hyperplanes is efficient in our context. MPS takes expected $O(n \log n)$ time (or $O(n)$ for finite precision) and uses $\Theta(n)$ deterministic space, where n is the output size. Exploiting the background grid and the uniformity of the sampling, constructing the Voronoi cell for a point is a constant-time and space operation. The overall complexity is the same as for MPS.

2 Algorithm

- **Protect** concave boundary features with random disks.
  - **Preprocess** sharp edges, reduce r for close edges, as needed.
- **Sample** the interior of the domain, until the set of disks is maximal.
- **Generate** Voronoi cells for sample points, trimmed by the boundary.
  - **Weld** vertices to eliminate small edges.

We describe sampling before we describe protecting, because it makes the exposition more clear. Protecting uses a form of sampling.

2.1 Sampling

We use the unbiased maximal Poisson-disk sampling algorithm described in Ebeida et al. [9]. An implicit background grid of cubes locates sample points and determines if they are disk-free. The cubes approximate the remaining disk-free area. We discard (the indices of) cubes that are known to be completely covered by disks. We successively refine the background grid, subdividing all uncovered cubes into eight. Keeping all the cubes the same size allows generating an unbiased candidate point in constant time, and, more importantly, reduces the memory needed.

Each cube $\mathcal{C}$ has diagonal length r and side length $r/\sqrt{3}$, and contains at most one point. (The exception is cubes cut by the domain boundary. In the current software these cubes have no sample points. In future versions a cube $\mathcal{C}$ might have one sample point for each connected component of $\mathcal{C} \cap \mathcal{D}$.)

2.2 Protect the Boundary

Reflex Edges

We pre-sample concave features of the boundary, in order to obtain star-shaped cells with the seed in the kernel. A concave edge is any edge of the

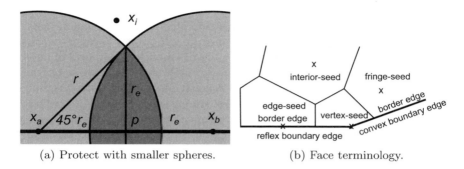

(a) Protect with smaller spheres.   (b) Face terminology.

Fig. 2. (a) Sampling reflex domain edges more densely, with radius $r_e = r/\sqrt{2}$, ensures that domain edges are completely inside the Voronoi cells for those samples. The same principle may be used to protect non-manifold facets. (b) 2d cartoon of 3d faces.

domain with a *reflex* dihedral angle, $\angle > 180°$. (Reflex with respect to the interior of the domain.) For each convex edge, we place a sample at both its domain vertices. We maximally Poisson sample the remainder of the edge with a smaller radius, $r_e = r/\sqrt{2}$. If a sample point x_i not on the edge is at least r from the edge samples, trigonometry shows that x_i must be at least $r/\sqrt{2}$ from the edge; see Figure 2(a). This ensures that every point of a convex edge is closest to one of its edge samples. The exceptions are a *sharp edge*, a convex edge that also makes a small angle with a second edge; and a *close edge*, a convex edge that is also closer to a disjoint domain facet than r.

Non-manifold Geometry

Non-manifold geometry can be protected using the same principles, but with a second $\sqrt{2}$ constant factor. We have not implement it yet. Non-manifold internal edges with no attached domain facets can be protected using the same radius as manifold boundary edges $r/\sqrt{2}$. For non-manifold facets, we may sample their attached edges and vertices with radius $r/2$, then the facets themselves with radius $r/\sqrt{2}$. This will result in all domain edge points being closest to an edge-seed, and all domain facet points closest to a bounding-edge seed or facet-seed. Voronoi cells for facet-seeds will be clipped into two, one for each side, but cells will agree on the domain facet. The penalty would be another factor of $\sqrt{2}$ in the worst-case aspect ratio in Section 3.

Sharp Features

If they are near a convex edge, sharp and close features may require extra care to obtain star-shaped cells. The extra work would be similar to what is done for primal meshes.

Reducing the sampling radius r is sufficient to handle close edges. Our plan to handle sharp edges is to first isolate their common vertex v. We may introduce points a fixed distance d from v, on the surface of a sphere of radius d. We place a sample at the intersection of this sphere and every edge of v, and also maximally sample the sphere in the interior of the domain. Depending on the smallest domain angle at v, some of these points might be close together, say distance r_v. The remainder of the domain can be sampled with radius $r' = \min_v r_v$ and $r'_e = r'/\sqrt{2}$ to obtain an aspect ratio bound on cells not dependent on feature size, or with the original radius and potentially degraded aspect ratio cells.

In the future, we may protect reflex and non-manifold features with spheres placed more randomly and in the interior of the domain, not on the boundary. This has some similarity to our "interior-disks" strategy for primal meshes [10]. The advantages would be improved aspect ratio and dihedral angle bounds, and a greater variation in dihedrals between cell facets and boundary facets. The disadvantage is the process is more complicated.

2.3 Generate Voronoi Cells

We use the locality of sample points in the background grid to generate Voronoi cells. We shall see in Section 3 that Voronoi cells sharing a facet have seeds that are at most $2r$ apart. The only relevant seeds are in cubes within a constant size template $\mathcal{T}$ of indices around the seed's cube. The Voronoi cell for any seed is constant size, and may be computed in constant time.

The template $\mathcal{T}$ is a $9 \times 9 \times 9$ grid of cubes centered at $\mathcal{C}$, trimmed to remove cubes that have no corner within $2r$ of the closest corner of $\mathcal{C}$. V is initialized to the bounding box of $\mathcal{T}$. The cubes contains pointers to the domain boundary faces (and their Voronoi cells) crossing them. We trim the bounding box by these boundary faces. We successively insert candidate seeds x_j from $\mathcal{T}$, and trim V by the perpendicular bisecting plane to $\overline{x_i x_j}$.

2.4 Weld Small Features

The weld tolerance w is a free user parameter; $w = r10^{-4}$ is reasonable. Voronoi vertices that are less than w apart are treated as one, on the fly as a cell containing them is generated. This will remove short edges and small angles. (Large angles do not occur because of the facet dihedral angle upper bound.)

3 Voronoi Cell Quality

See Figure 2(b). We call a Voronoi face a *border face* if it lies on the domain boundary, to distinguish it from unpartitioned domain boundary faces. A seed is a *vertex-seed* (with a vertex-cell) if the seed lies on a domain vertex; an *edge-seed* if it lies on a domain edge, necessarily reflex. Otherwise they are *fringe* if any of the cell's subfaces are border faces, or simply interior.

3.1 Aspect Ratio Bounds

Let *outradius* R_o be the radius of the *outsphere* S_o, the smallest enclosing sphere of a Voronoi cell. *Inradius* R_i is the radius of the *insphere* S_i, the largest contained sphere. The aspect ratio of a cell is $R_o/R_i > 1$. The centers of S_o and S_i are not in general x_i, but spheres centered at x_i provide bounds on R_o and R_i. Let $S_{p(R)}$ denote the sphere of radius R centered at p. We analyze cells prior to welding.

Lemma 1. $R_o \leq r$.

Proof. This follows trivially from maximal sampling. Since Voronoi cells are trimmed, all points p of a cell are in the domain. The domain is maximally sampled, so p is at most r away from some seed. By definition x_i is the closest one. $S_{x_i(r)}$ contains x_i's Voronoi cell.

Lemma 2. *For interior cells,* $R_i \geq r/2$.

Proof. Two seeds are at least distance r apart if at least one of them is interior. $S_{x_i(r/2)}$ is contained in x_i's Voronoi cell.

Corollary 1. *For edge-cells and vertex-cells* $S_{x_i(r_e/2)} \subset V$, *for fringe and interior cells* $S_{x_i(r/2)} \subset V$, *except where trimmed by the domain boundary.*

For non-interior cells, the smallest inradius might be driven by a small domain angle, or a small distance between disjoint faces on the domain. Let $\cdot_{\min}$ denote the smallest angle or feature over all the domain. Let the *local feature size lfs* [23] at point p be the smallest radius of a sphere at p containing two boundary faces that do not share a common subface. Let twice the *feature size* $fs_{\min}$ be the smallest distance between two disjoint boundary faces. ($fs_{\min} = \min lfs$, achieved at the midpoint of the smallest distance.) We consider several types of small angles. Let 2ϕ be a domain dihedral angle between two planar facets meeting at an edge; 2χ a domain angle between an edge and a facet, or two facets meeting at a vertex; and 2ψ a domain angle between two edges. Let 2ω be the aperture of the largest right-circular cone with apex at a domain vertex or point on an edge, inside the domain.

Claim. $\omega \geq \min(\phi_{\min}, 2\chi_{\min}/3, \psi_{\min})$.

The tighter dependence on χ occurs for an equilateral triangular cone, where several small angles together allow a smaller ω. In any event, ω is a domain-specific constant dependent on its small angles.

The following lemmas consider the above quantities as the limiting conditions on R_i in sequence. For curved domains, the smallest local angle would be the limiting quantity. In the proofs we assume $|x_i \partial \mathcal{D}| \leq r/2$, because otherwise Corollary 1 implies the bound from Lemma 2 applies.

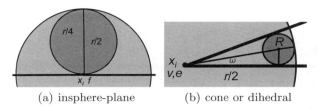

(a) insphere-plane   (b) cone or dihedral

Fig. 3. Minimum inspheres (a) near a domain plane and (b) near a vertex or edge.

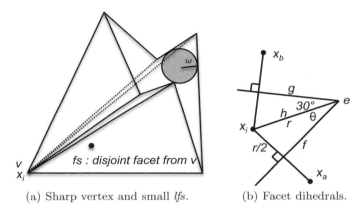

(a) Sharp vertex and small *lfs*.   (b) Facet dihedrals.

Fig. 4. (a) A minimum insphere limited by both a small cone angle ω and a small *lfs* from a nearby disjoint face. The largest insphere may be inside an image of the shown ω sphere, translated and scaled towards x_i so that it is closer to v than the disjoint face is to v. (b) The minimum dihedral angle between Voronoi facets.

Lemma 3. *For non-interior cells with exactly one border facet, $R_i \geq r/4$.*

Proof. See Figure 3(a), and Mitchell and Vavasis [19]. x_i lies inside the domain, so the border facet excludes at most half of $S_{x_i(r/2)}$. In particular the sphere with radius $r/4$, and diameter on the perpendicular to the border facet's plane through x_i, lies inside the $r/2$ sphere at x_i.

Corollary 2. *Edge-cells containing only superfacets of a reflex border edge also have $R_i \geq r/4$.*

Proof. Since reflex boundary edges are sampled more closely, adjacent edge-cells might limit the extent of the cell to $r_e/2$ in the direction of the domain edge. However, the planes at distance $r_e/2$ from x_i perpendicular to the domain edge do not intersect the radius $r/4$ sphere we constructed in Lemma 3.

Lemma 4. *For a non-interior cell containing only border edge e and superfacets with convex dihedral 2ϕ,*

$$R_i \geq \frac{r \sin \phi}{2(1 + \sin \phi)}$$

Proof. See Figure 3(b) or Mitchell [18] Figure 4.14. The smallest insphere occurs when x_i nearly lies on e. In this case $S_{x_i(r/2)}$ contains a sphere of radius R centered at point p with $\sin \phi = R/(r/2 - R)$.

Corollary 3. *For non-interior cell touching boundary faces sharing vertex v,*

$$R_i \geq \min \left(\frac{r_e}{4}, \frac{r_e \sin \omega}{2(1 + \sin \omega)} \right).$$

Proof. Here we must use r_e instead of r because some boundary edges of v might be reflex and be sampled with r_e.

Lemma 5. *For a non-interior cell touching two disjoint boundary faces,*

$$R_i \geq \min(r_e, fs) \min \left(\frac{1}{4}, \frac{\sin \omega}{2(1 + \sin \omega)} \right),$$

Proof. See Figure 4(a). Open sphere $S_{x_i(fs/2)}$ contains no disjoint faces, so one of the prior lemmas applies with r or r_e replaced by fs.

To see that these lemmas are tight, place x_i at the cone apex. Then place a second seed at distance r_e along a reflex edge, or a disjoint face at distance fs along the axis of the relevant cone. Combining the insphere and outsphere lemmas yields the following.

Theorem 1. *Prior to welding, interior Voronoi cells have aspect ratio $A \leq 2$, fringe and edge-cells $A \leq 4 \max(1, r/fs) \max (1, (1 + \sin \omega)/(2 \sin \omega))$, and vertex-cells*

$$A \leq 4 \max(\sqrt{2}, r/fs) \max \left(1, \frac{1 + \sin \omega}{2 \sin \omega} \right).$$

The worst case insphere and outsphere can be achieved simultaneously in many cases by achieving the maximum cell expanse r in directions not restricted by the boundary, roughly perpendicular to the ω cone axis. For example, the set of facets containing the common vertex in Figure 4(a) could be modified to extend further up and down, without increasing the insphere.

Welding increases R_o by at most w, and decreases R_i by at most w.

3.2 Dihedral Angle Bounds

Lemma 6. *Interior cells have interior dihedral angles in $[60°, 150°]$. Non-interior cells have border-internal facet dihedrals in $[30°, 150°]$, or $[20.7°, 159.7°]$ for cells near more than two border facets. Border-border facet dihedrals are determined by the domain.*

Proof. For interior cells the dihedral angle is equal to the supplement of $\angle x_a x_i x_b$. Each side of $\triangle x_a x_i x_b$ is between r and $2r$ by the maximal packing, and the circumcircle at most r, because the circumcenter is a point on the edge. From [8] Corollary 4, $\angle x_a x_i x_b$ is between $30°$ and $120°$. This shows $[60°, 150°]$ for interior cells.

For non-interior cells, let θ be "half" the dihedral; see Figure 4(b). For edge e, let f be its interior facets and g its border facet, and h the plane through e and x_i. Let $\theta = \angle fh$. By Lemma 1 the closest point to v on the line through e may be at most r away. The closest point of the plane through f as at least $r/2$ from x_i so $\sin\theta \geq 1/2$. The exception is if f is defined by two seeds on a reflex boundary edge, in which case we have $r_e/2$ and $\sin\theta \geq \sqrt{2}/4$. For non-internal cells θ is a tight lower bound on the dihedral because at worst the other facet through e might be coplanar with x_i. It cannot be less than θ because trimmed Voronoi cells are star-shaped with x_i in the kernel. Since reflex domain edges are protected, the supplement of θ is an upper bound on border-internal facet dihedrals. For interior cells 2θ is the minimum dihedral angle, which is another way to get the lower bound of $60°$.

For Voronoi meshes for fracture simulations, seeds on the boundary are undesirable because the angle between a domain facet and an adjacent cell facet is always exactly $90°$. For two dimensional meshing problems, this angle may be randomized by moving the mesh vertex on the boundary [2], but in higher dimensions this leads to non-planar facets.

Interior Voronoi facets are convex, but otherwise the smallest angle between edges may be arbitrarily close to zero. The largest angle between edges is bounded by the dihedral angle between faces; since this has an upper bound any facet with a large 2d-aspect ratio must have a small angle and a short edge. Short edges and small angles are resolved with welding, collapsing a short edge to a single vertex.

4 Experimental Results

Figures 1, 5, and 6 shows some initial example meshes. Because manifold domain facets are not sampled, they may have small-looking border facets. These facets are typically for seeds about r from the boundary; these fringe cells have aspect ratio between 1 and 2.

We report the aspect ratio and dihedral angles of convex cells achieved in practice. Most commercial software uses functions that capture features of aspect ratio but are easier to compute. Computing the aspect ratio is non-trivial, because the centers of the insphere and outsphere are not necessarily the seed. Miniball software computes the outsphere of the points of a cell [14]. For the insphere, a general solution is to compute the medial axis skeleton of each cell and consider the solution value at each medial axis vertex, or a voxel-based approximation. We are currently developing an efficient method for convex polyhedra based on walking the medial axis, traveling in directions where the insphere increases. For this paper we employed a brute force

Fig. 5. I-beam Voronoi meshes at three different resolutions, with cut-away along cell boundaries. At four domain vertices, two convex edges meet one concave edge. The non-convex, but star-shaped facets for the vertex- and edge-cells were subdivided into convex facets by the rendering engine.

approach for convex cells that computes the insphere for all combinations of four facets, and determines if that insphere is clipped by any other facets.

See Figures 7 and 8. Our quality example is a Voronoi mesh of the unit cube with 26,362 seeds (cells) and $r = 0.04$ spacing. 23,196 seeds are interior, and 3,166 are fringe. The cells have 163,808 vertices. There are no edge- and vertex-cells since the unit cube has 90° dihedrals.

The max and min quality values diverge from the theory due to a variation in the packing near the boundary: no background cubes that cut the boundary get a sample. We will remove this variation in future software versions.

5 Conclusion

In summary, we have demonstrated the ability to generate three dimensional polyhedral meshes as the clipped Voronoi cells of random points. The mesh is different from the dual of a boundary-fitted tetrahedral mesh. Voronoi cell quality is provably bounded because the point samples are the centers of uniform empty disks, and because the sampling is maximal.

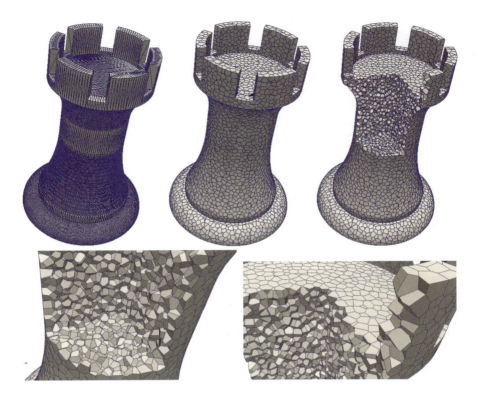

Fig. 6. GEO1.stl [15], based on a rook chess piece.

We have several planned improvements in the near-term. We plan to implement non-manifold (two-sided) internal boundary facets, edges, and vertices. These are all essential for crack simulations. We also plan to turn our research code into a user-ready tool, with a simplified interface for parameter selection. We seek the best-element-quality strategy for decomposing non-convex edge- and vertex-cells into convex cells for fracture analysis. We are researching alternatives to collapsing small edges based on seed location adjustment, because non-planar facets cause problems for the fracture simulations.

We plan to compare our running time and output quality to the alternatives. We are aware of no alternative that is solving the exact same problem, but we may compare to point-set Voronoi codes, primal meshing codes that also produce Voronoi cells, and centroidal-Voronoi tessellation codes.

We suggest the community engage in Voronoi mesh R&D in the following areas: graded meshes; quality metrics, both codes and theory; quality improvement strategies; and understanding the relationship between mesh quality and simulation quality. These have been important for primal meshes.

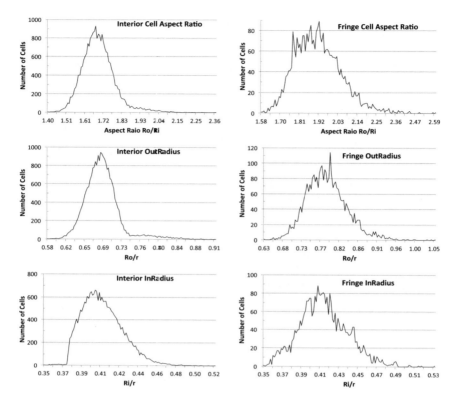

Fig. 7. Aspect ratios for a Voronoi mesh of a unit cube with $r = 0.04$ and 26k cells.

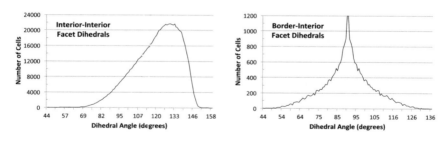

Fig. 8. Dihedral angles for a Voronoi mesh of a unit cube. The interior-interior dihedral distribution is the mirror image of the angle distribution of 2d MPS Delaunay meshes [8]. Border-interior dihedrals are symmetric because the dihedral on one side of an internal facet is the supplement of the dihedral on the other.

Acknowledgments

We thank Joseph Bishop for discussing the effect of random Voronoi meshes on fracture simulations with us. We thank Patrick Knupp for discussing Voronoi mesh quality with us and his helpful draft review. We thank the U.S. DOE, Office of Advanced Scientific Computing Research, SC-21, SciDAC-e, and Sandia's Computer Science Research Institute for supporting this work.

Sandia National Laboratories is a multi-program laboratory managed and operated by Sandia Corporation, a wholly owned subsidiary of Lockheed Martin Corporation, for the U.S. Department of Energy's National Nuclear Security Administration under contract DE-AC04-94AL85000.

References

1. Amenta, N.: Arbitrary dimensional convex hull, Voronoi diagram, Delaunay triangulation, http://www.geom.uiuc.edu/software/cglist/ch.html
2. Bishop, J.: Simulating the pervasive fracture of materials and structures using randomly close packed voronoi tessellations. Computational Mechanics 44, 455–471 (2009), 10.1007/s00466-009-0383-6
3. Bolander Jr., J.E., Saito, S.: Fracture analyses using spring networks with random geometry. Engineering Fracture Mechanics 61, 569–591 (1998)
4. Bondesson, L., Fahlén, J.: Mean and variance of vacancy for hard-core disc processes and applications. Scandinavian Journal of Statistics 30(4), 797–816 (2003)
5. Paul Chew, L.: Guaranteed-quality triangular meshes. Technical Report 89-983, Department of Computer Science, Cornell University (1989)
6. Brad Barber, C., Dobkin, D., Huhdanpaa, H.: Qhull (1995), http://www.qhull.org/
7. Du, Q., Faber, V., Gunzburger, M.: Centroidal Voronoi tessellations: Applications and algorithms. SIAM Review 41(4), 637–676 (1999)
8. Ebeida, M.S., Mitchell, S.A., Davidson, A.A., Patney, A., Knupp, P.M., Owens, J.D.: Efficient and good Delaunay meshes from random points. In: Proc. 2011 SIAM Conference on Geometric and Physical Modeling (GD/SPM11). Computer-Aided Design (2011)
9. Ebeida, M.S., Mitchell, S.A., Patney, A., Davidson, A.A., Owens, J.D.: Maximal Poisson-disk sampling with finite precision and linear complexity in fixed dimensions. In: ACM Transactions on Graphics (Proceedings of ACM SIGGRAPH-Asia 2011) (submitted 2011)
10. Ebeida, M.S., Patney, A., Mitchell, S.A., Davidson, A., Knupp, P.M., Owens, J.D.: Efficient maximal Poisson-disk sampling. In: ACM Transactions on Graphics (Proc. SIGGRAPH 2011), vol. 30(4) (2011)
11. Fortune, S.: Voronoi diagrams and Delaunay triangulations, pp. 193–233. World Scientific (1992), http://ect.bell-labs.com/who/sjf/Voronoi.tar
12. Fu, Y., Zhou, B.: Direct sampling on surfaces for high quality remeshing. Computer Aided Geometric Design 26(6), 711–723 (2009)
13. Gamito, M.N., Maddock, S.C.: Accurate multidimensional Poisson-disk sampling. ACM Transactions on Graphics 29(1), 1–19 (2009)

14. Gärtner, B.: Fast and Robust Smallest Enclosing Balls. In: Nešetřil, J. (ed.) ESA 1999. LNCS, vol. 1643, pp. 325–338. Springer, Heidelberg (1999), http://www.inf.ethz.ch/personal/gaertner/miniball.html
15. Johnson, J.: Geo1.stl (2008), http://www.3dvia.com/content/70FF9466784A5C6E
16. Knupp, P.: Algebraic mesh quality metrics. SIAM J. Sci. Comput. 23(1), 193–218 (2001)
17. Miller, G.L., Talmor, D., Teng, S.-H., Walkington, N., Wang, H.: Control volume meshes using sphere packing: Generation, refinement and coarsening. In: 5th International Meshing Roundtable, p. 4761 (1996)
18. Mitchell, S.A.: Mesh generation with provable quality bounds. Applied Math. Cornell PhD Thesis, Cornell CS Tech Report TR93-1327 (1993), http://ecommons.library.cornell.edu/handle/1813/6093
19. Mitchell, S.A., Vavasis, S.A.: An aspect ratio bound for triangulating a d-grid cut by a hyperplane. In: Proceedings of the 12th Annual Symposium on Computational Geometry, pp. 48–57. ACM (1996)
20. Morris, D.: topmod-test.stl (2010), http://www.3dvia.com/content/4D4234435567794B
21. Paoletti, S.: Polyhedral mesh optimization using the interpolation tensor. In: Proc. 11th International Meshing Roundtable, pp. 19–28 (2002)
22. Quey, R., Dawson, P.R., Barbe, F.: Large-scale 3D random polycrystals for the finite element method: Generation, meshing and remeshing. Computer Methods in Applied Mechanics and Engineering 200(17-20), 1729–1745 (2011)
23. Ruppert, J.: A Delaunay refinement algorithm for quality 2-dimensional mesh generation. J. Algorithms 18(3), 548–585 (1995)
24. Shewchuk, J.R.: Delaunay refinement algorithms for triangular mesh generation. Comp. Geom.: Theory and Applications 22, 21–741 (2002)
25. Shimada, K., Gossard, D.: Bubble mesh: Automated triangular meshing of non-manifold geometry by sphere packing. In: ACM Third Symposium on Solid Modeling and Applications, pp. 409–419. ACM (1995)
26. Shimada, K., Gossard, D.: Automatic triangular mesh generation of trimmed parametric surfaces for finite element analysis. Computer Aided Geometric Design 15(3), 199–222 (1998)
27. Hang, S.: Tetgen: A quality tetrahedral mesh generator and a 3D Delaunay triangulator (2005-2011), http://tetgen.berlios.de/
28. Spielman, D.A., Teng, S.-H., Üngör, A.: Parallel Delaunay refinement: Algorithms and analyses. Int. J. Comput. Geometry Appl. 17(1), 1–30 (2007)
29. Üngör, A.: Off-centers: A new type of Steiner points for computing size-optimal quality-guaranteed Delaunay triangulations. Comput. Geom. Theory Appl. 42, 109–118 (2009)
30. Wei, L.-Y.: Parallel Poisson disk sampling. ACM Transactions on Graphics 27(3), 1–20 (2008)
31. Yan, D.-M., Lévy, B., Liu, Y., Sun, F., Wang, W.: Isotropic remeshing with fast and exact computation of restricted Voronoi diagram. In: ACM/EG Symp. Geometry Processing / Computer Graphics Forum (2009)
32. Yan, D.-M., Wang, W., Lévy, B., Liu, Y.: Efficient Computation of 3D Clipped Voronoi Diagram. In: Mourrain, B., Schaefer, S., Xu, G. (eds.) GMP 2010. LNCS, vol. 6130, pp 269–282. Springer, Heidelberg (2010)

MESH OPTIMIZATION
(Tuesday Morning Parallel Session)

A Comparison of Mesh Morphing Methods for 3*D* Shape Optimization[⋆]

Matthew L. Staten[1], Steven J. Owen[1], Suzanne M. Shontz[2],
Andrew G. Salinger[1], and Todd S. Coffey[1]

[1] Sandia National Laboratories[⋆⋆], Albuquerque, NM, U.S.A.
   {mlstate,sjowen,agsalin,tscoffe}@sandia.gov
[2] The Pennsylvania State University, University Park, PA, U.S.A.
   shontz@cse.psu.edu

Summary. The ability to automatically morph an existing mesh to conform to geometry modifications is a necessary capability to enable rapid prototyping of design variations. This paper compares six methods for morphing hexahedral and tetrahedral meshes, including the previously published FEMWARP and LBWARP methods as well as four new methods. Element quality and performance results show that different methods are superior on different models. We recommend that designers of applications that use mesh morphing consider both the FEMWARP and a linear simplex based method.

Keywords: mesh morphing, mesh moving, mesh warping.

1 Introduction

The modeling and simulation process often involves many iterations of geometric design changes. Geometric parameters are driven automatically via an optimization procedure. The ability to automatically update an existing mesh to conform to a modified geometry is a necessary capability to enable rapid prototyping of many alternate geometric designs. In literature, this mesh update process is called mesh morphing [28, 21], mesh warping [17, 18], or mesh moving [22]. These algorithms first maintain constant mesh topology, while computing new locations for mesh nodes in order to conform to geometry changes. However, maintaining constant mesh topology has limitations based upon the magnitude of the geometric changes required. More advanced methods for tetrahedral meshes perform local mesh modifications once element quality degrades below a threshold [1, 4, 5, 14]. While

---

[⋆] The work of the third author is supported in part by NSF grant CNS-0720749 and NSF CAREER Award OCI-1054459.
[⋆⋆] Sandia National Laboratories is a multi-program laboratory managed and operated by Sandia Corporation, a wholly owned subsidiary of Lockheed Martin Corporation, for the U.S. Department of Energy's National Nuclear Security Administration under contract DE-AC04-94AL85000.

modifying mesh topology maintains element quality, local modification methods are generally intractable for hexahedral meshes. In addition, mesh topology changes also introduce noise into the gradient computations of the optimization. Extending the range in which the mesh topology may be reused through many iterations of the design process is the desired outcome of this research.

In this paper, we propose four new mesh morphing methods: 1. smoothing alone, 2. weighted residuals, 3. simplex-linear, and 4. simplex-natural neighbor. In addition, two existing methods, FEMWARP [17] and LBWARP [18], which have been previously published for tetrahedral meshes, are extended to hexahedral meshes. We compare these six methods with eight example meshes (four hex, four tet) with numbers of nodes varying from 11k to 136k. We assume the geometric topology of the models remains constant through all iterations of the optimization, which allows the geometric ownership of mesh nodes and elements to remain unchanged.

Ideally, the element quality resulting from any morphing method should be optimal. In practice the element quality from a morphed mesh can still be improved by using the morphed node locations as the initial condition for a post-processing step of optimization based smoothing [3, 11]. The results presented in this paper are generated without post-processing smoothing in order to compare the resulting element quality of the morphing methods.

2 Background

Mesh morphing methods can be categorized as either mesh-based or meshless. Mesh-based methods use the element topology of the mesh being morphed to define a computational space to compute new node locations. Meshless methods ignore the element topology, in favor of other algebraic relationships.

Numerous mesh-based techniques have been developed based on solving partial differential equations (PDEs) assuming the boundary deformation has been defined. For example, spring model approaches have been developed based on solving Laplace's equation, variable diffusion, and biharmonic PDEs [2, 9, 18, 22]. One of these methods, LBWARP [18], is used in the comparison in this paper. A finite element based method (FEMWARP) for triangle and tetrahedral meshes has also been presented [17], and is also part of the comparison in this paper. Several elasticity-based approaches have been developed as well [19, 23, 25]. An optimization-based approach to mesh warping based on the target matrix paradigm recently appeared in [13].

A meshless approach using radial basis interpolation functions has been proposed in [15]. Simplex based meshless approaches have been used for morphing surfaces meshes [24, 28]. In Section 3.3, we extend these simplex methods to morph volume meshes, and compare the results with other methods.

The computer graphics and geometry processing communities use morphing for real-time deformations for animations [10] and geometric mappings between 3D surfaces and the cooresponding 2D parameter space [7].

3 Mesh Morphing Methods

We begin with a geometric domain, $\Omega_G^n = \{G^r | r = 0, 1, 2, 3\}$ at iteration n, with geometry entities of dimension r. An existing finite element mesh, $\Omega_M^n = \{M^r | r = 0, 1, 2, 3\}$, also at iteration n, is assumed to have been generated and fully associated with Ω_G^n. We seek a transformation $\Omega_M^n \Rightarrow \Omega_M^{n+1}$ given a new Ω_G^{n+1} such that the element quality in Ω_M^{n+1} is maximized and usable. We assume the topology of Ω_G^n and Ω_G^{n+1} are identical, which allows the mesh topology of Ω_M^n and Ω_M^{n+1} to also be identical. Therefore, we seek only the nodal ($r = 0$) transformation $\left[M^0\right]^n \Rightarrow \left[M^0\right]^{n+1}$.

In defining the nodal transformation $\left[M^0\right]^n \Rightarrow \left[M^0\right]^{n+1}$ we address the nodes on each geometric entity type independently. For example, nodes are first transformed from vertices ($r = 0$) of $\left[G^0\right]^n$ to vertices of $\left[G^0\right]^{n+1}$, followed by curves ($r = 1$), surfaces ($r = 2$) and finally volumes ($r = 3$). In each step, the locations of nodes associated to lower dimension entities are assumed to be fixed. We introduce the notation $\mathbf{X}_r(e)$ to represent the location of entity e, which is either a geometric vertex or a mesh node. If e is a geometric vertex, r is omitted. If e is a mesh node, r represents the dimension of the geometric entity to which the node is associated.

The new location of node k associated to geometric vertex j is simply:

$$\mathbf{X}_0\left(\left[M_k^0\right]^{n+1}\right) = \mathbf{X}\left(\left[G_j^0\right]^{n+1}\right). \tag{1}$$

For curves, we make the assumption that the parametric location, t_k, on curve j, for any node k, remains constant through the transformation $n \Rightarrow n + 1$, ($t_k = t_k^n = t_k^{n+1}$). We define the location for any interior node k on curve j as:

$$\mathbf{X}_1\left(\left[M_k^0\right]^{n+1}\right) = \left[G_j^1\right]^{n+1}(t_k) \tag{2}$$

where $\left[G_j^1\right]^{n+1}(t_k)$ is a simple parametric evaluation of curve j at t_k.

For nodes associated to surfaces we use either the smoothing or weighted residual method described below. Any of the methods described below could be used to morph surface nodes if the surface is planar, or with a post-morph 3D surface projection. However, our focus is on 3D morphing of volume mesh nodes, so we have only implemented the smoothing and weighted residual methods for general 3D surface node transformations.

For nodes associated to volumes we use either the smoothing, weighted residual, simplex-linear, simplex-natural neighbor, FEMWARP, or LBWARP methods described below. Optionally, for structured hexahedral meshes, the interior node locations of the morphed models could be computed using transfinite interpolation. However, we seek methods which can be applied to unstructured hex mesh topologies.

3.1 Smoothing

Smoothing, a mesh-based method, utilizes smoothing techniques from the Mesquite [3] toolkit, and is used for morphing both surface and volume nodes ($r = 2, 3$). After the nodes on the bounding curves are morphed to their new positions, the interior surface mesh will often be inverted. Surface smoothing is then performed using the nodal coordinates established on the boundary curves as fixed. Mesquite's mean ratio, condition number [11] and untangling smoothing procedures are used adaptively based on local mesh quality. The same smoothing techniques are subsequently employed for volumes, where the node locations established on the boundary surfaces are fixed.

3.2 Weighted Residuals

Weighted residual, a meshless method, borrows from hexahedral sweeping, which morphs quad meshes from "source" surfaces to intermediate or "target" surfaces [12]. The weighted residual method is used for surface and volume morphing ($r = 2, 3$). The initial node locations, $\left[M^0\right]^n$, are first transformed using an affine transformation, $\mathbf{\Gamma}$, computed to minimize the function:

$$F(\mathbf{\Gamma}) = \min \sum \left[\mathbf{X_{r-1}}\left(\left[M_k^0\right]^{n+1}\right) - \mathbf{\Gamma} \cdot \mathbf{X_{r-1}}\left(\left[M_k^0\right]^n\right) \right] \quad (3)$$

for all nodes, k, on the domain boundary. A correction is then applied based upon the weighted sum of the nearby residual vectors at boundary nodes. The location of an interior node k, can be represented as:

$$\mathbf{X}_r\left(\left[M_k^0\right]^{n+1}\right) = \mathbf{\Gamma} \cdot \mathbf{X}_r\left(\left[M_k^0\right]^n\right) + \sum_{i=1}^{npts} w_i R_i, \quad (4)$$

$$R_i = \mathbf{X}_{r-1}\left(\left[M_i^0\right]^{n+1}\right) - \mathbf{\Gamma} \cdot \mathbf{X}_{r-1}\left(\left[M_i^0\right]^n\right), \text{and} \quad (5)$$

$$w_i = \frac{d_i^{-2}}{\sum_{j=1}^{npts} d_j^{-2}} \quad (6)$$

where R_i in (4) represents the minimization error in (3) for boundary node i. The weight w_i in (4), is the normalized inverse distance squared from node k to the nearest boundary nodes. Only a subset of the boundary nodes influence the weight w_i. We identify the $npts$ closest boundary nodes using a *kd-tree*. For the results in Section 4 we used $npts = 20$ for surface morphing and $npts = 80$ for volume morphing, however an adaptive method for determining the influencing nodes may be necessary.

3.3 Simplex-Linear Transformations

Simplex-linear is a meshless method based upon the BMSweep hex sweeping method [24], previously used for surface morphing [28]. In this study, simplex-linear was only implemented for 3D volume morphing, using weighted residual for surface morphing.

Simplex-linear creates a Delaunay tessellation, D^n, of the nodes on the fixed domain boundary. We use QMG[27] to generate D^n. No interior nodes are used. The enclosing tetrahedron, T_i^n, of each interior node i is determined. The barycentric coordinates, B_i, of $\mathbf{X}_3\left(\left[M_i^0\right]^n\right)$ with respect to T_i^n is computed. We require that the connectivity of D^n remains the same for

D^{n+1} and enforce the condition that the barycentic coordinates, B_i, for node i with respect to T_i^n and T_i^{n+1} will be the same. The location of interior node i at iteration $n+1$ is then:

$$\mathbf{X}_3\left(\left[M_i^0\right]^{n+1}\right) = \sum_{j=1}^{4} (B_i)_j \cdot \mathbf{X}_2\left(T_i^{n+1}\right)_j \qquad (7)$$

where $(B_i)_j$ is defined as the jth component of B_i and $\mathbf{X}_2\left(T_i^{n+1}\right)_j$ is the location of the jth node of tetrahedron T_i^{n+1}.

3.4 Simplex-Natural Neighbor Transformations

The simplex-natural neighbor method uses a similar vertex-based weighting scheme as the simplex-linear method described in Section 3.3, but generalizes the selection of weighting vertices. If we consider C_i^n, the set of all tetrahedron, $T_i^n \in D^n$, whose circumsphere contains $\mathbf{X}_3\left(\left[M_i^0\right]^n\right)$, then all vertices, $(N_i^n)_j | j = 1, 2, ...nvert$, of tetrahedron C_i^n, are used to compute the natural neighbor coordinates, $(\lambda_i)_j$ for $[M_i^0]^n$. Similar to equation 3.3 we can define the new location $\mathbf{X}_3\left(\left[M_i^0\right]^{n+1}\right)$ as a weighted average of $(N_i^{n+1})_j$ using weights $(\lambda_i)_j$,

$$\mathbf{X}_3\left(\left[M_i^0\right]^{n+1}\right) = \sum_{j=1}^{nvert} (\lambda_i)_j \cdot \mathbf{X}_2\left(N_i^{n+1}\right)_j \qquad (8)$$

where $(\lambda_i)_j$ are a function of the Voronoi volumes formed by temporarily inserting $[M_i^0]^n$ into D^n as described in [29] and [20]. Because the morphed locations of $[M_i^0]^n$ are influenced by the boundary points within its *natural neighborhood*, rather than only the four points defined by its enclosing tetrahedron, we hypothesize that the resulting morph quality will be improved.

3.5 Finite Elements

Finite element-based mesh warping (FEMWARP) was proposed by Baker [1] and developed by Shontz and Vavasis in [17] for tetrahedral meshes. We now adapt FEMWARP for hexahedral meshes. FEMWARP expresses the coordinates of each interior node, n_i, of the initial mesh as an affine combination of its neighbors with shape functions encapsulated in element stiffness matrices for each element. Extending FEMWARP to hexahedral meshes simply requires implementing hexahedral element stiffness matrices. We use standard bi-linear hex shape functions. Each n_i is then expressed as an affine combination of the nodes which share a common adjacent element with n_i.

The FEMWARP method proceeds as with tetrahedra. Let b and m be the numbers of boundary and interior nodes, $[M^0]^n$, in Ω_M^n, respectively. We form the $(m+b) \times (m+b)$ global stiffness matrix A by assembling the local element stiffness matrices for the boundary value problem $\triangle u = 0$ on $[G^0]^{n+1}$ with $u = u_0$ on $\partial [G^0]^{n+1}$. The result is a symmetric positive definite sparse global stiffness matrix where its nonzero entries correspond to pairs of neighboring nodes.

Next, let A_I denote the $m \times m$ submatrix of A whose rows and columns are indexed by interior nodes, and let A_B denote the $m \times b$ submatrix of A whose rows are indexed by interior nodes and whose columns are indexed by boundary nodes. Let X_I^n be the $m \times 3$ matrix consisting of x, y, z-coordinates of the interior nodes at step n:

$$X_I^n = \{\mathbf{X}_3\left(\left[M_1^0\right]^n\right), \mathbf{X}_3\left(\left[M_2^0\right]^n\right), \ldots, \mathbf{X}_3\left(\left[M_m^0\right]^n\right)\}^T \qquad (9)$$

and let X_B^n be the $b \times 3$ matrix consisting of x, y, z-coordinates of the boundary nodes at step n:

$$X_B^n = \{\mathbf{X}_2\left(\left[M_{m+1}^0\right]^n\right), \mathbf{X}_2\left(\left[M_{m+2}^0\right]^n\right), \ldots, \mathbf{X}_2\left(\left[M_{m+b}^0\right]^n\right)\}^T \qquad (10)$$

where interior nodes are numbered first. Then it follows that:

$$A_I X_I^n = -A_B X_B^n \qquad (11)$$

and

$$A_I X_I^{n+1} = -A_B X_B^{n+1}. \qquad (12)$$

In (11) and (12), the only unknown is X_I^{n+1}, which represents the morphed locations of the interior nodes of M^{n+1} and can be found by solving (12) as a linear system of equations.

3.6 Log Barrier

The log barrier-based mesh warping (LBWARP) technique was proposed by Shontz and Vavasis in [18] for tetrahedral meshes. We now adapt LBWARP for hexahedral meshes. Let $\mathbf{X}_3\left(\left[M_i^0\right]^n\right)$ denote the x, y, z coordinates of the ith interior node, n_i, in the initial mesh. In addition, let the x, y, z coordinates of n_i's adjacent nodes be given by $\{\mathbf{X}_{0,1,2,3}\left(\left[M_j^0\right]^n\right) : j \in N_i\}$, where N_i denotes the set of neighbors of n_i.

We define N_i as the set of all nodes which share an adjacent 3D element with n_i. For tetrahedra, this also means that n_i shares a common mesh edge with every n_j in N_i. However, this is not true for hexahedra because of the more complex hexahedral topology. We choose this definition of N_i to include more nodes in the interpolation and for consistency with FEMWARP.

In order to find the set of weights w_{ij}, where w_{ij} is the weight of node j on interior node i, we use the log barrier function from linear programming [16] to formulate the following optimization problem for each n_i:

$$\begin{array}{l} \max_{w_{ij}, j \in N_i} \quad \sum_{j \in N_i} \log(w_{ij}) \\ \text{subject to} \quad w_{ij} > 0 \\ \qquad \sum_{j \in N_i} w_{ij} = 1 \\ \mathbf{X}_3\left(\left[M_i^0\right]^n\right) = \sum_{j \in N_i} w_{ij} \mathbf{X}_{0,1,2,3}\left(\left[M_j^0\right]^n\right) \end{array} \qquad (13)$$

We solve for all w_{ij} using (13), a strictly convex optimization problem, for its unique optimum using the projected Newton method [30]. (See [18] for more details.)

Once the weights, w_{ij}, have been found, we assemble all w_{ij} into an $(m+b) \times (m+b)$ matrix A, and solve for the morphed interior node locations by decomposing it into A_I and A_B in the same manner used for FEMWARP. For LBWARP, A_I is an M-matrix.

4 Examples

We have implemented the six methods described in Section 3 using C++. For the FEMWARP and LBWARP methods, the Trilinos [26] implementation of the Amesos KLU sparse direct linear solver [6] was used to solve the linear systems. No matrix pre-conditioning was done.

Four example problems are illustrated in Figures 1, 3, 5, and 7. 3D morphing, using all six methods, was performed on a tetrahedral and a hexahedral mesh of each model, giving eight total test cases. To ensure a consistent baseline for the 3D methods, the boundary surface meshes of each example were morphed with the weighted residual method. Table 1 lists the number of nodes and elements in each example model. All test case data were run on an HP ProLiant linux workstation with two X5570 Intel quad cores with 24 GB RAM. The resulting quality and execution times were reported.

The tet meshes were generated with the GSH3D tet mesher [8]. The hex meshes on the bore and pipe models are completely structured and were generated with transfinite interpolation. In contrast, the hex meshes on the courier and canister models are unstructured, having been generated by decomposing the geometries into partitions each of which can be meshed with the pave and sweep method [24].

Table 1. Example Model Sizes

	Bore		Pipe		Canister		Courier	
	Hex	Tet	Hex	Tet	Hex	Tet	Hex	Tet
#nodes	11,904	21,859	11,520	17,174	136,462	83,562	101,817	136,106
#elems	15,190	109,535	8,532	81,304	128,269	472,924	83,934	699,867

Element quality is compared using the scaled Jacobian metric, which is the determinant of the elemental stiffness matrix. Normalization of element edges during formulation of stiffness matrices restricts the scaled Jacobian metric on the range [-1,1].

In each example, two sets of geometric parameters, $S_0 \in \{s_1^0, s_2^0, ...s_{nvar}^0\}$ and $S_F \in \{s_1^F, s_2^F, ...s_{nvar}^F\}$, define initial and final geometry configurations. Morphing is performed with N iterations between S_0 and S_F where the parameters, S_i, used for a given iteration i are defined as:

$$S_i = \frac{i}{N}(S_F - S_0), \qquad i \in \{0, 1, \ldots, N\}. \tag{14}$$

We used $N = 20$. Each model is morphed using both *relative* and *absolute* coordinates. With relative coordinates, the mesh is morphed from S_{i-1} to S_i, testing incremental morphing and iterative element quality degradation.

With absolute coordinates, the mesh is morphed from S_0 to S_i, testing how big a step each morphing method can take while maintaining element quality.

4.1 Bore Model

Figure 1 illustrates the "bore" model with its geometric parameters $S_0 \in \{h, r_1, r_2, r_3, \theta\}$ and $S_F \in \{h', r_1', r_2', r_3', \theta'\}$. The values used are $S_0 \in \{12.0, 2.0, 2.5, 3.0, 60°\}$ and $S_F \in \{12.0, 2.5, 3.5, 6.0, 30°\}$. The bore model tests scaling and rotations in each morphing method. The resulting element quality is illustrated in Figure 2. The hex mesh is a structured mesh generated with transfinite interpolation, while the tet mesh is completely unstructured. For this simple model, all morphing methods except smoothing provide near identical results for both relative and absolute morphing. The smoothing method does well for relative morphing, but quickly inverts element quality for absolute morphing. This suggests that smoothing should be restricted to small parametric changes. Table 2 lists the execution times for each method. It is notable that the execution times are roughly the same for the relative and absolute runs for all methods except for smoothing. Smoothing times more than double for absolute. This is because as the parametric changes get bigger, the displacement of the boundary nodes produces an inverted mesh that smoothing must work harder to untangle.

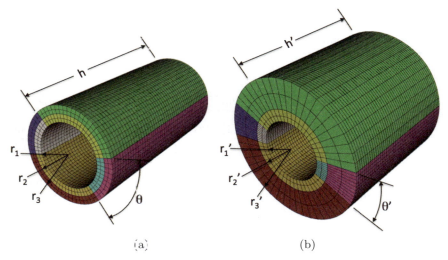

Fig. 1. Example problem "bore". (a) The parametric variables on the initial shape, (b) the final shape.

4.2 Pipe Model

Figure 3a illustrates the "pipe" model with its geometric parameters. Figures 3b and 3c illustrate the initial and ending geometric shapes. The pipe thickness and radius are decreased and the elbow radius is increased. The pipe model tests nonlinear stretching. Figure 4 again shows that the smoothing

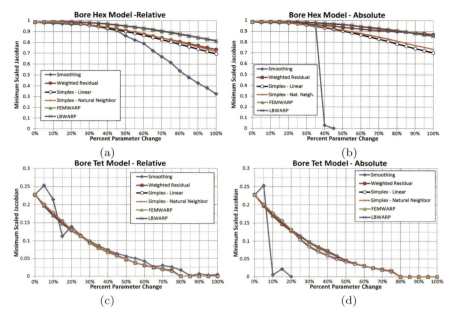

Fig. 2. Results of morphing on bore model. (a) Relative hex morph, (b) absolute hex morph, (c) relative tet morph, (d) absolute hex morph.

Table 2. Results for morphing bore model: CPU Time Per Step

Method	Ave Time Per Step (sec)			
	Hex-Relative	Hex-Absolute	Tet-Relative	Tet-Absolute
Smoothing	0.23	0.75	0.68	1.75
Weighted Residual	3.57	3.65	7.52	7.40
FEMWARP	0.30	0.29	0.56	0.58
LBWARP	1.55	1.36	1.97	1.96
Simplex-linear	0.33	0.40	0.39	0.36
Simplex-natural neighbor	0.46	0.38	0.94	0.56

method is unable to maintain a quality mesh for very long, especially in absolute morphing. For tet morphing the five other methods result in nearly the same element quality. Hex element quality from FEMWARP, simplex-linear, and simple-natural neighbor are roughly the same and superior to weighted residual, while LBWARP is somewhere in the middle.

Execution times for the pipe model are listed in Table 3. For this model, the FEMWARP, simplex-linear, and simplex-natural neighbor methods perform significantly better than LBWARP, smoothing, and weighted residual, while weighted residual is the most expensive method. Simplex-linear is by far the least expensive method.

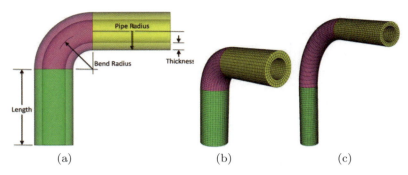

Fig. 3. Example problem "pipe". (a) The parametric variables, (b) the initial shape, (c) the final shape.

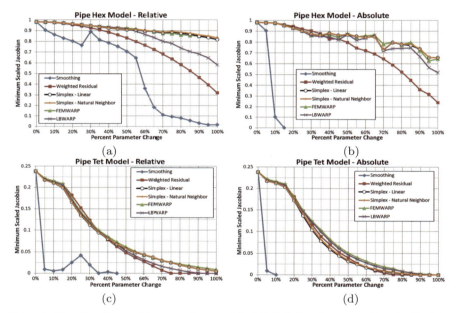

Fig. 4. Results of morphing on pipe model. (a) Relative hex morph, (b) absolute hex morph, (c) relative tet morph, (d) absolute hex morph.

Table 3. Results for morphing pipe model: CPU Time Per Step

Method	Ave Time Per Step (sec)			
	Hex-Relative	Hex-Absolute	Tet-Relative	Tet-Absolute
Smoothing	1.08	1.19	2.56	2.41
Weighted Residual	3.39	3.28	6.08	5.93
FEMWARP	0.31	0.32	0.55	0.56
LBWARP	1.53	1.50	1.88	1.85
Simplex-linear	0.16	0.28	0.31	0.28
Simplex-natural neighbor	0.29	0.35	0.53	0.54

4.3 Canister Model

Figure 5a illustrates the "canister" model which has only one geometric parameter (i.e. the offset of an inner cylindrical hole). Figures 5b and 5c illustrate the initial and ending geometric shapes. The canister model tests severe localized mesh warping, while the majority of the model remains unchanged. Figure 6 shows that for the canister model, FEMWARP and LBWARP provide the best quality. They maintain a positive scaled Jacobian mesh for longer than any other method. The exception is the smoothing algorithm, which performs well for hex-relative (Figure 6a), but poorly for hex-absolute and both tet morphs. The simplex and weighted residual methods also perform well, providing equivelant quality to FEMWARP and LBWARP for the first several steps of all four morphs, until eventually trailing off. Interestingly, simplex-linear and simplex-natural neighbor provide near identical results.

The increased model size of the canister model (~100k nodes) over the pipe and bore models (10k-20k nodes) is reflected in the execution times in Table 4. For hex, FEMWARP and LBWARP are by far the most expensive, followed by weighted residual. For tet, FEMWARP and LBWARP did not show the spike in computation time. The simplex-linear method is by far the most efficient method, several orders of magnitude faster than the others.

Table 4. Results for morphing canister model: CPU Time Per Step

Method	Ave Time Per Step (sec)			
	Hex-Relative	Hex-Absolute	Tet-Relative	Tet-Absolute
Smoothing	5.10	20.90	3.88	4.02
Weighted Residual	76.24	74.86	46.00	46.52
FEMWARP	215.48	216.39	31.92	31.81
LBWARP	265.14	266.96	43.35	44.50
Simplex-linear	0.55	0.83	0.57	0.69
Simplex-natural neighbor	54.25	52.88	23.43	23.48

4.4 Courier Model

Figure 7 illustrates the "courier" model. Eight parametric variables are modified to test full design optimization. For hex-relative, the simplex methods consistently provide the best quality, followed by FEMWARP. LBWARP is the first to invert the mesh. For hex-absolute and the tet morphs, FEMWARP and the simplex methods provide the best element quality.

Table 5 shows that simplex-linear is again the most efficient method. Interestingly, FEMWARP and LBWARP are also very efficient when compared to the Canister model results. It is clear that the efficiency of FEMWARP and LBWARP are model dependent, likely controlled by the conditioning of the sparse global system being solved.

4.5 Performance and Scaling

Twenty-five variations of a brick model were used to obtain scaling data. An $N \times 20 \times 20, N \in \{20, 30, \ldots 260\}$, hex mesh was fit to a $N \times 20 \times 20$ box. The dimensions of the box were then scaled to $N \times 20 \times 40$, and the mesh morphed to fit. Figures 9a and 9b illustrate the model for $N = 20$ and $N = 30$ respectively. Each $N \times 20 \times 20$ mesh was morphed to the $N \times 20 \times 40$ model with each of the six morphing methods, testing $3D$ affine scaling. Different scaling would likely result with different geometric changes such as torsion or shear.

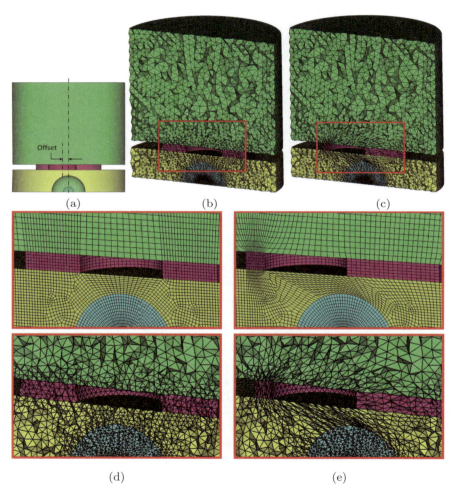

Fig. 5. Example problem "canister". (a) The offset parametric variable, (b) the initial shape, (c) the final shape, (d) zoom in of initial meshes, (e) zoom in of final meshes.

Table 5. Results for morphing courier model: CPU Time Per Step

Method	Ave Time Per Step (sec)			
	Hex-Relative	Hex-Absolute	Tet-Relative	Tet-Absolute
Smoothing	5.13	12.27	37.61	67.40
Weighted Residual	39.70	39.16	69.26	72.76
FEMWARP	4.75	4.89	14.92	14.78
LBWARP	18.55	18.71	28.44	28.49
Simplex-linear	1.07	1.50	1.95	2.55
Simplex-natural neighbor	6.69	7.45	13.23	12.67

The execution times are plotted in Figure 10. Figure 10 shows that simplex-linear, simplex-natural neighbor and weighted residual scale nearly linearly, while LBWARP, FEMWARP and smoothing scale to a power of ~1.50. The simplex methods scale the best and are the most efficient. LBWARP and FEMWARP are the most expensive since they require a solution to a global sparse matrix system.

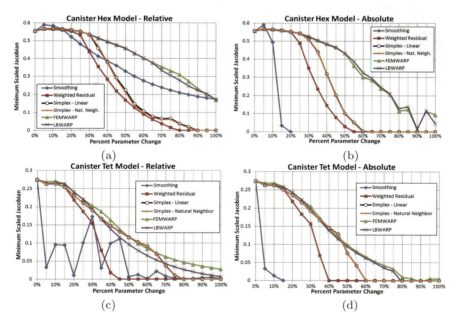

Fig. 6. Results of morphing on canister model. (a) Relative hex morph, (b) absolute hex morph, (c) relative tet morph, (d) absolute hex morph.

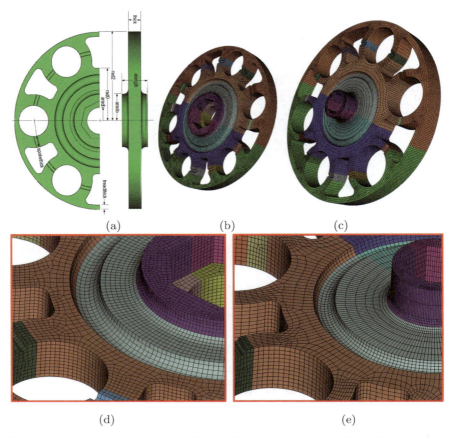

Fig. 7. Example problem "courier". (a) The parametric variables, (b) the initial shape, (c) the final shape, (d) zoom in of initial mesh, (e) zoom in of final mesh.

A Comparison of Mesh Morphing Methods for 3D Shape Optimization 307

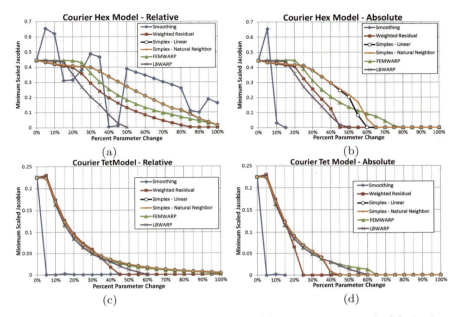

Fig. 8. Results of morphing on courier model. (a) Relative hex morph, (b) absolute hex morph, (c) relative tet morph, (d) absolute hex morph.

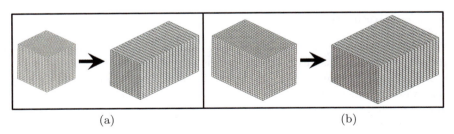

Fig. 9. Model for scaling study, (a) $N = 20$, (b) $N = 30$.

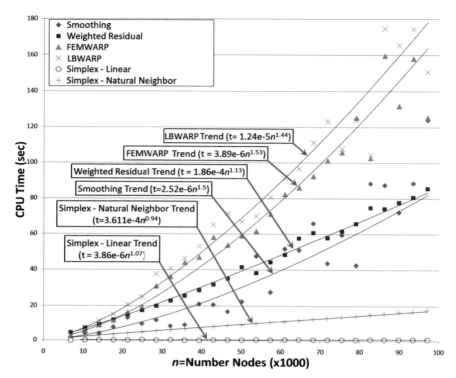

Fig. 10. Scaling (time per step).

5 Conclusions

We have compared six 3D mesh morphing methods on eight example models. Here we make conclusions based on the results of this comparison.

Smoothing: The smoothing method tracks small linear changes in geometry very well, as illustrated by the bore model. However, absolute (large step) morphing and nonlinear transformations, such as those in the pipe, courier, and canister models, proved impossible to capture. The morph of the model boundary produces a severely tangled mesh for large transformations, which a smoother cannot always untangle. Likely additional smoothing iterations are needed. However, in general, smoothing is not a reliable morphing method.

Weighted Residual: The weighted residual method performed on par with the other methods for simple models such as the bore model. However, as model complexity increased, weighted residual and smoothing are generally the first to produce poor mesh quality. Quality might be improved by increasing *npts* in equations (4) and (6), at the cost of decreased efficiency.

Simplex-Linear: This method proved effective for all example models including those with large transformations and complex geometry such as the courier and canister models. Performance and scaling is excellent due to the ability to localize the location of containing tets and the computation of barycentric coordinates. The most expensive part, however minimal, is the generation of the background Delaunay tesselation. However, there are good quality Delaunay tesselators available [27]. Even with the cost of generating the tesselation, simplex-linear is still by far the least computationally expensive method. Considering reasonable element quality and excellent performance, simplex-linear should be considered for any morphing implementation.

Simplex-Natural Neighbor: This method was included in the study in anticipation of improved element quality over simplex-linear by using a more judicious selection of weighting vertices. Using our implementation, both barycentric and natural neighbor weighting exhibited very similar results. As suggested by [29], augmenting the natural neighbor interpolant to use a nodal based gradient function may provide further improvement. However, with our current implementation, we observed little improvement in element quality and decreased performance over simplex-linear. Given that this method is more difficult to implement than simplex-linear, further studies would be needed before this method could be recommended over simplex-linear.

LBWARP: LBWARP provided excellent element quality on all test cases, except for the courier model with hexes. On the courier-hex model, element quality is still good, but less than the other methods. However, scaling of LBWARP is a concern. LBWARP requires solving an $n \times n$ sparse linear system, where n is the number of interior nodes being morphed. An efficient sparse linear solver is critical for any LBWARP implementation. Our example models show that LBWARP can morph meshes with ~100k elements in a few minutes. Computation times may become unacceptable as model sizes increase beyond a few hundred thousand nodes. Efficiency also varies significantly, for similarly sized models, as shown by a comparison of Tables 4 and 5. LBWARP is also complicated to implement. Tolerance and convergence issues must be considered carefully in the computation of the initial weights, w_{ij}, particularly with hexahedral meshes which tend to have large sets of coplanar element faces. Finally, LBWARP requires that the initial mesh be non-inverted (i.e. all elements being convex). Computation of initial weights, w_{ij}, requires that each point lie in the convex hull of its neighbors, which is not guaranteed in inverted meshes. The other five methods do not have this restriction.

FEMWARP: FEMWARP maintained the best element quality for all test cases and is easy to implement. Computation/assembly of local element stiffness matrices is the most difficult coding task. Although, FEMWARP requires solving an $n \times n$ global sparse system matrix, where n is the number of interior nodes, and efficiency varies significantly for similarly sized models, easy parallelization should minimize scaling concerns. We recommend FEMWARP based on its excellent element quality and easy of implementation.

In summary we recommend that both simplex-linear and FEMWARP should be considered when implementing a mesh morphing system. FEMWARP provides excellent element quality and easily extended to parallel for large models. The simplex-linear method also provides reasonable element quality, with excellent performance even for very large models. Additional studies would be required to identify the context in which the other morphing methods presented in this paper would be a reasonable option.

Finally, we observe that hexahedral meshes tend to maintain higher element quality longer than tetrahedral meshes during morphing, particularly with structued hex meshes, likely due to the more flexible element topology of hexahedral elements. For example, in the bore and pipe models, minimum element quality degrades only slightly for the hex meshes, while the quality of tet meshes on the same model become inverted (compare Figures 2a, 2b, 4a, and 4b with Figures 2c, 2d, 4c, and 4d). This same effect is seen to a lesser effect with the unstructured hex meshes on the canister and courier models. Maintaining high element quality during morphing is more important in hexahedral meshes because local remeshing of hexahedra is generally considered intractable, while straightforward for tetrahedral meshes.

References

1. Baker, T.J.: Mesh movement and metamorphosis. In: Proc. 10th Int. Meshing Roundtable, pp. 387–396 (2001)
2. Branets, L., Carey, G.F.: A local cell quality metric and variational grid smoothing algorithm. Eng. Comput. 21, 19–28 (2005)
3. Brewer, M., Diachin, L., Knupp, P., Leurent, T., Melander, D.: The Mesquite mesh quality improvement toolkit. In: Proc. 12th Int. Meshing Roundtable, pp. 239–250 (2003)
4. Cardoze, D., Miller, G., Olah, M., Phillips, T.: A Bézier-based moving mesh framework for simulation with elastic membranes. In: Proc. 13th Int. Meshing Roundtable, 71-80 (2004)
5. Dai, M., Schmidt, D.P.: Adaptive tetrahedral meshing in free-surface flow. J. Comp. Phys. 208, 228–252 (2005)
6. Davis, T.A., Natarajan, E.P.: Algorithm 8xx:KLU, a direct sparse solver for circuit simulation problems. ACM Transactions on Mathematical Software, 1-17 (2011)
7. Floater, M.S.: One-to-one piecewise linear mappings over triangulations. Math. Comp. 72, 685–696 (2003)
8. GSH3D, INRIA, http://www-roc.inria.fr/gamma/gamma/ghs3d/ghs.php
9. Helenbrook, B.T.: Mesh deformation using the biharmonic operator. Int. J. Num. Meth. Engr. 56, 1007–1021 (2003)
10. Jacobson, A., Baran, I., Popović, J., Sorkine, O.: Bounded Biharmonic Weights for Real-Time Deformation. To Appear In ACM Transactions On Graphics (Proc. Of ACM SIGGRAPH) 30(4) (2011),
    http://igl.ethz.ch/projects/bbw/
11. Knupp, P.: Achieving finite element mesh quality via optimization of the jacobian matrix norm and associated quantities, part I. Int. J. Num. Meth. Engr. 48, 401–420 (2000)

12. Knupp, P.: Next-generation sweep tool: a method for generating all-hex meshes on two-and-one-half dimensional geomtries. In: Proc. 7th Int. Meshing Roundtable, pp. 505–513 (1998)
13. Knupp, P.: Updating meshes on deforming domains: An application of the target-matrix paradigm. Commun. Num. Meth. Engr. 24(6), 467–476 (2007)
14. Li, R., Tang, T., Zhang, P.: Moving mesh methods in multiple dimensions based on harmonic maps. J. Comput. Phys. 170, 562–588 (2001)
15. Michler, A.K.: Aircraft control surface deflection using RBF-based mesh deformation. Int. J. Num. Meth. Engr. (2011), doi: 10.1002/nme.3208
16. Nocedal, J., Wright, S.J.: Numerical optimization, 2nd edn. Springer, Heidelberg (2006)
17. Shontz, S.M., Vavasis, S.A.: Analysis of and workarounds for element reversal for a finite element-based algorithm for warping triangular and tetrahedral meshes. BIT, Numerical Mathematics 50, 863–884 (2010)
18. Shontz, S.M., Vavasis, S.A.: A mesh warping algorithm based on weighted Laplacian smoothing. In: Proc. 12th Int. Meshing Roundtable, pp. 147–158 (2003)
19. Shontz, S.M., Vavasis, S.A.: A robust solution procedure for hyperelastic solids with large boundary deformation. Eng. Comput. (2011), doi: 10.1007/s00366-011-0225-y
20. Sibson, R.: A brief description of natural neighbor interpolation. In: Interpretting Multivariate Data, pp. 21–36. John Wiley and Sons, New York (1981)
21. Sigal, I.A., Hardisty, M.R., Whyne, C.M.: Mesh-morphing algorithms for specimen-specific finite element modeling. J. Biomech. 41(7), 1381–1389 (2008)
22. Stein, K., Tezduyar, T., Benney, R.: Mesh moving techniques for fluid-structure interactions with large displacements. Trans. ASME 2003 70, 58–63 (2003)
23. Stein, K., Tezduyar, T., Benney, R.: Automatic mesh update with the solid-extension mesh moving technique. Comput. Meth. Appl. M. 193, 2019–2032 (2004)
24. Staten, M.L., Canann, S.A., Owen, S.J.: BMSWEEP: locating interior nodes during sweeping. Eng. Comput. 15(3), 212–218 (1999)
25. Tezduyar, T., Behr, M., Mittal, S., Johnson, A.A.: Computation of unsteady incompressible flows with the finite element methods – Space-time formulations, iterative strategies and massively parallel implementations. New Methods in Transient Analysis, PVP-vol. 246/AMD-vol. 143, ASME, New York, 7-24 (1992)
26. The Trilinos Project, Sandia Nat. Lab., http://trilinos.sandia.gov/
27. Vavasis, S.: QMG: mesh generation and related software (2010), http://www.cs.cornell.edu/home/vavasis/qmg-home.html
28. Vurputoor, R., Mukherjee, N., Cabello, J., Hancock, M.: A mesh morphing technique for geometrically dissimilar tesselated surfaces. In: Proc. 16th Int. Meshing Roundtable, pp. 315–334 (2007)
29. Watson, D.F.: nngridr: an implementation of natural neighbor interpolation. University of Western Australia (1994)
30. Wright, S.J.: Primal-dual interior-point methods. SIAM (1997)

Families of Meshes Minimizing P_1 Interpolation Error

A. Agouzal[1], K. Lipnikov[2], and Y. Vassilevski[3]

[1] I.C.J. Universite Lyon1, France
    `agouzal@univ-lyon1.fr`
[2] Los Alamos National Laboratory, Los Alamos, NM 87545, U.S.A.
    `lipnikov@lanl.fr`
[3] Institute of Numerical Mathematics, Gubkina str. 8, Moscow 119333, Russia
    `yuri.vassilevski@gmail.com`

Summary. For a given function, we consider a problem of minimizing the P_1 interpolation error on a set of triangulations with a fixed number of triangles. The minimization problem is reformulated as a problem of generating a mesh which is quasi-uniform in a specially designed metric. For functions with indefinite Hessian, we show existence of a family of metrics with highly diverse properties. The family may include both anisotropic and isotropic metrics. A developed theory is verified with numerical examples.

1 Introduction

Let Ω_h be a conformal triangulation of a computational domain Ω and $\mathcal{I}_1(u)$ be a continuous piecewise linear Lagrange interpolant of a given function u. The interpolation error

$$e_h = u - \mathcal{I}_1(u)$$

depends on the triangulation. We consider the problem of minimizing this error on a set of triangulations with a fixed number of triangles. Methods developed for solution of this problem can be used to decrease significantly a discretization error in various applications, including complex fluid flows [12]. A theoretical basis for this phenomena exploits the fact that a discretization error can be bounded by the best interpolation error [9].

In many cases, an approximate solution to this minimization problem is sufficient. In [18], we have shown that there exists a sequence of meshes that provide asymptotically optimal reduction of the interpolation error in the $L^\infty(\Omega)$-norm. These meshes were called quasi-optimal. The theory of quasi-optimal meshes has been extended to the $L^p(\Omega)$-norm in [2, 19] and to the $W_2^p(\Omega)$-norm in [2, 3]. In this paper, we focus on the $L^\infty(\Omega)$-norm of the interpolation error, although an extension of our main result to the $L^p(\Omega)$-norm is possible.

A constructive approach to generating a quasi-optimal mesh is based on reformulating the minimization problem as a problem of building a mesh which is quasi-uniform in a metric field. The metric-based mesh generation has a long and successful history (see e.g., [5, 6, 7, 10, 12, 13, 17] and references therein). Many methods have been developed using a common sense that the mesh size should be small in regions of a strong solution gradient. A scalar metric proportional to a norm of the solution gradient is often called a monitor function [15]. It allows one to generate adaptive regular meshes for problems with isotropic solution singularities. A tensor metric derived from the Hessian of a solution is considered one of the best metrics nowadays [18, 1, 7, 16, 17]. It allows one not only to generate an adaptive mesh but also to stretch it in the direction where the gradient is small.

To the best of our knowledge, the first theoretical justification that a Hessian-based metric results in an approximate solution of the minimization problem has been done in [1, 18]. An upper and lower bounds have been derived there for functions with indefinite but nonsingular Hessians. Independently, a similar upper bound has been proved later in [8] for functions with definite Hessians. Upper bounds for P_k interpolation errors where $k > 1$ were proved in [16].

In [18, 7, 13] and similar papers, a Hessian is recovered from a discrete solution. Major disadvantage of this approach is that the recovered Hessian does not converge to the continuous one on a sequence of refined meshes. An alternative technology has been proposed in [2, 3, 4]. There, a metric is recovered from a posteriori error estimates prescribed to mesh edges. Both approaches allows one to implement an automatic (black-box) mesh adaptation. The primary goal of the first approach is to minimize the interpolation error. The second one tackles the discretization error and is potentially more beneficial in problems where the discretization and interpolation error differ significantly.

This paper extends the theory of quasi-optimal meshes for the P_1 interpolation problem. The main focus is on functions with indefinite Hessian. We show that in this case there exists a few families of quasi-optimal meshes with highly diverse properties. The developed theory and presented numerical examples demonstrate that one sequence may consist of isotropic meshes, while the other one includes anisotropic meshes. Moreover, the interpolation errors differ within 1-3% in both sequences.

The paper outline is as follows. In Section 2, we developed a theory of multiple metrics that result in quasi-optimal meshes. In Section 3, we verify the theoretical findings with numerical examples. Concluding remarks are collected in Section 4.

2 Analysis of the P_1 Interpolation Error

To analyze the interpolation error, $e_h = u - \mathcal{I}_1(u)$, we employ a divide and conquer approach. First, we prove error bounds for quadratic functions. Then, we extend them to $C^2(\Omega)$ functions.

2.1 Bounds for Quadratic Functions

Let $\Omega \subset \Re^2$ be a bounded polygonal domain and with $N(\Omega_h)$ triangles. Let $\mathcal{I}_1(u)$ be the continuous piecewise linear Lagrange interpolant of a given function u on mesh Ω_h and $\mathcal{I}_{1,\Delta}(u)$ be its restriction to triangle Δ.

Let us consider a triangle Δ with vertices $\mathbf{v}_i$, $i = 1, 2, 3$, edge vectors $\mathbf{e}_k = \mathbf{v}_i - \mathbf{v}_j$ and mid-edge points $\mathbf{c}_k$, $k = 6-i-j$, $1 \leq i < j \leq 3$. Let φ_i, $i = 1, 2, 3$, be linear functions on Δ associated with vertices $\mathbf{v}_i$, and $b_k = \varphi_i \varphi_j$ be quadratic bubble functions associated with edges $\mathbf{e}_k$. We define φ_i by requiring that $\varphi_i(\mathbf{v}_i) = 1$ and $\varphi_i(\mathbf{v}_j) = 0$ for $j \neq i$. Note that $0 \leq \varphi_i \leq 1$ and $0 \leq b_k \leq 1/4$ inside triangle Δ.

We start analysis of the interpolation error for a quadratic function u_2. The Hessian $\mathbb{H}_2$ of this function is constant. Since the local interpolation error $e_2 = u_2 - \mathcal{I}_{1,\Delta}(u_2)$ is zero at vertices of triangle Δ, we obtain easily the following Taylor formula:

$$e_2(\mathbf{x}) = -\frac{1}{2} \sum_{k=1}^{3} (\mathbb{H}_2 \mathbf{e}_k, \mathbf{e}_k) \, b_k(\mathbf{x}).$$

Thus,

$$\|e_2\|_{L^\infty(\Delta)} \leq \frac{1}{8} \sum_{k=1}^{3} |(\mathbb{H}_2 \mathbf{e}_k, \mathbf{e}_k)|. \tag{1}$$

The Hessian $\mathbb{H}_2$ is a symmetric matrix; therefore there exists a decomposition

$$\mathbb{H}_2 = \pm \mathbb{V}^T \mathbb{D} \mathbb{V} \tag{2}$$

where

$$\mathbb{D} = \begin{bmatrix} 1 & 0 \\ 0 & \epsilon \end{bmatrix}, \quad \epsilon = \mathrm{sgn}(\det(\mathbb{H}_2)).$$

In the sequel, it is sufficient to assume that the Hessian is either positive definite or indefinite; thus, we can consider only the plus sing in (2). The conclusions made for a positive definite Hessian will hold true for a negative one. The spectral module of $\mathbb{H}_2$ is defined as follows

$$|\mathbb{H}_2| = \mathbb{V}^T \mathbb{V}. \tag{3}$$

If $\epsilon = 1$, we obtain immediately that $\mathbb{H}_2 = |\mathbb{H}_2|$. Since $|\mathbb{H}_2|$ is positive definite, we can define a local metric as $\mathfrak{M}_\Delta = |\mathbb{H}_2|$. This approach has been extensively analyzed in the literature. It can be extended to general functions by approximating them locally as quadratic functions. The resulting piecewise constant metric $\mathfrak{M}$ allows one to generate a quasi-optimal mesh. In this paper, we look more closely at the case $\epsilon = -1$ when the Hessian is indefinite. We will show that in addition to (3), there exist other metrics that produce quasi-optimal meshes with drastically different properties.

In the approach developed in [18, 19, 2], the maximum norm of $e_{2,\Delta}$ is bounded from above by geometric quantities such as the length of edges $\mathbf{e}_k$ in metric $|\mathbb{H}_2|$:

$$\sum_{k=1}^{3} |(\mathbb{H}_2 \mathbf{e}_k, \mathbf{e}_k)| \leq \sum_{k=1}^{3} (|\mathbb{H}_2| \mathbf{e}_k, \mathbf{e}_k) = \sum_{k=1}^{3} (\mathbb{V}^T \mathbb{V} \mathbf{e}_k, \mathbf{e}_k). \tag{4}$$

In case $\epsilon = 1$, the above inequality becomes identity. Here, we try to improve the upper bound when $\epsilon = -1$ by exploiting the fact that the decomposition $\mathbb{H}_2 = \mathbb{V}^T \mathbb{D} \mathbb{V}$ is not unique. This results in the constrained minimization problem with respect to $\mathbb{V}$: Find $\mathfrak{M} = \mathbb{V}_o^T \mathbb{V}_o$ such that

$$\sum_{k=1}^{3} |(\mathbb{H}_2 \mathbf{e}_k, \mathbf{e}_k)| \leq \sum_{k=1}^{3} (\mathbb{V}_o^T \mathbb{V}_o \mathbf{e}_k, \mathbf{e}_k) = \inf_{\widetilde{\mathbb{V}}: \widetilde{\mathbb{V}}^T \mathbb{D} \widetilde{\mathbb{V}} = \mathbb{H}_2} \sum_{k=1}^{3} (\widetilde{\mathbb{V}}^T \widetilde{\mathbb{V}} \mathbf{e}_k, \mathbf{e}_k). \quad (5)$$

We denote by $\mathbb{V}_o$ the solution to this minimization problem. At this moment, we need to introduce additional notation and prove a technical result.

Let $[a \mid b]$ denote a 2×2 matrix with columns $a, b \in \Re^2$ and

$$\mathbb{Q} = \begin{bmatrix} 0 & -\epsilon \\ 1 & 0 \end{bmatrix}.$$

Lemma 1. *Let $\mathbb{H}_2 = \mathbb{V}^T \mathbb{D} \mathbb{V}$ and $\mathbb{H}_2 = \widetilde{\mathbb{V}}^T \mathbb{D} \widetilde{\mathbb{V}}$. Then, there exists a nonsingular matrix $\Phi = [\phi \mid \phi']$ such that $\widetilde{\mathbb{V}} = \Phi \mathbb{V}$. Moreover, the vector $\phi \in \Re^2$ satisfies $(\mathbb{D}\phi, \phi) = 1$ and $\phi' = \mathbb{Q} \phi$.*

Proof. The two decompositions imply that $\widetilde{\mathbb{V}} = \Phi \mathbb{V}$ and $\Phi^T \mathbb{D} \Phi = \mathbb{D}$, which in turn implies that $(\det(\Phi))^2 = 1$. Thus, Φ is nonsingular. Moreover, $(\mathbb{D}\phi, \phi) = 1$, $(\mathbb{D}\phi', \phi') = \epsilon$ and $(\mathbb{D}\phi, \phi') = 0$. Direct verification shows that $\phi' = \mathbb{Q} \phi$ gives the last two identities. This proves the assertion of the lemma. □

Immediate corollary of this lemma is that the local metric $\mathfrak{M}_\Delta$ defined by (3) is unique when $\mathbb{H}_2$ is a positive definite matrix. Since $(\phi, \phi') = 0$, Φ is an orthogonal matrix and

$$\widetilde{\mathbb{V}}^T \widetilde{\mathbb{V}} = \mathbb{V}^T \mathbb{V} = |\mathbb{H}_2| = \mathbb{H}_2.$$

Theorem 1. *Let Δ be a triangle with edges $\mathbf{e}_k$. Furthermore, let $\mathbb{H}_2$ be an indefinite matrix and $\mathbb{H}_2 = \mathbb{V}^T \mathbb{D} \mathbb{V}$ be one of the decompositions. Then, solution to the minimization problem (5) is*

$$\mathfrak{M}_O \equiv \mathbb{V}_o^T \mathbb{V}_o = \frac{1}{\sqrt{\mu^2 - \lambda^2}} \mathbb{V}^T \begin{bmatrix} \mu & -\lambda \\ -\lambda & \mu \end{bmatrix} \mathbb{V}, \quad (6)$$

where

$$\mu = \sum_{k=1}^{3} (\mathbb{V}\mathbf{e}_k, \mathbb{V}\mathbf{e}_k), \qquad \lambda = \sum_{k=1}^{3} (\mathbb{R}\mathbb{V}\mathbf{e}_k, \mathbb{V}\mathbf{e}_k), \quad (7)$$

$$\mathbb{R} = \begin{bmatrix} 0 & 1 \\ 1 & 0 \end{bmatrix}.$$

Proof. Using Lemma 1, we obtain the following representation of matrix $\widetilde{\mathbb{V}}^T \widetilde{\mathbb{V}}$:

$$\widetilde{\mathbb{V}}^T \widetilde{\mathbb{V}} = \mathbb{V}^T [\phi \mid \phi']^T [\phi \mid \phi'] \mathbb{V} = \|\phi\|^2 \mathbb{V}^T \mathbb{V} + (\phi, \phi') \mathbb{V}^T \mathbb{R} \mathbb{V}. \quad (8)$$

Let $\phi = [\phi_1, \phi_2]^T$, $\phi' = [\phi_2, \phi_1]^T$. Direct calculations and Lemma 1 give

$$(\phi, \phi)^2 - (\phi, \phi')^2 = \phi_1^4 + \phi_2^4 + 2\phi_1^2\phi_2^2 - 4\phi_1^2\phi_2^2 = (\mathbb{D}\phi, \phi)^2 = 1.$$

Therefore, there exists a number $z \in \Re^1$ such that $\|\phi\|^2 = \cosh(z)$ and $(\phi, \phi') = \sinh(z)$. Inserting this into (8), we obtain

$$\widetilde{\mathbb{V}}^T \widetilde{\mathbb{V}} = \cosh(z)\mathbb{V}^T\mathbb{V} + \sinh(z)\mathbb{V}^T\mathbb{R}\mathbb{V}. \tag{9}$$

Combining estimates (1) and (5) with representation (9), we obtain a one-dimensional minimization problem: Find $z_o \in \Re^1$ such that

$$z_o = \arg\inf_{\tilde{z} \in \Re^1} \left(\mu \cosh(\tilde{z}) + \lambda \sinh(\tilde{z}) \right), \tag{10}$$

where μ and λ are defined by (7).

Let $\mathbb{V}\mathbf{e}_k = [v_{1k}, v_{2k}]^T$. Note that this is a non-zero vector. Then, we have

$$\mu + \lambda = \sum_{k=1}^{3} \left((\mathbb{V}\mathbf{e}_k, \mathbb{V}\mathbf{e}_k) + (\mathbb{R}\mathbb{V}\mathbf{e}_k, \mathbb{V}\mathbf{e}_k) \right) = \sum_{k=1}^{3} \left(v_{1k}^2 + v_{2k}^2 + 2v_{1k}v_{2k} \right)$$

$$= \sum_{k=1}^{3} (v_{1k} + v_{2k})^2 > 0$$

and

$$\mu - \lambda = \sum_{k=1}^{3} (v_{1k} - v_{2k})^2 > 0.$$

We note that $\mu - \lambda \neq 0$ since the equality would imply that $v_{1k} = v_{2k}$, $k = 1, 2, 3$, and the vectors $\mathbf{e}_1, \mathbf{e}_2, \mathbf{e}_3$ are collinear which is possible only for degenerate triangle.

Minimization problem (10) has the explicit solution:

$$z_o = \frac{1}{2} \ln \frac{\mu - \lambda}{\mu + \lambda}.$$

The number z_o corresponds to the matrix $\mathbb{V}_o$ producing the metric

$$\mathfrak{M}_O \equiv \mathbb{V}_o^T \mathbb{V}_o = \frac{1}{\sqrt{\mu^2 - \lambda^2}} \left(\mu \mathbb{V}^T \mathbb{V} - \lambda \mathbb{V}^T \mathbb{R} \mathbb{V} \right) = \frac{1}{\sqrt{\mu^2 - \lambda^2}} \mathbb{V}^T \begin{bmatrix} \mu & -\lambda \\ -\lambda & \mu \end{bmatrix} \mathbb{V}.$$

This proves the theorem. □

Note that the metric $\mathfrak{M}_O$ differs from the metric

$$\mathfrak{M}_V = |\mathbb{H}_2| = \mathbb{V}^T \mathbb{V} \tag{11}$$

when $\lambda \neq 0$ although

$$\det \mathfrak{M}_V = \det \mathfrak{M}_O. \tag{12}$$

Thus, we derived two independent metrics $\mathfrak{M}_O$ and $\mathfrak{M}_V$ yielding bounds (4) and (5), respectively. The proof of the theorem implies there exist many metrics $\widehat{\mathbb{V}}^T \widehat{\mathbb{V}}$ produced by various values of z in formula (9). Hereafter, we use generation notation $\mathfrak{M}_\Delta$ to indicate any of these metrics, including the special metric $\mathfrak{M}_O$.

Combining (1) and (5), we get an upper bound on the interpolation error:

$$\|e_2\|_{L^\infty(\Delta)} \le \frac{1}{8} \sum_{k=1}^{3} \|\mathbf{e}_k\|_{\mathfrak{M}_\Delta}^2, \qquad \|\mathbf{e}_k\|_{\mathfrak{M}_\Delta} = (\mathfrak{M}_\Delta \mathbf{e}_k, \mathbf{e}_k)^{1/2}, \qquad (13)$$

where $\|\mathbf{e}_k\|_{\mathfrak{M}_\Delta}$ is the edge length measured in the constant tensor metric $\mathfrak{M}_\Delta$.

A lower bound for the interpolation error can be also expressed as a combination of geometric quantities associated with triangle Δ. In [11], the following estimate was shown

$$\max_{\mathbf{x}\in\Delta} |e_2(\mathbf{x})| \ge \max_{k=1,2,3} \max_{\mathbf{x}\in e_k} |e_2(\mathbf{x})| = \frac{1}{8} \max_{k=1,2,3} |(\mathbb{H}_2 \mathbf{e}_k, \mathbf{e}_k)| \ge \frac{4}{\sqrt{5}} |\hat{\Delta}|,$$

where $\hat{\Delta}$ the image of triangle Δ under the coordinate transformation $\hat{\mathbf{x}} = \mathbb{V}\mathbf{x}$. Note that

$$|\hat{\Delta}| = |\Delta| \det(\mathbb{V}) = |\Delta| \sqrt{\det(\mathbb{V}^T \mathbb{V})} = |\Delta| \sqrt{\det(\mathfrak{M}_\Delta)} = |\Delta|_{\mathfrak{M}_\Delta},$$

where $|\Delta|$ is the area of Δ and $|\Delta|_{\mathfrak{M}_\Delta}$ is its area in metric $\mathfrak{M}_\Delta$. Thus, we have immediately that

$$\|e_2\|_{L^\infty(\Delta)} \ge \frac{4}{\sqrt{5}} |\Delta|_{\mathfrak{M}_\Delta}. \qquad (14)$$

This implies that the interpolation error $\|e_2\|_{L^\infty(\Delta)}$ is controlled from above and below by geometric quantities associated with triangle Δ. We proved the following theorem.

Theorem 2. *Let u_2 be a quadratic function with a nonsingular Hessian $\mathbb{H}_2 = \mathbb{V}^T \mathbb{D} \mathbb{V}$ and e_2 denote its linear interpolation error on triangle Δ. Let $\mathfrak{M}_\Delta = \mathbb{V}^T \mathbb{V}$. Then,*

$$\frac{4}{\sqrt{5}} |\Delta|_{\mathfrak{M}_\Delta} \le \|e_2\|_{L^\infty(\Delta)} \le \frac{1}{8} \sum_{k=1}^{3} \|\mathbf{e}_k\|_{\mathfrak{M}_\Delta}^2. \qquad (15)$$

If $\det(\mathbb{H}_2) < 0$, the specific metric $\mathfrak{M}_O$ provides the best upper bound in (15) whereas $|\Delta|_{\mathfrak{M}_\Delta}$ is the same for any metric (9).

2.2 Illustrative Example

In order to illustrate diversity of metrics generated by (8), we consider the bilinear function $u_2 = xy$ with the indefinite Hessian

$$\mathbb{H}_2 = \begin{bmatrix} 0 & 1 \\ 1 & 0 \end{bmatrix}.$$

It is shown in [11] that for a quadratic function with an indefinite Hessian, the interpolation achieves maximum on the boundary of triangle Δ:

$$\|e_2\|_{L^\infty(\Delta)} = \frac{1}{8} \max_{k=1,2,3} |(\mathbb{H}_2 \mathbf{e}_k, \mathbf{e}_k)|.$$

Let us apply this result to various meshes with a characteristic mesh size h schematically shown in Fig. 1. Consider a triangle Δ_1 from the isotropic mesh, e.g. the one with vertices $\mathbf{v}_1 = [0,0]^T$, $\mathbf{v}_2 = [h,h]^T$, and $\mathbf{v}_3 = [0,h]^T$. The interpolation error on this triangle is

$$\|e_2\|_{L^\infty(\Delta_1)} = |(\mathbb{H}_2 \mathbf{e}_3, \mathbf{e}_3)| = \frac{1}{4}h^2 = \frac{1}{2}|\Delta_1|.$$

Due to the mesh structure, the interpolation error is the same for all triangles.

Consider a triangle Δ_2 from the first anisotropic mesh, e.g the one with vertices $\mathbf{v}_1 = [0,0]^T$, $\mathbf{v}_2 = [h,1]^T$, and $\mathbf{v}_3 = [0,1]^T$. The interpolation error on this triangle is

$$\|e_2\|_{L^\infty(\Delta_2)} = |(\mathbb{H}_2 \mathbf{e}_3, \mathbf{e}_3)| = \frac{1}{4}h = \frac{1}{2}|\Delta_2|.$$

Again, the interpolation error is the same for all triangles in this mesh. Exactly the same interpolation error holds true for the triangles in the second anisotropic mesh, e.g. for triangle Δ_3 with vertices $\mathbf{v}_1 = [0,0]^T$, $\mathbf{v}_2 = [1,0]^T$, and $\mathbf{v}_3 = [1,h]^T$.

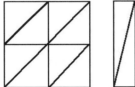

Fig. 1. Illustration of one isotropic and two anisotropic meshes.

Let all three meshes cover the unit square and have the same number of triangles, N. Then, the interpolation error equals to $1/(2N)$ in all three examples. These three meshes represent three different families of meshes yielding the optimal (i.e. reciprocal to N) reduction of the interpolation error. Thus these meshes belong to three different families of quasi-optimal meshes. One family contains shape regular meshes; while the other two contain anisotropic meshes stretched in x and y directions, respectively.

Consider the following decomposition of Hessian $\mathbb{H}_2$:

$$\mathbb{H}_2 = \mathbb{V}^T \begin{bmatrix} 1 & 0 \\ 0 & -1 \end{bmatrix} \mathbb{V}, \quad \mathbb{V} = \frac{1}{\sqrt{2}} \begin{bmatrix} 1 & 1 \\ 1 & -1 \end{bmatrix}.$$

Using formula (11), we obtain the isotropic metric $\mathfrak{M}_V = |\mathbb{H}_2| = \mathbb{I}$. Only triangles from the isotropic meshes are shape-regular in metric $\mathfrak{M}_V$. However, using formulas (6)-(7), we obtain

$$\mathfrak{M}_O(\Delta_1) = \begin{bmatrix} 1 & 0 \\ 0 & 1 \end{bmatrix}, \qquad \mathfrak{M}_O(\Delta_2) = \begin{bmatrix} h^{-1} & 0 \\ 0 & h \end{bmatrix}, \qquad \mathfrak{M}_O(\Delta_3) = \begin{bmatrix} h & 0 \\ 0 & h^{-1} \end{bmatrix}.$$

Note that the triangles Δ_k are shape-regular in the respective metrics $\mathfrak{M}_O(\Delta_k)$, $k = 1, 2, 3$.

2.3 Bounds for C^2-Functions

Let u be a continuous function and $\mathcal{I}_{2,\Delta}(u)$ be its quadratic Lagrange interpolant on triangle Δ. In Theorem 2, we derived the geometric representation of the L^∞-norm of $e_{2,\Delta} = \mathcal{I}_{2,\Delta}(u) - \mathcal{I}_{1,\Delta}(u)$. It was shown in [2] that the norm of $e_{2,\Delta}$ provides a good approximation for the corresponding norm of the true error $e_\Delta = u - \mathcal{I}_{1,\Delta}(u)$. For completeness, we formulate this result in the next lemma. Let $\mathcal{F}$ be a space of symmetric 2×2 matrices. We define the following norm:

$$|||\partial\Delta|||^2_{|\mathbb{H}|} = \sum_{k=1}^{3} |||\mathbf{e}_k|||^2_{|\mathbb{H}|}, \qquad |||\mathbf{e}_k|||^2_{|\mathbb{H}|} = \max_{\mathbf{x} \in \Delta} (|\mathbb{H}(\mathbf{x})|\mathbf{e}_k, \mathbf{e}_k). \tag{16}$$

Lemma 2. *[2] Let $u \in C^2(\bar{\Delta})$. Then*

$$\frac{3}{4}\|e_{2,\Delta}\|_{L^\infty(\Delta)} \leq \|e_\Delta\|_{L^\infty(\Delta)} \leq \|e_{2,\Delta}\|_{L^\infty(\Delta)} + \frac{1}{4}\inf_{\mathbb{F} \in \mathcal{F}} |||\partial\Delta|||^2_{|\mathbb{H}-\mathbb{F}|}. \tag{17}$$

The second term in the right inequality is the typical for a contemporary posteriori error analysis. It depends on the triangle and particular features of function u. In many cases it is essentially smaller than $\|e_{2,\Delta}\|_{L^\infty(\Delta)}$. For instance, a quasi-optimal mesh Ω_h generated by $\mathfrak{M}_\Delta = \mathfrak{M}_V = |\mathbb{H}_2|$ is characterized by balance between volume and perimeter of its triangles:

$$|\Delta|_{\mathfrak{M}_\Delta} \simeq |||\partial\Delta|||^2_{\mathfrak{M}_\Delta} \qquad \forall \Delta \in \Omega_h,$$

where $a \simeq b$ means existence of a constant C independent of the mesh and the triangle such that $C^{-1} a \leq b \leq C a$. Hereafter, C denotes a generic constant. Using (14) and the fact that $\mathbb{H}_2$ is a constant Hessian, we obtain

$$\|e_{2,\Delta}\|_{L^\infty(\Delta)} \geq C^{-1}\|\partial\Delta\|^2_{\mathfrak{M}_\Delta} = C^{-1}\||\partial\Delta\||^2_{\mathfrak{M}_\Delta}.$$

Therefore, for a function with a nonsingular Hessian, we obtain

$$\frac{\inf_{\mathbb{F} \in \mathcal{F}} |||\partial\Delta|||^2_{|\mathbb{H}-\mathbb{F}|}}{\|e_{2,\Delta}\|_{L^\infty(\Delta)}} \leq C \frac{\inf_{\mathbb{F} \in \mathcal{F}} |||\partial\Delta|||^2_{|\mathbb{H}-\mathbb{F}|}}{|||\partial\Delta|||^2_{\mathfrak{M}_\Delta}} = C \frac{\inf_{\mathbb{F} \in \mathcal{F}} |||\partial\Delta|||^2_{|\mathbb{H}-\mathbb{F}|}}{|||\partial\Delta|||^2_{|\mathbb{H}_2|}} = o(1).$$

This argument justifies usage of the quadratic Lagrange interpolant for the derivation of the optimal metric for functions $u \in C^2(\bar{\Delta})$ with nonsingular Hessians on quasi-optimal meshes.

The local analysis is naturally extended to triangulations. Let $\mathfrak{M}$ be a piecewise constant metric composed of local metrics $\mathfrak{M}_\Delta$. Let $N(\Omega_h)$ be the number of triangles in mesh Ω_h. If Ω_h is a quasi-uniform mesh with respect to metric $\mathfrak{M}$, then all triangles have approximately the same area measured in this metric:

$$N(\Omega_h)^{-1}|\Omega|_\mathfrak{M} \simeq |\Delta|_{\mathfrak{M}_\Delta} \simeq |\partial\Delta|^2_{\mathfrak{M}_\Delta} \qquad \forall \Delta \in \Omega_h.$$

Thus, the following error estimates is held

$$\|e\|_{L^\infty(\Omega)} = \max_{\Delta \in \Omega_h} \|e\|_{L^\infty(\Delta)} \leq C \max_{\Delta \in \Omega_h} |\Delta|_{\mathfrak{M}_\Delta} \leq C|\Omega|_\mathfrak{M} N(\Omega_h)^{-1},$$

which implies the asymptotically optimal error reduction and proves quasi-optimality of $\mathfrak{M}$-quasi-uniform meshes.

3 Numerical Experiments

Generation of a quasi-optimal mesh for a given function u is in general an iterative process. First, we generate an initial mesh with a desirable number of elements and calculate a piecewise constant metric $\mathfrak{M}$. Second, we generate a new mesh which is quasi-uniform in metric $\mathfrak{M}$. Also, we require that the number of triangles in the new mesh is approximately the same as that in the initial mesh. To generate a $\mathfrak{M}$-quasi-uniform mesh, we use a sequence of local mesh modifications described in [18] and implemented in the publicly available package Ani2D (sourceforge.net/projects/ani2d). Frequently, the initial mesh is not related to the function u which results in a large interpolation error and a non-optimal metric $\mathfrak{M}$. In this case, the two-step adaptation process can be repeated (see Algorithm 1). A few iterations may be required until the interpolation error is stabilized. The number of iterations depends on smoothness of function u.

Algorithm 1. Adaptive mesh generation

1: Generate an initial mesh Ω_h and compute the metric $\mathfrak{M}$.
2: **loop**
3:     Generate a $\mathfrak{M}$-quasi-uniform mesh Ω_h with the prescribed number triangles.
4:     Recompute the metric $\mathfrak{M}$.
5:     If Ω_h is $\mathfrak{M}$-quasi-uniform, then exit the loop
6: **end loop**

In practice, Algorithm 1 convergences faster and results in a smoother mesh when the metric is continuous. To define a continuous metric we use a method of shifts. For every node $\mathbf{v}_i$ in Ω_h, we define the superelement σ_i as the union of all triangles sharing $\mathbf{v}_i$. Then, $\mathfrak{M}(\mathbf{v}_i)$ is defined as one of the metrics in σ_i with the largest determinant. This method always chooses the worst metric in the superelement. Once the metric is computed at nodes of each triangle, the metric is linearly interpolated inside triangles.

We consider three examples of functions with indefinite Hessians. Their isolines are presented in Fig.2. In the experiments, we study the asymptotic behavior of the P_1 interpolation error for two families of quasi-optimal meshes based on the local metrics $\mathfrak{M}_V = \mathbb{V}^T \mathbb{V}$ and $\mathfrak{M}_O = \mathbb{V}_o^T \mathbb{V}_o$, respectively.

Fig. 2. From left to right: Isolines of functions $u^{(1)}$, $u^{(2)}$, and $u^{(3)}$.

Let Ω be the unit square in all examples. The first function is the canonical hyperbolic function,

$$u^{(1)}(x,y) = (x - 0.5)^2 - (y - 0.5)^2,$$

with the constant indefinite Hessian $\mathbb{H}_2 = \text{diag}\{2, -2\}$. Isolines of this function are shown on the left picture in Fig. 2. Fig. 3 shows first two meshes in two sequences of quasi-optimal meshes with approximately 4000 and 8000 triangles. The meshes in the top row are quasi-uniform in metric $\mathfrak{M}$ generated by local metrics $\mathfrak{M}_O$. The meshes in the bottom row are quasi-uniform in metric $\mathfrak{M}$ generated by local metrics $\mathfrak{M}_V$. Obviously that both sequences are different, one contains strongly anisotropic meshes stretched along bisectors of four quadrants, while the other one contains isotropic meshes. The data in Table 1 also confirm that. The second and the fifth columns in this table show maximal ratio of the circumscribed radius R to the inscribed radius r across all triangles in the mesh. The interpolation errors are proportional to $N(\Omega_h)^{-1}$ which is the optimal error reduction. Thus, both sequences contain quasi-optimal meshes. Moreover, the errors in two sequences differ by 1-3% only.

The second function is

$$u^{(2)}(x,y) = (x + \sin(\pi x))^2 - (y + \sin(\pi x))^2.$$

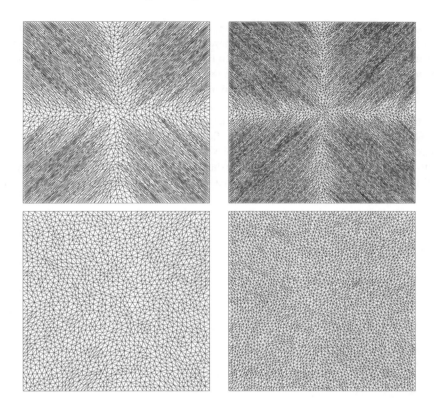

Fig. 3. Example 1. Quasi-optimal meshes with approximately 4000 (left column) and 8000 (right column) triangles. The top and bottom rows correspond to two different families of meshes.

Table 1. Example 1. Interpolation error on two families of quasi-optimal meshes.

$\mathfrak{M}_\Delta = \mathfrak{M}_O$			$\mathfrak{M}_\Delta = \mathfrak{M}_V$		
$N(\Omega_h)$	R/r	$\|e_h\|_{L^\infty(\Omega)}$	$N(\Omega_h)$	R/r	$\|e_h\|_{L^\infty(\Omega)}$
4073	834.2	7.974e-05	3930	15.1	8.195e-05
8267	768.1	4.054e-05	7885	24.6	4.064e-05
16589	1121.	2.047e-05	15813	18.7	2.040e-05
33118	2259.	1.053e-05	31786	27.9	1.018e-05
66557	4794.	5.233e-06	63373	46.2	5.126e-06
rate		0.991			0.936

Isolines of this function are shown on the middle picture in Fig. 2. The Hessian of this function is indefinite almost everywhere in the computational domain except for a parabola-shaped region around point $(0.6, 0.4)$. This region can be identified on the top-right picture in Fig. 4 as the region where the mesh is isotropic. Each row in this figure shows two meshes with approximately 4000 and 8000 triangles. The top and bottom rows correspond to local metrics $\mathfrak{M}_O$ and $\mathfrak{M}_V$, respectively. Clearly, the metric (6) results in more stretched meshes.

The interpolation errors presented in Table 2 verify that both metrics result in quasi-optimal meshes. The error decrease is again proportional to $N(\Omega_h)^{-1}$, i.e. both sequences contain quasi-optimal meshes. The values of the maximal anisotropy ratio, R/r, confirm visual impression that one sequence of meshes is much more

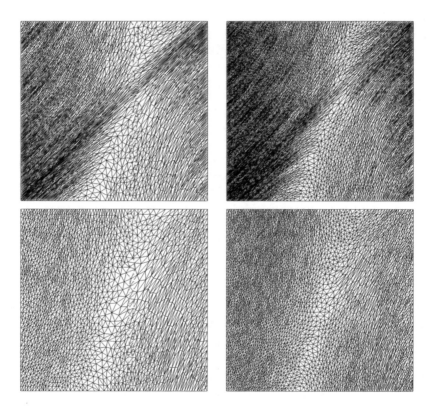

Fig. 4. Example 2. Quasi-optimal meshes with approximately 4000 (left column) and 8000 (right column) triangles. The top and bottom rows correspond to two different families of meshes.

Table 2. Example 2. Interpolation error on two families of quasi-optimal meshes.

$\mathfrak{M}_\Delta = \mathfrak{M}_O$			$\mathfrak{M}_\Delta = \mathfrak{M}_V$		
$N(\Omega_h)$	R/r	$\|e_h\|_{L^\infty(\Omega)}$	$N(\Omega_h)$	R/r	$\|e_h\|_{L^\infty(\Omega)}$
4070	2489.	9.523e-04	3934	62.0	8.818e-04
8097	1492.	5.393e-04	7839	82.6	3.977e-04
16269	937.1	2.281e-04	15761	203.	2.376e-04
32881	2704.	1.209e-04	31478	190.	1.374e-04
65147	4142.	6.427e-05	63031	239.	5.837e-05
rate		0.991			0.936

stretched than the other. Thus, for this example, there exist at least two different families of quasi-optimal meshes. There is no conclusive evidence that one sequence of meshes gives a consistently smaller error.

The third function

$$u^{(3)}(x,y) = \frac{(x-0.5)^2 - (y+0.2)^2}{((x-0.5)^2 + (y+0.2)^2)^2}$$

satisfies the Laplace equation. Its isolines are shown on the right picture in Fig. 2. Fig. 5 is identical to Figs. 3 and 4. The top and bottom rows present two first meshes in the different sequences of quasi-optimal meshes. One observes the presence of two anisotropic jet-like structures on the top pictures.

The interpolation errors presented in Table 3 verify that both metrics result in quasi-optimal meshes. The errors are again proportional to $N(\Omega_h)^{-1}$ which is the optimal error reduction. As in the first example, they are within 1-3% of one another which indicates that neither of the sequences is preferable for minimizing the interpolation error.

Table 3. Example 3. Interpolation error on two families of quasi-optimal meshes.

$\mathfrak{M}_\Delta = \mathfrak{M}_O$			$\mathfrak{M}_\Delta = \mathfrak{M}_V$		
$N(\Omega_h)$	R/r	$\|e_h\|_{L^\infty(\Omega)}$	$N(\Omega_h)$	R/r	$\|e_h\|_{L^\infty(\Omega)}$
4003	380.1	2.643e-03	3946	19.0	2.562e-03
8087	353.2	1.285e-03	7872	12.8	1.295e-03
16069	652.0	6.502e-04	15776	10.9	6.457e-04
32206	955.7	3.283e-04	31579	11.8	3.254e-04
64478	5291.	1.647e-04	62970	26.3	1.620e-04
rate		0.996			0.996

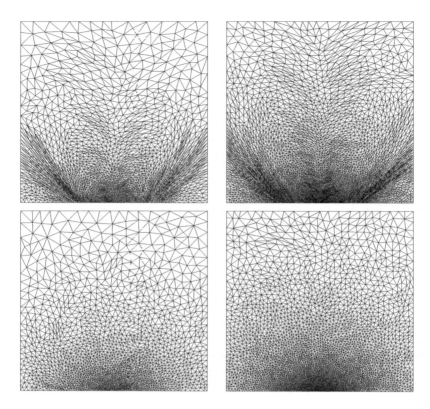

Fig. 5. Example 3. Quasi-optimal meshes with approximately 4000 (left column) and 8000 (right column) triangles. The top and bottom rows correspond to two different families of meshes.

4 Conclusion

We extended a theory of optimal meshes that minimize the P_1 interpolation error for a given function u. We have shown that the exist families of quasi-optimal meshes giving approximately the same interpolation error for a fixed number of triangles. At the same time, the properties of this meshes may vary significantly, one mesh can be isotropic, while the other one is anisotropic. These quasi-optimal meshes are generated using different metric fields. Formally, one of the metrics can be referred to as optimal since it provides the sharpest bound for the local interpolation error.

The existing metric-based mesh generation technology can produce any of the available families of quasi-optimal meshes. The result depends on selection of an initial mesh for the adaptive iterations. In the future, we shall analyze this phenomenon in more details.

References

1. Agouzal, A., Lipnikov, K., Vassilevski, Y.: Adaptive generation of quasi-optimal tetrahedral meshes. East-West J. Numer. Math. 7, 223–244 (1999)
2. Agouzal, A., Lipnikov, K., Vassilevski, Y.: Hessian-free metric-based mesh adaptation via geometry of interpolation error. Comp. Math. Math. Phys. 50, 124–138 (2010)
3. Agouzal, A., Vassilevski, Y.: Minimization of gradient errors of piecewise linear interpolation on simplicial meshes. Comp. Meth. Appl. Mech. Engnr. 199, 2195–2203 (2010)
4. Agouzal, A., Lipnikov, K., Vassilevski, Y.: Edge-based a posteriori error estimators for generating quasi-optimal simplicial meshes. Math. Model. Nat. Phenom. 5, 91–96 (2010)
5. Agouzal, A., Lipnikov, K., Vassilevski, Y.: On optimal convergence rate of finite element solutions of boundary value problems on adaptive anisotropic meshes. Math. Comput. Simul. (in Press, 2011)
6. Borouchaki, H., Hecht, F., Frey, P.J.: Mesh gradation control. Inter. J. Numer. Meth. Engrg. 43, 1143–1165 (1998)
7. Buscaglia, G.C., Dari, D.A.: Anisotropic mesh optimization and its application in adaptivity. Int. J. Numer. Meth. Eng. 40, 4119–4136 (1997)
8. Chen, L., Sun, P., Xu, J.: Optimal anisotropic meshes for minimizing interpolation errors in L^p-norm. Mathematics of Computation 76, 179–204 (2007)
9. Ciarlet, P.G.: The finite element method for elliptic problems. North-Holland (1978)
10. Coupez, T.: Generation de maillage et adaptation de maillage par optimisation locale. Revue Europeenne Des Elements Finis 49, 403–423 (2000) (in French)
11. D'Azevedo, E.: Optimal triangular mesh generation by coordinate transformation. SIAM J. Sci. Stat. Comput. 12, 755–786 (1991)
12. Farrell, P.E., Maddison, J.R.: Conservative interpolation between volume meshes by local Galerkin projection. Comput. Methods Appl. Mech. Engrg. 200, 89–100 (2011)
13. Fortin, M., Vallet, M.-G., Dompierre, J., Bourgault, Y., Habashi, W.G.: Anisotropic mesh adaptation: theory, validation and applications Computational Fluid dynamics, pp. 174–180. John Wiley & Sons Ltd (1996)
14. George, P.L.: Automatic mesh generation: Applications to Finite Element Methods. John Wiley & Sons, Inc., New York (1991)
15. Huang, W., Sun, W.: Variational mesh adaptation II: Error estimates and monitor functions. J. Comput. Phys. 184, 619–648 (2003)
16. Huang, W.: Metric tensors for anisotropic mesh generation. J. Comput. Phys. 204, 633–665 (2005)
17. Loseille, A., Alauzet, F.: Optimal 3D highly anisotropic mesh adaptation based on the continuous mesh framework. In: Clark, B.W. (ed.) Proc. of 18th Int. Meshing Roundtable, pp. 575–594 (2009)
18. Vassilevski, Y., Lipnikov, K.: Adaptive algorithm for generation of quasi-optimal meshes. Comp. Math. Math. Phys. 39, 1532–1551 (1999)
19. Vassilevski, Y., Agouzal, A.: An unified asymptotic analysis of interpolation errors for optimal meshes. Doklady Mathematics 72, 879–882 (2005)

A Log-Barrier Method for Mesh Quality Improvement

Shankar P. Sastry[1], Suzanne M. Shontz[2], and Stephen A. Vavasis[3]

[1] Department of Computer Science and Engineering, The Pennsylvania State University, University Park, PA, U.S.A.
`sps210@cse.psu.edu`
[2] Department of Computer Science and Engineering, The Pennsylvania State University, University Park, PA, U.S.A.
`shontz@cse.psu.edu`
[3] Department of Combinatorics and Optimization, University of Waterloo, Waterloo, ON, Canada
`vavasis@math.uwaterloo.ca`

Summary. The presence of a few poor-quality mesh elements can negatively affect the stability and efficiency of a finite element solver and the accuracy of the associated partial differential equation solution. We propose a mesh quality improvement method that improves the quality of the worst elements. Mesh quality improvement of the worst elements can be formulated as a nonsmooth unconstrained optimization problem, which can be reformulated as a smooth constrained optimization problem. Our technique solves the latter problem using a log-barrier interior point method and uses the gradient of the objective function to efficiently converge to a stationary point. The technique can be used with convex or nonconvex quality metrics. The method uses a logarithmic barrier function and performs global mesh quality improvement. Our method usually yields better quality meshes than existing methods for improvement of the worst quality elements, such as the active set, pattern search, and multidirectional search mesh quality improvement methods.

Keywords: mesh quality improvement, interior point method, log-barrier method.

1 Introduction

High-quality meshes are necessary for the stability [1] and efficiency of finite element (FE) solvers and for the accuracy [2, 3, 4] of the solution of the associated partial differential equations (PDEs). Poor quality elements affect the conditioning of the linear system that arises from the PDE and the mesh [1]. Therefore, mesh quality improvement algorithms are used to improve the quality of mesh elements when the initial qualities of the elements obtained from mesh generators are poor or after mesh warping [5, 6, 7]. There are numerous geometric mesh quality improvement algorithms which are effective in improving the average mesh quality [8, 9, 10, 11, 12]. However, even a few poor quality elements can negatively affect the entire finite element analysis due to the resulting ill-conditioned matrices that hinder the efficiency and the accuracy of the finite element solvers [13]. Therefore, we focus on the development of an algorithm to improve the worst quality mesh element. Our algorithm solves a smooth constrained optimization problem to improve the worst quality elements and can be classified as an interior point method [14].

Other algorithms have been developed for improvement of the quality of the worst element. Freitag and Plassmann developed an active set algorithm [15] for mesh quality improvement. However, their algorithm requires the the objective function specifying the mesh quality metrics to be convex. Park and Shontz developed two derivative-free optimization algorithms, namely the pattern search and multidirectional search methods [16], for mesh quality improvement. These algorithms do not use the gradient to optimize the mesh quality; thus, we expect their rate of convergence to be slow. They are described in more detail in Section 2.

Quality improvement of worst quality mesh elements is a nonsmooth unconstrained optimization problem. In Section 3, we describe the problem statement and show its reformulation into a constrained optimization problem. We solve this unconstrained optimization problem using an interior point method.

We develop a log-barrier interior point method [14] that seeks to improve the quality of the worst element in a mesh. Our method overcomes the disadvantages posed by the other algorithms presented above by employing a logarithmic barrier term, which is a function of the quality of the worst element. Though derived from classical optimization theory, the log-barrier method in our context has the following natural interpretation. On each iteration, the gradient of the log-barrier function points in a direction that is a weighted combination of the directions that improve each individual element. The weights are selected automatically in such a way that elements with the worst qualities have the highest weights (see (6) below). Therefore, the method globally updates vertex positions but concentrates on the improvement of the worst elements. In addition, the method has the built-in feature that the line search for maximizing the objective function will automatically prevent inversion, since the objective function (weighted element qualities) tends to minus infinity as inversion is approached. Our interior point method solves the primal formulation of the constrained optimization problem and can be used on both convex and nonconvex objective functions. The algorithm is presented in Section 4, and its characteristics are discussed in Section 5.

We run numerical experiments to assess the efficiency of our algorithm by comparing it against existing algorithms such as the active set, pattern search, and multidirectional search methods. Numerical experiments and their results are discussed in Section 6. We give our concluding remarks and indicate future research directions in Section 7.

2 Related Work

2.1 Derivative-Free Algorithms for Mesh Quality Improvement

Park and Shontz developed pattern search (PS) and multidirectional search (MDS) mesh quality improvement algorithms [16] to improve the worst quality mesh element. Their derivative-free algorithms do not compute the gradient of the objective function, but instead use function evaluations to move the vertices. These algorithms also employ local mesh quality improvement, as pattern search techniques are not efficient on large problems. The following objective function is maximized in order to improve the worst element mesh quality:

$$f(x) = \min_{1 \leq i \leq m} q_i(x), \qquad (1)$$

where m is the number of elements, and $q_i(x)$ is a mesh quality metric.

Pattern Search (PS) Method. The PS method moves a mesh vertex in one of a pre-defined (usually orthogonal) set of directions. The direction and distance by which each vertex is moved is determined by evaluating objective function at the pattern points. A backtracking line search is used to untangle an element if vertex movement deems this necessary.

Multidirectional Search (MDS) Method. The MDS method uses search directions given by a simplex, i.e., a triangle (2D) or a tetrahedron (3D). The simplex is expanded, contracted, and/or reflected in order to determine the optimal position for a vertex. A backtracking line search is used to untangle an element, if needed, in a similar manner as for the PS method.

2.2 Active Set Method for Mesh Quality Improvement

Freitag and Plassmann developed an active set mesh quality improvement algorithm [15], which maximizes the quality of triangular or tetrahedral mesh elements. To guarantee convergence, the relevant objective function formed by quality metrics should possess convex level sets. Examples of such quality metrics include the minimum of the sine of the angles of the triangle in 2D and the aspect ratio quality metric in 3D for individual submeshes. This method employs a local quality improvement technique, where individual submeshes are optimized by moving one vertex at a time.

For each submesh, the objective function is defined as the quality of the worst element. The objective function described above in (1) is maximized in order to improve quality of the worst element in the mesh. They define *active value* as the current value of the objective function due to the vertex placement, x. They define *active set*, denoted by A, as a set of those functions that result in the active value. The nonsmooth optimization problem of improving the quality of the worst element is solved using the steepest descent algorithm [14]. Here, the gradient and hence the descent direction is obtained by considering all possible convex combinations of active set gradients, $\nabla_x (q_i(x)) = g_i(x)$, and choosing the one that solves

$$\min_x \bar{g}^T \bar{g}, \text{ where } \bar{g} = \sum_{i \in A} \beta_i g_i(x),$$

$$\text{s.t.} \sum_{i \in A} \beta_i = 1, \beta_i \geq 0.$$

A backtracking line search technique is used to determine the points at which the active set changes, and the vertex is moved to the point that results in the best quality improvement.

3 Mesh Quality Improvement Problem Formulation

In this section, we mathematically formulate our mesh quality improvement problem. Our problem involves the description of the quality metrics used to measure the mesh quality and the objective function that is optimized in order to improve the mesh quality. We use two quality metrics, i.e., one smooth and one nonsmooth, to demonstrate the effectiveness of our algorithm.

Aspect Ratio Quality Metric. We define the quality of tetrahedral mesh element i as

$$q_i(x) = \left(\frac{l_1^2 + l_2^2 + \cdots + l_6^2}{6} \right)^{\frac{3}{2}} / \left(\text{vol} \times \frac{12}{\sqrt{2}} \right),$$

where x are the vertex positions, vol is the unsigned volume of the ith tetrahedron, and l_j is the length of side j of the tetrahedron [17]. The range of this quality metric is from 1 to ∞, where 1 is the quality of an regular tetrahedron. The quality tends to infinity as the tetrahedron becomes degenerate.

Nonsmooth Aspect Ratio Quality Metric. We define the quality of the ith tetrahedral mesh element as

$$q_i(x) = \frac{\sqrt{2}}{12} \frac{l_{\max}^3}{\text{vol}},$$

where $l_{\max}$ is the length of the longest edge of the tetrahedron. The range of this quality metric is from 1 to ∞, where 1 is the quality of an equilateral tetrahedron. The quality tends to infinity as the tetrahedron becomes degenerate.

Objective Function. The problem of improving the worst quality element can be expressed as

$$\min_x \left(\max_{i \in [1,m]} q_i(x) \right).$$

However, our algorithm is designed for maximization of a quality metric. Thus, the objective function is modified by taking the reciprocal of the quality metric as follows:

$$\max_x \left(\min_{i \in [1,m]} \frac{1}{q_i(x)} \right).$$

This can be reformulated as a constrained optimization problem

$$\max_x t \text{ subject to } t < \frac{1}{q_i(x)}, \forall i \in [1,m]. \qquad (2)$$

Note that all of these formulations improve the quality of the worst element, as minimizing a function is equivalent to maximizing its reciprocal.

4 A Log-Barrier Method for Mesh Quality Improvement

In this section, we describe our log-barrier method for mesh quality improvement. We develop our algorithm based on interior point methods, which can be used to solve constrained optimization problems [14]. Our method uses a logarithmic barrier term, which emphasizes the improvement of poor quality elements, and solves the constrained optimization problem using the gradient of the objective function.

4.1 Interior Point Methods

Interior point methods are a class of methods used to solve constrained optimization problems. For a constrained optimization problem, the objective function, $f(x)$, is maximized while respecting the constraint, $g(x) < 0$. When interior point methods are employed, the constraint is added as logarithmic barrier term to the objective function. The new objective function, $F(x, \mu)$, is given by:

$$F(x, \mu) = f(x) + \mu \log(-g(x)),$$

where $\mu > 0$. As we will describe in Section 5.1, μ is chosen such that it enables the satisfaction of the Karush Kuhn Tucker (KKT) conditions (i.e., first order, necessary conditions) [14]. The modified objective function is iteratively maximized using an unconstrained optimization algorithm. On every iteration, μ is reduced so that the barrier term is eventually negligible, and the original objective function, $f(x)$, is maximized subject to the constraint. Psuedocode for a typical interior point method is presented in Algorithm 1.

Algorithm 1. Interior Point Method

Input: $f(x), g(x)$, tol.
Output: $\max_x f(x)$ such that $g(x) < 0$.
Initialize $\mu_k > 0$ and x_0 such that $g(x_0) < 0$.
while $\mu_k \geq$ tol **do**
    Maximize $f(x) + \mu_k \log(-g(x))$ using any gradient-based optimization algorithm.
    Decrease μ_k towards 0.
end while

4.2 The Log-Barrier Method for Mesh Quality Improvement

We seek to improve the worst mesh element quality. A quantity, t, is defined, which is a function of the worst quality element such that $t < \min_i \frac{1}{q_i(x)}$, where $q_i(x)$ is the quality of the ith element. When t is maximized, the quality of the worst element is improved. The expression $t - \frac{1}{q_i(x)} < 0$ is used as a constraint. We iteratively maximize

$$t + \mu_k \sum_{i=1}^{m} \log\left(\frac{1}{q_i(x)} - t\right) \tag{3}$$

to improve the quality of the worst element in the mesh. After every iteration, μ is brought closer to 0. We modify t such that $\frac{d}{dt}(F(x,t,\mu)) \approx 0$. For a fixed μ and x, the objective function is strictly concave in t. Therefore, setting its derivative to 0 corresponds to globally maximizing the objective w.r.t. t. The log-barrier method for mesh quality improvement is shown in Algorithm 2.

Algorithm 2. Interior Point Method for Mesh Quality Improvement

Initialize μ_k and $t < 1/q_i(x)$ for all $i \in [1, m]$ where $q_i(\cdot)$ is the quality metric function.
while not converged **do**
    Maximize $F(x, t, \mu_k) = t + \mu_k \sum_{i=1}^{m} \log\left(\frac{1}{q_i(x)} - t\right)$, where t and μ are held constant, using the nonlinear conjugate gradient method.
    Decrease μ_k towards 0.
    Update t to a new value such that $\frac{d}{dt}(F(x,t,\mu)) \approx 0$.
end while

Log Barrier Term for Nonsmooth Quality Metrics. In our paper, the nonsmooth aspect ratio quality metric is defined using the longest edge of a tetrahedron. Our method handles the metric using each of the edges in the tetrahedron to compute six qualities for the tetrahedron and uses them as additive terms in the log barrier function. Since each individual term (for each edge in the tetrahedron) is smooth, the resulting log barrier function is also smooth.

5 Characteristics of the Log-Barrier Method for Mesh Quality Improvement

In this section, we show that the set of first-order necessary conditions, i.e., the KKT conditions [14], are satisfied for a solution of our constrained optimization problem. This fact is well known in the optimization community; we are including it here for the sake of completeness and also to provide more detail about the properties of the objective function and the algorithm. In addition, we examine the monotonicity of our algorithm.

5.1 Satisfaction of the KKT Conditions

For a solution, x^*, of a constrained optimization problem, the gradient of the Lagrangian vanishes at the solution, i.e., $\nabla_x L(x^*, t^*, \lambda^*) = 0$. For (2), the Lagrangian is given by

$$L(x, t, \lambda) = t + \sum_{i=1}^{m} \lambda_i \left(\frac{1}{q_i(x)} - t \right).$$

Hence, its gradient is given by

$$\nabla_x L(x, t, \lambda) = \sum_{i=1}^{m} \lambda_i \nabla_x \left(\frac{1}{q_i(x)} \right). \tag{4}$$

The nonlinear conjugate gradient (CG) step in the log barrier method computes x such that the gradient of the objective function given by (3), i.e., $\nabla_x F(x, t, \mu_k)$, vanishes. Thus,

$$\nabla_x F(x, t, \mu_k) = \mu_k \nabla_x \sum_{i=1}^{m} \log \left(\frac{1}{q_i(x)} - t \right) \tag{5}$$

$$= \mu_k \sum_{i=1}^{m} \frac{1}{\left(\frac{1}{q_i(x)} - t \right)} \nabla_x \left(\frac{1}{q_i(x)} \right). \tag{6}$$

From Equations (4) and (6), we see that, if λ_i satisfies

$$\lambda_i^* = \frac{\mu_k^*}{\frac{1}{q_i(x^*)} - t^*}, \tag{7}$$

then the solution obtained by our method satisfies the stationarity requirement of the KKT conditions. The stationarity conditions are satisfied, as

$$\nabla_x L(x^*, t^*, \lambda^*) = \sum_{i=1}^{m} \lambda_i^* \nabla_x \left(\frac{1}{q_i(x^*)} \right) = 0.$$

Primal feasibility is also satisfied, since

$$\frac{1}{q_i(x^*)} - t^* \geq 0.$$

Dual feasibility is satisfied if

$$\lambda_i^* \geq 0.$$

From (7), and since $\mu_k > 0$ and $\frac{1}{q_i(x^*)} - t^* > 0$ at the solution, we have

$$\lambda_i^* \geq 0.$$

The complementarity condition requires that

$$\lambda_i^* \left(\frac{1}{q_i(x^*)} - t^* \right) = 0.$$

Substituting for λ_i^*, we see that

$$\lambda_i^* \left(\frac{1}{q_i(x^*)} - t^* \right) = \mu_k.$$

The log-barrier method drives μ_k to 0 as $k \to \infty$. Thus, the complementarity condition is also satisfied. Therefore, our log-barrier method converges to stationary points. Our implementation explicitly checks that the line search exploration moves the vertices in an ascent direction.

5.2 Monotonicity

In our algorithm, the optimization method maximizes the objective function given by (3),

$$F(x, t, \mu_k) = t + \mu_k \sum_{i=1}^{m} \log \left(\frac{1}{q_i(x)} - t \right),$$

on every iteration. Because t and $\mu_k > 0$ are constants for a given iteration, the maximization of the objective function is equivalent to maximization of the sum of the logarithmic terms. This is equivalent to maximizing the product of the terms (without taking their logarithms).

For simplicity of the analysis, let us now examine the monotonicity of our algorithm when employed on a patch having only two elements. If we plot the qualities of the two elements on the X and Y axes, we obtain hyperbolic contours representing the objective function as shown in Fig. 1. In Fig. 1, P represents a patch with near-equal qualities of the elements, and Q represents a patch with unequal element qualities. The symbols a, b, c, and d represent the paths the patches can take in order to maximize the objective function.

A Log-Barrier Method for Mesh Quality Improvement

Ideally, P should take path b, and Q should take path d so that the qualities of both the elements improve. In many cases, this is not possible, as improving the quality of one of the elements decreases the quality of the other. In the near-equal case, if P takes path a, the quality of the worst element decreases. Thus, we see that our algorithm may not monotonically increase the quality of the worst element in the mesh. For the unequal case, path c also improves the quality of the worst element.

Fig. 2 shows how the nonsmooth objective function for maximizing the worst quality element in Fig. 2(a) is converted to a smooth objective function in Fig. 2(b). In Fig. 2(a), the nonsmooth aspect ratio is plotted for a patch with a free vertex in the square formed by the diagonal from $(0, 0)$ to $(1, 1)$. Other vertices are on the perimeter of the square at $(0, 0)$, $(0, 0.5)$, $(0, 1)$, $(0.5, 1)$, $(1, 1)$, $(1, 0.5)$, $(1, 0)$, and $(0.5, 0)$. Note that the contours are nonsmooth when plotting the worst quality element in the patch. For illustration purposes, we chose the point $(0.1, 0.1)$, where the function is nonsmooth and set t in the log-barrier objective function as some quantity less than the worst element quality in the patch with the free vertex at $(0.1, 0.1)$. When the contours of the objective function are plotted, we see in Fig. 2(b) that they are smooth. Our algorithm moves the free vertex at $(0.1, 0.1)$ to a point close to $(0.5, 0.5)$.

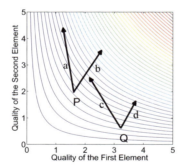

Fig. 1. Illustration of possible nonmonotonicity in the convergence of our algorithm. The X and Y axes represent the qualities of two elements in a patch, and the hyperbolic contours represent the sum of the two qualities on a logarithmic scale (which is maximized in an iteration). P and Q are possible locations of qualities of the patch. The symbols a, b, c, and d are the possible paths our algorithm can take. Although the objective function is maximized, notice that the quality of the worst element may not improve in all cases.

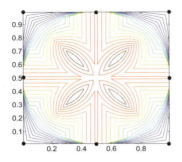
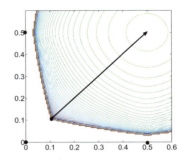

(a) Contours of the mesh quality

(b) Contours of the objective function

Fig. 2. Contours of the worst element quality and log-barrier objective function for a patch with eight vertices on the perimeter. (a) Contours of the nonsmooth aspect ratio quality metric. They are nonsmooth at points where two worst quality elements are present. (b) Contours of the log-barrier objective function with the free vertex at $(0.1, 0.1)$. Notice how smooth the contours are.

6 Numerical Experiments

In this section, we describe the numerical experiments that we designed to evaluate the performance of our method. We implemented our algorithm, the PS, and the MDS methods in the Mesquite Mesh Quality Improvement Toolkit Version 2.0.0 [18]. The active set method was already implemented in Mesquite. For each of the methods, the movement of the surface vertices was enabled. For the star meshes, the gradient of the objective function for the boundary vertices was projected onto the respective planes. For the sphere meshes, if the boundary vertices moved away from the surface, they were snapped back onto the surface.

Star and sphere (Fig. 3) meshes of various sizes were constructed using CUBIT [19]. In order to test our algorithm on challenging meshes, 50% of the vertices in the original meshes were randomly perturbed. The following three experiments were performed.

6.1 Effect of Parameters on Algorithmic Performance

For our first experiment, the following set of parameters were modified to determine their effect on the performance of the mesh quality improvement. Three variants of the nonlinear conjugate gradient method, i.e., the Fletcher-Reeves, Polak-Ribière, and Hestenes-Stiefel variants were used to improve the mesh quality in the inner loop. Two, four, and eight CG iterations per outer iteration were used in each of the experiments. The parameter μ was reduced to 90%, 60%, and 30% of its value after every outer iteration. We used the largest star mesh with approximately 1.012 million elements. We

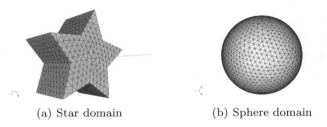

(a) Star domain   (b) Sphere domain

Fig. 3. Unperturbed coarse meshes on the two domains representative of actual meshes considered in this paper.

maximized the reciprocal of the aspect ratio quality metric for each of the subexperiments. The subexperiments were carried out for all combinations of these parameters. The numerical experiments were run until the quality of the worst element did not improve for five successive iterations.

Hestenes-Stiefel Conjugate Gradient Method. For this subexperiment, the Hestenes-Stiefel CG method was used for mesh quality improvement. The results from all the parameter combinations are shown in Fig. 4. Fig. 4(a) shows the plot of mesh quality verses time when μ was reduced to 30% of its value in every outer iteration for a set of CG iterations per outer iteration. It can be clearly seen that eight CG iterations for every outer iteration give the best mesh quality improvement. The results indicate that eight and four iterations of Hestenes-Stiefel CG give the best performance for the 60% and 90% cases, respectively.

Fletcher-Reeves and Polak-Ribière Conjugate Gradient Methods. We repeated the above subexperiment using the Fletcher-Reeves and Polak-Ribière CG methods. The results for two iterations of Polak-Ribière CG and four iterations of Fletcher-Reeves CG are show in Figs. 5(a), (b), and (c). For all the numerical experiments we conducted, it was seen that four Fletcher-Reeves CG iterations per outer iteration returned the best performance Polak-Ribière CG method gave best performance when two CG iterations were carried out for every outer iteration.

Parameters for Best Performance of Log-Barrier Method. Fig. 5 shows a summary of the best results obtained for the above subexperiments. Figs. 5(a), (b), and (c) show the best performance of each of the CG variants for μ being reduced to 30%, 60%, and 90% of its value after every outer iteration, respectively. Fig. 5(d) summarizes the best performing results from Figs. 5(a), (b), and (c). For each of the cases, the Hestenes-Stiefel CG variant gives the best performance. When μ is reduced to 90% of its value after four iterations of the Hestenes-Stiefel CG algorithm, maximum quality improvement is seen. Thus, we use this set of parameter values in our subsequent experiments, described in Sections 6.3 and 6.4, in which we compare the performance of our methods with existing algorithms for worst element mesh quality improvement including the active set, PS, and MDS methods.

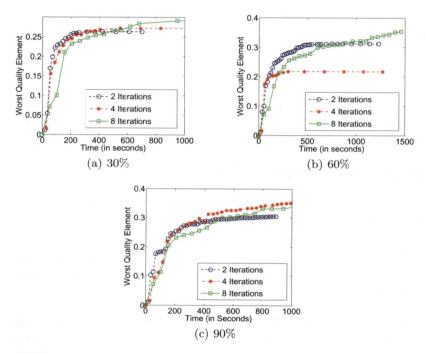

Fig. 4. Results obtained by using the Hestenes-Stiefel CG on the star mesh. The reciprocal of the aspect ratio is maximized using the log-barrier method. The percentages refer to the factor to which μ is reduced after every outer iteration. The number of iterations refer to the number of CG iterations per outer iteration.

6.2 Scalability

For this experiment, the reciprocal of the aspect ratio metric was maximized to improve the quality of the perturbed meshes using all the methods described in Section 2. Two inner iterations were carried out for every outer iteration each the method. The log-barrier method employed the Hestenes-Stiefel conjugate gradient algorithm [14] in the inner loop. After every outer iteration, μ was reduced to 90% of its value in the previous iteration. In order to accurately estimate the time per iteration, our implementation was executed for 50 iterations on each star mesh.

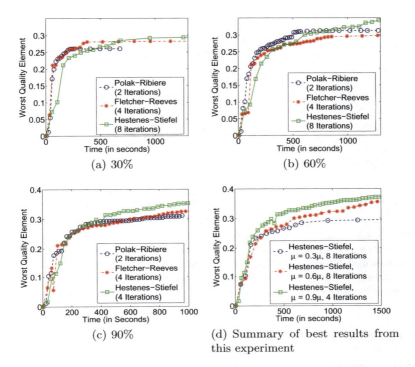

Fig. 5. Results obtained by using variants of CG on the star mesh. The reciprocal of the aspect ratio is maximized using the log barrier method. The percentages refer to the factor to which μ is reduced after every outer iteration. The number of iterations refer to the number of CG iterations per outer iteration.

The results from the scalability experiment are shown in Fig. 6, where the time per outer iteration scales linearly (for each method) with the number of elements in the mesh, as the time required to compute the gradient and move the vertices is directly proportional to the mesh size. We determined the order of convergence as a function of the problem size for our method. The order of convergence, α, is given by $T = k * m^{\alpha}$, where T is the time to convergence, m is the number of mesh elements, and $k > 0$ is a constant. In order to determine α, a least squares fit was computed by taking the logarithm of both sides. The value of α was found to be 0.9946.

6.3 Comparison with Existing Mesh Quality Improvement Methods for the Aspect Ratio Quality Metric

For this experiment, we used the largest meshes for each domain containing approximately 1.012 million and 1.014 million elements in the star and the sphere mesh, respectively. The reciprocal of the aspect ratio quality metric

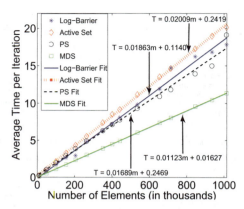

Fig. 6. Scalability experiment results. The data and the linear regression fits are shown. In the equations, T refers to average time per iteration, and m refers to number of elements (in thousands). The reciprocal of the aspect ratio quality metric was maximized. The average time per iteration for the first 50 outer iterations of all the methods were used to compute the least squares fit. For the log-barrier method, two Hestenes-Stiefel CG iterations were carried out per outer iteration.

was maximized by the four algorithms previously discussed: the log-barrier, active set, PS, and MDS methods. Several experiments were conducted to find the values of the various parameters in each of the algorithms resulting in the best performance (measured as the worst element quality at the time by which our algorithm converged). The numerical experiments were carried out until the worst element quality remained the same for five iterations. We present results for the best performance for each of the algorithms.

The results for the mesh quality improvement on the star mesh are shown in Fig. 7(a). Our method improved the mesh quality by the greatest amount when compared to the other methods. It was followed by the MDS, PS, and then the active set method. A closer inspection of the plot reveals that, in the initial iterations, the active set method was the fastest method to significantly improve the worst quality. The method was followed by the MDS and PS methods. During the initial phase, our method was slower than the rest.

The slow convergence of the other three methods, despite their initial performance, may have been caused by the slow propagation of unequal patches due to their use of local mesh quality improvement. The active set method was able to quickly improve the worst quality by a significant amount within two iterations but became stagnant afterward. The active set method moves every vertex to the optimal location with respect to the patch.

The optimal vertex locations are approximately determined in each iteration for the PS and MDS methods. Thus, unequal patches are present throughout the mesh. This enables steady improvement of the worst element quality in MDS and PS. In MDS, we noticed a behavior similar to the active set method where the worst element quality was constant for seven iterations and then the method converged to a mesh with a slightly better quality.

The results for the sphere mesh are shown in Fig. 7(b). As in the earlier case, our method was able to improve the mesh quality by the greatest amount. Here the PS method was very competitive and converged faster than our method, but to a lower optimal value. The MDS and active set methods also converged to a lower optimal value.

Table 1. Number of vertices and elements in the star meshes with their initial and final qualities after 50 outer iterations of quality improvement using the log-barrier method. The objective function that is formed from the reciprocal of the aspect ratio quality metric is maximized. The aspect ratio metric is a smooth, convex quality metric.

# Vertices	# Elements	Initial Worst Quality	Final Worst Quality
2,128	9,099	1.8544e-03	5.2991e-02
9,501	48,219	8.9960e-04	5.6172e-02
21,972	99,684	4.6232e-04	2.0447e-02
29,096	153,780	1.2875e-05	3.4908e-02
38,163	204,612	4.2422e-05	4.0148e-02
48,880	263,602	5.7753e-04	4.5985e-02
64,673	350,303	9.4858e-06	3.3304e-02
73,617	400,522	3.1302e-05	3.4584e-02
80,926	440,711	6.1325e-05	2.9847e-02
97,981	535,921	7.2535e-05	2.3580e-02
119,137	654,606	4.5933e-05	4.0360e-02
129,952	714,495	7.9804e-06	2.3341e-02
152,929	844,425	4.9885e-06	3.1510e-02
169,024	935,178	4.9250e-06	1.4332e-02
182,760	1,012,632	2.3352e-04	3.0506e-02

6.4 Comparison with Existing Mesh Quality Improvement Methods for the Nonsmooth Aspect Ratio Quality Metric

Through this experiment, we demonstrate that our algorithm is also efficient for mesh quality improvement when a nonsmooth or nonconvex quality metric is used to define the objective function. For this experiment, the objective function that was formed from the reciprocal of the nonsmooth aspect ratio quality metric was maximized by the log barrier, PS, and MDS methods. The active set method is designed to be used with a convex objective function, and yields a tangled mesh when used with a nonconvex quality metric. Thus, we have shown only the results for the other three methods in Fig. 8(a).

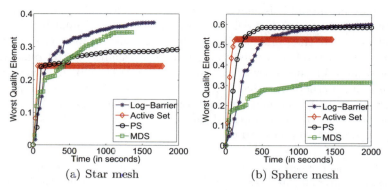

Fig. 7. Results from the experiment that compares the mesh quality improvement algorithms. The aspect ratio quality metric was improved in the meshes.

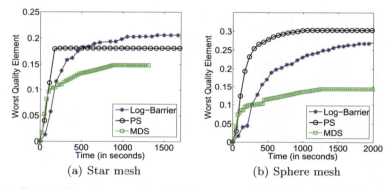

Fig. 8. Results from the experiment that compares the mesh quality improvement algorithms. The nonsmooth aspect ratio quality metric was improved in the meshes. The active set method is designed to be used with a convex objective function, and it yields a tangled mesh when used with a nonconvex quality metric, and hence its performance is not shown here.

The numerical experiments were carried out until the quality of the worst element did not improve for five iterations. It can be clearly seen in Fig. 8(a) that our method yields a better quality mesh than the other methods. In Fig. 8(b), it can be seen that our log-barrier method converged to a mesh of somewhat lower quality than the PS method. We seek to determine the line search parameter values that can make our method converge to a mesh with better quality.

We also found that the PS and MDS methods were very sensitive to parameter changes, but our log-barrier method and the active set method are not as sensitive. We have shown the effectiveness of our method compared to other existing methods. Our method can also improve the quality of a mesh even when it is assessed using a nonsmooth, nonconvex metric.

7 Conclusions and Future Work

We have presented a log barrier interior point method for improving the worst quality elements in a finite element mesh. We reformulated the nonsmooth problem of maximizing the quality of the worst element as a smooth constrained optimization problem, which is solved using a log barrier interior point method. Our method uses a log barrier function whose gradient places a greater emphasis on poor quality elements in the mesh and performs global mesh quality improvement. Our method usually yields a better quality mesh than other existing worst quality mesh improvement methods and scales roughly linearly with the problem size.

We have used the conjugate gradient method to perform mesh quality improvement in the inner loop. We plan to use Newton's method for the objective function maximization instead of the nonlinear CG method because it uses second-order information which may lead to faster convergence. The constrained optimization problem can also be solved using primal-dual Newton-based methods. An example of such a method includes the Mehrotra predictor-corrector method [20]. We plan to explore such methods to determine the most efficient solver for mesh quality improvement. We will also conduct research on other barrier functions that may make our method more efficient.

Our method can be naturally extended to handle mesh untangling by appropriately choosing the barrier term in the objective function. In this case, the barrier term should place the highest emphasis on inverted elements, medium emphasis on poor quality elements, and the least emphasis on good quality elements. Finally, by combining these approaches, our method may be used for simultaneous untangling and quality improvement.

Acknowledgments

The work of the first two authors was supported in part by NSF grants CNS-0720749 and NSF CAREER Award OCI-1054459. The work of the third author was supported in part by a Discovery grant from NSERC (Canada) and a grant from the U.S. Air Force Office of Scientific Research.

References

1. Fried, E.: Condition of finite element matrices generated from nonuniform meshes. AIAA Journal 10, 219–221 (1972)
2. Babuska, I., Suri, M.: The p and h-p versions of the finite element method, basic principles, and properties. SIAM Review 35, 579–632 (1994)
3. Berzins, M.: Solution-based mesh quality for triangular and tetrahedral meshes. In: Proc. of the 6th International Meshing Roundtable, Sandia National Laboratories, pp. 427–436 (1997)

4. Berzins, M.: Mesh quality - Geometry, error estimates, or both? In: Proc. of the 7th International Meshing Roundtable, Sandia National Laboratories, pp. 229–237 (1998)
5. Shontz, S., Vavasis, S.: Analysis of and workarounds for element reversal for a finite element-based algorithm for warping triangular and tetrahedral meshes. BIT, Numerical Mathematics 50, 863–884 (2010)
6. Shontz, S., Vavasis, S.: A robust solution procedure for hyperelastic solids with large boundary deformation. Engineering with Computers (2011), http://www.springerlink.com, doi: 10.1007/s00366-011-0225-y
7. Knupp, P.: Updating meshes on deforming domains: An application of the target-matrix paradigm. Commun. Num. Meth. Engr. 24, 467–476 (2007)
8. Knupp, P.: Matrix norms and the condition number: A general framework to improve mesh quality via node-movement. In: Proc. of the 8th International Meshing Roundtable, pp. 13–22 (1999)
9. Knupp, P., Freitag, L.: Tetrahedral mesh improvement via optimization of the element condition number. Int. J. Numer. Meth. Eng. 53, 1377–1391 (2002)
10. Amenta, N., Bern, M., Eppstein, D.: Optimal point placement for mesh smoothing. In: Proc. of the 8th ACM-SIAM Symposium on Discrete Algorithms, pp. 528–537 (1997)
11. Munson, T.: Mesh shape-quality optimization using the inverse mean-ratio metric. Mathematical Programming 110, 561–590 (2007)
12. Plaza, A., Suárez, J., Padrón, M., Falcón, S., Amieiro, D.: Mesh quality improvement and other properties in the four-triangles longest-edge partition. Comput. Aided Geom. D. 21(4), 353–369 (2004)
13. Shewchuk, J.: What is a good linear element? Interpolation, conditioning, and quality measures. In: Proc. of the 11th International Meshing Roundtable, Sandia National Laboratories, pp. 115–126 (2002)
14. Nocedal, J., Wright, S.: Numerical Optimization, 2nd edn. Springer, New York (2006)
15. Freitag, L., Plassmann, P.: Local optimization-based simplicial mesh untangling and improvement. Int. J. Numer. Meth. Eng. 49, 109–125 (2000)
16. Park, J., Shontz, S.: Two derivative-free optimization algorithms for mesh quality improvement. In: Proc. of the 2010 International Conference on Computational Science., vol. 1, pp. 387–396 (2010)
17. Parthasarathy, V., Graichen, C., Hathaway, A.: A comparison of tetrahedron quality measures. Finite Elem. Anal. Des. 15, 255–261 (1994)
18. Brewer, M., Frietag Diachin, L., Knupp, P., Laurent, T., Melander, D.: The mesquite mesh quality improvement toolkit. In: Proc. of the 12th International Meshing Roundtable, Sandia National Laboratories, pp. 239–250 (2003)
19. CUBIT generation and mesh generation toolkit, http://cubit.sandia.gov/
20. Mehrotra, S.: On the implementation of a primal-dual interior point method. SIAM J. Optimiz. 2(4), 575–601 (1992)

A Novel Geometric Flow-Driven Approach for Quality Improvement of Segmented Tetrahedral Meshes

Juelin Leng[1], Yongjie Zhang[2], and Guoliang Xu[1,*]

[1] LSEC, Institute of Computational Mathematics, Academy of Mathematics and System Sciences, Chinese Academy of Sciences, Beijing 100190, China
lengjl@lsec.cc.ac.cn, xuguo@lsec.cc.ac.cn
[2] Department of Mechanical Engineering, Carnegie Mellon University, USA
jessicaz@andrew.cmu.edu

Summary. This paper presents an efficient and novel geometric flow-driven method for mesh optimization of segmented tetrahedral meshes with non-manifold boundary surfaces. The presented method is composed of geometric optimization and topological transformation techniques, so that both location and topology of vertices are optimized. Non-manifold boundary can be divided into manifold surface patches having common boundary curves with each other. We adopt the averaged curvature flow to fair boundary curves with shape preserved, and the averaged mean curvature flow to fair surface patches with the property of volume-preserving. Meanwhile, boundary meshes are regularized by adjusting curve nodes and surface nodes along tangent directions. Locations of interior nodes are optimized by minimizing an energy functional which reflects the mesh quality. In addition, face-swapping and edge-removal operations are applied to eliminate poorly-shaped elements. Finally, we validate the presented method on several application examples, and the results demonstrate that mesh quality is improved significantly.

Keywords: Segmented tetrahedral mesh, quality improvement, geometric flow-driven, optimization-based mesh smoothing, shape-preserving.

1 Introduction

Unstructured tetrahedral meshes for complex three dimensional domains have been recognized as indispensable tools in various application fields, including computer graphics, finite element simulations, and partial differential equations. Since mesh quality is an extremely critical factor influencing the stability, convergence, and accuracy of the numerical solution, tremendous efforts have been made to achieve better mesh quality. However, it is still a challenging problem to generate quality meshes for complicated structures with non-manifold boundaries. In finite element analysis, the research objects are

---

* Corresponding author.

often segmented into multiple regions with respect to different physical attributes, chemical attributes, or material properties. Thus, quality segmented meshes, with conforming non-manifold boundaries, are needed for partitioned regions. In this paper, we focus on quality improvement of segmented tetrahedral meshes with boundary surface smoothed and shape preserved.

The advancing front technique [2], octree methods [3, 17], and Voronoi Delaunay-based methods [5, 14] are well studied techniques in unstructured mesh generation. Several mesh generation methods [24, 26] for domains with segmented regions have been developed in recent years. Unfortunately, these techniques cannot avoid the existence of distorted elements efficiently. Therefore, a post processing step is necessary to improve the overall quality of the meshes produced by automatic mesh generators.

The existing methods for mesh improvement fall into three typical categories [7, 14]: vertex insertion, topological transformation, and geometric optimization (also called mesh smoothing). It is intuitive to eliminate poor elements by adding vertices into meshes. Hence, vertex insertion methods are powerful ways to improve mesh quality. However, adding vertices will increase the number of mesh elements, which is not we would like to see. In topological transformation, several operations are implemented to reconnect vertices such that a set of adjacent elements are replaced by another set of elements with higher quality. Operations like edge/face swapping [7] and edge/multiface removal [13, 22] are usually local, easy to implement, and effective in removing poor elements. The effect for mesh quality improvement is also limited since reconnections are considered within small regions. To alleviate the limitation, a new reconnection way was proposed in [19] for relatively larger polyhedron composed of 20 to 40 tetrahedra. There are mainly two types of mesh smoothing methods, Laplacian smoothing and optimization-based smoothing. Laplacian smoothing [1] is simple and inexpensive, but it does not guarantee an improvement of the mesh in quality metrics and also results in degraded or inverted elements. Thus, various optimization-based methods were proposed. In these approaches, the objective function is based on a quality metric such as solid angle [21], dihedral angle [22], Jacobian matrix [9], or condition number [12]. In addition, Chen et al [15, 16] defined the interpolation error as the quality metric based on the concept of optimal Delaunay triangulation.

Surface smoothing is an important step of mesh improvement, since the generated meshes are often bumpy and irregular on boundary surfaces. During smoothing, surface features should be well preserved rather than be treated as noise and smoothed. Therefore, shape-preserving approaches were developed rapidly. Geometric flows [8] have the powerful ability to preserve features and reduce volume shrinkage. The surface diffusion flow which keeps the object volume invariant was used in [18, 25] to remove noise. Moreover, surface fitting and curvature information were applied to surface smoothing in [23].

In this paper, we present an efficient and novel geometric flow-driven method for mesh optimization of segmented tetrahedral meshes with non-manifold surfaces. Vertices of the original meshes are classified into four types: fixed vertices, curve vertices, surface vertices and interior vertices. Different vertices are handled by different strategies. For curve vertices, the averaged curvature flow for shape preserving is used to fair curves, and vertices are also modified along the tangential direction to achieve equi-distribution on curves. The averaged mean curvature flow, with the property of volume-preserving, is selected to remove bumpiness of surfaces by moving surface vertices along the normal direction. Meanwhile, an optimization objective function is defined to regularize triangular surface meshes. Interior vertices are regularized by an optimization-based method, with the objective function reflecting mesh quality. These approaches can improve the overall mesh quality efficiently, but some poorly-shaped elements still exist, because some vertices have bad valence. Hence, supplement operations like face-swapping and edge-removal are joined into the mesh improvement process. Our experiment results demonstrate that the presented method improves mesh quality significantly and preserves surface boundary features efficiently.

The remainder of this paper is organized as follows: in section 2, we introduce the problem description and preparation work for mesh improvement; section 3 presents quality improvement algorithms and implementation details; several application examples are given in section 4 to demonstrate the effectiveness of the presented method; and conclusion is drawn in the end.

2 Problem Description and Preparation Work

This section gives the description of quality improvement problem for segmented tetrahedral meshes. Before performing the mesh optimization algorithms, some preparation work are needed. We classify all the vertices into four types and select proper quality metrics to measure the mesh quality.

2.1 Problem Description

Suppose we are given a tetrahedral mesh T in $\mathbb{R}^3$, which is partitioned (segmented) into several volumetric components $\{T_i\}$. Fig. 1(a) shows a simple example of a segmented tetrahedral mesh with three components displayed by different colors. Our goal is to modify the mesh to regularize all the tetrahedra as much as possible. The given mesh is always bumpy and irregular on the boundary. Hence, we also aim to fair and regularize the boundary mesh with shape preserved.

For a given tetrahedral mesh, a position vector and a boundary indicator are given for each vertex, where the boundary indicator indicates if it lies on the boundary or not. Each tetrahedron is expressed by a list of its vertices in order. From the last vertex point of view, if the remaining three vertices are in the counterclockwise order, we say this tetrahedron is positive. If the ordering of a tetrahedron reverses this rule, then it is considered to be negative or inverted. In the given tetrahedral mesh, all tetrahedra are assumed to be positive. In addition, a component index is provided for each tetrahedron indicating which component it belongs to. If a tetrahedron has a triangular face on the boundary, then two component indices are given as well, indicating which pair of components share this triangle.

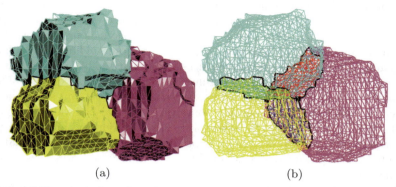

Fig. 1. (a) Tetrahedral mesh with three components; (b) the boundary mesh of (a).

2.2 Vertex Classification

Due to the complexity of non-manifold boundary, we classify the mesh vertices into four groups, such that different improvement strategies can be applied to each vertex group. Before vertex classification, we first introduce the concepts of boundary surface patches, boundary curves, and corner vertices.

Boundary surface patches: The common surface shared by any two components is referred to a boundary surface patch. Besides, the exterior boundary of each component is regarded as a boundary surface patch as well. As shown in Fig. 1(b), we use different colors to represent six boundary surface patches.

Boundary curves: The common curve shared by any two boundary surface patches is referred to a boundary curve, which is marked black in Fig. 1(b).

Corner vertices: The common vertex of any two boundary curves is referred to a corner vertex.

Then, we categorize the vertices into the following four groups:

Interior vertices: Interior vertices are vertices inside one volumetric component.

Surface vertices: Surface vertices are manifold vertices on boundary surface patches, which can move along the normal direction to remove noise, and move along the tangent direction to improve the aspect ratio of elements.

Curve vertices: Curve vertices are vertices located on boundary curves excluding end points. Curve vertices can only move along the tangent direction of the boundary curve during regularization.

Fixed vertices: Fixed vertices are end points of boundary curves and other non-manifold vertices, which are fixed during the mesh improvement process.

2.3 Quality Metrics of Tetrahedral Meshes

A number of quality metrics have been used to measure the quality of tetrahedral meshes, such as the longest-to-shortest edge length ratio [17], the minimum dihedral angle [22] or solid angle [21], and the element condition number [12]. Here, we choose the element aspect ratio introduced by Liu and Joe [4] to measure the teterahedron quality,

$$Q = \frac{8 \cdot 3^{\frac{5}{2}} V}{(\sum_{j=1}^{6} e_j^2)^{\frac{3}{2}}}, \quad (1)$$

where e_j are six edge lengths of one tetrahedron and V is the volume. If the oriented tetrahedron is positive, then the aspect ratio $Q \in [0, 1]$, and the quality gets better as Q is closer to 1.

Fig. 2 shows several examples of tetrahedra with poor quality. Tetrahedron (a) and (b) with one or two very short edges, can be recognized by all the above quality metrics. Four vertices of tetrahedra (c) and (d) are almost coplanar, but the ratios of the longest-to-shortest edge length are good. Tetrahedron (e) is slender but the minimum dihedral is away from 0 and π, so it cannot be detected by using the minimum dihedral angle as a quality metric. In contrast, the element aspect ratio Q can detect all the poor elements successfully.

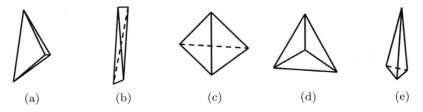

Fig. 2. Examples of poorly-shaped elements.

3 Quality Improvement Algorithm and Implementation

Our quality improvement algorithm for segmented tetrahedral meshes is composed of four steps:

1. Boundary curve fairing and regularization by adjusting curve vertices.
2. Boundary surface patch fairing and regularization by adjusting surface vertices.
3. Volume mesh regularization by relocating the interior vertices.
4. Topology improvement.

To fair a curve/surface mesh, we relocate the vertices such that the curve/surface is as smooth as possible. To regularize a curve/surface/volume mesh, we relocate the mesh vertices so that each element of the mesh has an optimal shape in a certain sense.

The first three steps are geometric optimization for different groups of vertices, which are implemented and carried out iteratively. Then the topological transformations are used to optimize the connection of vertices. Geometric optimization and topological transformation are also performed iteratively, until the desirable result is achieved. In the following, we explain each step of the quality improvement algorithm in detail.

3.1 Curve Fairing by Averaged Curvature Flow

Let $[\mathbf{x}_0 \mathbf{x}_1 \cdots \mathbf{x}_n]$ be a boundary curve of the mesh, which is actually a polygonal line. To fair this curve with shape preserved, we construct the following averaged curvature flow

$$\frac{d\mathbf{x}_i}{dt} = [\|\mathbf{h}_i(t)\| - h(t)]\, \mathbf{n}_i(t), \quad i = 1, \cdots, n-1, \tag{2}$$

where

$$\mathbf{n}_i(t) = \frac{\mathbf{h}_i(t)}{\|\mathbf{h}_i(t)\|}, \quad \mathbf{h}_i(t) = \frac{\mathbf{t}_{i+1} - \mathbf{t}_i}{s_i}, \quad \mathbf{t}_i = \frac{\mathbf{x}_i - \mathbf{x}_{i-1}}{\|\mathbf{x}_i - \mathbf{x}_{i-1}\|}, \tag{3}$$

and

$$h(t) = \sum_{i=1}^{n-1} s_i \|\mathbf{h}_i\| \Big/ \sum_{i=1}^{n-1} s_i, \quad s_i = \frac{\|\mathbf{x}_i - \mathbf{x}_{i-1}\| + \|\mathbf{x}_i - \mathbf{x}_{i+1}\|}{2}. \tag{4}$$

The equation can be solved using the explicit Euler scheme

$$\mathbf{x}_i^{(k+1)} = \mathbf{x}_i^{(k)} + \tau\, [\|\mathbf{h}_i(t_k)\| - h(t_k)]\, \mathbf{n}_i(t_k), \quad i = 1, \cdots, n-1, \tag{5}$$

where τ is a temporal step-size, $\mathbf{x}_i^{(0)} = \mathbf{x}_i$, and $\mathbf{x}_0^{(k)} = \mathbf{x}_0^{(k+1)} = \mathbf{x}_0$, $\mathbf{x}_n^{(k)} = \mathbf{x}_n^{(k+1)} = \mathbf{x}_n$. $\mathbf{h}_i(t_k)$, $h(t_k)$ and $\mathbf{n}_i(t_k)$ are defined in (3)–(4), taking $\mathbf{x}_i = \mathbf{x}_i^{(k)}$, $i = 0, \cdots, n$.

Remark 1. To ensure that the vertex relocation will not result in inverted tetrahedra, we perform an explicit check. During each iteration step, if the relocation of vertices inverts any tetrahedron, we reduce the step-size by a predefined factor 0.618, until no inverted tetrahedron is produced. Similarly, this check will be used in the following algorithm.

3.2 Curve Regularization

Let $L = \sum_{i=1}^{n} \|x_i - x_{i-1}\|$. Then L is the total length of the polygonal line $[\mathbf{x}_0 \mathbf{x}_1 \cdots \mathbf{x}_n]$. We intend to regularize the curve such that vertices are uniformly distributed. Therefore we construct the following energy functional

$$\mathcal{E}(\mathcal{C}) = \frac{1}{2}\sum_{i=1}^{n}(\|\mathbf{x}_i - \mathbf{x}_{i-1}\| - h)^2, \tag{6}$$

where $h = \frac{L}{n}$. At each free vertex $\mathbf{x}_i$ of the curve, we vary $\mathbf{x}_i$ as $\mathbf{x}_i \to \mathbf{x}_i + \epsilon_i \Phi_i$, $\Phi_i \in \mathbb{R}^3$, $i = 1, \cdots, n-1$. Then $\mathcal{E}(\mathcal{C})$ can be denoted as $\mathcal{E}(\mathcal{C}, \epsilon_i)$ and

$$\left.\frac{\partial \mathcal{E}(\mathcal{C}, \epsilon_i)}{\partial \epsilon_i}\right|_{\epsilon_i=0} = (\|\mathbf{x}_{i+1} - \mathbf{x}_i\| - h)\frac{\Phi_i^T(\mathbf{x}_i - \mathbf{x}_{i+1})}{\|\mathbf{x}_i - \mathbf{x}_{i+1}\|}$$
$$+ (\|\mathbf{x}_i - \mathbf{x}_{i-1}\| - h)\frac{\Phi_i^T(\mathbf{x}_i - \mathbf{x}_{i-1})}{\|\mathbf{x}_i - \mathbf{x}_{i-1}\|}.$$

Let $\mathbf{e}_i$ be the unit tangential direction at $\mathbf{x}_i$, then we construct a set of L^2-gradient flows as follows:

$$\frac{d\mathbf{x}_i}{dt} + (\|\mathbf{x}_{i+1} - \mathbf{x}_i\| - h)\frac{\mathbf{e}_i \mathbf{e}_i^T(\mathbf{x}_i - \mathbf{x}_{i+1})}{\|\mathbf{x}_i - \mathbf{x}_{i+1}\|} + (\|\mathbf{x}_i - \mathbf{x}_{i-1}\| - h)\frac{\mathbf{e}_i \mathbf{e}_i^T(\mathbf{x}_i - \mathbf{x}_{i-1})}{\|\mathbf{x}_i - \mathbf{x}_{i-1}\|} = \mathbf{0}, \tag{7}$$

$i = 1, \cdots, n-1$. The discretization of (7) can be written as

$$\frac{\mathbf{x}_i^{(k+1)} - \mathbf{x}_i^{(k)}}{\tau} + (\|\mathbf{x}_{i+1}^{(k)} - \mathbf{x}_i^{(k)}\| - h)\frac{\mathbf{e}_i \mathbf{e}_i^T(\mathbf{x}_i^{(k)} - \mathbf{x}_{i+1}^{(k)})}{\|\mathbf{x}_i^{(k)} - \mathbf{x}_{i+1}^{(k)}\|}$$
$$+ (\|\mathbf{x}_i^{(k)} - \mathbf{x}_{i-1}^{(k)}\| - h)\frac{\mathbf{e}_i \mathbf{e}_i^T(\mathbf{x}_i^{(k)} - \mathbf{x}_{i-1}^{(k)})}{\|\mathbf{x}_i^{(k)} - \mathbf{x}_{i-1}^{(k)}\|} = 0. \tag{8}$$

The initial value is chosen as $\mathbf{x}_i^{(0)} = \mathbf{x}_i$. Each $\mathbf{e}_i$ is obtained by computing the unit tangent direction of a fitting quadratic curve with respect to $\mathbf{x}_{i-1}, \mathbf{x}_i$ and $\mathbf{x}_{i+1}$.

3.3 Surface Mesh Fairing by Averaged Mean Curvature Flow

To regularize a partitioned tetrahedral mesh, it is pre-requested that the volume of each component should be preserved. It is well known that for a compact (closed and finite) smooth surface, the averaged mean curvature flow is volume preserving. However, the problem here is different because the boundary surface consists of several surface patches with fixed boundary curves. Hence, the volume preserving property of the averaged mean curvature flow needs to be re-considered.

Let M_0 be a piece of compact orientable surface in $\mathbb{R}^3$ with boundary denoted as Γ. A curvature driven geometric evolution consists of finding a family $M = \{M(t) : t \geq 0\}$ of smooth orientable surfaces in $\mathbb{R}^3$ which evolve according to the flow equation

$$\frac{\partial \mathbf{x}}{\partial t} = V_n(\mathbf{x})\mathbf{n}(\mathbf{x}), \quad M(0) = M_0, \quad \partial M(t) = \Gamma. \tag{9}$$

Here $\mathbf{x}(t)$ is a surface point on $M(t)$, $V_n(\mathbf{x})$ denotes the normal velocity of $M(t)$, and $\mathbf{n}(\mathbf{x})$ stands for the unit normal of the surface at $\mathbf{x}(t)$.

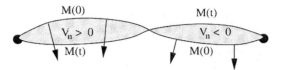

Fig. 3. The directional area between the curves $M(0)$ and $M(t)$. The area of the region with normal velocity $V_n > 0$ (or $V_n < 0$).

Theorem 1. *Let $V(t)$ denote the (directional) volume of the region enclosed by $M(0)$ and $M(t)$ (see Fig. 3 for 2D curve case). Then we have*

$$\frac{dV(t)}{dt} = \frac{1}{3}\int_{M(t)} V_n dA. \tag{10}$$

Proof. Let $\mathcal{S}$ be a closed smooth surface and V be the volume enclosed by $\mathcal{S}$. Then we have (see [20]),

$$V = \frac{1}{3}\int_{\mathcal{S}} \mathbf{x}^T \mathbf{n}\, dA. \tag{11}$$

By taking derivative with respect to t, we have

$$\begin{aligned}\frac{dV(t)}{dt} &= \frac{1}{3}\frac{d}{dt}\left[\int_{M(t)} \mathbf{x}^T \mathbf{n}\, dA + \int_{M(0)} \mathbf{x}^T \mathbf{n}\, dA\right]\\ &= \frac{1}{3}\frac{d}{dt}\int_{M(t)} \mathbf{x}^T \mathbf{n}\sqrt{g}\, du dv\\ &= \frac{1}{3}\int_{M(t)}\left[\frac{d\mathbf{x}^T}{dt}\mathbf{n}\sqrt{g} + \mathbf{x}^T \frac{d(\mathbf{n}\sqrt{g})}{dt}\right] du dv.\end{aligned} \tag{12}$$

Since the flow is a normal motion of the surface without tangential movement, hence $\frac{d\mathbf{x}_u}{dt} = \frac{d\mathbf{x}_v}{dt} = \mathbf{0}$. We have $\frac{d(\mathbf{n}\sqrt{g})}{dt} = \mathbf{0}$. Substituting (9) into (12), we obtain

$$\frac{dV(t)}{dt} = \frac{1}{3}\int_{M(t)} V_n dA. \tag{13}$$

□

In (9), if we take $V_n = H(t) - h(t)$, where $h(t) = \int_{M(t)} H dA / \int_{M(t)} dA$, then we have the **Averaged Mean Curvature Flow** [10] (AMCF)

$$\frac{\partial \mathbf{x}}{\partial t} = [H(\mathbf{x}) - h(t)]\mathbf{n}(\mathbf{x}), \quad M(0) = M_0, \quad \partial M(t) = \Gamma. \tag{14}$$

The existence proof of the global solution to this flow can be found in Escher and Simonett's paper [6]. It follows from (10) that

$$\frac{dV(t)}{dt} = \frac{1}{3}\left(\int_{M(t)} H dA - h(t)\int_{M(t)} dA\right) = 0.$$

Hence the averaged mean curvature flow is volume preserving, and its steady solution depends upon the initial surface.

Let M be a triangular surface patch and $\{\mathbf{x}_i\}_{i=1}^N$ be its free vertex set. For a vertex $\mathbf{x}_i$ with valence n, $N(i) = \{i_1, i_2, \cdots, i_n\}$ denotes the index set of one-ring neighbors of $\mathbf{x}_i$. Equation (14) is solved for the triangular mesh M using an explicit discretization method, where the discrete approximation of the mean curvature vector, mean curvature and surface normal are required. These approximations can be found in [8, 20]. Moreover, to compute $h(t)$, the integration $\int_{M(t)} H dA$ can be discretized as $\sum_{i=1}^N [H(\mathbf{x}_i) A_M(\mathbf{x}_i)]$, where $A_M(\mathbf{x}_i)$, as shown in Fig. 4, denotes the area represented by $\mathbf{x}_i$.

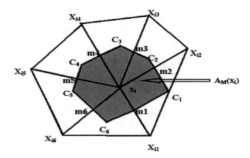

Fig. 4. Area represented by $\mathbf{x}_i$. $\{\mathbf{m}_j\}_{j=1}^6$ are midpoints of edges $[\mathbf{x}_i \mathbf{x}_{i_j}]$. $\mathbf{c}_j$ is the circumcenter point for the triangle $[\mathbf{x}_{i_j} \mathbf{x}_{i_{j+1}} \mathbf{x}_i]$ if the triangle is non-obtuse; if the triangle is obtuse, $\mathbf{c}_j$ is chosen to be the midpoint of the longest edge.

3.4 Surface Mesh Regularization

Suppose $\mathcal{S}$ is a piece of triangular surface patch. We intend to regularize it with fixed boundary. Let m be the triangle number of of $\mathcal{S}$, and A be the total area of all the triangles. Let $h = \frac{2}{3^{\frac{1}{4}}} \left(\frac{A}{m}\right)^{\frac{1}{2}}$. To make vertices equally distributed, we define

$$\mathcal{E}(\mathcal{S}) = \frac{1}{2} \sum_{i=1}^N \sum_{j \in N(i)} (\|\mathbf{x}_j - \mathbf{x}_i\| - h)^2. \tag{15}$$

At each free vertex $\mathbf{x}_i$, we vary $\mathbf{x}_i$ as $\mathbf{x}_i \to \mathbf{x}_i + \epsilon_i \Phi_i$, where $\Phi_i \in \mathbb{R}^3$, $i = 1, \cdots, N$. Then $\mathcal{E}(\mathcal{S})$ can be denoted as $\mathcal{E}(\mathcal{S}, \epsilon_i)$, and

$$\left.\frac{\partial \mathcal{E}(\mathcal{S}, \epsilon_i)}{\partial \epsilon_i}\right|_{\epsilon_i = 0} = \sum_{j \in N(i)} (\|\mathbf{x}_j - \mathbf{x}_i\| - h) \frac{\Phi_i^T (\mathbf{x}_i - \mathbf{x}_j)}{\|\mathbf{x}_i - \mathbf{x}_j\|}.$$

Let $\mathbf{e}_i^{(1)}$ and $\mathbf{e}_i^{(2)}$ be two unit orthogonal tangential directions at $\mathbf{x}_i$. Then we construct two sets of L^2-gradient flows as follows:

$$\frac{d\mathbf{x}_i}{dt} + \sum_{j \in N(i)} (\|\mathbf{x}_j - \mathbf{x}_i\| - h) \frac{\mathbf{e}_i^{(l)} (\mathbf{e}_i^{(l)})^T (\mathbf{x}_i - \mathbf{x}_j)}{\|\mathbf{x}_i - \mathbf{x}_j\|} = 0, \quad i = 1, \cdots, N, \tag{16}$$

where $l = 1, 2$. Equation (16) is solved iteratively by an explicit Euler scheme for the unknown $\mathbf{x}_i$, $i = 1, \cdots, N$. Tangential directions $\mathbf{e}_i^{(1)}$ and $\mathbf{e}_i^{(2)}$ can be computed from the limit surface of Loop's subdivision [11] or the quadratic fitting surface [20]. The discretization of (16) can be written as

$$\frac{\mathbf{x}_i^{(k+1)} - \mathbf{x}_i^{(k)}}{\tau} + \sum_{j \in N(i)} (\|\mathbf{x}_j^{(k)} - \mathbf{x}_i^{(k)}\| - h) \frac{\mathbf{e}_i^{(l)}(\mathbf{e}_i^{(l)})^T(\mathbf{x}_i^{(k)} - \mathbf{x}_j^{(k)})}{\|\mathbf{x}_i^{(k)} - \mathbf{x}_j^{(k)}\|} = 0,$$

with $i = 1, \cdots, N$, and $l = 1, 2$. The initial value is chosen as $\mathbf{x}_i^{(0)} = \mathbf{x}_i$. $\mathbf{e}_i^{(1)}$ and $\mathbf{e}_i^{(2)}$ are computed using the data at step k.

Remark 2. In the energy functional (15), h is defined as a global constant with respect to the average area of all triangles. If the mesh is adaptive to various density of distribution, then the local h_i can be used for each free vertex $\mathbf{x}_i$, and the energy functional is replaced by $\mathcal{E}(\mathcal{S}) = \frac{1}{2} \sum_{i=1}^{N} \sum_{j \in N(i)} (\|\mathbf{x}_j - \mathbf{x}_i\| - h_i)^2$. h_i can be chosen as $\frac{2}{3^{1/4}} \cdot A_i$, where A_i is the average area of triangles surrounding $\mathbf{x}_i$. Similarly, in the curve regularization energy functional (6), h can be replaced by a local h_i for adaptive meshes as well.

In addition, both the global h and the local h_i are updated during the iteration process of curve regularization and surface regularization.

3.5 Volume Mesh Regularization

Let M be a tetrahedral mesh for one component, and $\{\mathbf{x}_i\}_{i=1}^N$ be its interior vertex set. For a vertex $\mathbf{x}_i$ with tetrahedron valence n, let $N(i) = \{i_1, i_2, \cdots, i_n\}$ be the index set of its one-ring tetrahedron neighbors.

Let $\mathcal{T}$ be the set of all tetrahedra in M. Define the energy functional as

$$\mathcal{E}(M) = \mathcal{E}(\mathbf{x}_1, \cdots, \mathbf{x}_N) = \frac{1}{2} \sum_{\tau \in \mathcal{T}} (Q_\tau - 1)^2, \quad (17)$$

where $Q_\tau = \frac{(\sum_{j=1}^{6} e_{\tau,j}^2)^{\frac{3}{2}}}{8 \cdot 3^{\frac{5}{2}} V_\tau}$ is a quality metric, V_τ is the volume of tetrahedron τ, and $e_{\tau,j}$ ($j = 1, \cdots, 6$) are six edge lengths of τ. Note that Q_τ^{-1} is just the quality metric given in section 2.3, and $Q_\tau = 1$ if and only if τ is a regular tetrahedron. Hence, the overall mesh quality is improved as the energy functional $\mathcal{E}(M)$ reduces. At each interior vertex $\mathbf{x}_i$, we vary $\mathbf{x}_i$ as $\mathbf{x}_i \to \mathbf{x}_i + \epsilon_i \Phi_i$, where $\Phi_i \in \mathbb{R}^3$, $i = 1, \cdots, N$. Then

$$\left.\frac{\partial \mathcal{E}(M, \epsilon_i)}{\partial \epsilon_i}\right|_{\epsilon_i=0} = \left.\frac{\partial \mathcal{E}(\mathbf{x}_1, \cdots, \mathbf{x}_N, \epsilon_i)}{\partial \epsilon_i}\right|_{\epsilon_i=0} = \sum_{\tau \in N(i)} (Q_\tau(\epsilon_i) - 1) \left.\frac{\partial Q_\tau(\epsilon_i)}{\partial \epsilon_i}\right|_{\epsilon_i=0},$$

where

$$\left.\frac{\partial Q_\tau(\epsilon_i)}{\partial \epsilon_i}\right|_{\epsilon_i=0} = \frac{3(\sum_{j=1}^{6} e_{\tau,j}^2)^{\frac{1}{2}} \sum_{k=1}^{3} \Phi_i^T(\mathbf{x}_i - \mathbf{x}_{\tau,k})}{8 \cdot 3^{\frac{5}{2}} V_\tau} - \frac{(\sum_{j=1}^{6} e_{\tau,j}^2)^{\frac{3}{2}} \left.\frac{\partial V_\tau(\epsilon_i)}{\partial \epsilon_i}\right|_{\epsilon_i=0}}{8 \cdot 3^{\frac{5}{2}} V_\tau^2}.$$

Here, τ is an adjacent tetrahedron of $\mathbf{x}_i$, $\mathbf{x}_{\tau,k}$ ($k = 1, 2, 3$) are the three other vertices of τ besides $\mathbf{x}_i$.

Suppose $\mathbf{x}_i, \mathbf{x}_{\tau,1}, \mathbf{x}_{\tau,2}$ and $\mathbf{x}_{\tau,3}$ are in the positive order, so that

$$\frac{1}{6}\begin{vmatrix} 1 & x_i & y_i & z_i \\ 1 & x_{\tau,1} & y_{\tau,1} & z_{\tau,1} \\ 1 & x_{\tau,2} & y_{\tau,2} & z_{\tau,2} \\ 1 & x_{\tau,3} & y_{\tau,3} & z_{\tau,3} \end{vmatrix} = V_\tau,$$

where $\mathbf{x}_i = (x_i, y_i, z_i)^T$, $\mathbf{x}_{\tau,k} = (x_{\tau,k}, y_{\tau,k}, z_{\tau,k})^T$, $k = 1, 2, 3$. Then

$$\left.\frac{\partial V_\tau(\epsilon_i)}{\partial \epsilon_i}\right|_{\epsilon_i=0} = -\frac{1}{6}\Phi_i^T \left(\begin{vmatrix} 1 & y_{\tau,1} & z_{\tau,1} \\ 1 & y_{\tau,2} & z_{\tau,2} \\ 1 & y_{\tau,3} & z_{\tau,3} \end{vmatrix}, \begin{vmatrix} 1 & z_{\tau,1} & x_{\tau,1} \\ 1 & z_{\tau,2} & x_{\tau,2} \\ 1 & z_{\tau,3} & x_{\tau,3} \end{vmatrix}, \begin{vmatrix} 1 & x_{\tau,1} & y_{\tau,1} \\ 1 & x_{\tau,2} & y_{\tau,2} \\ 1 & x_{\tau,3} & y_{\tau,3} \end{vmatrix}\right)^T$$

$$\triangleq -\frac{1}{6}\Phi_i^T \Psi_{\tau,i}.$$

Therefore,

$$\left.\frac{\partial \mathcal{E}(M, \epsilon_i)}{\partial \epsilon_i}\right|_{\epsilon_i=0}$$

$$= \sum_{\tau \in N(i)} (Q_\tau - 1)\Phi_i^T \left(\frac{3(\sum_{j=1}^6 e_{\tau,j}^2)^{\frac{1}{2}} \sum_{k=1}^3 (\mathbf{x}_i - \mathbf{x}_{\tau,k})}{8 \cdot 3^{\frac{5}{2}} V_\tau} + \frac{(\sum_{j=1}^6 e_{\tau,j}^2)^{\frac{3}{2}} \Psi_{\tau,i}}{48 \cdot 3^{\frac{5}{2}} V_\tau^2} \right)$$

$$\triangleq \Phi_i^T \mathbf{d}_i,$$

where $\mathbf{d}_i$ is the gradient direction of $\mathcal{E}(M)$ with respect to $\mathbf{x}_i$. Then, we get the following discrete scheme:

$$\mathbf{x}_i^{(k+1)} = \mathbf{x}_i^{(k)} - \alpha_k \mathbf{d}_i^{(k)}.$$

Here, α_k is the step size in the gradient direction $\mathbf{d}_i^{(k)}$, which is properly selected such that the energy functional decreases and the worst quality among neighboring tetrahedra of $\mathbf{x}_i$ is also improved.

3.6 Topological Transformation

Most experiments show that, even after geometric optimization, some less-ideal elements still remain in the mesh. Thus, reconnection for mesh vertices should be considered. Face swapping, as the most popular topological operation, is used in our method to further improve the tetrahedral mesh. Basic operations for face swapping are shown in Fig. 5. These operations are simple, but it is critical to choose a proper algorithm.

Algorithm 1 presents our face swapping scheme for improving segmented tetrahedral mesh. Here, if the boundary is not destroyed and no inverted tetrahedron is produced, we say the operation is legal.

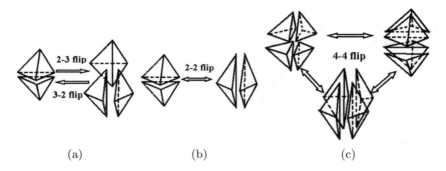

Fig. 5. Face swapping operations. (a) 2-3 and 3-2 flip; (b) 2-2 flip; (c) 4-4 flip.

Algorithm 1. *Face swapping*

1. *Compute the quality metric Q for all tetrahedra in the segmented mesh.*
2. *For each tetrahedron τ with $Q < \varepsilon$, where $\varepsilon \in (0,1)$ is a given threshold, perform the following steps to find the optimal operation to improve the quality.*
    a. *Set $f_i = -1.0, i = 1, \cdots, 4$, and $e_i = -1.0, i = 1, \cdots, 6$.*
    b. *Check each face of the tetrahedron. If the 2-3 flip operation is legal for removing face i, then set f_i as the worst quality of the three new tetrahedra.*
    c. *Check each edge of the tetrahedron.*
    If the valence of edge i is 3 and the 3-2 flip is legal, then set e_i as the worst quality of the two new tetrahedra.
    If the edge is valence 4 and the 4-4 flip is legal, then set e_i as the worst quality if the operation performs.
    If the edge with valence 2 is on an exterior boundary and the 2-2 flip is legal, then set e_i as the worst quality if the 2-2 flip performs.
    d. *If*
    $$\max\{\max_{i=1,\cdots,4}\{f_i\}, \max_{i=1,\cdots,6}\{e_i\}\} > Q,$$
    perform the corresponding operation such that the worst quality reaches the maximum, and then update the quality Q of all the new tetrahedra.
3. *Go back to step 1 until no operation can be performed or reach the given loop steps.*

It is well known that edge swapping is a simple and efficient way to eliminate sliver triangles for triangular meshes. For tetrahedral meshes, we use edge removal operation [13] to swap boundary edges. Fig. 6 shows the process of removing boundary edge $[AB]$. The gray area is an interior boundary. After removing boundary edge $[AB]$, vertices C and D are connected to generate two new boundary triangles. The next step is to find an optimal triangulation for the two polygons that maximize the quality of the worst tetrahedron. The edge removal operation is implemented by performing a sequence of 2-3 flips followed by a single 4-4 flip. For the exterior boundary edge removal, the last 4-4 flip in the interior case is replaced by a 2-2 flip.

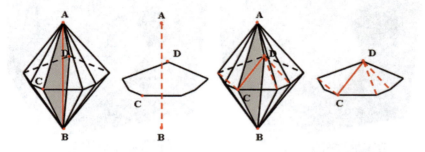

Fig. 6. An edge removal transformation.

4 Application Examples

In this section, we present several examples to show the efficiency of our new method. The input segmented meshes were generated by an octree-based iso-contouring method in [26].

4.1 Titanium Alloy with 52 Grains

The first example is a representative volume element of titanium alloy with 52 grains. The original tetrahedral mesh shown in Fig. 7(a) consists of 512,191 nodes and 3,000,564 tetrahedra. There are 3,678 elements with the quality value Q below 0.2, and the lowest value is 0.0002. Since the outline of the mesh is a cube, we treat the eight corners as fixed vertices, and treat vertices on the cube edges as curve points which can only move along the edge.

We improve the given mesh using the presented method and get the improved mesh displayed in Fig. 7(b), with each color representing a different grain. It is clear that planar boundary curves are smoothed and vertices are regularly distributed on curves. Moreover, sliver triangles near boundary curves in Fig. 7(a) are eliminated. Fig. 8 shows the improvement of interior boundary surface patches. Compared to the original mesh, both smoothness and regularity of the improved mesh are desirable. In addition, meshes of some internal grains are displayed in Fig. 9.

The mesh quality statistics before and after improvement are compared in Table 1, which shows a remarkable improvement of mesh quality using our method. The minimum quality value increases to 0.1346, and the average quality increases from 0.8721 to 0.8909. The mesh is overall optimized with good elements increased and poor elements reduced. Furthermore, six dihedral angles for all tetrahedra are calculated to measure the mesh quality,

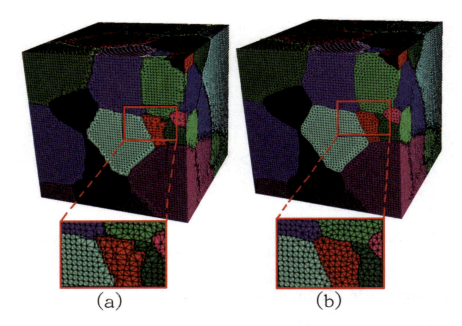

Fig. 7. (a) The original tetrahedral mesh; (b) the improved tetrahedral mesh.

and the histogram of dihedral angles is plotted in Fig. 10 (Left). The black line denotes the original mesh with dihedral angles ranging from $0.009°$ to $179.98°$, while the red line denotes the improved mesh with dihedral angles ranging from $5.83°$ to $153.43°$. The read line has a gap near $10°$ with the percentage value equals to zero. The original mesh is generated by an octree-based method which has structured interior elements, that explains why two peaks appear in the chart.

Table 1. Quality comparison

Quality value	min	max	average	0-0.2	0.2-0.4	0.4-0.6	0.6-0.8	0.8-1.0
Original mesh	0.0002	0.9994	0.8721	3,678	36,190	128,830	284,705	2,547,161
Improved mesh	0.1346	0.9998	0.8909	7	835	26,305	381,840	2,591,614

4.2 Brain Model with 41 Components

The fairing and regularization methods for boundary curves and surfaces introduced in section 3 can be used to optimize triangular meshes as well. Here, we give an example for improving triangular boundary meshes of a brain model. The brain model is made up of 41 components, and the original

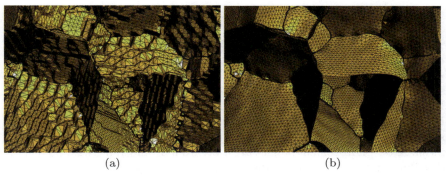

Fig. 8. Internal structure of the boundary mesh. (a) The original boundary mesh; (b) the smoothed boundary mesh.

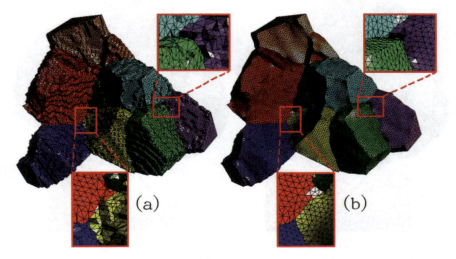

Fig. 9. Meshes of internal grains. (a) The original mesh; (b) the improved mesh.

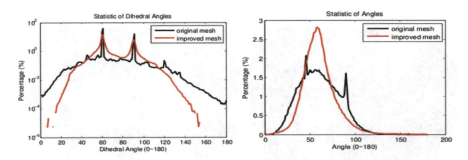

Fig. 10. Left: Dihedral angle statistics of the tetrahedral mesh in Fig. 7 (52-grain titanium alloy). Right: Angle statistics of the triangular mesh in Fig. 11 (brain).

triangular mesh consists of 97,295 nodes and 211,222 triangles. We apply our algorithms to optimize the mesh.

Fig. 11(a) and (b) show the exterior boundary meshes for the original mesh and the improved mesh, respectively. Surface patches belong to different components are represented with different colors. It can be seen that, surface noise is removed successfully and triangles are much more regular. Fig. 12 shows the improvement performance on interior surface patches. The histogram of mesh angles is plotted in Fig. 10 (Right), with the black line standing for the original mesh and the red line standing for the improved mesh. After improvement, angles are away from 0° and 180°, with its density distribution close to a normal distribution.

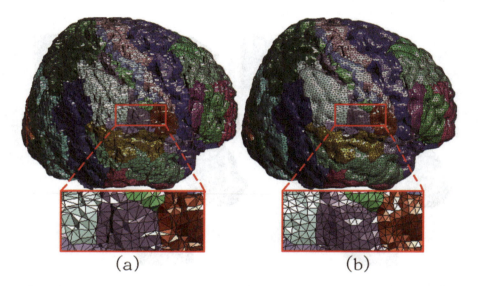

Fig. 11. (a) The original triangular mesh; (b) the improved triangular mesh.

Fig. 12. Cross section of brain. (a) The original mesh; (b) the smoothed mesh.

5 Conclusion

We have presented a novel geometric flow-driven method for quality improvement of segmented tetrahedral meshes, with volume-preserving for each component. At first, mesh vertices are classified into four groups, and each group of vertices is relocated by various geometric optimization strategies. Moreover, face-swapping and edge-removal operations are applied to eliminate poorly-shaped elements by changing the topology of vertices. Finally, we validate the presented method on several examples. Experiment results indicate that our method is capable of significantly improving the quality of segmented tetrahedral meshes and efficiently preserving boundary features.

Acknowledgement. Juelin Leng and Guoliang Xu were supported in part by NSFC under the grant 60773165, NSFC key project under the grant 10990013 and Funds for Creative Research Groups of China (grant No. 11021101). Yongjie Zhang was supported in part by a NSF/DOD-MRSEC seed grant. The authors are grateful to Jin Qian for preparing the tetrahedral mesh data.

References

1. Field, D.: Laplacian Smoothing and Delaunay Triangulation. Communications in Applied Numerical Methods 4, 709–712 (1988)
2. Lohner, R., Parikh, P.: Generation of Three-dimensional Unstructured Grids by the Advancing-front Method. International Journal for Numerical Methods in Fluids 8, 1135–1149 (1988)
3. Shephard, M.S., Georges, M.K.: Automatic Three-dimensional Mesh Generation Technique by the Finite Element Octree Technique. International Journal for Numerical Methods in Engineering 32, 709–749 (1991)
4. Liu, A., Joe, B.: Relationship Between Tetrahedron Quality Measures. BIT 34, 268–287 (1994)
5. Borouchaki, H., Lo, S.H.: Fast Delaunay Triangulation in Three Dimensions. Computer Methods in Applied Mechanics and Engineering 128, 153–167 (1995)
6. Escher, J., Simonett, G.: The Volume Preserving Mean Curvature Flow Near Spheres. Proceedings of the American Mathematical Society 126(9), 2789–2796 (1998)
7. Freitag, L.S., Ollivier-Gooch, C.: Tetrahedral Mesh Improvement Using Face Swapping and Smoothing. International Journal for Numerical Methods in Engineering 40(21), 3979–4002 (1998)
8. Desbrun, M., Meyer, M., Schröder, P., Barr, A.H.: Implicit Fairing of Irregular Meshes Using Diffusion and Curvature Flow. In: SIGGRAPH 1999, Los Angeles, USA, pp. 317–324 (1999)
9. Knupp, P.M.: A Framework for Volume Mesh Optimization and the Condition Number of the Jacobian Matrix. International Journal For Numerical Methods In Engineering 48(8), 1165–1185 (2000)
10. Sapiro, G.: Geometric Partial Differential Equations and Image Analysis. Cambridge University Press (2001)

11. Bajaj, C., Xu, G., Warren, J.: Acoustics Scattering on Arbitrary Manifold Surfaces. In: Proceedings of Geometric Modeling and Processing, Theory and Application, Japan, pp. 73–82 (2002)
12. Freitag, L.A., Knupp, P.M.: Tetrahedral Mesh Improvement via Optimization of the Element Condition Number. International Journal for Numerical Methods in Engineering 53, 1377–1391 (2002)
13. Shewchuk, J.R.: Two Discrete Optimization Algorithms for the Topological Improvement of Tetrahedral Meshes (2002) (unpublished manuscript)
14. Du, Q., Wang, D.: Tetrahedral Mesh Generation and Optimization Based on Centroidal Voronoi Tessellations. International Journal on Numerical Methods in Engineering 56(9), 1355–1373 (2003)
15. Chen, L., Xu, J.: Optimal Delaunay Triangulations. Journal of Computational Mathematics 22(2), 299–308 (2004)
16. Chen, L.: Mesh Smoothing Schemes Based on Optimal Delaunay Triangulations. In: Proceedings of 13th International Meshing Roundtable, pp. 109–120 (2004)
17. Zhang, Y., Bajaj, C., Sohn, B.S.: 3D Finite Element Meshing from Imaging Data. Computer Methods in Applied Mechanics and Engineering 194(48-49), 5083–5106 (2005)
18. Zhang, Y., Bajaj, C., Xu, G.: Surface Smoothing and Quality Improvement of Quadrilateral/Hexahedral Meshes with Geometric Flow. In: 14th International Meshing Roundtable, pp. 449–468 (2005)
19. Liu, J., Sun, S.: Small Polyhedron Reconnection: A New Way to Eliminate Poorly-shaped Tetrahedra. In: Proceedings of the 15th International Meshing Roundtable, pp. 241–257 (2006)
20. Xu, G.: Geometric Partial Differential Equation Methods in Computational Geometry. Scientific Publishing Press (2008)
21. Ghadyan, H.R.: Tetrahedral Meshes: Generation, Boundary Recovery and Quality Enhancements (2009)
22. Misztal, M.K., Brentzen, J.A., Anton, F., Erleben, K.: Tetrahedral Mesh Improvement Using Multi-face Retriangulation. In: Proceedings of the 18th International Meshing Roundtable, pp. 539–555 (2009)
23. Wang, J., Yu, Z.: A Novel Method for Surface Mesh Smoothing: Applications in Biomedical Modeling. In: Proceedings of the 18th International Meshing Roundtable, pp. 195–210 (2009)
24. Lederman, C., Joshi, A., Dinov, I., Van Horn, J.D., Vese, L., Toga, A.: Tetrahedral Mesh Generation for Medical Images with Multiple Regions using Active Surfaces. In: 2010 IEEE International Symposium on Biomedical Imaging From Nano to Macro, pp. 436–439 (2010)
25. Qian, J., Zhang, Y., Wang, W., Lewis, A.C., Siddiq Qidwai, M.A., Geltmacher, A.B.: Quality Improvement of Non-manifold Hexahedral Meshes for Critical Feature Determination of Microstructure Materials. International Journal for Numerical Methods in Engineering 82(11), 1406–1423 (2010)
26. Zhang, Y., Hughes, T., Bajaj, C.: An Automatic 3D Mesh Generation Method for Domains with Multiple Material. Computer Methods in Applied Mechanics and Engineering 199(5-8), 405–415 (2010)

Defining Quality Measures for High-Order Planar Triangles and Curved Mesh Generation

Xevi Roca[1,2], Abel Gargallo-Peiró[1], and Josep Sarrate[1]

[1] Laboratori de Càlcul Numèric (LaCàN),
Universitat Politècnica de Catalunya, Barcelona 08034, Spain
{xevi.roca,abel.gargallo,jose.sarrate}@upc.edu

[2] Department of Aeronautics and Astronautics
Massachusetts Institute of Technology, Cambridge, MA 02139, USA
xeviroca@mit.edu

Summary. We present a technique to extend any Jacobian based quality measure for linear elements to high-order isoparametric planar triangles of any interpolation degree. The extended quality measure is obtained as the inverse of the distortion of the high-order element with respect to an ideal element. To measure the high-order distortion, we integrate on the curved element the inverse of the Jacobian based quality measure. Thus, we can proof that if the Jacobian based quality is invariant under a particular affine mapping, then the resulting quality measure is also invariant under that mapping. In addition, we check that the quality measure detects non-valid and low-quality high-order elements. Finally, we present and test an approach to generate curved meshes by minimizing the high-order distortion measure of the elements.

Keywords: High-order quality, high-order mesh generation, mesh optimization, curved elements.

1 Introduction

In the last decades several computational methods have been widely used to solve partial differential equations (PDE) in applied sciences and engineering. Some of these methods allow the use of unstructured meshes, such as the finite element method (FEM), the finite volume method (FVM), and the discontinuous Galerkin method (DG). The unstructured methods have been proven to be very successful to solve PDE in complex domains (geometry flexibility). To solve a PDE with these methods, an unstructured mesh of the domain is generated. Then, a linear system is created by assembling the contributions of each mesh element to the system matrix. These contributions can be computed by integrating directly in the physical element or by changing the variable and integrating in a reference element.

To apply the reference element approach, it is required to use a differentiable, invertible and smooth mapping (diffeomorphism) from the reference element to the mesh element. Hence, the mapping has to be expressed by means of differentiable functions and the mesh elements have to be valid (non-folded) and present high-quality (regular shape). If one element is invalid then the determinant of the mapping Jacobian presents non-positive values. These non-positive determinant values invalidate the change of variable, and therefore, the obtained solution. Moreover, if one element has low quality then the element is distorted respect a regular element. Thus, the approximation accuracy is degraded and the solution may be polluted by the introduced error [1]. In summary, quality measures have to be used to assess the validity and quality of a given mesh.

Quality measures also have an alternative and significant application. They allow the use of optimization based techniques to repair non-valid meshes (untangle) and to improve the mesh quality (smooth) by maximizing the quality of the mesh elements. This technique allows the generation of high-order meshes with a posteriori approach [2, 3, 4, 5, 6, 7]. That is, it allows the generation of meshes that might contain inverted or low-quality elements, and then untangle and smooth them a posteriori to ensure and enhance the mesh quality. Specifically, a high-order mesh can be obtained by generating first a linear mesh. Second, the linear mesh is converted to a high-order mesh by adding additional nodes and by curving the boundary elements. Finally, the converted mesh is untangled and smoothed to remove the non-valid (folded) and low-quality (distorted) elements. However, the application of this approach together with a mesh quality optimization has been hampered by the absence of quality measures for high-order iso-parametric elements with degree superior than two. Note that the capability of generating valid high-order meshes is of the major importance for the high-order methods community.

The main contribution of this work is to present a technique that allows extending any Jacobian based quality measure for linear elements to high-order iso-parametric planar triangles of any interpolation degree. The proposed approach is compared with other related work in Section 2. Similarly to the linear elements technique, we measure the deviation of the physical element respect an ideal element. Specifically, we integrate the selected Jacobian based distortion measure in the curved element. Then, the quality measure for high-order elements is defined as the inverse of this distortion measure. The resulting quality inherits some of the properties of the original linear quality measures, Section 3. We also check that the proposed measure detects non-valid and low-quality elements for different initial Jacobian based quality measures, Section 4. To assess the applicability of the proposed measures, we overview a technique to optimize high-order meshes by minimizing the proposed distortion measure, Section 5. Finally, we apply this optimization technique to untangle and smooth several high-order triangular meshes, Section 6.

2 Related Work

In this work we propose an extension of quality measures for linear elements that allows determining the quality of iso-parametric elements of any interpolation degree. There are several previous works that determine the distortion or the quality of non-linear iso-parametric elements but only for quadratic degree [8, 9, 10, 11, 12, 13]. We would like to highlight that we share a similar formulation to the one proposed before by Branets and Carey in [10]. However, they extend only one distortion measure to quadratic elements, while we can extend any Jacobian based distortion measure to any interpolation degree. A different approach to extend Jacobian based disortion measures was previously proposed by Knupp [14]. The main difference is that we propose to integrate the distortion measure on the curved element, instead of computing the minimum, maximum or the mean on a set of sampling points. In addition, we also present numerical tests and mesh optimizations beyond the quadratic case.

The proposed extension of quality measures also allows the detection of non-positive values of the Jacobian determinant of the reference mapping for any interpolation degree. If the quality is strictly positive, we can ensure that the reference mapping is a local diffeomorphism. Other techniques to detect non-positive Jacobian determinants have been proposed before for B-spline based mappings [2, 3, 4, 5, 6] and quadratic iso-parametric elements [15, 16, 17]. Note that checking that the Jacobian determinant is strictly positive is not a guarantee for the global invertibility of the reference mapping for non-linear elements. That is, it is also required to check that the image of the reference mapping is simply connected [18]. In this work we do not study the conditions on the coordinates of the element nodes that ensure simply connected images, and therefore, global invertibility. It is important to point out that these conditions have been studied only in 2D for quadratic iso-parametric elements in [19, 20].

Finally, the proposed measures allow untangling and smoothing non-valid and low-quality high-order meshes. The optimization technique is a generalization of the methods for linear elements of Knupp [21] and Escobar *et al.* [22]. A similar optimization method has been presented previously by Branets and Carey for quadratic elements [10]. The main application of the untangling and smoothing method is to generate curved meshes by means of the named a posteriori approach. We have to remark that the a posteriori approach has been previously used with success by Shephard and co-workers for B-spline mappings [2, 3, 4, 5, 6] (instead of iso-parametric elements), and by Persson and Peraire by means of a nonlinear elasticity problem [7] (instead of optimizing a quality measure).

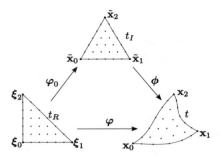

Fig. 1. Mappings between the reference, the ideal and the physical elements.

3 Quality Measures for High-Order Triangles

In order to determine the quality of a high-order triangular element t, we generalize the Jacobian based quality measures for linear elements [23, 24], Section 3.1. To this end, we consider a mapping ϕ from the ideal element t_I to the physical element t. To determine this mapping, we consider two high-order isoparametric mappings (Section 3.1), φ and φ_0, from a reference element t_R to t and t_I, respectively. Figure 1 presents the generalized diagram of mappings between the reference, the ideal and the physical high-order elements. This setting allows extending a Jacobian based distortion measure to high-order elements, Section 3.2. The proposed high-order quality measure inherits some properties of the initial Jacobian based quality measure, Section 3.3. In addition, the proposed definitions allow detecting invalid and low quality elements for different initial Jacobian based quality measures, Section 4.

3.1 Preliminaries

Jacobian Based Quality Metrics

For linear elements, the three mappings presented in Figure 1 are affine. In particular, the mapping between the reference and the ideal element is:

$$\varphi_0 : t_R \longrightarrow t_I \\ \boldsymbol{\xi} \longmapsto \tilde{\mathbf{x}} = \mathbf{W}\boldsymbol{\xi} + \tilde{\mathbf{x}}_0, \tag{1}$$

where

$$\mathbf{W} = (\tilde{\mathbf{x}}_1 - \tilde{\mathbf{x}}_0 \quad \tilde{\mathbf{x}}_2 - \tilde{\mathbf{x}}_0) = \begin{pmatrix} \tilde{x}_1 - \tilde{x}_0 & \tilde{x}_2 - \tilde{x}_0 \\ \tilde{y}_1 - \tilde{y}_0 & \tilde{y}_2 - \tilde{y}_0 \end{pmatrix} \tag{2}$$

is a constant matrix. The ideal element is chosen to be a valid and properly oriented element. Thus, φ_0^{-1} exists and is affine, since $\mathbf{W}$ is not singular. Similarly, the mapping between the reference and the physical triangle is defined as:

Table 1. Algebraic distortion measures for linear elements

Name	Distortion measure $\eta(\mathbf{S})$
Shape measure	$\eta(\mathbf{S}) = \dfrac{\|\mathbf{S}\|^2}{d \cdot \sigma(\mathbf{S})^{2/d}}$
Oddy et al. measure	$\eta(\mathbf{S}) = \dfrac{3}{d} \sigma^{-4/d}(\mathbf{S}) \left(\|\mathbf{S}^T\mathbf{S}\|^2 - \dfrac{1}{3}\|\mathbf{S}\|^4 \right)$
Condition number	$\eta(\mathbf{S}) = \dfrac{1}{2}\|\mathbf{S}\| \cdot \|\mathbf{S}^{-1}\|$

$$\varphi : t_R \longrightarrow t \\ \boldsymbol{\xi} \longmapsto \mathbf{x} = \mathbf{A}\boldsymbol{\xi} + \mathbf{x}_0, \quad (3)$$

where

$$\mathbf{A} = (\mathbf{x}_1 - \mathbf{x}_0 \quad \mathbf{x}_2 - \mathbf{x}_0).$$

Hence, φ is also an affine mapping with a constant Jacobian matrix $\mathbf{A}$. Finally, a mapping between the ideal and the physical element is determined by

$$\boldsymbol{\phi} = \varphi \circ \varphi_0^{-1}. \quad (4)$$

Note that $\boldsymbol{\phi}$ is also an affine mapping, since φ_0^{-1} and φ are so. Moreover, the Jacobian of $\boldsymbol{\phi}$ is constant and can be written as

$$\mathbf{S} := \mathbf{D}\boldsymbol{\phi} = \mathbf{A} \cdot \mathbf{W}^{-1}. \quad (5)$$

For linear elements it is usual to define a distortion measure in terms of the Jacobian matrix (5). These distortion measures, herein denoted by $\eta(\mathbf{S})$, quantify a specific type of distortion of the physical element in a range scale $[1, \infty)$. In addition, the quality measures of the physical elements are defined as the inverse of these distortion measures:

$$q(\mathbf{S}) = \frac{1}{\eta(\mathbf{S})}. \quad (6)$$

Several distortion measures for linear triangles have been proposed in literature, see [23]. In Table 1 we present three distortion measures that we use to test the proposed high-order quality measure, Section 4. In Table 1, parameter d is the number of spatial dimensions, $\sigma(\mathbf{S})$ is a function of the determinant of $\mathbf{S}$, and $\|\mathbf{S}\| = \sqrt{\operatorname{tr}(\mathbf{S}^t\mathbf{S})}$ is its Frobenius norm. In general $\sigma(\mathbf{S}) := \det \mathbf{S}$. However, to compute the quality of an element we set $\sigma(\mathbf{S}) = \frac{1}{2}(\det \mathbf{S} + |\det \mathbf{S}|)$ to assign a null quality to inverted elements ($\det \mathbf{S} < 0$).

Nodal High-Order Triangles

Let t be a nodal high-order element of order p determined by n_p nodes with coordinates $\mathbf{x}_i \in \mathbb{R}^d$, for $i = 1, \ldots, n_p$. Given a reference element t_R with nodes $\boldsymbol{\xi}_j \in \mathbb{R}^d$, $j = 1, \ldots, n_p$, we consider the basis $\{N_i\}_{i=1,\ldots,n_p}$ of nodal shape functions (Lagrange interpolation) of order p. In this basis, the high-order isoparametric mapping from t_R to t can be expressed as:

$$\varphi : t_R \subset \mathbb{R}^d \longrightarrow t \subset \mathbb{R}^d$$
$$\boldsymbol{\xi} \longmapsto \mathbf{x} = \varphi(\boldsymbol{\xi}; \mathbf{x}_1, \ldots, \mathbf{x}_{n_p}) = \sum_{i=1}^{n_p} \mathbf{x}_i N_i(\boldsymbol{\xi}), \quad (7)$$

where $\boldsymbol{\xi} = (\xi^1, \ldots, \xi^d)^T$ and $\mathbf{x} = (x^1, \ldots, x^d)^T$. Note that the shape functions $\{N_i\}_{i=1,\ldots,n_p}$ depend on the selection of $\boldsymbol{\xi}_j$, for $j = 1, \ldots, n_p$. In addition, they form a partition of the unity on t_R, and hold that $N_i(\boldsymbol{\xi}_j) = \delta_{ij}$, for $i, j = 1, \ldots, n_p$.

In this paper we focus on nodal high-order triangular elements of order p. Hence, the number of nodes n_p is $\frac{(p+1)(p+2)}{2}$, and the number of spatial dimensions d is 2. For this 2-dimensional case, the Jacobian of the isoparametric mapping (7) is:

$$\mathbf{D}\varphi(\boldsymbol{\xi}; \mathbf{x}_1, \ldots, \mathbf{x}_{n_p}) = \begin{pmatrix} \sum_{i=1}^{n_p} x_i^1 \frac{\partial N_i}{\partial \xi^1}(\boldsymbol{\xi}) & \sum_{i=1}^{n_p} x_i^1 \frac{\partial N_i}{\partial \xi^2}(\boldsymbol{\xi}) \\ \sum_{i=1}^{n_p} x_i^2 \frac{\partial N_i}{\partial \xi^1}(\boldsymbol{\xi}) & \sum_{i=1}^{n_p} x_i^2 \frac{\partial N_i}{\partial \xi^2}(\boldsymbol{\xi}) \end{pmatrix}.$$

3.2 Definitions

To define the high-order distortion measure of the physical element, we have to select first the equilateral ideal element t_I and a distribution of points. Herein, we choose a straight-sided triangle as the ideal element. In addition, we map the chosen distribution on the reference element (e.g. equi-distributed or Fekete points) to determine the distribution on the ideal element. Note that the mapping φ_0 is affine and its Jacobian matrix is given by equation (2). However, the mapping φ between the reference and the physical element, see Equation (7), can be not affine. Hence, $\phi = \varphi \circ \varphi_0^{-1}$ is in general not affine, and the Jacobian matrix is not constant. The expression of the Jacobian is:

$$\begin{aligned}
\mathbf{D}\phi(\tilde{\mathbf{x}}; \mathbf{x}_1, \ldots, \mathbf{x}_{n_p}) &= \mathbf{D}(\varphi(\cdot; \mathbf{x}_1, \ldots, \mathbf{x}_{n_p}) \circ \varphi_0^{-1})(\tilde{\mathbf{x}}) \\
&= \mathbf{D}\varphi(\varphi_0^{-1}(\tilde{\mathbf{x}}); \mathbf{x}_1, \ldots, \mathbf{x}_{n_p}) \cdot \mathbf{D}\varphi_0^{-1}(\tilde{\mathbf{x}}) \quad (8) \\
&= \mathbf{D}\varphi(\varphi_0^{-1}(\tilde{\mathbf{x}}); \mathbf{x}_1, \ldots, \mathbf{x}_{n_p}) \cdot \mathbf{W}^{-1},
\end{aligned}$$

where $\tilde{\mathbf{x}}$ is a point on the ideal triangle. Note that, according to (8), the local variation between the ideal and the physical triangles depends on $\tilde{\mathbf{x}}$ and also on the physical configuration of the high-order element $\mathbf{x}_1, \ldots, \mathbf{x}_{n_p}$.

Similar to the linear element case, we want to define a distortion measure based on the the Jacobian matrix of ϕ, see Equation (6). However, we cannot apply directly this approach because the Jacobian is not constant. Nevertheless, the Jacobian allows measuring the local deviation between the ideal and the physical element. Thus, we can obtain an elemental distortion measure by integrating the Jacobian based distortion measure on the whole physical element.

Definition 1. *The **high-order distortion measure** for a high-order element with nodes* $\mathbf{x}_1, \ldots, \mathbf{x}_{n_p}$ *is*

$$\hat{\eta}_r(\mathbf{x}_1, \ldots, \mathbf{x}_{n_p}) := \left(\frac{1}{|t|} \int_t \eta^r (\mathbf{D}\phi(\phi^{-1}(\mathbf{x}); \mathbf{x}_1, \ldots, \mathbf{x}_{n_p})) \, d\mathbf{x} \right)^{\frac{1}{r}}, \qquad (9)$$

where η is a distortion measure for linear elements based on the Jacobian matrix (see Table 1), $|t|$ is the area of the physical triangle, and r is a real number greater or equal to one.

Remark 1. Taking into account the change of variable determined by the isoparametric mapping φ and expression (8), we compute the distortion as:

$$\hat{\eta}_r(\mathbf{x}_1, \ldots, \mathbf{x}_{n_p}) = \left(\frac{1}{|t|} \int_{t_R} \eta^r \left(\mathbf{D}\varphi(\boldsymbol{\xi}) \cdot \mathbf{W}^{-1} \right) \cdot |\det \mathbf{D}\varphi(\boldsymbol{\xi})| \, d\boldsymbol{\xi} \right)^{\frac{1}{r}}.$$

In practical applications, we approximate the value of this integral with the symmetrical numerical quadrature for high-order triangles proposed in [25].

Definition 2. *The **high-order quality measure** for a high-order element with nodes* $\mathbf{x}_1, \ldots, \mathbf{x}_{n_p}$ *is*

$$\hat{q}_r(\mathbf{x}_1, \ldots, \mathbf{x}_{n_p}) := \frac{1}{\hat{\eta}_r(\mathbf{x}_1, \ldots, \mathbf{x}_{n_p})}. \qquad (10)$$

3.3 Properties

The proposed high-order distortion and quality measures, Definitions 1 and 2, present several properties. First, it is straightforward to prove that the linear case is just a particular case of these generalizations. That is, $\hat{\eta}_r(\mathbf{x}_1, \mathbf{x}_2, \mathbf{x}_3) = \eta(\mathbf{S}(\mathbf{x}_1, \mathbf{x}_2, \mathbf{x}_3))$ and $\hat{q}_r(\mathbf{x}_1, \mathbf{x}_2, \mathbf{x}_3) = q(\mathbf{S}(\mathbf{x}_1, \mathbf{x}_2, \mathbf{x}_3))$. Second, $\hat{\eta}_r$ and $\hat{q}_r$ maintain the image range of their respective linear distortion and quality measures. In particular, let q be a quality measure for linear elements with image range $[0, 1]$. Then, $\hat{q}_r$ is a quality measure for high-order elements with image range $[0, 1]$. Finally, we prove that $\hat{\eta}_r$ and $\hat{q}_r$ inherit the geometric properties of the Jacobian based distortion measure η.

Proposition 1. *If η is invariant under an affine mapping ψ, then $\hat{\eta}_r$ is also invariant under ψ.*

Proof. The affine mapping ψ can be written as $\psi(\mathbf{x}) := \mathbf{A}\mathbf{x} + \mathbf{b}$, where $\mathbf{A}$ is the linear mapping, and $\mathbf{b}$ is the translation vector. Thus, the transformation of the high-order element by the mapping ψ is the isoparametric mapping for the points $\psi(\mathbf{x}_i)$, $i = 1, \ldots, n_p$:

$$\begin{aligned}
\psi(\boldsymbol{\phi}(\tilde{\mathbf{x}}; \mathbf{x}_1, \ldots, \mathbf{x}_{n_p})) &= \mathbf{A} \cdot \boldsymbol{\phi}(\tilde{\mathbf{x}}; \mathbf{x}_1, \ldots, \mathbf{x}_{n_p}) + \mathbf{b} \\
&\stackrel{\mathbf{A} \text{ is linear}}{=} \sum_{i=1}^{n_p} \mathbf{A}\mathbf{x}_i N_i(\boldsymbol{\varphi}_0^{-1}(\tilde{\mathbf{x}})) + \mathbf{b} \\
&\stackrel{\text{Partit. unity}}{=} \sum_{i=1}^{n_p} \mathbf{A}\mathbf{x}_i N_i(\boldsymbol{\varphi}_0^{-1}(\tilde{\mathbf{x}})) + \sum_{i=1}^{n_p} \mathbf{b} N_i(\boldsymbol{\varphi}_0^{-1}(\tilde{\mathbf{x}})) \\
&= \sum_{i=1}^{n_p} (\mathbf{A}\mathbf{x}_i + \mathbf{b}) N_i(\boldsymbol{\varphi}_0^{-1}(\tilde{\mathbf{x}})) \\
&= \boldsymbol{\phi}(\tilde{\mathbf{x}}; \psi(\mathbf{x}_1), \ldots, \psi(\mathbf{x}_{n_p})).
\end{aligned}$$

Thus, the Jacobian for the transformed element is

$$\begin{aligned}
\mathbf{D}(\boldsymbol{\phi}(\tilde{\mathbf{x}}; \psi(\mathbf{x}_1), \ldots, \psi(\mathbf{x}_{n_p}))) &= \mathbf{D}(\mathbf{A}\boldsymbol{\phi}(\tilde{\mathbf{x}}; \mathbf{x}_1, \ldots, \mathbf{x}_{n_p}) + \mathbf{b}) \\
&\stackrel{\mathbf{b} \text{ is constant}}{=} \mathbf{D}(\mathbf{A}\boldsymbol{\phi}(\tilde{\mathbf{x}}; \mathbf{x}_1, \ldots, \mathbf{x}_{n_p})) \quad (11)\\
&= \mathbf{A} \cdot \mathbf{D}\boldsymbol{\phi}(\tilde{\mathbf{x}}; \mathbf{x}_1, \ldots, \mathbf{x}_{n_p}).
\end{aligned}$$

Finally, we can prove the invariance of $\hat{\eta}_r$ under ψ:

$$\begin{aligned}
\hat{\eta}_r(\psi(\mathbf{x}_1), \ldots, \psi(\mathbf{x}_{n_p})) &= \left(\frac{1}{|t|} \int_t \eta^r(\mathbf{D}\boldsymbol{\phi}(\boldsymbol{\phi}^{-1}(\mathbf{x}); \psi(\mathbf{x}_1), \ldots, \psi(\mathbf{x}_{n_p}))) \, \mathrm{d}\mathbf{x}\right)^{\frac{1}{r}} \\
&\stackrel{\text{by Eq. (11)}}{=} \left(\frac{1}{|t|} \int_t \eta^r(\mathbf{A} \cdot \mathbf{D}\boldsymbol{\phi}(\boldsymbol{\phi}^{-1}(\mathbf{x}); \mathbf{x}_1, \ldots, \mathbf{x}_{n_p})) \, \mathrm{d}\mathbf{x}\right)^{\frac{1}{r}} \\
&\stackrel{\eta \text{ is invariant}}{=} \left(\frac{1}{|t|} \int_t \eta^r(\mathbf{D}\boldsymbol{\phi}(\boldsymbol{\phi}^{-1}(\mathbf{x}); \mathbf{x}_1, \ldots, \mathbf{x}_{n_p})) \, \mathrm{d}\mathbf{x}\right)^{\frac{1}{r}} \\
&= \hat{\eta}_r(\mathbf{x}_1, \ldots, \mathbf{x}_{n_p}). \qquad \square
\end{aligned}$$

Corollary 1. *If a Jacobian based distortion measure for linear elements fulfills any of the following properties:*

- *translation-free,*
- *scale-free,*
- *rotation-free,*
- *symmetry-free,*

then the proposed high-order distortion and quality measures, Definitions 1 and 2, also hold the same properties.

Proof. Since $\hat{q}_r$ is defined as the inverse of $\hat{\eta}_r$, we only have to prove the previous properties for $\hat{\eta}_r$, Equation (9). All the *translations, scalings, rotations, symmetries*, and their compositions are affine mappings. Therefore, by Proposition 1, we have that $\hat{\eta}_r$ inherits the invariance under translation, or scaling, or rotation, or symmetry that η could have. $\qquad \square$

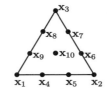

Fig. 2. Mesh composed by a triangle of order three.

Table 2. Locations of the free node for displacements restricted to one direction.

Free node	Location 1 (bue)	Location 2 (red)	Location 3 (green)
x_3	$(-1, \sqrt{3}/2)$	$(1/2, \sqrt{3}/2)$	$(2, \sqrt{3}/2)$
x_4	$(1/3, -3/2)$	$(1/3, 0)$	$(1/3, 3/2)$
x_{10}	$(1/2, -1)$	$(1/2, \sqrt{3}/6)$	$(1/2, 1.5)$

4 Behavior of the High-Order Quality Measure

In this section we illustrate the behavior of the proposed quality measure for high-order elements. Using Equation (9), we compute the high-order distortion measure for the three algebraic distortion measures presented in Table 1. Then, we use Equation (10) to evaluate the corresponding high-order quality measures. Specifically, we apply three tests to a triangle of order three with nodes located in an equispaced configuration, see Figure 2:

$$x_1 = (0,0), \quad x_2 = (1,0), \quad x_3 = \left(\tfrac{1}{2}, \tfrac{\sqrt{3}}{2}\right), \quad x_4 = \left(\tfrac{1}{3}, 0\right), \quad x_5 = \left(\tfrac{2}{3}, 0\right),$$
$$x_6 = \left(\tfrac{5}{6}, \tfrac{\sqrt{3}}{6}\right), \quad x_7 = \left(\tfrac{2}{3}, \tfrac{\sqrt{3}}{3}\right), \quad x_8 = \left(\tfrac{1}{3}, \tfrac{\sqrt{3}}{3}\right), \quad x_9 = \left(\tfrac{1}{6}, \tfrac{\sqrt{3}}{6}\right), \quad x_{10} = \left(\tfrac{1}{2}, \tfrac{\sqrt{3}}{6}\right).$$

In each test we consider a free node (keeping the rest of nodes fixed in the equispaced ideal configuration) and compute the quality of the high-order element when the node moves in $\mathbb{R}^2$. The free nodes are: the vertex node x_3, the edge node x_4, and the face node x_{10}. Figure 3 shows the contour plots of the previous high-order qualities for each test.

To visualize the configuration of the high-order triangle and to analyze in more detail the behavior of each high-order quality measure, we now restrict the displacement of the free nodes to one direction: vertex node x_3 moves along the x direction, and edge node x_4 and face node x_{10} move along the y direction. For each test, Figures 4(a), 4(b) and 4(c) display the configurations of the high order elements corresponding to the location of the free node presented in Table 2.

Figures 4(d), 4(e) and 4(f) plot the three high-order quality measures based on the linear distortion measures presented in Table 1 when:

- vertex node $x_3 = (x_3, \sqrt{3}/2)$ moves along the x direction, $x_3 \in [-2, 3]$;
- edge node $x_4 = (1/3, y_4)$ moves along the y direction, $y_4 \in [-2, 2]$;
- face node $x_{10} = (1/2, y_{10})$ moves along the y direction, $y_{10} \in [-3/2, 2]$.

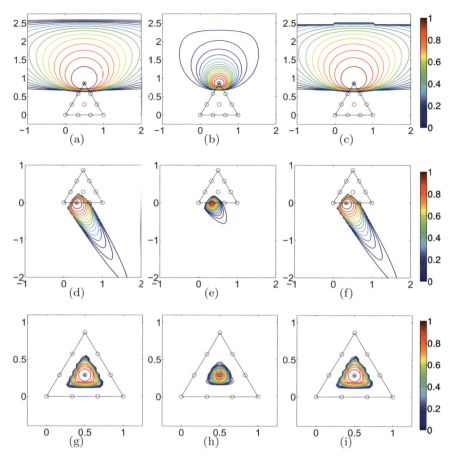

Fig. 3. Level sets for the three high-order quality measures (in columns: shape, Oddy and condition number) when the free node is: (a) to (c) the vertex node x_3; (d) to (f) the edge node x_4; and (g) to (i) the face node x_{10}.

As expected, in Figure 3 we realize that the three high-quality measures have similar behavior. Moreover, all of them define the same feasible region. However, the Oddy high-order quality is more strict and tends to zero faster than the other two measures. In these tests, the high-order quality measure detects all the non-valid configurations. Specifically, it detects tangled elements due to crossed edges or folded areas. Several conclusions can be drawn from Figure 4. From Figures 4(a), 4(b) and 4(c) we realize that moving away a node from its ideal location induces oscillations in the representation of the high-order element, even if the boundary of the element does not change as in Figure 4(c). Hence, tangled elements can appear, see for instance Figures

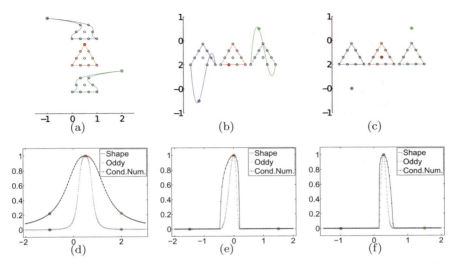

Fig. 4. Configurations and high-order qualities for the three tests. (a) and (b) vertex node $\mathbf{x}_3$ moves on the x direction, (c) and (d) edge node $\mathbf{x}_4$ moves on the y direction; and (e) and (f) face node $\mathbf{x}_{10}$ moves on the y direction.

4(b) and 4(c). From Figures 4(d), 4(e) and 4(f) we first realize that the defined measure properly detects when the high-order element folds and gets tangled. In Figures 4(e) and 4(f) all the measures detect the same tangling positions, where the quality achieves the zero value. Moreover, in all cases, the three measures detect the proper ideal configurations, with quality equals to 1. Note that in Figure 4(d), the three high-order quality measures do not degenerate, despite the Oddy measure decreases faster than the other two. Finally, Figures 3 and 4 show that vertex nodes have larger feasible regions than edge or face nodes.

5 Application to High-Order Mesh Optimization

One of the main problems in high-order mesh generation is to ensure that all the mesh elements are valid (untangled) and regular (smooth). Similarly to the methods proposed for linear meshes [21, 22], we propose to untangle and smooth a high-order mesh by minimizing the distortion of the elements, Definition 1. In this way, we are maximizing the mesh quality since the distortion is the inverse of the quality, Definition 2. For a free node $\mathbf{x}_0$, we define the local value of the objective function as the integral of the distortion measure on all the elements that contain $\mathbf{x}_0$:

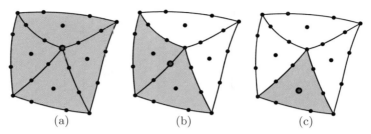

Fig. 5. Local patch for the element-based objective function for a node on: (a) a vertex node, (b) an edge, and (c) a face.

$$f(\mathbf{x}_0) := \left(\frac{1}{\sum_{i=1}^{N} |t_i|} \sum_{i=1}^{N} \int_{t_i} \eta^r \left(\mathbf{D}\phi_i(\phi^{-1}(\mathbf{x})) \right) d\mathbf{x} \right)^{1/r} \quad (12)$$

where N is the number of elements t_i that contain $\mathbf{x}_0$. Recall that $\phi_i(\tilde{\mathbf{x}}) \equiv \phi_i(\tilde{\mathbf{x}}; \mathbf{x}_1^i, \ldots, \mathbf{x}_0, \ldots, \mathbf{x}_{n_p}^i)$ denotes the mapping between the ideal triangle t_I, and the physical triangle t_i, with nodes $\mathbf{x}_1^i, \ldots, \mathbf{x}_0, \ldots, \mathbf{x}_{n_p}^i$. For all the results presented in this work, we set the value of r to 2. Note that the number of neighboring elements N depends on the position of the node inside the element. That is, if the node is on: a vertex, an edge or a face. In Figure 5 we illustrate the local patch (grey region) around a free node (red circle) depending on its location inside the element.

Finally, to minimize the objective function we perform a local non-linear minimization with updates. This approach is similar to the Gauss-Seidel technique for solving linear systems. Specifically, we loop on all the nodes that are not on the boundary of the mesh. Then, in each step we move the position of one node, and we fix the other ones. To this end, we minimize the local objective function of the node and we update the node location to the optimum. Then, we repeat the process again for the next node.

6 Numerical Examples

In this section we apply a high-order quality measure to untangle and smooth non-valid high-order meshes. To this end, we use a high-order distortion based on the shape measure, see Table 1. The initial high-order meshes, and therefore the initial curved boundary approximation, have been computed using the ez4u meshing environment [26, 27, 28]. Mesh quality statistics for the examples are presented in Table 3.

Circular Ring. In the first example we generate four meshes of orders 3, 4, 5 and 10 for a circular ring, see Figure 6. The four meshes are composed by 24 elements. The number of nodes depends on the selected order: 126 nodes for order 3, 216 nodes for order 4, 330 nodes for order 5, and 1260 nodes for order 10. All the initial meshes have the same straight inner edges and only differ on the degree of the polynomial approximation of the boundary. Figure 6(a) shows the initial mesh for order 3 displaying also the quality of its elements. Note that the inner edges of this mesh are straight. Therefore, several tangled elements appear at the inner boundary. Figure 6(b) shows a detail of the upper-right inner boundary of this initial mesh, where a tangled element with null quality appears. It also displays an equispaced sub-grid (in gray) inside each element to visualize the distortion of the mapping between the reference element and the physical one.

Figure 6 shows the initial non-valid meshes and the final optimized meshes. To perform the optimization, we use an equidistributed set of points on the reference and ideal element. In this way, we can plot a straight-sided structured grid on the reference element and map it to the physical element. This sub-grid helps to visualize the behavior of the isoparametric mapping. However, it is well known that for equidistributed interpolation points the Lebesgue constant grows with the interpolation degree. Therefore, the minimum quality for $p = 10$ is slightly worse than for lower degrees, see Table 3. To amend this issue, the interpolation points have to be placed to improve the value the Lebesgue constant. To this end, in practical applications we use an approximated Fekete distribution on both the reference and ideal element. For instance, we use these point distribution in the following example.

Barcelona Harbor. In this example we generate a high-order mesh for computing the wave agitation inside the Barcelona (Spain) harbor. The physical problem that is studied is the wave propagation in highly reflective coastal areas. The final goal is to obtain the wave amplification factor for an incident wave of height one. The Barcelona harbor contains several small geometric features (10 m length) compared to the total extension of the domain (12 km), requiring fine computational meshes if linear elements are used. On top of that, high-order elements are needed in order to reduce the numerical dispersion error, commonly associated with the propagation of high frequency waves in presence of numerous reflections. Using a mesh composed by 2.4 millions of linear elements an erroneous solution without physical meaning is obtained. However, using a high-order mesh of order 7 composed by 32802 elements (803649 nodes), the dispersion error can be reduced obtaining an accurate solution. Figure 7 shows the wave amplification factor for the Barcelona harbor when the angle between the incident wave and the x-axis is 43 degrees and the period is 6 seconds [29].

To generate a high-order mesh for this problem, and analogously to the previous example, we first create a triangular mesh composed by elements with curved edges on the boundary of the domain and with straight edges in the interior. Figure 8 shows the initial and smoothed meshes, displaying also the high-order quality of the elements, for the four areas marked in Figure 7. We apply the optimization procedure using a Fekete distribution of nodes on the reference and ideal element. Figures 8(a) to 8(d) present the four

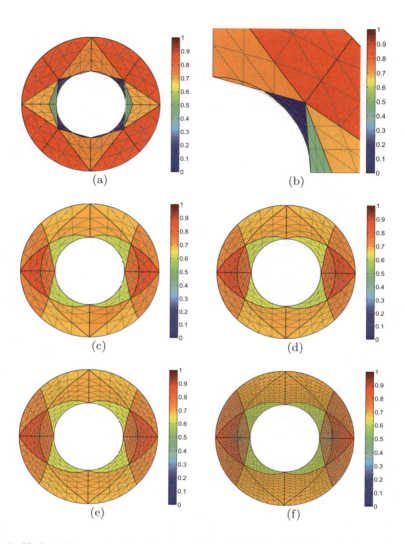

Fig. 6. High-order meshes for the ring: (a) and (b) the initial mesh ($p = 3$); and smoothed and untangled meshes for (c) $p = 3$, (d) $p = 4$, (e) $p = 5$, and (f) $p = 10$.

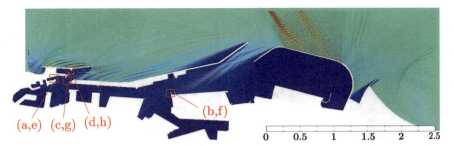

Fig. 7. Wave amplification factor on the Barcelona harbor for an incident wave of height equal to 1 [29]. The solution is obtained on a high-order mesh of order 7.

Table 3. High-order quality statistics for the circular ring and the Barcelona harbor: interpolation order (p); minimum, maximum, mean and standard deviation of the quality; number of inverted elements (#inv); and number of inner nodes on faces (N_f), edges (N_e) and vertices (N_v).

Meshes	p	Min.	Max.	Mean	Std. dev.	#inv	N_f	N_e	N_v
Initial ring	3	0.00	0.87	0.60	0.31	4	24	84	18
Smoothed ring	3	0.61	0.85	0.72	0.08	0	24	84	18
Smoothed ring	4	0.61	0.85	0.72	0.08	0	72	126	18
Smoothed ring	5	0.61	0.85	0.72	0.08	0	144	168	18
Smoothed ring	10	0.59	0.87	0.72	0.08	0	864	378	18
Initial harbor	7	0.00	1.00	0.91	0.08	3	492030	295218	16401
Smoothed harbor	7	0.36	1.00	0.91	0.02	0	492030	295218	16401

selected details of the initial mesh. Note that the three first details contain non-valid elements. Figures 8(e) to 8(h) show the four selected details of the smoothed mesh. The final mesh is composed by valid and high-quality elements. Specifically, on the boundary we obtain well shaped elements with curved edges, whereas inner elements tend to have straight edges.

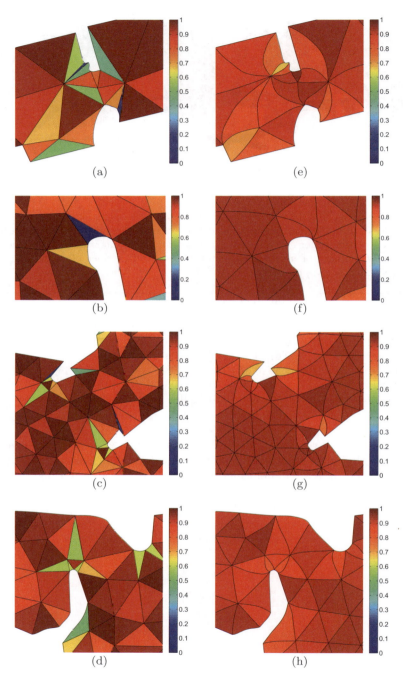

Fig. 8. Details of a high-order mesh for the Barcelona harbour: (a) to (d) details of the initial mesh, (e) to (h) details of the smoothed mesh.

7 Concluding Remarks and Future Work

In this work we have presented a technique to define quality measures for nodal high-order triangular elements of any interpolation degree. The quality allows the generalization of quality measures for linear triangles that are based on the Jacobian of an affine mapping. In addition, the generalization inherits the properties of the linear quality measure such as being: invariant under translation, scaling and orthogonal transformations. To assess the reliability of the technique, we have extended and tested three quality measures for linear triangles. The tests show that the obtained high-order quality measures detect invalid and low-quality configurations. Finally, we have shown the applicability of the proposed method by developing a technique to optimize high-order triangular meshes. This technique repairs non-valid elements (untangles) and improves low-quality elements (smooths). The numerical examples show optimized meshes that are fully untangled and composed by higher quality elements. Thus, we can generate high-order triangular meshes that are valid for a finite element simulation on a curved domain. To show this claim, we have included a high-fidelity solution of the Berkhoff equation on a curved mesh of the Barcelona harbor. The curved triangular mesh has been optimized with the proposed technique.

Our long term goal is to develop a mesh optimization tool that untangles and smoothes high-order meshes in 2D and 3D. In this sense, this work presents several limitations that should be investigated and solved in the near future. First, we have only presented quality measures for high-order triangles. Thus, we would like to extend the proposed technique to define quality measures for high-order quadrilaterals, tetrahedra and hexahedra. Second, we have used only the shape quality metric to optimize high-order triangular meshes. Therefore, we would like to compare the results obtained with several quality measures. Third, detailing the implementation of our optimization technique was out of the scope of this paper. However, we would like to detail the formulation and implementation of our procedure in a future work. Finally, in this work we have only ensured local invertibility. Thus, further research is needed to determine the conditions on the coordinates of nodes that ensure global invertibility. To this end, we would like to extend to high-order elements the conditions currently stated only for quadratic elements.

References

1. Shewchuk, J.: What is a good linear finite element? interpolation, conditioning, anisotropy, and quality measures (2002) (Preprint)
2. Dey, S., Shephard, M., Flaherty, J.: Geometry representation issues associated with p-version finite element computations. Comput. Method. Appl. M 150(1–4), 39–55 (1997)

3. Dey, S., O'Bara, R., Shephard, M.: Curvilinear mesh generation in 3d. Comput Aided Design 33, 199–209 (2001)
4. Luo, X., Shephard, M., Remacle, J., O'Bara, R., Beall, M., Szabó, B., Actis, R.: P-version mesh generation issues. In: 11th IMR, Citeseer, pp. 343–354 (2002)
5. Luo, X., Shephard, M., O'Bara, R., Nastasia, R., Beall, M.: Automatic p-version mesh generation for curved domains. Eng. Comput. 20(3), 273–285 (2004)
6. Shephard, M., Flaherty, J., Jansen, K., Li, X., Luo, X., Chevaugeon, N., Remacle, J., Beall, M., O'Bara, R.: Adaptive mesh generation for curved domains. Appl. Numer. Math. 52(2-3), 251–271 (2005)
7. Persson, P., Peraire, J.: Curved mesh generation and mesh refinement using lagrangian solid mechanics. In: AIAA Proceedings (2009)
8. Yuan, K.Y., Huang, Y., Pian, T.: Inverse mapping and distortion measures for quadrilaterals with curved boundaries. Int. J. Numer. Meth. Eng. 37(5), 861–875 (1994)
9. Chen, Z., Tristano, J., Kwok, W.: Combined laplacian and optimization-based smoothing for quadratic mixed surface meshes. In: 12th IMR (2003)
10. Branets, L., Carey, G.: Extension of a mesh quality metric for elements with a curved boundary edge or surface. J. Comput. Inf. Sci. Eng. 5(4), 302–308 (2005)
11. Salem, A., Canann, S., Saigal, S.: Robust distortion metric for quadratic triangular 2d finite elements. Appl. Mech. Div. ASME 220, 73–80 (1997)
12. Salem, A., Canann, S., Saigal, S.: Mid-node admissible spaces for quadratic triangular arbitrarily curved 2d finite elements. Int. J. Numer. Meth. Eng. 50(2), 253–272 (2001)
13. Salem, A., Saigal, S., Canann, S.: Mid-node admissible space for 3d quadratic tetrahedral finite elements. Eng. Comput. 17(1), 39–54 (2001)
14. Knupp, P.: Label-invariant mesh quality metrics. In: 18th IMR (2009)
15. Mitchell, A., Phillips, G., Wachspress, E.: Forbidden shapes in the finite element method. IMA J. Appl. Math. 8(2), 260 (1971)
16. Field, D.: Algorithms for determining invertible two-and three-dimensional quadratic isoparametric finite element transformations. Int. J. Numer Meth. Eng. 19(6), 789–802 (1983)
17. Baart, M., Mulder, E.: A note on invertible two-dimensional quadratic finite element transformations. Commun. Appl. Numer. M 3(6), 535–539 (1987)
18. de la Vallée Poussin, C.: Cours d'analyse infinitésimale, vol. 1. Gauthier-Villars (1921)
19. Baart, M., McLeod, R.: Quadratic transformations of triangular finite elements in two dimensions. IMA J. Numer. Anal. 6(4), 475 (1986)
20. Frey, A., Hall, C., Porsching, T.: Some results on the global inversion of bilinear and quadratic isoparametric finite element transformations. Math. Comput. 32(143), 725–749 (1978)
21. Knupp, P.: A method for hexahedral mesh shape optimization. Int. J. Numer. Meth. Eng. 58(2), 319–332 (2003)
22. Escobar, J., Rodríguez, E., Montenegro, R., Montero, G., González-Yuste, J.: Simultaneous untangling and smoothing of tetrahedral meshes. Comput. Method. Appl. M 192(25), 2775–2787 (2003)
23. Knupp, P.: Algebraic mesh quality metrics. SIAM J. Sci. Comput. 23(1), 193–218 (2002)
24. Knupp, P.: Algebraic mesh quality metrics for unstructured initial meshes. Finite Elem. Anal. Des. 39(3), 217–241 (2003)

25. Wandzurat, S., Xiao, H.: Symmetric quadrature rules on a triangle. Comput. Math. Appl. 45(12), 1829–1840 (2003)
26. Roca, X., Ruiz-Gironés, E., Sarrate, J.: ez4u: Mesh generation environment (2010), http://www-lacan.upc.edu/ez4u.htm
27. Roca, X., Sarrate, J., Ruiz-Gironés, E.: A graphical modeling and mesh generation environment for simulations based on boundary representation data. In: CMNE (2007)
28. Roca, X.: Paving the path towards automatic hexahedral mesh generation. PhD thesis, Universitat Politècnica de Catalunya (2009)
29. Huerta, A., Giorgiani, G., Modesto, D.: Adaptive cdg and hdg computations. In: 16th FEF (2011)

CAD REPAIR and DECOMPOSITION

(Tuesday Afternoon Parallel Session)

Hybrid Approach for Repair of Geometry with Complex Topology

Amitesh Kumar[1] and Alan M. Shih[2]

[1] University of Alabama at Birmingham, AL, USA
   amitesh@uab.edu
[2] University of Alabama at Birmingham, AL, USA
   ashih@uab.edu

Summary. A discrete geometry can have artifacts such as holes, intersections, non-manifold edges, mesh fragment among other defects depending upon its origin. These kinds of defects sometime cause the geometry to be unsuitable for any further use in computational simulation in absence of a satisfactory geometry repair technique. There are two main approaches to geometry repair, surface based and volume based. Surface based approaches, in general, provide better quality results when they work but require that the input model already satisfies certain quality requirements to be able to guarantee a valid output. Many of these requirements cannot even be met or checked automatically. Volume based approaches, in general, can guarantee watertightness but they usually significantly change the underlying model in this process and are computationally more expensive.

A hybrid approach for mesh repair, combining surface based approach and a two step volume based approach is being presented in this paper. The two steps in the volume based approach are heat diffusion solution as the first step and Poisson surface reconstruction from oriented points in 3D space as the second step. This approach presents a reliable method for the repair of those defective discrete surface geometries which otherwise could not be completely repaired using existing surface-based techniques due to geometric and topological complexities presented as holes, isles, intersections and small overlaps.

Keywords: geometry repair, volume based approach, volumetric diffusion.

1 Introduction

A preferred approach for geometry creation is to do so with mesh generation tools directly. This approach in theory can reduce the conversion errors which are introduced during geometry translation between various Computer-Aided Design (CAD) packages and the mesh generation tools. However, most of the mesh generation tools do not have sophisticated solid modeling capabilities as the CAD systems do. As a result, geometries for most of the sophisticated real-world applications are first produced on CAD systems and then imported into the mesh generation tools. Geometry can also come from other sources in discrete forms. For example, scanned data using range scanning devices such as a laser scanner, can produce fairly accurate geometric model defined by point clouds, which can be turned into discrete elements. In the image-based, patient-specific biomedical applications, geometry can be produced by reconstructing from the segmented contours on each image slice. Regardless of the source of the geometry data, or the form in which they are represented (parametrically or discretely), the geometries obtained can have many defects due to the data conversion errors or ambiguities in the process. One of the major defects encountered in many geometries is the lack of water-tightness. A surface mesh, or geometry, is considered to be not watertight or non-manifold in the following two cases:

1. It has edges which are shared by only one polygon, i.e. the edges lie on the boundary. The occurrences of a set of connected boundary edges create a hole in the surface.
2. It has edges which are shared by more than two polygons. This kind of non-manifold mostly occurs in CAD applications due to improper stitching of surfaces patched together to generate a desired geometric model.

The most common type of mesh defects or artifacts encountered are holes or isles, singular vertex, handle, gaps, overlaps, inconsistent orientation, complex edges and intersections [1]. A number of research papers have tried different approaches in an attempt to address this issue using various automated and intelligent methods. Those approaches broadly fall in two main categories: volume-based repair methods and surface-based repair methods.

This research work tries to address surface defects of the type of topologically simple as well as complex holes in a discrete geometry, by using diffusion equation as a system of governing equation to obtain a convergent solution in the volumetric domain. That would help us in generating a set of well placed, regularly sampled and correctly oriented points in the areas of discontinuity, which are later used in surface reconstruction using Poisson's Surface Reconstruction technique [29], [31]. In this work, a robust and fully automatic hybrid methods is being presented which utilizes the strengths of both surface based and volume based techniques and can handle dirty geometry with holes, isles and intersections.

2 Previous Works

2.1 Surface Based Methods

Surface based repair methods ([2]–[12]) operate directly on the input data and hence need to explicitly identify and resolve artifacts on the surface. As an example, narrow gaps could be removed by snapping boundary elements (vertices and edges) onto each other or by stitching triangle strips in between the gap. Holes could also be closed by a triangulation that minimizes a certain error term. Intersections could be located and resolved by explicitly splitting edges and triangles although robustly resolving intersections is usually a very expensive process in terms of computational cost due to numerical accuracies. Surface based repair methods only minimally perturb the input model and are able to preserve the model structure in areas that are away from artifacts. In particular, structure that is encoded in the connectivity of the input (e.g. curvature lines) or material properties that are associated with triangles or vertices are usually well preserved in the areas away from the artifact. Furthermore, these algorithms introduce only a limited number of additional triangles [1].

These class of methods, however, usually require that the input model already satisfies certain quality requirements to be able to guarantee a valid output. Many of these requirements cannot be guaranteed or even be checked automatically hence these algorithms are rarely fully automatic and require manual post-processing. Other artifacts, like gaps between two closed connected components of the input model that are geometrically close to each other, are quite difficult to identify. As a result a guaranteed repair using only Surface based technique is not always possible.

Turk and Levoy [2] proposed a mesh zippering algorithm tailored to fuse range images using surface-based approach. Barequet and Kumar [3] and Barequet and Sharir [4] describe the use of an interactive system that closes small gaps, generated by CAD programs while joining the surfaces by stitching and triangulation within the hole. Borodin et al. [6] propose a progressive gap closing algorithm that works by vertex edge contraction accompanied with insertion of vertices on the boundary edges and progressively contracting the edge. Leipa [7] describes a method for filling holes by a weight-based hole triangulation, mesh refinement based on the Delaunay criterion and mesh fairing based on energy minimization. Jun [8] describes an algorithm based on stitching a planar projection of a complex hole in 2D space and projecting the stitched patch back in 3D space. Branch et al. [9] suggest a method for filling holes in triangular meshes using a local radial basis function. Pernot et al. [10] describe a method to fill holes by first cleaning up the geometry by removing unwanted triangles and then filling the holes with a patch of the disk topology by inserting a point in the middle and connecting all the nodes on the hole boundary with it.

The usability of most of these algorithms is constrained by their assumptions related with shapes, sizes, or sources of the holes and other defects on the mesh surface.

2.2 Volume Based Methods

The key to all volume based methods lies in converting a surface model into an intermediate volumetric representation. Volume based techniques are then used to classify and orient all voxels of the volumetric representation representing whether the particular voxel lies inside, outside or on the surface of the geometry. This classification determines and defines the topology and geometry of the reconstructed model which is extracted out of the intermediate volume representation. Due to their very nature, volumetric representations do not allow for artifacts like intersections, holes, gaps, overlaps or inconsistent normal orientation [1]. Volumetric algorithms are typically fully automatic and produce watertight models and depending on the type of volume and they can often be implemented in a very robust manner. Surface mesh from the intermediate volumetric representation is usually extracted using a surface extraction technique like Marching Cubes [13].

Volume based approaches ([14] – [20]) to mesh repair also produce some undesirable effects. The conversion to and from a volume leads to a resampling of the model. It often introduces aliasing artifacts, loss of model features and destroys any structure that might have been present in the connectivity of the input model [1]. The number of triangles in the output of a volumetric algorithm is usually much higher than that of the input model and thus has to be decimated in a post-processing step. The quality of the output triangles often degrades in this process as well. As a result, quality of the repaired mesh needs to be improved as a post-processing step to the volume based geometry repair techniques. Finally, volumetric representations are quite memory intensive so it is hard to run them at high resolutions.

Curless and Levoy [14] proposed a hole filling method based on volumetric diffusion optimized for range scanning devices. Murali and Funkhouser [15] use Spatial subdivision using BSP-Tree and determination of solid regions using region adjacency relationship to construct a set of polygons from the solid regions. Davis et al. [16] presented a method of in which the voxels of the volumetric representation of a surface mesh is classified as inside or outside with the help of distance map generated using line of sight information. Nooruddin and Turk [17] suggested the use of parity count and ray stabbing to repair the intermediate volumetric representation. Ju [18] presented a method for generating signs of voxels using octree. Bischoff et al. [19] proposed an improved volumetric technique to repair arbitrary polygonal soup using adaptive octree. Podolak and Rusinkiewicz [20] described a method using atomic volumes based on hierarchical non-intersecting graph representation in 3D space.

3 Our Work

Our work is based on a hybrid approach to mesh Repair which uses both Surface and volume based techniques.

The surface based technique which is being used in our hybrid approach is closely related to our previous work [11] and [12] with some improvements. The main features of it are:

- Support for non-manifold mesh.
- Explicit identification of holes and their sorting based on their sizes in terms of number of edges.

- Octree based search for locating points, edges and triangles.
- Sizing and refinement controlled by point insertion at centroid and edge swapping based on Delaunay criteria.
- Hole patching using localized NURBS based surface definition.
- Hole patching only supported for discrete geometries with simple topologies. The presence of isles in the hole region is neither detected nor supported in the surface repair mode.

Our surface based technique, in general, would produce good results when it would work. However as is the case with all other surface based techniques, they are not robust and require that the input model already satisfies certain quality requirements to be able to guarantee a valid output. Many of these requirements cannot even be met or checked automatically for all the input geometries. As a result when the surface based techniques fail we use the output of the surface based technique as an input to our volume based repair algorithm.

The work most closely related with our volume based technique appears in Davis *et al.* [16]. However our method is different in many ways. The main features of our volume based technique are as following:

- No assumption regarding source of model and hence no additional requirements such as line of sight information.
- Identification of independent and non-intersecting solution columns in 3D space.
- Convergent Diffusions equation based solution in each of the solution columns.
- Extraction of surface patch in the hole region away from the original input surface.
- Extraction of point and their normals from input geometries and surface fragments, in case of dirty geometries, and the extracted patch.
- Poisson surface reconstruction to get a watertight output model using the extracted oriented point set.

The above mentioned steps would be discussed in the next few sections. We would be using three discrete geometries namely a modified Stanford Bunny [35] with a few extra holes, Laurent Hand [36] and Chinese Lion [37] to demonstrate the results of our work.

3.1 Voxelization of Discrete Geometry

The first step in volume based repair method is to convert the surface mesh into a volumetric mesh. In this research, for simplicity, we have used a Cartesian grid to represent the data. The voxelization is only performed in non-intersection regions near the surface defects. We call each of these non-intersecting region as a *"solution column"*. Each of the solution column is represented as a uniform Cartesian grid. A Cartesian grid can be represented as a block of 3D tiles as shown in figure 1(a). The voxel size is a function of the average edge lengths of the triangles in the column so the points and triangles generated with this method have the same sample density as those in its neighborhood in the surface mesh. The tiles are padded with an extra layer of ghost cells along the boundary as shown in figure 1(b). The memory allocation is completely dynamic and the information exchange between the contiguous tiles is completely hidden from the user with an abstraction layer. This kind of memory allocation for representation of the Cartesian mesh has two major benefits:

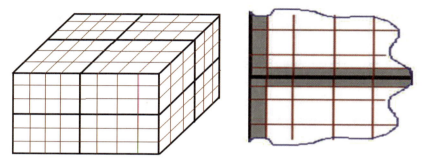

(a) A Cartesian Grid composed of multiple blocks in 3D

(b) Ghost cells at the interface of two tiles shown in gray color in a 2D Cartesian grid

Fig. 1. Description of the cartesian solution space

1. Depending on the cache size and the size of each tile, this configuration may speed up the diffusion solver performance due to cache effect despite the cost associated with information exchange between ghost cells at the end of every iteration.
2. The solver requires an exact temporary copy of the tile in the intermediate step to compute and transfer data. If we have a small tile size then only a small intermediate amount of memory would be needed. This would prevent duplication of larger tile, that would instead need to fit the whole model.

The information exchange between ghost cells is completely hidden from the user. Hence the user does not experience any differences while using the APIs compared to the situation when whole block is composed of a single tile. The discrete geometry is composed of individual triangles. Intersection of each triangle with the *voxels* of the Cartesian grid lying within its bounding box is checked using AABB Triangle-Box intersection algorithm [32], [33]. All the *voxels* found intersecting with triangles are masked as "*model voxels*" with a static value of '0'. All the voxels adjacent to the models with '0' value along the positive normal of the triangle plane are masked as *"heated"* and given a static positive value of "+1" while all the voxels in the opposite direction of the positive normal are masked as *"cold"* and given a negative value of "-1". This process creates a voxelized representation of the discrete geometry having a neutral value ("0") sandwiched between hot and cold side. The rest of the voxels are dynamic and can change values during the solution. In our implementation, we assume that the input discrete geometry along with all present input surface fragments, have consistent orientation with respect to each other, for the purpose of defining the boundary conditions. The input geometry may not have normal information present. As a result normal generation is performed at every vertex of the input geometry in order to sign those unsigned voxels of the cartesian grid column which are touching the embedded geometry. In the hybrid approach being presented in this work, parts of the original input geometry along with all the surface fragments as well as the output surface patches from surface based approach are embedded in the voxelized non-intersecting cartesian grid columns prior to generating a diffusion equation based solution in those columns. Figure 2(a) shows non-intersecting solution columns on Stanford Bunny.

3.2 Diffusion Equation

Diffusion is a time-dependent process, constituted by random motion of given entities and causing the statistical distribution of these entities to spread in space. The concept of diffusion is tied to notion of mass transfer, driven by a concentration gradient. The diffusion equations can be obtained easily from Continuity equation when combined with the Fick's first law, which assumes that the flux of the diffusing material in any part of the system is proportional to the local density gradient [34]. Diffusion equation is a partial differential equation continuous in both space and time which describes density fluctuations in a material undergoing diffusion. Diffusion equation can be simplified and written as,

$$\frac{\partial \phi}{\partial t} = \alpha \Delta^2 \phi + S \tag{1}$$

Where α is a constant and S is a source term. In our formulation we assume that there are no source terms, Hence the equation becomes,

$$\frac{\partial \phi}{\partial t} = \alpha \Delta^2 \phi \tag{2}$$

 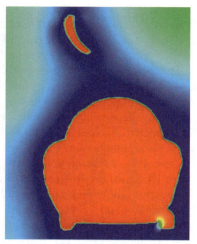

(a) Non-intersecting diffusion equation solution columns

(b) Diffusion solution on the whole bunny as one column

Fig. 2. Diffusion Solutions for the Stanford Bunny

Using *forward difference* scheme in time and *central difference* scheme in space based on Taylor's series expansion, the diffusion equation for a Cartesian grid simplifies to,

$$\phi_{i,j,k}^{n+1} = \phi_{i,j,k}^{n} + \alpha \Delta t \left[\begin{array}{c} \left(\frac{\phi_{i+1,j,k}^{n} - \phi_{i,j,k}^{n}}{\Delta x^2} - \frac{\phi_{i,j,k}^{n} - \phi_{i-1,j,k}^{n}}{\Delta x^2} \right) \\ + \left(\frac{\phi_{i,j+1,k}^{n} - \phi_{i,j,k}^{n}}{\Delta y^2} - \frac{\phi_{i,j,k}^{n} - \phi_{i,j-1,k}^{n}}{\Delta y^2} \right) \\ + \left(\frac{\phi_{i,j,k+1}^{n} - \phi_{i,j,k}^{n}}{\Delta z^2} - \frac{\phi_{i,j,k}^{n} - \phi_{i,j,k-1}^{n}}{\Delta z^2} \right) \end{array} \right] \qquad (3)$$

where $\Delta x, \Delta y$ and Δz are the grid spacings of a Cartesian grid in x, y and z directions respectively. Equation (3) provides the explicit numerical solution which is central difference in space and forward difference in the time domain for the diffusion equation given in equation (2).

After performing stability analysis on our numerical scheme using Von Neumanns analysis and Courant–Friedrichs–Lewy (CFL) condition for the stability of the numerical scheme in three dimensions (3D), we further come with the inequality defining the values for the constants given in equation (3).

$$0 \leq \alpha \Delta t < \frac{\Delta x_{min}^2}{6} \qquad (4)$$

Equation (4) also implies that the scheme is numerically stable for a value of $\alpha \Delta t$ satisfying the above given condition.

We define the Cartesian grid with the analogy of a curved thin plate in 3D space whose one side is heated while the other side is cold. Diffusion equations presented earlier are used to find a steady state solution. For illustration purpose, figure 2(b) shows the rendering of one slice of Stanford bunny after 120 iterations when the whole model is used as one solution column. However in practice, the whole model is rarely, if ever, embedded as whole in one single solution column as represented in 2(a).

$$l_2norm_{max}^{n+1} = max\left(\sqrt{(\phi_{i,j,k}^{n+1} - \phi_{i,j,k}^n)^2}\right) \quad (5)$$

$$l_2norm_{avg}^{n+1} = \frac{1}{i \times j \times k} \sum \sqrt{(\phi_{i,j,k}^{n+1} - \phi_{i,j,k}^n)^2} \quad (6)$$

$$\%l_2norm_{change}^{n+1} = \frac{abs\left(l_2norm_{avg}^{n+1} - l_2norm_{avg}^n\right)}{l_2norm_{avg}^{n+1}} \times 100.0 \quad (7)$$

Equations (5), (6) and (7) define three measures of change for each successive iteration for our solution. These are $l_2norm_{max}^{n+1}$, $l_2norm_{avg}^{n+1}$ and $\%l_2norm_{change}^{n+1}$ respectively. In our case, we use $\%l_2norm_{change}^{n+1}$ as our measure of change and use an arbitrary number (0.5%) or 1000 iterations, whichever happens earlier, to determine whether or not convergence has been reached for a particular solution column. Figures 3(a) and 3(b) show convergence plots of columns 2 and 3 for the Stanford Bunny. In both the cases the plots are asymptotic and seem to converge within a few iterations. However looking at the log plots shown in corresponding figures 4(a) and 4(b), it becomes clear that actual convergence is reached much later.

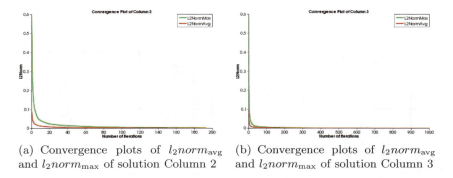

(a) Convergence plots of l_2norm_{avg} and l_2norm_{max} of solution Column 2

(b) Convergence plots of l_2norm_{avg} and l_2norm_{max} of solution Column 3

Fig. 3. Convergence plots of two solution columns of Stanford Bunny

Explicit schemes are known to be notoriously slow for convergence. The convergence becomes slower as we refine the grid to a finer resolution. To circumvent this problem, we provide the capability to be able to use a primitive algebraic interpolation approach using coarse grid to initialize the flow field for the fine grid. This allows the overall solution field to be initialized using

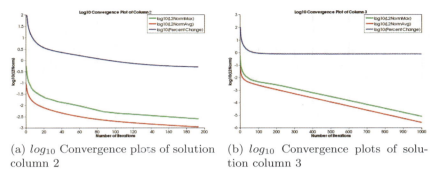

(a) log_{10} Convergence plots of solution column 2

(b) log_{10} Convergence plots of solution column 3

Fig. 4. log_{10} Convergence plots of two solution columns of Stanford Bunny

the preliminary solution from the coarser grid, which can reduce the time needed to convergence. First the solver is run on a coarse grid for a number of iterations and then its solution is interpolated onto the finer Cartesian grid as initial values for faster convergence. It should however be noted that, we are not using algebraic interpolation in the examples being presented in this work.

3.3 Extraction of Consistently Oriented Surfaces and PointSet

Zero-set of the numerical solution for the diffusion equation as described in the previous section provides a closed surface. This zero-set surface is otherwise only open at the places where it intersects with the extremities of the Cartesian grid. Extraction of the surfaces using a suitable contouring method

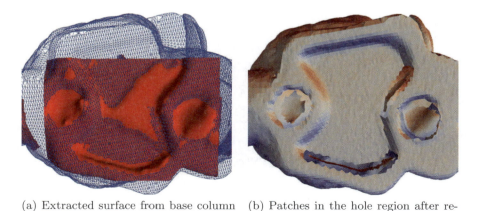

(a) Extracted surface from base column in red

(b) Patches in the hole region after removal of triangles

Fig. 5. Extracted surface and patches after diffusion equation solution in the base column

 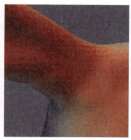 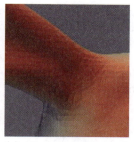

(a) Surface around a hole in Stanford Bunny (b) Reconstruction without taking into account voxel shift (c) Reconstruction after taking into account voxel shift

Fig. 6. Shift in the position of reconstructed surface due to voxelization

should provide us with the desired zero-set surface. In our implementation we have used Marching Cubes algorithm [13] due to ease of use and availability. Once the surfaces are extracted, All those triangles are removed which intersect with those voxels which embed input surface mesh. This leaves us with only those triangles which lie within the hole regions in the original surface mesh as a patch. Parts of input geometry is embedded in a number of solution columns in voxels, while surface extraction contouring is done by evaluating values on the voxel corners. As a result, it was observed that conversion of surface mesh to a volume mesh and subsequent contouring causes a noticeable shift of about half a voxel in the position of the extracted surfaces in comparison with the original input surface. If left uncorrected, the combination of original input surface and extracted surface patch would produce a noticeable bump in the output reconstructed surface as shown in figures 6(a), 6(b) and 6(c). The shift is corrected by shifting the points on the extracted surface by half voxel in the direction away from the positive normal. Furthermore, normals are calculated and generated at every points in the extracted patches as well as input surface and mesh fragments to create a pointset. The pointset loosely represents a surface mesh with well sampled points everywhere including hole regions and is used as an input for surface reconstructions.

4 Surface Reconstruction and Results

Surface reconstruction and surface fitting from point samples is a well studied problem in computer graphics and has applications in a number of areas including reverse engineering. Reconstruction itself is a very challenging area due to uneven sampling of points, noisy data, scan misregistration among other problems [29]. There are a number of schemes for surface reconstruction based on implicit form among other techniques. These implicit surface fitting methods are either global or local in nature. Global fitting methods commonly define the implicit function as the sum of radial basis functions (RBFs) centered at the points. Local fitting methods consider subsets of nearby points at a time. These methods are well studied and have been compared in a number of papers including but not limited to [23, 24, 25, 26, 27, 28, 29, 30] among others.

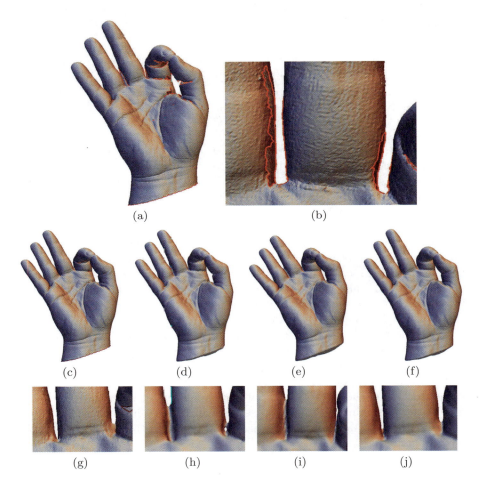

Fig. 7. Mesh repair on Laurent's Hand [36] in different configurations: (a) Original model of Laurent's hand, (b) closeup of model around fingers, (c) repair based on surface approach, (d) repair based on volume approach, (e) repair based on Poisson's reconstruction, (f) repair based on hybrid approach, (g) closeup after repair based on surface approach, (h) closeup after repair based on volume approach, (i) closeup after repair based on Poisson's reconstruction and (j) closeup after repair based on hybrid approach

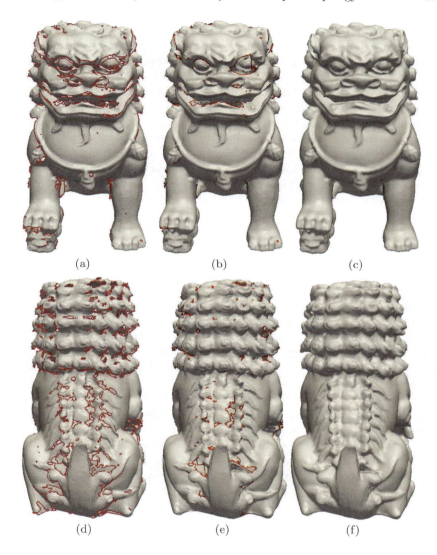

Fig. 8. Chinese Lion (a), (d): Front and back of original model , (b), (e): Front and back of model after mesh repair based on surface approach and (c), (f): Front and back of model after mesh repair based on hybrid approach

In this paper, we do not try to create any new method for surface reconstruction, but we would be using an existing well established method for it. We chose Poisson Surface Reconstruction as described in [29] as its open source implementation is widely available [31]. Although Poisson surface reconstruction provides a watertight surface, however its accuracy is dependent on the sampling of input points. We, in this method, try to provide a well sampled pointset with correctly oriented normals as an input to the Poisson reconstruction so that we could get an output surface which is not only smooth, well behaved but also is feature sensitive due to our hybrid approach. We use [31] implementation of Poisson reconstruction at octree refinement level 10 in our examples. Figures 7(a) - 7(j) show a visual comparison of results obtained on Laurent Hand model. Figure 7(a) shows the original model with discontinuities. Figure 7(b) shows a closeup image around the middle two fingers. Figure 7(c) shows the mesh repair using only surface based approach. We can clearly see in figure 7(g) that although the repair approximates the surface well but it fails to repair all the discontinuities. Figures 7(d), 7(e) and 7(f) shows the result after volume based repair followed by Poisson Surface reconstruction, only Poisson reconstruction and hybrid approach which uses both surface and volume based techniques, respectively. Looking at the closeup images it can be noticed that fingers in volume based approach 7(h) and only Poisson based approach 7(i) are fused together which is clearly an undesirable outcome. Figure 7(j) shows the closeup image from the hybrid approach where middle two fingers are not fused together and hence is a better result.

Figures 8(d), 8(e) and 8(f) show the front side of Chinese Lion [37] as original model, model after repair based on surface based technique and repair based on hybrid approach, respectively. Similarly, figures 8(a), 8(b) and 8(c) show the backside of Chinese Lion [37] as original model, model after repair based on surface based technique and repair based on hybrid approach, respectively. Lines in red show discontinuities present in the model.

The input models being used and presented in this section are of high resolution and have high level of details and geometric complexities. The input Chinese Lion model had a number of isles or small surface fragments along with a large number of holes. The results presented in this section demonstrate that merely using either surface or volume based techniques may not be sufficient as only one of those in isolation may not be able to repair the model completely or they may generate a result which may not be reliable or satisfactory. Combining surface and volume approach based techniques as a hybrid approach helps us in preserving the details as well as helps us in providing better results in that process. This was especially demonstrated using Laurent Hand model in figures 7(a) - 7(j).

5 Conclusion

Surface based methods explicitly try to identify surface artifacts prior to operating on them and require that input geometry meets some mesh quality conditions. This is sometimes not feasible causing the mesh repair techniques to fail. However the surface based methods have some compelling advantages over the volume based methods, which emphasize that we should try to employ surface based techniques wherever possible to get better quality results compared to when we are only using volume based methods for geometry repair.

Volume based approaches like the ones described in this research can be used to repair the models with artifacts that surface based models otherwise cannot robustly handle. They however also pose some potential problems. The conversion to and from a volume leads to resampling of the model. It often introduces aliasing artifacts, loss of model features and destroys any structure that might have been present in the connectivity of the input model. The focus of this work is to address the need to develop a method to obtain a watertight geometry from a geometric model that could have the presence of holes of complex topology. Despite all their shortcomings, volumetric algorithms can solve some problems in geometry repair robustly which can not be handled by surface based approaches alone. As a result we have presented a hybrid method for repairing a discrete geometry by combining surface and volume based methods, which in many cases provides superior results when compared with either Surface or volume based techniques in isolation. We have presented a number of repaired models in support of the results presented in this work.

Acknowledgement

We would like to express our gratitude towards Stanford 3D Scanning Repository [35] and AIM@SHAPE Shape Repository for putting test models in public domain, some of which were used in this paper to present and validate our results. We would also like to express our gratitude towards Laurent Saboret *et al.* [36], [37] of INRIA for making "Laurent Hand" and "Chinese Lion" models publicly available.

References

1. Botsch, M., Pauly, M., Kobbelt, L., Alliez, P., Lévy, B., Bischoff, S., Rössl, C.: Geometric modeling based on polygonal meshes. In: ACM SIGGRAPH 2007 Courses - International Conference on Computer Graphics and Interactive Techniques (2007)
2. Turk, G., Levoy, M.: Zippered polygon meshes from range images. In: Proceedings of ACM SIGGRAPH 1994, pp. 311–318 (1994)
3. Barequet, G., Sharir, M.: Filling gaps in the boundary of a polyhedron. Computer Aided Geometric Design 12, 207–229 (1995)

4. Barequet, G., Kumar, S.: Repairing CAD Models. Proceedings of IEEE Visualization, 363–370 (1997)
5. Guéziec, A., Taubin, G., Lazarus, F., Horn, B.: Cutting and Stitching: Converting Computer sets of Polygons to Manifold Surfaces. IEEE Transactions on Visualization and Graphics 7(2), 136–151 (2001)
6. Borodin, P., Novotni, M., Klein, R.: Progressive Gap Closing for mesh repairing. In: Vince, J., Earnshaw, R. (eds.) Advances in Modelling, Animation and Rendering, pp. 201–213. Springer, Heidelberg (2002)
7. Liepa, P.: Filling Holes in Meshes. In: Proceedings of the 2003 Eurographics/ACM SIGGRAPH Symposium on Geometry processing, Eurographics Association, pp. 200–205 (2003)
8. Jun, Y.: A Piecewise Hole Filling Algorithm in Reverse Engineering. Computer-Aided Design 37, 263–270 (2005)
9. Branch, J., Prieto, F., Boulanger, P.: A Hole-Filling Algorithm for Triangular Meshes using Local Radial Basis Function. In: Proceedings of the 15th International Meshing Roundtable, pp. 411–431. Springer, Heidelberg (2006)
10. Pernot, J.P., Moraru, G., Vernon, P.: Filling holes in meshes using a mechanical model to simulate the curvature variation minimization. Computer and Graphics 30(6), 892–902 (2006)
11. Kumar, A., Shih, M.A., Ito, Y., Ross, H.D., Soni, K.B.: A Hole-Filling Algorithm Using Non-Uniform Rational B-Splines. In: Proceedings of 16th International Meshing Roundtable, pp. 169–182. Springer, Heidelberg (2007), ISBN: 978-3-540-75102-1
12. Kumar, A., Ito, Y., Yu, T., Ross, H.D., Shih, M.A.: A novel hole patching algorithm for discrete geometry using non-uniform rational B-spline. International Journal for Numerical Methods in Engineering (2011),
http://onlinelibrary.wiley.com/doi/10.1002/nme.3157/abstract
13. Lorenson, E.W., Cline, E.H.: Marching Cubes: A high resolution 3D surface construction algorithm. ACM Computer Graphics 21(4) (1987)
14. Curless, B., Levoy, M.: A volumetric method for building complex models from range images. Computer Graphics 30, 303–312 (1996)
15. Murali, M.T., Funkhouser, A.T.: Consistent solid and boundary representations from arbitrary polygonal data. In: Proceedings of Symposium on Interactive 3D Graphics, pp. 155–162 (1997)
16. Davis, J., Marschner, S.R., Garr, M., Levoy, M.: Filling holes in complex surfaces using volumetric diffusion. In: Proceedings of First International Symposium on 3D Data Processing, Visualization, Transmission, pp. 428–861 (2002)
17. Nooruddin, F.S., Turk, G.: Simplification and repair of polygonal models using volumetric techniques. IEEE Transactions on Visualization and Computer Graphics 9(2), 191–205 (2003)
18. Ju, T.: Robust repair of polygonal models. In: Proceedings of ACM SIGGRAPH 2004; ACM Transactions on Graphics 23, 888–895 (2004)
19. Bischoff, S., Pavic, D., Kobbelt, L.: Automatic restoration of polygon models. Transactions on Graphics 24(4), 1332–1352 (2005)
20. Podolak, J., Rusinkiewicz, S.: Atomic volumes for mesh completion. In: Symposium on Geometry Processing (2005)
21. Hoppe, H., DeRose, T., DuChamp, T., HalStead, M., Jin, H., McDonald, J., Schweitzer, J., Stuetzle, W.: Piecewise Smooth Surface Reconstruction. In: Proceedings of ACM SIGGRAPH 1994, New York, NY, pp. 295–302 (1994)

22. Amenta, N., Bern, M.: Surface Reconstruction by Vornoi Filtering. Discrete & Computational Geometry 22(4), 481–504 (1999)
23. Amenta, N., Hoi, S.C., Kolluri, R.: The power crust, unions of balls, and the medial axis transform. Computational Geometry: Theory and Applications 19, 127–153 (2001)
24. Carr, C.J., Beatson, K.R., Cherrie, B.J., Mitchell, J.T., Evans, R.T., Fright, R.W., McCallum, C.B.: Reconstruction and representation of 3D objects with Radial Basis functions. In: Proceedings of ACM SIGGRAPH 2001, pp. 67–76 (2001)
25. Dey, K.T., Goswami, S.: Tight Cocone: A Water-tight Surface Reconstructor. Journal of Computing and Information Science in Engineering 3(4), 302–307 (2003)
26. Dey, K.T., Goswami, S.: Provable surface reconstruction from noisy samples. Computational Geometry 35(1-2), 124–141 (2004)
27. Shen, C., O'Brien, F.J., Shewchuk, R.J.: Interpolating and approximating implicit surfaces from polygon soup. In: Proceedings of ACM SIGGRAPH 2004, pp. 896–904 (2004)
28. Casciola, G., Lazzaro, D., Montefusco, B.L., Morigi, S.: Fast surface reconstruction and hole filling using positive definite radial basis functions. Journal of Numerical Algorithms 39, 289–305 (2005)
29. Kazhdan, M., Bolitho, M., Hoppe, H.: Poisson Surface Reconstruction. In: Proceedings of the fourth Eurographics Symposium on Geometry processing (SGP 2006), Switzerland, pp. 61–70 (2006)
30. Mullen, P., Goes, D.F., Desbrun, M., Cohen-Steiner, D., Alliez, P.: Signing the Unsigned: Robust Surface Reconstruction from Raw Pointsets. In: Eurographics Symposium on Geometry Processing 2010, vol. 29(5), pp. 1733–1741 (2010)
31. Doria, D., Gelas, A.: Poisson Surface Reconstruction for VTK. The VTK Journal (2010), http://www.insight-journal.org/download/viewpdf/718/2/download
32. Akenine-Möller, T.: Fast 3D triangle-box overlap testing. In: Fujii, J. (ed.) ACM SIGGRAPH 2005 Courses, SIGGRAPH 2005. ACM, New York (2005)
33. Akenine-Möller, T.: AABB-triangle overlap test code (2010), Source code is located at, http://jgt.akpeters.com/papers/AkenineMoller01/tribox.html
34. See Diffusion Equation on Wikipedia (2011), http://en.wikipedia.org/wiki/Diffusion_equation
35. Stanford Bunny at The Stanford 3D Scanning Repository(2011), http://graphics.stanford.edu/data/3Dscanrep/
36. Saboret, L., Attene, M., Alliez, P.: Laurent Hand at AIM@SHAPE Shape Repository (2011), http://shapes.aim-at-shape.net/viewgroup.php?id=785
37. Saboret, L., Attene, M., Alliez, P.: Chinese Lion at AIM@SHAPE Shape Repository (2011), http://shapes.aim-at-shape.net/viewgroup.php?id=783

A Surface-Wrapping Algorithm with Hole Detection Based on the Heat Diffusion Equation

Franjo Juretić[1] and Norbert Putz[2]

[1] AVL-AST d.o.o, Avenija Dubrovnik 10/III, Zagreb, Croatia
`franjo.juretic@avl.com`
[2] AVL GmbH, Alte Poststrasse 152, Graz, Austria
`norbert.putz@avl.com`

Summary. This paper presents a method for detecting holes during the surface wrapping process causing surface leaks into the volume parts that shall not be meshed. The method solves a heat diffusion equation, and the holes are detected as regions of high temperature gradients. It can detect both holes with open edges and semantic holes due to some missing parts. The sensitivity of the method is controlled via user-adjustable parameter representing the ratio between the volume that shall not be meshed and the area of the hole. The potential of the method is presented on complex engineering examples.

Keywords: mesh generation, surface wrapping, hole detection, watertight surface generation, octree mesh.

1 Introduction

Computer Aided Engineering (CAE) tools have become a widely accepted in the process of developing new products in almost any engineering discipline. Such tools enable engineers to perform simulations and virtually test their designs even before the first prototype is made. However, these tools often require either surface or volume meshes to perform the desired simulations. These days most tools for generation of volume meshes require high quality surface meshes as the input into the meshing process, and the surface meshes are generated from the underlying CAD model. Such surface meshes models often contain errors such as misplaced parts by a small tolerance, missing parts, open edges, non-manifold connections, overlaps, etc. [19]. These errors often occur during the process of converting the data from CAD format to the other, and sometimes are caused by the user of the CAD software. It is common that the analysis shall be performed before some parts are designed, causing semantic holes in the model which also have to be tackled during the geometry preparation phase. In order to reduce the manual work required for repairing the surface mesh, many methods for repairing surface meshes have emerged over the last decade.

Shrink wrapping is a popular technique for re-meshing bad quality surfaces [2, 5, 14, 18, 13]. Kobbelt et. al. [13] have developed a concept for shrink-wrapping starting from a coarse wrapper triangulation which is subjected to refinement, projecting and smoothing until the until the wrapper surface is a satisfactory representation on the original surface mesh. However, the method cannot capture feature edges, and refinement process does not give satisfactory results in regions where the normals of the wrapper surface do not fit the normals of the input surface. Lee et. al. [14] have developed the method which generates the initial wrapper surface from the underlying cartesian mesh, and the quality of the wrapper surface is not influenced by the quality of the initial surface in regions where the problems like, overlaps, gaps, holes, etc. are smaller than the cell size in the cartesian mesh, Their method is based on the concept of inside-out volume mesh generated developed by Wang et. al. [21]. The methods [2, 5, 18] are not described in the scientific literature. However, large holes often have a disastrous effect on the quality of the generated wrapper surface, because the wrapper surface leaks into the volume that shall not be meshed. There are two major types of holes causing this undesired behaviour. The first major type are holes with open boundary edges occurring due to missing surface segment or occlusion of surface triangles. Sanchez et. al. [16] developed the methodology classifying open boundary holes based on the shape of the boundary curve. Bendels et. al. developed an algorithm for detecting holes in the point set surfaces. The algorithms developed for closing open holes are a mature subject [3, 4, 8, 7, 22, 23]. [3, 4, 8] use Radial Basis Functions to interpolate the missing surface parts, [22, 23] interpolate the surface with the help of the Level Set Method, and the method developed by Davis et. al. [7] solves a local heat diffusion problem to get the smooth interpolation of the missing surface parts. The second major group of holes are semantic holes which are often a consequence of missing parts in an assembly. Recently Schilling et. al. [17] have developed a method for closing sematic holes based on the geometric properties of the surface and complex holes must be detected by the user.

This paper presents the methodology for surface wrapping based on the underlying octree mesh, together with the method for detection of large holes during the surface wrapping process, which can cause the surface leaks into the unwanted parts of the original geometry. It is based on solving the heat equation on the underlying octree mesh, and is not sensitive whether the hole that causes the leak is a hole with boundary edges or a semantic hole due to some missing and/or misplaced parts. The procedure is presented in Sect. 2, and its potential is shown in Sect. 3. Conclusions and suggestions for future work are given in Sect. 4.

2 Methodology

The first phase of the procedure is the generation of the octree from the input surface mesh with marked feature edges and the settings file containing information about the requested cell size, refinement regions and smoothing options. The octree is refined based on:

- Global cell size - the size of the octree root box is adjusted to meet the requested cell size.
- User-defined regions - the boxes intersected by the selected surface elements are refined, and the number of refinements is specified by the user.
- Curvature - is estimated using the methods in [11] and the octree boxes containing surface elements are refined based in case the mean curvature radius is smaller than the diagonal of the octree box.
- Proximity - is controlled by refining the octree boxes which contain two distinct surface parts within a ball centred at the centre of the octree leaf with the radius equal to its diagonal length.

The initial wrapped surface is generated using the Marching Cubes Method [15] at the interface between the outer octree leaves and the octree leaves intersected by the input surface mesh. This surface mesh is a watertight manifold, and its geometry does not fit the input surface. An example of the surface mesh with a semantic hole at the front and an open hole at the back side is shown in Fig. 1, and the initial wrapped surface is given in Fig. 2.

The geometry of the initial template is adjusted to fit the original surface in a series of steps. The surface vertices are gradually moved towards the original surface in an iterative procedure which combines both Laplacian smoothing and projections onto the nearest surface location in a single iteration:

$$\mathbf{x}_{smooth} = \frac{\sum_{i=0}^{i=N-1} \mathbf{x}_i}{N}, \tag{1}$$

$$\mathbf{x}_{projected} = \min \|\mathbf{x}_{smooth} - \mathbf{x}_{projected}\| \tag{2}$$

$$\mathbf{x}_{new} = \alpha \mathbf{x}_{projected} + (1-\alpha)\mathbf{x}, \tag{3}$$

where α is the under-relaxation factor by default set to 0.5, N is the number of neighbouring vertices, $\mathbf{x}_i$ are the neighbouring vertices of a given mesh vertex, $\mathbf{x}_{smooth}$ is the position after Laplacian smoothing, $\mathbf{x}_{projected}$ is the position after projecting the vertex, and $\mathbf{x}_{new}$ is the final position of the vertex after the current iteration. This procedure smooths out the stairs in the original surface template and brings it closer to the original surface. The final iteration consists only of the projecting step, see equation (2), to speed up the convergence towards the original surface mesh.

Feature edges and corners are captured by segmenting the input surface into regions consisting of surface elements, connected over manifold edges, which are bounded either by selected feature edges, non-manifold edges, or boundary edges. Consequently, the triangles in the wrapped surface are

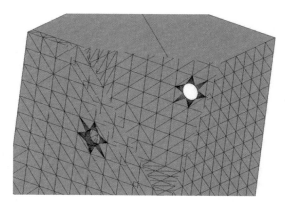

Fig. 1. Surface mesh with semantic and open holes

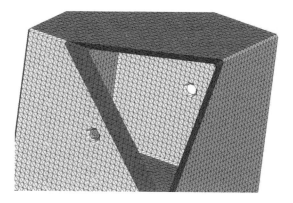

Fig. 2. Initial wrapper surface

assigned a region by projecting the centre of each triangle onto the input surface, and copying the region assigned to the nearest surface element. The vertices are classified as edge vertices if the number of regions at the vertex is greater than one. In case a triangle has two edges assigned to the same feature edge, it is assigned to the other region and the third edge becomes the feature edge. This operation is performed as long as the problem persists. The selected edge vertices are projected onto the nearest location on the feature edges of the input surface, and the edges which do not exist in the input surface are recovered in an iterative manner by finding the nearest vertices in all regions the vertex is in, and by moving it to the average position of these vertices. The procedure terminates once the displacements are smaller than the prescribed tolerance.

The quality of the surface mesh is improved in a smoothing procedure consisting of non-shrinking Laplacian smoothing [20] and optimisation-based

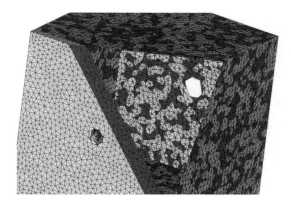

Fig. 3. Final wrapped surface with leaks through the holes

smoothing [9], by using the optimisation-based smoothing for regions consisting of triangles with small angles and folded patches. Figure 3 shows an example of the wrapped surface, and the regions with multiple shading indicate that the surface has leaked inside the volume. For complex geometries, it is often very difficult to find such problematic holes, and they often have complex shapes which impose high requirements on the hole detection algorithm, especially in case of semantic holes. Sect. 2.1 presents the methodology developed for finding the locations where these problematic holes are located, in order to provide the hint to the user where to look for holes, and reduce the effort spent in searching for the problematic holes.

2.1 Heat Diffusion Equation

The heat diffusion equation is a mathematical representation of the balance between the heat production within a solid body and the heat fluxes through the boundaries of the body. In case the heat production is not in balance with the fluxes through the outer boundary, it results in temperature changes within the domain. In addition, the above applies for any closed surface within the domain and the volume enclosed within that surface. The method presented in this paper considers the steady-state equation where the heat production is balanced with the boundary heat fluxes. The steady-state heat diffusion equation has the following form:

$$-\nabla \cdot \nabla T(\mathbf{x}) = S(\mathbf{x}), \tag{4}$$

where $T(\mathbf{x})$ is the temperature field and $S(\mathbf{x})$ is the heat source term, constant in space, and can be set to any value; the default is 1. The equation is solved on the volume mesh consisting of outer octree boxes with the following boundary conditions:

$$T = 0 \tag{5}$$

at the outer boundaries of the octree space, and

$$\mathbf{n} \cdot \nabla T(\mathbf{x}) = 0 \qquad (6)$$

otherwise, where $\mathbf{n}$ is the unit normal vector of a given octree mesh face. Since all internal walls originating from surface mesh are considered adiabatic, the heat generated inside the volume that shall not be meshed can be transported out of the volume only through the holes in the surface mesh. In addition, this equation has an interesting property that at any location in the domain it is possible to find the path to the outer boundary by travelling in the opposite direction of the temperature gradient. If the location is selected inside the volume that shall not be meshed, this path shall pass through the hole in the vicinity with the smallest resistance. An example of the octree mesh, coloured in dark grey, and the surface mesh mesh, coloured in light grey, is shown in Fig. 4. The equation is solved by using a face-based cell-centered finite volume approach, [12, 6]. The heat fluxes at the internal faces of the octree mesh are approximated as follows [10]:

$$\mathbf{n} \cdot \nabla T(\mathbf{x}) = \frac{T_N - T_P}{\frac{l_N}{2} - \frac{l_P}{2}}, \qquad (7)$$

where l_N and l_P are the sizes of the neighbouring octree leaf and the currently treated octree leaf, respectively, with a constraint that the index of

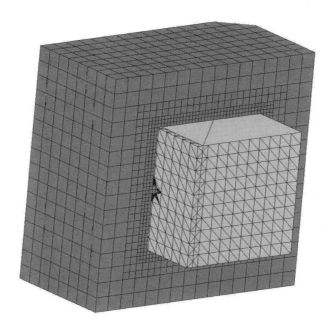

Fig. 4. Octree mesh for the cube with holes

the neighbouring leaf is higher than the index of the current octree leaf. The discretisation process results in a symmetric system matrix which is solved using an iterative multigrid solver.

2.2 Leak Detection

Once the solution of the heat equation is available, surface leaks are detected based on heat flux intensity, which is high near holes, because the heat generated inside the volume that shall not be meshed can only be transported out of the volume through the holes, since other boundaries are considered adiabatic. Therefore, the leaking holes are detected based on the balance between the heat flux through the holes and the heat production inside the volume that the surface mesh leaks into, which can be mathematically expressed as:

$$-\int_{A_{hole}} \nabla T \cdot \partial \mathbf{A} = \int_{V_{inside}} S(\mathbf{x})\partial V, \tag{8}$$

where $\int_{A_{hole}} \nabla T \cdot \partial \mathbf{A}$ represents the heat flux through the hole and $\int_{V_{inside}} S(\mathbf{x})\partial V$ represents the heat generated inside the volume which shall not be meshed. Equation 8 is approximated with the following expression:

$$-A_{hole} \nabla T = S(\mathbf{x}) V_{inside}, \tag{9}$$

where A_{hole} represents the surface area of the hole, and V_{inside} is the volume inside the hole that shall not be meshed. The gradient of the solution is approximated using the least squares fit [10]:

$$\mathbf{G} \nabla T_P = \sum_f \frac{\mathbf{d}(T_N - T_P)}{|\mathbf{d}|^2}, \tag{10}$$

where the matrix G is defined as:

$$\mathbf{G} = \sum_f \frac{\mathbf{dd}}{|\mathbf{d}|^2}, \tag{11}$$

where $\sum_f$ denotes the summation over the faces of each control volume, $\mathbf{d}$ is the vector connecting the centroids of the current octree leaf and the neighbour N over a given face.

Please note that holes can be detected as regions of high gradients, since the flux through the hole is proportional to the temperature gradient according to the Fourier's Law of heat conduction, see Fig. 5. In order to filter out such holes from the rest of the features, a non-dimensional sensitivity parameter is implemented:

$$\psi = \frac{V_{inside}}{A_{hole}\, l_{octreeSpace}}, \tag{12}$$

where $l_{octreeSpace}$ is the edge length of the root octree box, set to be fifty percent larger than the maximum edge length of the surface bounding box.

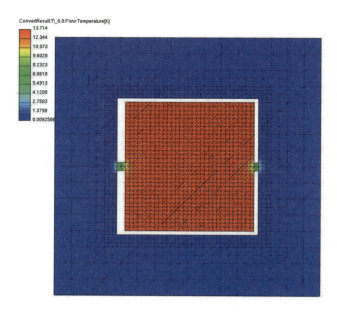

Fig. 5. Temperature field with high gradients near the holes

ψ can be considered as a ratio between the volume inside the volume that shall not be meshes and the area of the hole. Finally, the holes are detected by marking the surface elements contained inside the octree leaves which are neighbours of the outer octree leaves with the gradient higher than the critical value:

$$(\nabla T)_{crit} = \psi S(\mathbf{x}) l_{octreeSpace}. \tag{13}$$

Please note that the increase of ψ in (13) makes the method less sensitive and its reduction makes the method more sensitive. In practice, the value shall be greater than $\psi >= \frac{1}{6}$ in order not to mark all octree boxes and all surface elements as holes.

In the present study, the holes are closed manually by clicking on mesh vertices and generating the triangulations of selected polygons, and the method for detecting holes only provides hints to the user where to look for holes. Once the holes are closed, the user can start the next iteration of the wrapping process from the beginning, and continue the process until all surface leaks are resolved.

3 Examples

3.1 Engine

The aim of this exercise is to generate a watertight envelope of the outer boundaries of the engine which can be used for acoustics simulation. The input surface mesh, depicted in Fig. 6, consists of thirty-eight disconnected parts with many holes and open boundary edges. The final surface mesh, without any leaks, was generated generated with uniform cell size of 1mm after seven iterations of the wrapping process, and is depicted in Fig. 7. Each iteration lasted approximately thirty minutes, five minutes is needed to run the surface wrapper, fifteen seconds for the hole detection, and the rest of the is time needed for manual surface repair.

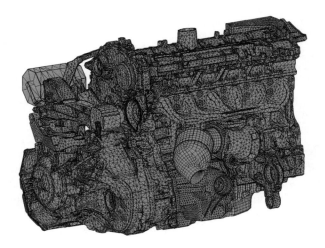

Fig. 6. Initial surface mesh for the engine example (321219 elements)

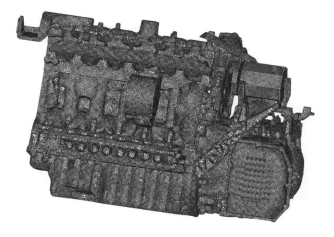

Fig. 7. Wrapped surface mesh after 7 iterations (484672 triangles)

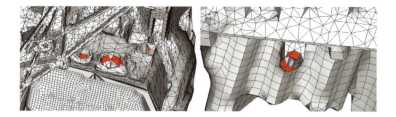

Fig. 8. Detected holes for the engine example. In left, are semantic holes due to missing parts. In right, is a hole with open boundary edges

Table 1. Values of the sensitivity parameter ψ with the number of wrapping iterations for the engine example.

Iteration	1	2	3	4	5	6	7
ψ	6	4	2	2	1	1	1

Figure 8 shows examples of detected holes during the first iteration, which are displayed as red surface facets in the input surface mesh. All holes are closed manually in the input surface mesh, and the detection algorithm provides hints where to look for holes. The sensitivity parameter ψ has decreased from six in the first iteration down to one in the seventh iteration and the intermediate values are given in the Table. 1. The sensitivity parameter is decreased during the process to detect the holes with smaller and smaller volumes that shall not be meshed. In general, it is up to the user to adjust the sensitivity parameter until the method starts providing useful hints where the problematic holes are.

3.2 Tractor

The input surface mesh, shown in Fig. 9, is used to show the performance of the method for the underhood analysis. The mesh consists of 1.4 million elements distributed in eighty disconnected regions with many open boundary edges and self-intersecting parts.

The wrapped surface after the first iteration has leaking regions above the wheels next to the cabin, see Fig. 10, which are caused by the holes in the passage underneath the cabin. Figure 11 shows the region under the cabin, and the triangles near the detected holes are coloured in red. The sensitivity parameter is set to $\psi = 6$. The resulting output of the hole detection algorithm is dependent on the resolution of the initial surface mesh, and therefore the detected hole stretches underneath the whole cabin due long triangles with high aspect ratio that are found there. Despite the high aspect ratio triangles, it was possible to spot the holes and close them manually. The final wrapped surface, given in Fig. 12, was achieved after the second iteration and the

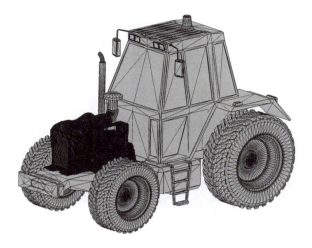

Fig. 9. Initial surface mesh with feature edges for the tractor example

Fig. 10. Leaks after the first iteration for the tractor example

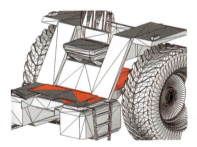

Fig. 11. Surface parts with marked elements

whole process lasted approximately 1 hour. Figure 12 shows that the edges at the intersection between the cabin and the underhood are resolved in the final mesh, even though they are not present in the input surface mesh, and it is the consequence of the iterative edge capturing algorithm.

Fig. 12. Final surface mesh for the tractor example (1563336 triangles) is given in the top figure. At the bottom, zoom of the region near the wheel.

4 Conclusions and Future Work

The paper presents the method for surface wrapping with a method for detecting holes based on the intensity of heat fluxes in various regions of the underlying octree volume mesh. The method is capable of detecting both holes with open edges and the semantic holes occurring due to some missing or misplaced parts. The heat conduction equation is solved on the octree mesh and the resolution of the octree mesh corresponds to the resolution of the newly generated wrapped surface. Its sensitivity is controlled by the user via a non-dimensional parameter. The higher values of the parameter make the method less sensitive and the lower values increase the sensitivity with a tendency to select all outside facets in the surface mesh. The method can be further improved by implementing an automatic hole closing algorithm into the wrapping process in order to reduce the number of iterations needed to close holes that affect the wrapping process. In addition, the results of this study can be further improved by implementing an algorithm for automatic adjustment of the sensitivity parameter and a filtering algorithm for removing the regions which are not topological holes but are selected during the process.

References

1. Bendels, G.H., Schnabel, R., Klein, R.: Detecting Holes in Point Set Surfaces. In: Proceedings: 14th International Conference in Central Europe on Computer Graphics, Visualization and Computer Vision (2006)
2. Beta CAE Systems USA. Meshing and Assembly (2010),
http://www.ansa-usa.com/products/ansa/meshing-and-assembly
3. Branch, J., Prieto, F., Boulanger, P.: A Hole-Filling Algorithm for Triangular Meshes Using Local Radial Basis Function. In: Proceedings: 15th International Meshing Roundtable, pp. 411–431 (2006)
4. Carr, J.C., Beatson, R.K., Cherrie, J.B., Mitchell, T.J., Fright, W.R., McCallum, B.C., Evans, T.R.: Reconstruction and representation of 3D objects with Radial Basis Functions. In: Proceedings: SIGRAPH 2001: Proceedings of the 28th Annual Conference on Computer Graphics and Interactive Techniques, pp. 67–76 (2001)
5. CD-Adapco, Star CCM+ (2011),
http://www.cd-adapco.com/products/star_ccm_plus/robustness.html
6. Davidson, L.: A pressure correction methods for unstructured meshes with arbitrary control volumes. Int. J. Numer. Meth. Fluids 22, 265–281 (1996)
7. Davis, J., Marschner, S., Garr, M., Levoy, M.: Filling holes in in complex surfaces using volumetric diffusion. In: Proceedings: First International Symposium on 3D Data Processing, Visualization and Transmission, vol. 11 (2002)
8. Dinh, H.Q., Turk, G., Slabaugh, G.: Reconstructing Surfaces Using Anisotropic Basis Functions. In: Proceedings: International Conference on Computer Vision, pp. 606–613 (2001)
9. Escobar, J.M., Rodriguez, E., Montenegro, R., Montero, G., Gonzalez-Yuste, J.M.: Simultaneous untangling and smoothing of tetrahedral meshes. Comput. Methods Appl. Mech. Engrg 192, 2775–2787 (2003)
10. Ferziger, J., Perić, M.: Computational Methods for Fluid Dynamics. Springer, Heidelberg (1996)
11. Garimella, R.V., Swartz, B.K.: Curvature Estimation for Unstructured Triangulations of Surfaces. Technical Report LA-UR-03-8240, Los Alamos National Laboratory (2003)
12. Jasak, H.: Error analysis and estimation in the Finite Volume Method with applications to fluid flows. PhD Thesis, Imperial College, University of London, London (1996)
13. Kobbelt, L.P., Vorsatz, J., Labsik, U., Seidel, H.P.: A Shrink Wrapping Approach to Remeshing Polygonal Surfaces. Comput. Graph. Forum 18, 119–130 (1999)

14. Lee, Y.K., Lim, C.K., Ghazilam, H., Vardhan, H., Eklund, E.: Surface Mesh Generation for Dirty Geometries by Shrink Wrapping using Cartesian Grid Approach. In: Proceedings: 15th International Meshing Roundtable, pp. 393–410 (2006)
15. Lorensen, W.E., Cline, H.E.: Marching cubes: A high resolution 3D surface construction algorithm. Computer Graphics 21, 163–169 (1987)
16. Sanchez, G.T., Branch, J.W., Atencio, P.: A Metric for Automatic Hole Characterization. In: Proceedings: 19th International Meshing Roundtable, pp. 195–208 (2010)
17. Schilling, A., Bidmon, K., Sommer, O., Ertl, T.: Filling Arbitrary Holes in Finite Element Models. In: Proceedings: 17th International Meshing Roundtable, pp. 231–248 (2008)
18. Sharc Ltd, New wrapping technology in Harpoon (2006), http://www.sharc.co.uk/html/notes_wrap.htm
19. Veleba, D., Felkel, P.: Survey of errors in surface representation and their detection and correction. In: WSCG 2007: Proceedings of the 15th International Conference in Central Europe on Computer Graphics, Visualization and Computer Vision, Plzen-Bory, Czech Republic (2007)
20. Vollmer, J., Mencl, R., Müller, H.: Improved Laplacian Smoothing of Noisy Surface Meshes. Comput. Graph. Forum, 131–138 (1999)
21. Wang, Z.J., Srinivasan, K.: An adaptive Cartesian grid generation method for 'Dirty' geometry. Int. J. Numer. Meth. Fluids 39, 703–717 (2002)
22. Whitaker, R.T.: A Level-Set Approach to 3D Reconstruction From Range Data. Int. J. Computer Vision 29, 203–231 (1998)
23. Zhao, H.K., Osher, S., Fedkiw, R.: Fast Surface Reconstruction Using the Level Set Method. In: Proceedings: IEEE Workshop on Variational and Level Set Methods in Computer Vision (VLSM 2001), pp. 194–201 (2001)

Mesh and CAD Repair Based on Parametrizations with Radial Basis Functions

Cécile Piret, Jean-François Remacle, and Emilie Marchandise

Université catholique de Louvain, Institute of Mechanics,
Materials and Civil Engineering (iMMC), Place du Levant 1,
1348 Louvain-la-Neuve, Belgium
cecile.piret@uclouvain.be, jean-francois.remacle@uclouvain.be,
emilie.marchandise@uclouvain.be

Summary. The goal of this paper is to present a new repair process that includes both model/mesh repair and mesh generation. The repair algorithm is based on an initial mesh of the CAD that may contain topological and geometrical errors. This initial mesh is then remeshed by computing a discrete parametrization with radial basis functions (RBF's).

[34] showed that a discrete parametrization can be computed by solving PDE's on an initial correct triangulation using finite elements. Paradoxically, the meshless character of the RBF's makes it an attractive numerical method for solving the PDE's for the parametrization in the case where the initial mesh contains errors.

In this work, we implement the Orthogonal Gradients method which was recently introduced in [32], as a technique to solve PDE's on arbitrary surfaces with RBF's. We will implement the low order version of the algorithm, which already gives great results in this context.

Different examples show that the presented method is able to deal with errors such as gaps, overlaps, T-joints and simple holes and that the resulting meshes are of high quality. Moreover, the presented algorithm can be used as a hole-filling algorithm to repair meshes with undesirable holes. The overall procedure is implemented in the open-source mesh generator Gmsh [18].

Keywords: geometry processing, hole filling algorithm, radial basis functions, RBF, surface remeshing, surface parametrization, STL file format, surface mapping, harmonic map, Orthogonal Gradients method.

1 Introduction

Using CAD data for finite element analysis has become the actual standard in the engineering practice. Yet, geometries that come out of design offices are not free of problems: slivers, cross-overs, surfaces with multiples unnecessary patches, super-small model entities and many other issues that are encountered in the CAD data make the meshing process a nightmare. Those *dirty geometries* are still the cause of time consuming repair processes. The same kind of issues are present when dealing with STL triangulations as the input geometry: the mesh may be noisy, self-intersecting, not watertight, with T-junctions[1] and have undesirable holes. Figure 1 gives an example of such dirty CAD models or STL triangulations that need to be repaired.

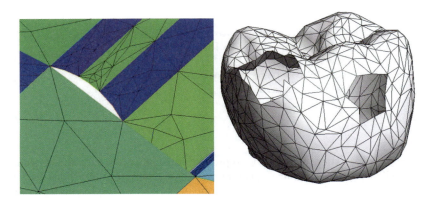

Fig. 1. Two examples for which a cad and mesh repair algorithm is needed. The figure on the left shows an initial triangulation of a dirty CAD model with topological errors: gaps (holes), overlaps and T-junctions. The right figure displays an STL triangulation of a tooth that contains undesirable holes.

There are two approaches for cleaning dirty geometries: one acts on the CAD model and one acts on the mesh.

The first approach corrects the geometry directly by using point and edge merging algorithms [5, 19, 38]. Those approaches thus provide specific tools for model correction that are controlled by the user [31, 37]. Presently, there are also many commercial software modules that claim to be able to perform automatic geometry healing. However, these third party software modules can only rectify common geometry problems and a successful or unsuccessful outcome is possible. Thus there is yet no absolute solution for geometry/mesh healing of CAD models.

Another approach is that of correcting an initial triangulation of the model through the addition of triangles and different stitching procedures [2, 1, 29]. In the same vein, Nooruddin and Turk [30] proposed a method to repair polygonal meshes using volumetric techniques. Unfortunately, those algorithms do not consider geometric intersections and inconsistent topology may be present. Also based on an initial triangulation, Wu et al. [40] suggested some specific hole-filling algorithms that employ RBF's as an interpolation

---

[1] A T-Junction is an intersection of two or more faces in a mesh where the vertex of one face lies on the edge or interior of another face.

technique to construct an implicit surface patch and to intersect this implicit surface with the existing triangulation. However, the intersection method to reconstruct the mesh is quite complex.

In this paper, an original alternative approach is presented. The repair process includes both model/mesh repair and mesh generation. The remeshing procedure relies on an discrete parametrization. Surface parametrization techniques [41, 27, 22] originate mainly from the computer graphics community: they have been used extensively for applying textures onto surfaces [4, 24] and have become a very useful and efficient tool for many mesh processing applications [22, 36, 9]. In the context of remeshing procedures, the initial surface is parametrized onto a surface in $\mathcal{R}^2$, the surface is meshed using any standard planar mesh generation procedure and the new triangulation is then mapped back to the original surface [7, 28]. We showed in [34, 26, 25] that harmonic maps can be computed efficiently by solving partial differential equations (PDE's) on the initial triangulation with finite elements.

In the context of CAD and mesh repair, the initial triangulation may contain topological errors such as holes, T-junctions and overlaps that make numerical techniques such as finite elements fail. The meshless character of the RBF's makes it then an attractive numerical method for solving those PDE's. Although the RBF method has been used as an interpolation technique since the 1970s, it is only in the 1990s that it was introduced as a technique to solve PDE's [20, 21]. Its high accuracy and meshless character have made it the method of choice for problems set on complicated geometries [8]. Often overlooked for having poor stability and high complexity issues, the method has finally gained acknowledgement. A number of studies showed the method's great potential by solving full-scale geophysical applications, and by showing that RBF's could compete with the most trusted numerical techniques [10, 12, 11, 39, 35]. Although a good part of the RBF literature deals with surface reconstruction [6, 3], no technique has been developed to solve PDE's on them, until [32], which provides the very first methods, based on RBF's, for solving PDE's on completely arbitrary surfaces. In this work, we implement the RBF's Orthogonal Gradients method of [32] which relies on a level set representation of the initial surface.

The paper is organized as follows. In section 2, we present the PDE's for solving the parametrization. In section 3, we show how to solve the PDE's with RBF's Orthogonal Gradients method centered on the mesh vertices of the initial triangulation. Next, we explain in section 4 how to find the inverse mapping in order to be able to project the new points on the 3D surface (CAD patches or discrete surface). Finally, results are presented to validate the method and to reveal the efficiency and accuracy of our proposed algorithm.

2 Discrete Parametrization

The discrete parametrization aims at computing the discrete mapping $\mathbf{u}(\mathbf{x})$ that maps every mesh vertex $\mathbf{x}$ of an initial triangulation of a surface Γ to a point $\mathbf{u}$ of Γ' embedded in $\mathcal{R}^2$:

$$\mathbf{x} = \{x, y, z\} \in \Gamma \subset \mathcal{R}^3 \mapsto \mathbf{u}(\mathbf{x}) = \{u, v\} \in \Gamma' \subset \mathcal{R}^2 \tag{1}$$

In the remainder of this paper, we restrict ourselves to the parametrization of non-closed surfaces meshes of zero genus. If the surface mesh is closed, of genus greater that zero, or non-manifold, we partition the initial mesh into different mesh patches [26, 25] and compute a discrete parametrization for each of those mesh patches with the presented method.

In this work, we have chosen a harmonic mapping for the parametrization [28, 7, 34]. Harmonic maps $\mathbf{u}(\mathbf{x})$ can be computed by solving two Laplace's equations on the 3D surface Γ:

$$\nabla^2 u(\mathbf{x}) = 0, \quad \nabla^2 v(\mathbf{x}) = 0, \quad \forall\, \mathbf{x} \in \Gamma \tag{2}$$

with appropriate Dirichlet boundary condition for one of the boundaries of the surface Γ,

$$u(l) = \cos(2\pi l/T)\ ,\quad v(l) = \sin(2\pi l/T), \tag{3}$$

and with Neumann boundary conditions for the other boundaries. In (3), l denotes the curvilinear abscissa of a point along the boundary of total length T.

We will show in the next section how to compute the discrete parametrization using RBF's. Figure 2 shows a triangulation of a CAD model with topological and geometrical errors such as gaps, overlaps and T-junctions. Those gaps presented in the initial mesh are magnified during the parametrization procedure. The T-junctions are also visible in both the 3D space and the parametric space. With the presented approach, there is no need to identify those T-junctions and gaps. The mesh points of the dirty initial triangulation are directly used to compute a discrete parametrization $\mathbf{u}(\mathbf{x})$ with RBF's. The remeshing procedure will then be performed in the parametric space given this mapping $\mathbf{u}(\mathbf{x})$ and the parametric description of the 4 CAD patches $\mathbf{u}_j^{CAD}$ visible on the left figure 2. For the remeshing of the holes, RBF interpolations will be used.

3 Harmonic Map with Radial Basis Functions

In the case where the initial triangulation contains topological and geometrical errors (such as in Fig. 2), standard numerical methods based on meshes

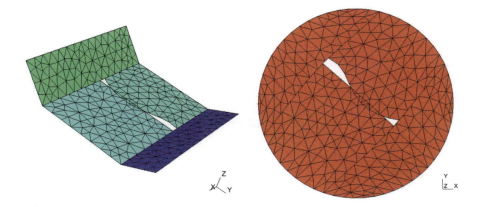

Fig. 2. Initial triangulation of a CAD model with topological and geometrical errors such as gaps, overlaps and T-junctions (left) and its discrete parametrization $\mathbf{u}(\mathbf{x})$ computed with RBF's (right). One can see clearly the T-junctions, holes and overlaps of the initial mesh in the parametric space.

Table 1. Definitions of the most commonly used C_∞ radial functions.

Name of RBF	Abbreviation	Definition
Multiquadric	MQ	$\sqrt{1+(\varepsilon r)^2}$
Inverse multiquadric	IMQ	$\dfrac{1}{\sqrt{1+(\varepsilon r)^2}}$
Inverse quadratic	IQ	$\dfrac{1}{1+(\varepsilon r)^2}$
Gaussian	GA	$e^{-(\varepsilon r)^2}$

such as finite elements, or finite volumes fail to compute the solutions to Laplace's equation (2). The meshless character of the RBF's makes it then an attractive numerical method for solving the two Laplace's equations (2) whose solution is the harmonic map.

Different techniques have been recently suggested to solve PDE's on arbitrary surfaces with RBF's [32]. In this work, we use [32]'s simplest version of the Orthogonal Gradients method, which relies on a level set representation of the initial surface. Let M be the number of points on the initial triangulation (points on the surface $\varGamma$ in Figures 3 and 4).

In this work, we use the MQ (Table 1) radial function $\phi(r)$ to represent the solutions (u,v) of the Laplace's equations on $\varGamma$.[2]

---

[2] In the remainder of this section, we develop the Laplacian of u. The Laplacian of v can be developed in a similar way.

$$u(\mathbf{x}) = \sum_{i=1}^{N} \lambda_i \, \phi(\|\mathbf{x} - \mathbf{x}_i\|), \quad \text{with} \quad \phi(r) = \sqrt{1 + \epsilon^2 r^2} \tag{4}$$

where $\|\cdot\|$ denotes the Euclidean norm, $\mathbf{x}_i$ are the set of N data points, λ_i are the RBF's expansion coefficients, and ϵ is the shape parameter. All C_∞ smooth radial functions yield a similar accuracy, and any of them could be used.

Let us find a matrix D that discretizes, via an RBF representation, a continuous differential operator L (for the Laplacian operator, we have $L = \mathcal{L} = \partial_{xx} + \partial_{yy} + \partial_{zz}$). We can rewrite the RBF interpolation (4) in matrix form for the N points:

$$A\Lambda = U, \quad \text{with} \quad A_{i,j} = \phi(\|\mathbf{x}_i - \mathbf{x}_j\|), \quad U_i = u(\mathbf{x}_i) \tag{5}$$

where $A_{i,j}$ is the RBF interpolation matrix of size $N \times N$ and Λ is the vector of expansion coefficients λ_i. Analytically applying the differential operator to the radial function interpolation $u(\mathbf{x})$ gives

$$u^L(\mathbf{x}_k) = \sum_{j=1}^{N} \lambda_j \underbrace{L\phi(\|\mathbf{x} - \mathbf{x}_j\|)_{\mathbf{x}=\mathbf{x}_k}}_{B_{k,j}^L}, \tag{6}$$

where u^L is the value of the differential operator applied to u at each $\mathbf{x}_k$. Thus we have $B^L \Lambda = U^L$ in matrix form, where matrix B is of size $K \times N$, and K is the number of points on which we wish to compute the differentiation. Eliminating the expansion coefficient vector Λ leads to:

$$U^L = \underbrace{B^L A^{-1}}_{D^L} U. \tag{7}$$

where D^L (of size $K \times N$) is the matrix of differentiation, i.e the discretization of L. Note that if L is the identity operator I, then D^I is an interpolation matrix that interpolates the values of $u(\mathbf{x}_i)$ at points $\mathbf{x}_k$.

We wish now to compute the Laplacian operator of a function defined on the 3D surface Γ by using the Orthogonal Gradients method of Piret [32]. This method is loosely inspired from the closest point method [23] in that the goal is to make that function constant in the direction normal to the surface Γ. If that is the case, the Laplacian contribution from the normal direction vanishes and leaves only the Laplacian contribution from the tangential direction, i.e. the surface Laplacian.

In order to have a reliable RBF representation of the surface Γ and to reliably compute the direction that is normal to the surface, it is now quite usual to introduce additional points, on either side of the surface, and to define the level set distance function as $s(\mathbf{x}) = 0$ on Γ, $s(\mathbf{x}) = \pm 1$ on $\Gamma^\pm$, where $\Gamma^\pm$ are surfaces that surround and that are parallel to Γ (as in Figures 3 and 4) (see for example [3, 8]) The alternative to introducing additional points is to define a level surface $s(\mathbf{x}) = c$ ($c \neq 0$ to avoid the trivial RBF expansion) on Γ through the original nodes only. However, it is known that RBFs do a poor job of interpolating constant values [16]. The interpolated level-set would risk to be self-intersecting, which would invalidate any result. Adding nodes on either sides of Γ makes the distance function interpolation much more stable. We thus extend the original M points of the surface inward and outward as follows:

$$\mathbf{x}_i^e = \mathbf{x}_i \pm \delta \mathbf{n}(\mathbf{x}), \qquad (8)$$

where $\mathbf{n}(\mathbf{x})$ is the unit normal at point $\mathbf{x}_i$ and δ is the offset parameter. The normals are computed from the differentiation of an RBF expansion for smooth function $\sigma(\mathbf{x})$ that is given a value of $\sigma(\mathbf{x}) = 1$ on the M points on the surface and a value of $\sigma = 2$ on an additional point located far away from the surface[3]:

$$\mathbf{n} = \frac{\nabla \sigma}{\|\nabla \sigma\|}. \qquad (9)$$

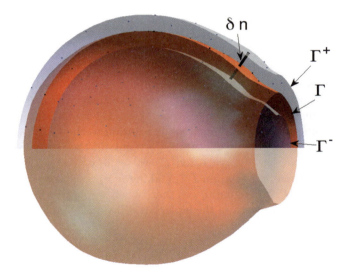

Fig. 3. Illustration of the RBF Orthogonal Gradients method. M points are uniformly distributed on the main surface Γ (in red). They are the mesh vertices of the initial triangulation. At each point, the normal $\mathbf{n}$ to the surface is computed and 2 new points are obtained at distance δ from the surface, one on either side of Γ. One can see these $2M$ additional points on the two grey layers (which we call Γ^+ and Γ^-) that surround Γ.

---

[3] This additional point guarantees that the normal is always well defined even on planar surface parallel to the coordinate axis.

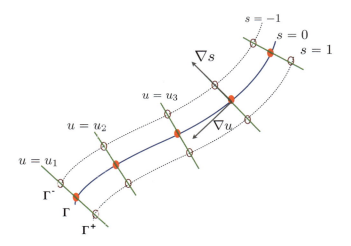

Fig. 4. Schema of the RBF Orthogonal Gradients method.

A schema of the RBF Orthogonal Gradients method is displayed in Figure 4. M points (full red dots) are uniformly distributed on Γ. At each point, the normal $\mathbf{n}$ to the surface is computed and two new points are obtained at distance δ from the surface, one on either side of Γ (empty red dots). We define two RBF expansions on these $3M$ nodes, one for the level set distance function $s(\mathbf{x})$ and one for the solution to the PDE $u(\mathbf{x})$. We set $s(\mathbf{x}) = 0$ on Γ, $s(\mathbf{x}) = \pm 1$ on $\Gamma^{\pm}$, thus ∇s points in the normal direction to Γ. Moreover, as we let $u(\mathbf{x_i}) = u_i$ on Γ, we set $u(\mathbf{x_i^{\pm}}) = u_i$, where $\mathbf{x_i^{\pm}}$ are the points corresponding to $\mathbf{x_i}$ on $\Gamma^{\pm}$, making $u(\mathbf{x})$ constant in the direction that is normal to Γ. This guarantees that, when we apply the Laplacian to $u(\mathbf{x})$, the normal component of the laplacian will vanish, and only the surface laplacian will remain.

We have thus a set of $N = 3M$ points that we use to compute the differentiation matrix for the Laplacian $D^{\mathcal{L}}$. This differentiation matrix is of size $M \times 3M$. In order to restrict the Laplacian operator to the surface Γ, we have to cancel the derivative of u in the direction normal to the surface:

$$\nabla_{\mathbf{n}} u = 0, \quad \nabla_{\mathbf{n}} \cdot \nabla_{\mathbf{n}} u = 0 \tag{10}$$

so that we only keep the restriction of the Laplacian on the surface Γ.

$$\nabla^2 u = (\nabla_{\mathbf{n}} + \nabla_{\mathbf{t}}) \cdot (\nabla_{\mathbf{n}} + \nabla_{\mathbf{t}}) u = \nabla_{\mathbf{t}} \cdot \nabla_{\mathbf{t}} u, \tag{11}$$

In (11), n and t denote respectively the surface normal and tangential directions.

There are two versions of the RBF Orthogonal Gradients method. The first version is less accurate but very easy to implement. This method assigns a constant value of $u(\mathbf{x})$ in the normal direction to the surface, i.e at the inward and outward points of $\mathbf{x}_i$, i.e $u(\mathbf{x}_i) = u(\mathbf{x}_i^e)$.

If the initial approximation of the normals (9) (which we use to compute the additional $2M$ points, thus before the level set distance function has been defined) is not accurate enough, the extended points $\mathbf{x}_i^e$ need to be corrected in order to guarantee the cancelation of the derivatives in the normal direction. The orthogonality condition is then $u(\mathbf{x}_i) = u(\mathbf{x}_i^{ec})$, where $u(\mathbf{x}_i^{ec})$ can be computed from (7) as $U^{ec} = D^I U$, so that we come up with the following linear system of size $3M \times 3M$ to solve:

$$\begin{pmatrix} D^{\mathcal{L}} \\ I_\Gamma - D^I \\ I_\Gamma - D^I \end{pmatrix} U = \bar{0}, \tag{12}$$

where I_Γ is an $M \times 3M$ matrix whose entries are zero everywhere except for the entries (i,i) (corresponding to points $\mathbf{x}_i$ on Γ) that have a value of 1.

The second version of the RBF Orthogonal Gradients method is more accurate, and we will implement it as part of our future work. Dirichlet boundary conditions (3) on the boundary nodes of $\mathbf{x}_i$ are then applied directly on the linear system (12) by setting the entries of the corresponding line to zero, the diagonal term to 1 and the right hand side to the cosine value (3). This algorithm has a global and a local version. In the global version, the derivatives are computed using all the nodes, while in the local version, the derivatives at a particular point are computed using only the points in a cluster of neighbors. Although the local method loses the potential for spectral accuracy in the solution representations, it still is a high order method and it has a significantly reduced complexity compared to the global method, since the differentiation matrices are now sparse.

The only two parameters that need to be properly defined are the shape parameter ϵ defined in (4) and the offset parameter δ defined in (8). Both of these parameters have a direct impact on both the accuracy and the conditioning of the differentiation matrix. A lot of work has been done on the search for an 'optimal' shape parameter ϵ. Since the topology and the nodes' distribution and density also impact the conditioning, finding a formula for this 'optimal' ϵ is near impossible. As ϵ gets smaller, the accuracy improves but the conditioning worsens [8, 17]. Unless one uses one of the algorithms that bypass the small shape parameter conditioning issue [15, 14, 13], the shape parameter will need to be set large enough to have a good conditioning. Since we know the smallest distance between two nodes ($d_{\min}$), and that the nodes are uniformly distributed, we can 'normalize' the shape parameter as $\epsilon = \frac{\epsilon^*}{d_{\min}}$. ϵ^* will vary with the total number of nodes (global method) or with the number of nodes in a cluster (local method). A lot less work has been made in finding an optimal δ. However, the accuracy seems less sensitive to its value than to the value of the shape parameter. In order to avoid level sets crossing, we set $\delta = d_{min} \times \delta^*$, where $0 < \delta^* < 1$. For our examples, which feature relatively small sets of nodes, we set $\delta^* = 0.33$ and $\epsilon^* = M/300$. If M were to be much larger, ϵ^* would need to grow exponentially with M.

4 Inverse Mapping

Once the harmonic map has been computed, it can be used as input for planar surface meshers (delaunay, frontal, etc.) to produce high quality triangulations [34]. It is this remeshing step performed in the parametric space that will remove the overlaps, intersections or T-junctions. The only information we need to provide to the planar meshers is the inverse map $\mathbf{x}(\mathbf{u})$ and the Jacobians of the mapping $\mathbf{x}_u$ and $\mathbf{x}_v$. The jacobians are then used to build the metric tensor (or first fundamental form)

$$\mathbf{M} = \mathbf{x}_{,\mathbf{u}}^T \mathbf{x}_{,\mathbf{u}} = \begin{bmatrix} \mathbf{x}_{,u} \cdot \mathbf{x}_{,u} & \mathbf{x}_{,u} \cdot \mathbf{x}_{,v} \\ \mathbf{x}_{,v} \cdot \mathbf{x}_{,u} & \mathbf{x}_{,v} \cdot \mathbf{x}_{,v} \end{bmatrix} \qquad (13)$$

in order to compute edge lengths, angles and areas.

In this section, we explain how to compute the inverse mapping. We already have the solution of the harmonic map at the M points of the mesh: $\mathbf{u}_i(\mathbf{x}_i)$. The planar mesh algorithm needs to insert a new point $\mathbf{u}_p$ in the parametric space and needs therefore to know $\mathbf{x}_p(\mathbf{u}_p)$ as well as the mesh metric in order to have edge length and angle measures.

Figure 5 shows the remeshing procedure in the parametric space as well as the resulting mesh for the CAD model of Fig. 2. As can be seen from Fig. 5, the presented algorithm is also able to keep some specified shape features such as topological edges.

Let suppose that our first mesh is obtained by meshing different CAD patches that have a given parametric description $\mathbf{u}^{CAD}$. In the case the new mesh point to be inserted by the meshing algorithm, lies within a mesh triangle T_j, then the inverse mapping can be found as follows [34].

1. Find a triangle T_j of the parametric space Γ' that contains point $\mathbf{u}$; Note that in the case the parametrization contains overlaps or intersections that come from the dirty model (see right Fig. 2), then any of the triangles containing the point $\mathbf{u}$ can be chosen;
2. Compute local coordinates $\xi = (\xi, \eta)$ of point $\mathbf{u}$ inside triangle T_j;
3. Compute the parametric coordinate of the CAD patch as follows:
   $\mathbf{u}^{CAD}(\xi, \eta) = (1 - \xi - \eta)\mathbf{u}_1^{CAD} + \xi \mathbf{u}_2^{CAD} + \eta \mathbf{u}_3^{CAD}$;
4. Use the CAD model to obtain the inverse mapping $\mathbf{x}(\mathbf{u}^{CAD})$ and the derivatives of this mapping.

Suppose now that the point to be inserted does not lie within any parametrized triangle or suppose that the triangulation does not have an underlying CAD model (stl triangulations obtained from image segmentation). In this case, the inverse mapping $\mathbf{x}_p$ for a point $\mathbf{u}_p = (u_p, v_p)$ is computed with a local RBF interpolation of the form (7):

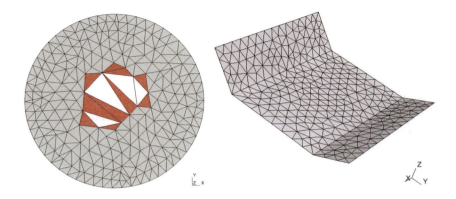

Fig. 5. Starting from a initial triangulation of a CAD model with topological and geometrical errors and from its parametrization (see the Fig. 2), we compute with a planar mesh generator the new mesh. The left figure shows an intermediate step of the frontal mesh algorithm (the grey triangles are the resolved layers, the red the active layers and the white the waiting layers) in the parametric space and the right figure shows the resulting mesh in the 3D space.

$$X^L = \underbrace{B_{\mathbf{u}}^L A_{\mathbf{u}}^{-1}}_{D_{\mathbf{u}}^L} X, \quad (14)$$

where X is taken as the vector of the closest P points $\mathbf{x}_i$. Those closest points are found by computing the P closest[4] parametrized points $\mathbf{u}_i$ to $\mathbf{u}_p$ in the set of the M parametrized points and by taking the corresponding points in the 3D space $\mathbf{x}_i$. In 14, X^L is the inverse mapping we are looking for (of size $K = 1$) and $D_{\mathbf{u}}^L$ is the matrix of differentiation in the parametric space of size $K \times P$ (i.e. with the radial basis functions written in the parametric space: $\phi(\mathbf{u} - \mathbf{u}_i)$). From (14), we obtain easily the inverse quantities needed by the planar meshes. Indeed, when $L = I$, we get the inverse map $\mathbf{x}_p(\mathbf{x}_u)$, when $L = \partial_u$, we obtain the Jacobian $\mathbf{x}_{u,p}$ and when $L = \partial_v$, we have the Jacobian $\mathbf{x}_{v,p}$.

From Fig. 5, we can see that the proposed method can be used as a hole-filling algorithm. Indeed, thanks to the RBF interpolation, we are able to find new mesh vertices $\mathbf{x}_p$ corresponding to parametrized points $\mathbf{u}_p$ that are located inside undesirable holes.

5 Results

Two examples aim to show that the presented algorithm can be seen as a CAD and mesh repair algorithm.

---

[4] A fast kdtree method is used for computing those closest points.

The first example shows a raw mesh model of a face (Fig. 6). The triangulation is made of 1048 triangles, contains different self-intersection triangles (see red triangles in Fig. 6) and has two undesirable holes. It follows that the parametrization cannot be computed with a finite element Laplacian. The parametrization computed with RBF's is shown in Fig. 7. A Frontal-Delaunay algorithm [33] was used for the remeshing in the parametric space. We first compare the quality of the initial and final mesh by computing for every triangle the aspect ratio κ that is the ratio between the inscribed and circumscribed radius of the triangle [18, 34]. Parameter κ is such that an equilateral triangle has $\kappa = 1$ and a degenerated flat triangle has $\kappa = 0$. The initial triangulation of the face has $\bar{\kappa} = 0.69$ and $\kappa_{min} = 0.01$, while the new mesh made of 2705 triangles has $\bar{\kappa} = 0.94$ and $\kappa_{min} = 0.36$. The L_2 Hausdorff distance (normalized by the bounding box) between the initial triangulation and the new mesh is only 0.0007. This small error show that the underlying shape is predicted with good geometric fidelity. The total time for remeshing the face is $1s^5$, and 25% of that time is for computing the parametrization with RBFs.

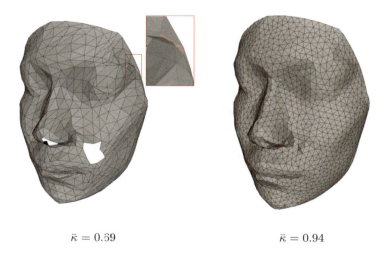

$\bar{\kappa} = 0.69$ $\bar{\kappa} = 0.94$

Fig. 6. Dirty raw mesh model of the face (left) and the new mesh (right) and the average quality $\bar{\kappa}$ of the meshes. The red triangles on the left are the self-intersecting triangles.

In the next example we consider a dirty CAD model from a rocket reinforcement of a vehicle.

Most of the time, a straightforward meshing of the patches of a clean CAD does not give a suitable mesh for finite element analysis. An efficient

---

[5] On a MACBOOK PRO clocked at 2.4 GHz.

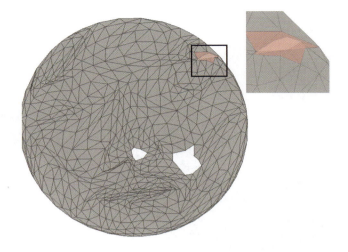

Fig. 7. The parametrized mesh computed with RBFs. The red triangles are the self-intersecting triangles.

manner to build a high quality mesh for those CAD models is then to build from the initial CAD mesh a cross-patch parametrization that enables the remeshing of merged patches. Indeed, as most surface mesh algorithms mesh model faces individually, mesh points are generated on the bounding edges of those patches and if thin patches exist in the model they will result in the creation of small distorted triangles with very small angles. Those low quality elements present in the surface mesh will often hinder the convergence of the FE simulations on those surface meshes. An efficient manner to build a high quality mesh for those CAD models is then to build from the initial CAD mesh a *cross-patch parametrization* that enables the remeshing of merged patches. The new mesh is then build in the cross-patch parametric space and the new points are projected back onto the CAD patches using the parametric representation of the patches $\mathbf{u}^{CAD}$ (e.g. NURBS). Now, if the CAD is dirty such as the example of Fig. 8, standard techniques for computing cross-patch parametrizations will fail and the method we present in this paper can be considered as an original approach for creating a finite element mesh based of the dirty CAD that does that bypasses the geometrical repair. The dirty CAD model of Fig. 8 is given in IGES format, and includes 141 NURBS surfaces. The CAD model contains a lot of gaps, overlaps, redundant surfaces, T-joints and patches with reverted normals. This CAD model has so many topological and geometrical errors that none our available commercial packages that claim to perform geometrical CAD healing was able to repair the model.

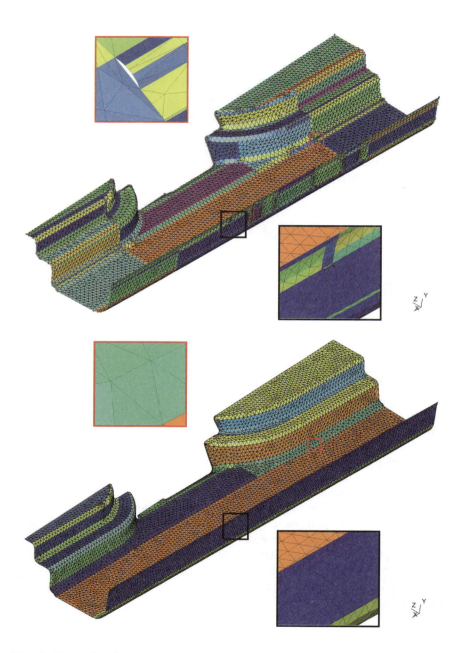

Fig. 8. Example of an initial triangulation of a CAD model from the automotive industry that contains a lot of gaps, overlaps and T-joints. The bottom figure shows the new mesh that is suitable for finite element simulations.

Seventeen groups of patches (also called *compounds*) have been created and a cross-patch parametrization has been computed with the presented method for every of those compounds. This is implemented quite simply in Gmsh by creating a .geo file that reads:

```
Mesh.RemeshParametrization=2; // (2) for  rbf
Merge "CAD.iges";
Mesh.CharacteristicLengthFactor = 0.1;
Compound Surface(20000) = {34:65,118} ;
```

The new mesh (Fig. 8 bottom) is obtained with a planar Delaunay mesh algorithm with a given uniform mesh size field. This new mesh is made of 13216 triangles that have a high mean quality of $\bar{\kappa} = 0.94$. The total time for meshing the compounds is $30s$. This example also demonstrates that our method works with non-uniformly distributed nodes. The minimum distance between two nodes is $1.e-3$ while the maximum distance is 5.2.

6 Conclusion

In this paper, we presented a brand new approach for repairing and generating meshes, which paradoxically finds its strength in the meshless character of the RBF method on which our approach is based. We showed that our algorithm gives excellent results in repairing serious topological and geometrical errors such as holes, reversed normals, overlaps and T-junctions, non-manifold vertices.

The approach makes use of the RBF Orthogonal Gradients method, recently introduced in [32], to solve PDEs on arbitrary surfaces using RBF's. We use this method to solve Laplace's equation (with Dirichlet boundary conditions on a closed boundary curve of the surface) to obtain a discrete parametrization. Next, the surface is remeshed in the parametric space with a computed inverse mapping. We showed that the presented algorithm can be seen as a CAD and mesh repair algorithm but also a global quality remeshing algorithm, and a hole filling-algorithm.

Our technique can be further improved by using the higher order version of the RBF Orthogonal Gradients method, or by computing conformal mappings instead of harmonic mappings. Indeed, as conformal mappings preserve angles, they will enable us to generate also high quality quadrilateral meshes. Moreover, they require only two Dirichlet boundary conditions for the solution of the mapping. Also different algorithmic advances involving hierarchical and fast-multipole like methods combined with interpolatory filters can be used to reduce the computational cost associated to the RBF interpolation.

Acknowledgements

The rocket reinforcement model of Fig. 8 is provided courtesy of ArcelorMittal and the two raw mesh models are provided courtesy of IMATI by the AIM@SHAPE Shape Repository [6].

References

1. Barequet, G.: Using geometric hashing to repair cad objects. In: IEEE Comput. Science and Engineering, pp. 22–28 (1997)
2. Barequet, G., Sharir, M.: Filling gaps in the boundary of a polyhedron. Computer Aided Geom. Design 12, 207–229 (1995)
3. Beatson, R.K , Cherrie, J.B., McLennan, T.J., et al.: Surface reconstruction via smoothest restricted range approximation. Geometric Modeling and Computing 46, 41–52 (2004)
4. Bennis, C., Vézien, J.-M., Iglésias, G.: Piecewise surface flattening for non-distorted texture mapping. In: ACM SIGGRAPH Computer Graphics, pp. 237–246 (1991)
5. Butlin, G., Stops, C.: Cad data repair. In: Proceedings of the 5th International Meshing Roundtable, Pennsylvania, USA, pp. 7–12 (1996)
6. Carr, J.C., Beatson, R.K., McCallum, B.C., et al.: Smooth surface reconstruction from noisy range data. In: GRAPHITE 2003, pp. 119–126 (2003)
7. Eck, M., DeRose, T., Duchamp, T., Hoppe, H., Lounsbery, M., Stuetzle, W.: Multiresolution analysis of arbitrary meshes. In: SIGGRAPH 1995: Proceedings of the 22nd Annual Conference on Computer Graphics and Interactive Techniques, pp. 173–182 (1995)
8. Fasshauer, G.E.: Meshfree approximation methods with MATLAB. In: Interdisciplinary Mathematical Sciences. World Scientific (2007)
9. Floater, M.S., Hormann, K.: Surface parameterization: a tutorial and survey. In: Advances in Multiresolution for Geometric Modelling, pp. 157–186. Springer, Heidelberg (2005)
10. Flyer, N., Fornberg, B.: Radial basis functions: Developments and applications to planetary scale flows. Computers and Fluids 46, 23–32 (2011)
11. Flyer, N., Wright, G.-B.: Transport schemes on a sphere using radial basis functions. J. Comp. Phys. 226, 1059–1084 (2007)
12. Flyer, N., Wright, G.-B.: A radial basis function method for the shallow water equations on a sphere. Proc. Roy. Soc. 465, 1949–1976 (2009)
13. Fornberg, B., Larsson, E., Flyer, N.: Stable computations with gaussian radial basis functions. SIAM J. Scientific Computing 33(2), 869–892 (2011)
14. Fornberg, B., Piret, C.: A stable algorithm for flat radial basis functions on a sphere. SIAM J. Scientific Computing 30(1), 60–80 (2007)
15. Fornberg, B., Piret, C.: On choosing a radial basis function and a shape parameter when solving a convective pde on a sphere. Journal of Computational Physics 227, 2758–2780 (2008)
16. Fornberg, B., Wright, G.: Larsson E. Some observations regarding interpolants in the limit of flat radial basis functions. Computers and Mathematics with Applications 47, 37–55 (2004)

---

[6] http://shapes.aim-at-shape.net/

17. Fornberg, B., Wright, G.-B.: Stable computation of multiquadric interpolants for all values of the shape parameter. Comp. Math. Applic. 48, 853–867 (2004)
18. Geuzaine, C., Remacle, J.-F.: Gmsh: a three-dimensional finite element mesh generator with built-in pre- and post-processing facilities. International Journal for Numerical Methods in Engineering 79(11), 1309–1331 (2009)
19. Jones, M.R., Price, M.A., Butlin, G.: Geometry management support for automeshing. In: Proceedings, 4th International Meshing Roundtable, pp. 153–164 (1995)
20. Kansa, E.J.: Multiquadrics, a scattered data approximation scheme with applications to computational fluid-dynamics (i): Surface approximations and partial derivative estimates. Comput. Math. Appl. 19, 127–145 (1990)
21. Kansa, E.J.: Multiquadrics: a scattered data approximation scheme with applications to computational fluid-dynamics (ii): Solutions to parabolic, hyperbolic and elliptic partial differential equations. Comput. Math. Appl. 19, 147–161 (1990)
22. Levy, B., Petitjean, S., Ray, N., Maillot, J.: Least squares conformal maps for automatic texture atlas generation. In: Computer Graphics (Proceedings of SIGGRAPH 2002), pp. 362–371 (2002)
23. Macdonald, C.B., Ruuth, S.J.: The implicit closest point method for the numerical solution of partial differential equations on surfaces. Siam Journal on Scientific Computing 31, 4330–4350 (2009)
24. Maillot, J., Yahia, H., Verroust, A.: Interactive texture mapping. In: Proceedings of ACM SIGGRAPH 1993, pp. 27–34 (1993)
25. Marchandise, E., Carton de Wiart, C., Vos, W.G., Geuzaine, C., Remacle, J.-F.: High quality surface remeshing using harmonic maps. Part II: Surfaces with high genus and of large aspect ratio. International Journal for Numerical Methods in Engineering 86(11), 1303–1321 (2011)
26. Marchandise, E., Compère, G., Willemet, M., Bricteux, G., Geuzaine, C., Remacle, J.-F.: Quality meshing based on stl triangulations for biomedical simulations. International Journal for Numerical Methods in Biomedical Engineering 83, 876–889 (2010)
27. Marcum, D.L.: Efficient generation of high-quality unstructured surface and volume grids. Engineering with Computers 17, 211–233 (2001)
28. Marcum, D.L., Gaither, A.: Unstructured surface grid generation using global mapping and physical space approximation. In: Proceedings, 8th International Meshing Roundtable, pp. 397–406 (1999)
29. Morvan, S.M., Fadel, G.M.: Ives: an interactive virtual environment for the correction of .stl files. In: Proceedings of Conference on Virtual Design, University of California, Irvine (1996)
30. Nourrudin, F.S., Turk, G.: Simplification and repair of polygonal models using volumetric techniques. IEEE Transactions on Visualisation and Computer Graphics 9(2), 191–205 (2003)
31. Peterson, N.A., Kyle, K.: Detecting translation errors in cad surfaces and preparing geometries for mesh generation. Technical Report UCRL-JC-144019, Center for Applied Scientific Computing, Lawrence Livermore National Labs, Livermore, CA 94551 (2001)
32. Piret, C.: The orthogonal gradients method: a radial basis function solution method for solving partial differential equations on arbitrary surfaces. Journal of Computational Physics (in preparation, 2011)

33. Rebay, S.: Efficient unstructured mesh generation by means of delaunay triangulation and bowyer-watson algorithm. Journal of Computational Physics 106, 25–138 (1993)
34. Remacle, J.-F., Geuzaine, C., Compère, G., Marchandise, E.: High quality surface remeshing using harmonic maps. International Journal for Numerical Methods in Engineering 83, 403–425 (2010)
35. Schmidt, J., Piret, C., Zhang, N., et al.: Modeling of tsunami waves and atmospheric swirling flows with graphics processing unit (gpu) and radial basis functions (rbf). Concurrency Comput.: Pract. Exp. 22, 1813–1835 (2010)
36. Sheffer, A., Praun, E., Rose, K.: Mesh parameterization methods and their applications. Found. Trends. Comput. Graph. Vis. 2(2), 105–171 (2006)
37. Sheng, X., Meier, I.R.: Generating topological structures for surface models. IEEE Comput. Graphics Appl. 15(6), 35–41 (1995)
38. Steinbrenner, J.P., Wyman, N.J., Chawner, J.R.: Procedural cad model edge tolerance negotiation for surface meshing. Engineering with Computers 17, 315–325 (2001)
39. Wright, G.-B., Flyer, N., Yuen, D.A.: A hybrid radial basis function - pseudospectral method for thermal convection in a 3d spherical shell. Geochemistry Geophysics Geosystems 11(7), 1–18 (2010)
40. Wu, X.J., Wang, M.Y.: B Han. An automatic hole-filling algorithm for polygonal meshes. Computer-Aided Design and Applications 5(6), 889–899 (2008)
41. Zheng, Y., Weatherill, N.P., Hassan, O.: Topology abstraction of surface models for three-dimensional grid generation. Engineering with Computers 17, 28–38 (2001)

Automated Two-Dimensional Multi-block Meshing Using the Medial Object

Jeremy Gould[1], David Martineau[1], and Robin Fairey[2]

[1] Aircraft Research Association Ltd, Bedford, UK
 jgould@ara.co.uk
[2] TranscenData Europe Ltd, Cambridge, UK
 rmf2@transcendata.com

Summary. This paper describes the automatic generation of a structured-dominant mesh based on an automatic block decomposition obtained using the medial object. The mesh generator will produce a mesh for an arbitrary two-dimensional geometry, where quadrilateral-dominant or triangular meshing is used when structured meshing fails. The domain is automatically partitioned into blocks (sub-regions), and an appropriate meshing algorithm for meshing each block is then automatically selected. It will also generate conformal interfaces or hanging interfaces between adjacent blocks as required.

Keywords: Medial object, automated block decomposition, mesh generation, structured-dominant mesh, two-dimensional mesh.

1 Introduction

For aerodynamic analysis, it is generally accepted that structured meshes provide greater accuracy and efficiency compared to unstructured meshes. However, considerable manual effort is still required to generate the block decomposition of the flow domain necessary for structured meshing. The work described in this paper automates the generation of a structured-dominant mesh through generation of a block decomposition using the medial object. The mesh generator will produce a structured-dominant mesh for an arbitrary two-dimensional geometry, where quadrilateral-dominant or triangular meshing is used if structured meshing is not possible. The process will automatically partition the domain into blocks (sub-regions), and then select automatically an appropriate meshing algorithm for each block. It will generate conformal interfaces or hanging interfaces between adjacent blocks where required.

The paper provides details of the design of the two-dimensional mesh generation process and is organized as follows. Section 2 describes requirements for the block decomposition of a domain. Section 3 provides an account of how the medial object is used to generate a block decomposition of the flow domain. The meshing process starts with the meshing of block boundary curves, which is described in Section 4. Section 5 describes a solution of the auto-interval assignment problem, which occurs when generating conformal structured meshes in blocks. Section 6 provides an account of the mesh selection strategy used to automatically choose a meshing algorithm for different blocks, and also to automatically place non-matched interfaces between certain blocks. Sections 7 and 8 gives details of the algorithms employed for meshing of block boundaries and the blocks themselves. Finally, Section 9 demonstrates the application of the mesh generator to three representative aerofoil configurations.

2 Block Decomposition Requirements

A block decomposition of the meshing domain is a partition of the domain into non-overlapping regions which cover the whole domain. The overall aim of the block decomposition strategy used for the mesh generator is to decompose the domain into well-shaped regions suitable for structured meshing wherever possible. The requirements are as follows:

1. Blocks should ideally be four-sided and failing this, three-sided.
2. The block decomposition should be conformal. This means that each bounding curve of a block is shared by exactly two blocks unless the curve is part of the geometry or is part of the far-field boundary in which case it belongs to just one block.
3. Also, if a block vertex lies on a bounding curve of the geometry it should be shared by only two quadrilateral blocks, unless the vertex coincides with a discontinuity of the geometry.
4. If a block boundary curve which is not part of the geometry intersects the geometry then it should do so orthogonally.

Ideally, the viscous boundary layer adjacent to the aerodynamic surfaces should be captured within a region of good quality structured mesh. This implies that the block topology should feature quadrilateral blocks adjacent to the aerodynamic surfaces which extend sufficiently far to capture the likely extent of the boundary layer. Additionally the blocking around geometric features, such as trailing edges, should use an appropriate local topology (C-type, O-type or H-type).

3 Block Decomposition Using the Medial Object

The medial object (MO) of a region in two dimensions is defined as the set of centres of all maximal inscribed circles in the region [1,2,3,4]. These are circles contained within the region which are not strictly contained within any other circle inside the region.

Automated Two-Dimensional Multi-block Meshing Using the Medial Object

The block decomposition algorithm uses a shelling based approach and is capable of automatically subdividing a domain into blocks which are mostly four sided, and mostly have angles close to ninety degrees.

Two dimensional shelling is the process of creating an offset profile or shell that is everywhere a fixed distance from the boundary of a trimmed CAD curve. The two dimensional medial object is instrumental in generating shells of this kind.

Building blocks from a shell requires some modifications above and beyond a true 'offset' operation, in which all points on the shell are a fixed distance from the original surface boundary. These include:

- Corner squaring.
- Split generation for most of the shell.
- Additional split generation using the medial axis more directly in narrow gap regions.
- Nested shell construction.

Corner squaring is the principal modification to the shelling process. It creates square corner pieces in place of the arcs that appear at convex corners of the surface being shelled. The motivation for this is to create four-sided blocks, and avoid exclusively O-type topologies (Fig. 1). For two dimensional planar cases, this is a relatively simple process using the tangents of the shell on either side of the corner. Care must be taken however, and the new vertex we are creating to become the fourth corner of the square must be examined to ensure it does not intersect other shell geometry. The two dimensional medial object is again helpful in carrying out this test and determining a better corner placement (Fig. 2).

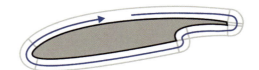

Fig. 1. O-type topology from unmodified shell

Fig. 2. Corner squaring

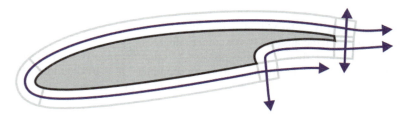

Fig. 3. H-type topology from shell with squared corners

The corner squaring process removes most of the three-sided blocks, and also gives a locally H-type topology (Fig. 3). This technique works for general two-dimensional profiles, and is not specific to aerofoil geometries. Some three sided blocks remain, in highly concave regions or where the shell intersects itself in models with multiple element aerofoils.

Split generation is the splitting of a shell into blocks. A shell is divided into blocks using straight line segments (orthogonal to the geometry if the block is adjacent to the geometry). This is always safe to do, by construction of the two dimensional medial object.

Where a narrow gap occurs, for example between two elements of a multi-element aerofoil, the available space is not wide enough to accommodate two full shell thicknesses, and so the two parts of the shell must join together and be reduced in height. In these regions, the two-dimensional MO will run through the middle of the gap, and sections of the MO can be used directly as block boundary geometry (Fig. 4). This technique also applies when there is detail on the surface boundary that is completely contained within the shell thickness.

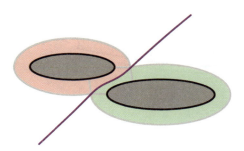

Fig. 4. Medial object used as block boundary

Nested shell construction proceeds by building one layer of shell, before generating a completely new two dimensional medial object on the remaining domain. This is feasible since each medial object computation is relatively cheap.

4 Meshing Block Boundary Curves

The meshing of blocks and block boundaries requires specification of a target mesh density which is provided as input.

As well as the spacing information, refinement of the background spacing is desirable during meshing in order to improve mesh quality. At block boundaries between structured and unstructured zones, there is potential for a large mismatch in spacing, due to the propagation of the curve discretisation through a set of topologically parallel curves. This potential problem is avoided by generating additional mesh sizing information along the boundary of structured zones adjacent to any unstructured zones.

Each boundary curve of a block has an associated marker indicating the nature of that curve. The marker distinguishes between curves belonging to the geometry, curves on the far-field boundary, curves capturing a special flow feature (e.g. a wake plane), or curves in the interior of the domain. These markers are important for meshing because they determine which blocks will be meshed with a geometrical-growth (Navier-Stokes) meshing algorithm.

The first phase of the meshing process involves the generation of an initial discretisation of the curves bounding each block in the input geometry. The initial meshing is required to make subsequent decisions about how to modify the block topology if required. The initial curve meshing is therefore performed independently of any constraints which may be imposed later, due to structured blocks for example. The curves forming the boundaries of each block are meshed initially using either a standard spacing-based discretisation, or a Navier-Stokes algorithm.

The choice between the above two approaches is based on whether the curve is connected to the end of a curve that belongs to the geometry or to a curve belonging to a special flow feature, which is not itself of one of these types. If it is, then it is meshed using the Navier-Stokes algorithm, otherwise the spacing-based method is used.

Viscous meshes are generated directly, as opposed to generating a mesh based on the Euler background spacing field and then obtaining the viscous mesh through a subsequent refinement process.

5 Auto-interval Assignment

Certain meshing algorithms impose constraints on the numbers of nodes in the meshes of the boundary curves of a block. For example, for a four-sided block to have a structured mesh, it is required that the opposite edges are discretised with the same number of nodes. A quadrilateral meshing algorithm requires an even number of intervals around the whole boundary mesh of the block. The problem of auto-interval assignment is the problem of simultaneously satisfying the constraints for each block so that a conformal mesh can be generated for the whole domain.

An approach to solving the interval assignment problem for structured meshing was proposed by Mitchell in [5]. In Mitchell's paper the auto-interval assignment problem is stated as a mixed-integer linear program. This method involves tackling the determination of all the intervals simultaneously through the definition of constraints, which depend on the meshing algorithm to be applied to each block. However, only the constraint for structured blocks is applicable in this work, and although the linear programming approach is still applicable, a simpler and more robust approach was chosen. The technique used here is based on the notion of curve clan introduced in [6], and provides a way of obtaining solutions to the interval assignment problem for structured meshing based on the spacing specified for the domain.

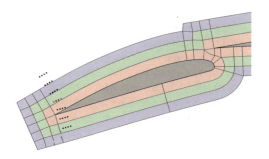

Fig. 5. A curve clan.

A *curve clan* is a set of topologically parallel boundary curves of four-sided blocks in a block decomposition of a domain. The term topologically parallel means that any two curves in a clan are related by a sequence of opposite edges of neighbouring four-sided blocks. The definition is extended in this work to include three-sided blocks by identifying a pair of opposite curves which share the same boundary condition and have the closest match in terms of target mesh density based on the background spacing. If three-sided blocks are allowed, then the determination of curve clans for a given block decomposition is not entirely topological, but depends also on the spacing. In Fig. 5, the set of dashed curves form a curve clan.

Collecting together all such curves into a curve clan, and then meshing all curves in the clan together, will ensure that the requirement for structured meshes mentioned above can be met. Furthermore, curve clans can be meshed independently of each other, and therefore curve meshing is parallelizable at the clan level.

Any discretisation of the curves in the curve clan would be a solution to the interval assignment problem described earlier, as long as all the curves in the clan share this discretisation. For the purposes of this prototype, the discretisation is selected from the curve in the clan with the most nodes in its curve mesh, based on the initial meshing of the curves using either spacing-based or Navier-Stokes mode, as described above. This curve will be called the *clan leader* of the curve clan. The remaining curves in the clan are then meshed by mapping the mesh of the clan leader on to each of the other curves in the clan.

The idea of *curve dependency* is used to facilitate the meshing of one curve (*dependent curve*) by copying the parameterisation of the mesh of another curve (*master curve*). In the case of curve clans, the clan leader will be the master curve for all the other curves in the clan.

6 Mesh Algorithm Selection and Non-conformal Interfaces

The strategy used to select a suitable meshing algorithm for each block in the decomposition of the domain makes use of curve clans and curve dependencies.

A key concept introduced in the new mesh generator is that of a hierarchy of mesh generation algorithms. The aim of this is to provide the best quality meshes without sacrificing robustness.

One of the main aims of the prototype mesh generator is to generate structured-dominant meshes, in order to exploit the improved accuracy and potential efficiency benefits of structured meshes in the flow solver. Hence, structured meshing is the first algorithm in the meshing algorithm hierarchy. A necessary condition that a block must satisfy for it to be given a structured mesh is that it should have three or four bounding curves. Thus blocks with five sides or more are not meshed with a structured meshing algorithm.

The hierarchical meshing approach works by attempting to mesh each block with the "best" meshing algorithm possible, subject to certain constraints. In the two-dimensional meshing prototype, these constraints are purely topological; however the constraints could well include geometric considerations. If structured meshing fails for a block, then an attempt is made to mesh the block using the unstructured quadrilateral-dominant meshing algorithm. If this fails, then triangular meshing will be used.

Experience of this approach in the current work has shown that success criteria need to be given more consideration. In the current implementation, surface mesh quality metrics, including self-intersection checks, are only conducted after all the surface meshes have been generated. It is also clear from the testing that sometimes, a good quality unstructured quad-dominant mesh would be preferable to a poor quality structured mesh, and that the strictly hierarchical meshing algorithm selection described above is not necessarily optimal. This highlights the need for quantifiable definitions of acceptable mesh quality metrics to enable automatic selection of the most appropriate algorithm. However, the approach has successfully demonstrated the improved robustness through the ability to resort to alternative meshing algorithms in the event of meshing failure.

If structured meshing is used for the majority of the blocks in the flow domain, with conformal interfaces between blocks, it can give rise to the problem of high density structured mesh being propagated out to regions with a low target mesh density. This problem is typical in conventional multi-block structured meshing tools. However, in the prototype mesh generator, two possible remedies for this situation are allowed. The first is to select an unstructured meshing algorithm for a particular block. This effectively splits into two the curve clans running in both directions through the block, allowing them to be meshed with different discretisations. The second approach is to introduce a non-conformal interface in the block topology. Again, this splits a curve clan into two separate clans, but only in one direction. The non-conformal interface is implemented by replacing a single curve in the underlying geometry with two topologically distinct, but physically coincident curves.

Fig. 6. (a) A hanging interface, (b) a non-matched interface.

There are two types of non-conformal interface used in the block topology. The first type is a *hanging interface* where one curve is a refinement or coarsening of the other (Fig. 6a). This introduces a new type of dependency between the curves. The second type is a *non-matched* interface (Fig. 6b) where the two coincident curves can have completely independent discretisations, and have no dependency on each other.

A hanging or non-matched interface can be used to generate structured mesh which conforms more closely to the target background spacing, but the mesh will of course, no longer be conformal.

The automatic block meshing algorithm selection strategy initially assumes that all three and four-sided blocks will be meshed using a structured algorithm, and all other blocks will be meshed using an unstructured algorithm. Block boundary curves are selected to be hanging or non-matched interfaces or structured/unstructured interfaces by analyzing curve clans. The idea is to compare spacing along the different curves in a curve clan. A curve is chosen to be a non-conformal interface when there is a large decrease in spacing between that curve and the next or previous curve in the clan. This is implemented as follows. Recall that all the boundary curves of the block decomposition are given an initial mesh. These meshes are used to calculate the ratio of the number of nodes in the mesh of each clan curve to the number of nodes in the mesh of the clan leader. A sequence of Booleans is generated for the curve clan by comparing the sequence of ratios to a configurable threshold value. A ratio above the threshold is mapped to true (T), otherwise it is mapped to false (F). Setting every curve with a false value to be a hanging interface may lead to too many hanging interfaces. Therefore, a form of smoothing is applied, which replaces the patterns of the form TFT and FTF in the sequence of true/false values associated with a clan by the patterns TTT and FFF, respectively. After the smoothing operation interface curves are selected by searching for patterns of the form FTT, TTF, FFT and TFF. If either pattern is found, then the interface curve will be the curve in the clan corresponding to the middle symbol. Only one split is allowed in each clan.

Once the decisions have been made about where to introduce unstructured blocks or non-conformal interfaces, the curve clans are re-determined for the modified topology. The above process is then repeated for the new set of curve clans, and more interfaces introduced if necessary. This cycle is repeated until no new non-conformal interfaces or unstructured blocks are deemed necessary.

7 Block Interface Meshing

As mentioned above, block boundaries are initially meshed using either standard spacing-based discretisation, or an advancing-layer style discretisation. The advancing-layer curve meshing algorithm is based on the following parameters: first cell height, number of linear layers and growth rate. The algorithm uses a geometric expansion of the projected edge length which terminates when the projected edge length matches the background spacing. The remainder of the curve is then meshed using the conventional spacing-based discretisation technique.

Following the initial curve meshing, the curves are re-meshed based on the constraints imposed by the blocking topology and the choice of meshing algorithms and block interface types. The re-meshing of the curves is based entirely upon the curve dependencies. The re-meshing involves an iterative loop over the curves, only meshing a curve if it either has no dependency, or if its *master* curve (i.e. the curve upon which it is dependant) has already been re-meshed. The curve dependency can take one of two forms: *direct* or *hanging*.

In the case of a direct dependency, the parametric distribution is copied from the master curve. A small amount of Laplacian smoothing is applied to the curve distribution as this was seen to improve the mesh quality in certain situations, for example, smoothing out the Navier-Stokes refinement away from the aerofoil geometry.

At a hanging interface, two possibilities exist for generation of the dependent curve mesh: refinement of the master mesh or coarsening of the master mesh. Refinement of a master mesh is the simplest operation, as it simply involves subdividing each segment of the master mesh into a specified number of edges. Coarsening of a master mesh is slightly more involved. The coarse mesh is generated through sub-sampling of the master mesh using a specified coarsening factor, however, the master mesh nodes must then be projected onto the segments of the coarse mesh in order to ensure conservation of volume at the interface and avoid creation of gaps or overlaps in the mesh.

Experience of applying hanging interfaces to realistic aerofoil geometries revealed that a fixed refinement or coarsening factor is not necessarily desirable, since the background spacing can vary considerably along the block boundary. To account for this, a modified algorithm was implemented for curve coarsening based on a variable coarsening factor. This modified algorithm uses a rather different approach to the curve re-meshing. The curve is first meshed using the standard sourcing-based approach, and then the nodes are snapped to the closest nodes in the master mesh. As with the fixed-ratio coarsening, the master mesh nodes must then be projected onto the segments of the coarse mesh.

Initially, meshing of non-matched curve interfaces might seem trivial, since the two curves are simply meshed independently from one another. However, it is essential that volume is conserved at the interface, and that there are no gaps or overlaps in the mesh. For the implementation of the prototype, non-matched interfaces were therefore only allowed at straight block boundaries.

8 Block Meshing

The main algorithm used to generate structured meshes is linear transfinite interpolation. The method extends to triangular blocks, by collapsing one of the boundary edges of the block.

For a rectangular block, linear transfinite interpolation will give a perfectly rectilinear surface mesh. However, as the shape of the block deviates from being rectangular the quality of the mesh will deteriorate. For such cases, the mesh lines attached to the aerofoil will not be orthogonal to the aerofoil surface. For particularly badly shaped blocks, linear transfinite interpolation may not be one-to-one which will cause 'folds' in the surface mesh.

For this prototype meshing capability, an alternative algorithm was implemented for structured meshing of "viscous" blocks. The alternative exploits the fact that, by construction, the blocks adjacent to the surface of the aerofoil geometry will have straight edges in the direction perpendicular to the aerofoil surface. This allows the structured mesh generation to be reduced to a linear interpolation of the distributions along the perpendicular block edges.

Hermite transfinite interpolation [7] was also used. This technique is able to generate a mesh for a non-rectangular block which has mesh curves orthogonal to the boundary curves of the block. This method was modified to allow normal vectors to be specified only along a wall-type boundary curve, where orthogonality of the mesh is most critical.

9 Results and Discussion

The mesh generator has been demonstrated on three test cases: a clean wing case, a high-lift aerofoil, and a clean aerofoil with a leading edge ice shape. Viscous meshes suitable for RANS simulation of nominally attached flows have been generated for all cases.

The first test case corresponds to the computation of flow around a single element two-dimensional aerofoil section at the transonic cruise condition. The test case used is the RAE 2822 aerofoil section, which exhibits regions of concave and convex geometry. The test case has a sharp (discontinuous) trailing edge. The block topology and mesh generated for the RAE 2822 aerofoil are shown in Fig. 7a and Fig. 7b, respectively, based on six layers of shelling. The block topology and mesh demonstrate the desirable C-topology around the aerofoil resulting from the corner squaring process.

A series of meshes were generated for the RAE 2822 configuration using various options to accommodate changes in target mesh density:

a. Default mesh generated with structured blocks wherever possible and only conformal interfaces
b. Structured-dominant mesh with hanging interfaces of fixed ratio.
c. Structured-dominant mesh with hanging interfaces of variable ratio.
d. Structured-dominant mesh with non-matched interfaces.
e. Completely conformal mesh generated using largely structured meshing, but allowing use of unstructured blocks to accommodate changes in density.

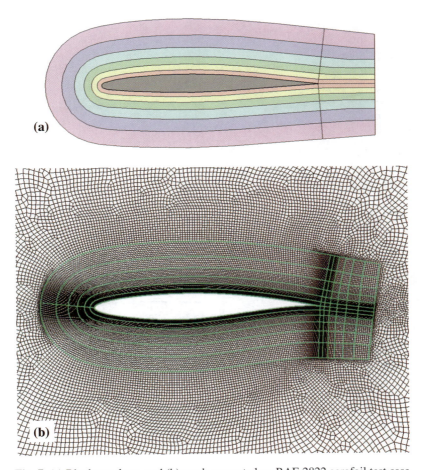

Fig. 7. (a) Block topology, and (b) mesh generated on RAE 2822 aerofoil test case.

The different meshes corresponding to these alternative strategies are shown in Fig. 8 (a) to (e) and illustrate how the variation in target mesh density can be handled in a variety of ways.

The second test case is the L1T2 geometry, a multi-element high-lift configuration at take-off or landing. This system involves three elements, a leading-edge slat, main element and trailing-edge flap. Fig. 9 shows the detail of the medial object computed on the L1T2 test case, which includes segments passing through the gaps between the elements, and also segments which approach the cove region of the slat and main element.

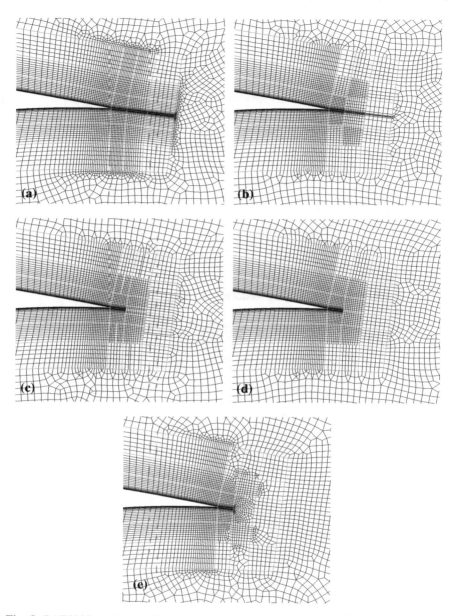

Fig. 8. RAE2822 mesh at trailing edge of aerofoil with (a) no interfaces, (b) fixed ratio hanging interfaces, (c) variable ratio hanging interfaces, (d) non-matched interfaces, and (e) unstructured blocks.

The block topology for this case is shown in Fig. 10, illustrating how the medial object is used to form part of the block boundaries in gap regions. Various details of the structured-dominant mesh generated for the L1T2 test case are shown in Fig. 11. The effect of using only conformal interfaces in this mesh can be seen in the propagation of the refined mesh downstream from each of the trailing edges of the three elements.

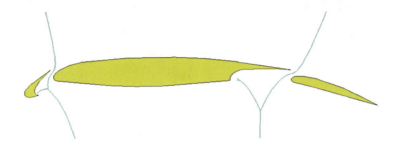

Fig. 9. Detail of the medial object close to the L1T2 geometry

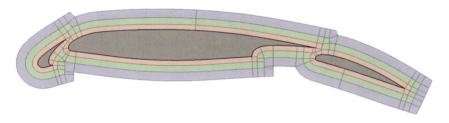

Fig. 10. Block topology generated on the L1T2 test case

The final test case corresponds to a clean aerofoil configuration with leading edge ice shape. The test case used is the C6 configuration from the NATO-RTO Workshop on Ice Accretion Simulation in 2000 at CIRA, Italy [8]. The medial object and block topology generated for the C6 icing case are shown in Fig. 12. The shelling offset is relatively small compared with the chord length, in order to capture the detail of the ice shape.

Detail of the mesh on the C6 icing test case is shown in Fig. 13a and 13b. In this case triangular meshing was used to generate the mesh in the far-field region of the flow domain as a result of a failure in the quad-dominant advancing-front algorithm, demonstrating the hierarchical meshing approach. The detail of the mesh in Fig. 13b demonstrates the potential of this automatic blocking and mesh generation process to handle viscous mesh generation around complex geometric features.

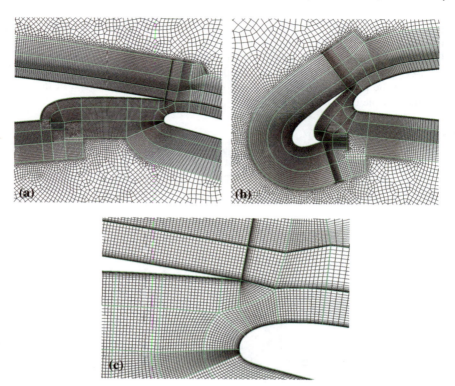

Fig. 11. Detail of the mesh on the L1T2 test case at (a) region around the slat, (b) cove region of main element, and (c) gap between main element and flap.

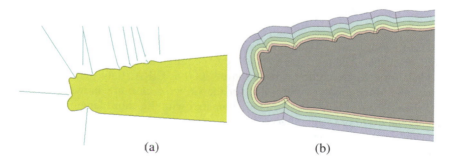

Fig. 12. Detail of the (a) medial object, and (b) block topology at the leading edge of the C6 test case.

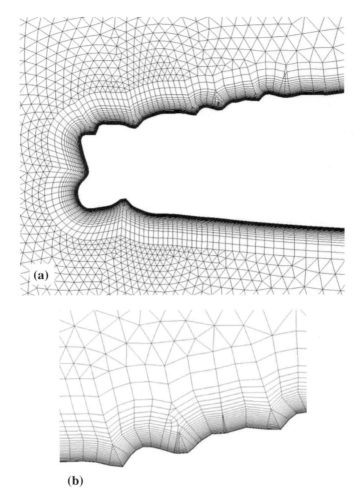

Fig. 13. Mesh on the C6 icing test case around (a) the leading edge, and (b) a detail of the ice shape.

10 Conclusions

A new prototype mesh generator has been presented which will produce a structured-dominant mesh for an arbitrary two-dimensional geometry. The process will automatically partition the domain into sub-regions or blocks using the medial object. An appropriate meshing algorithm is automatically selected to generate each block mesh. Quadrilateral-dominant or triangular meshing is used if structured meshing is not possible. The process will generate conformal interfaces or hanging interfaces between adjacent blocks where required.

Acknowledgement. The work reported here has been carried out under the ANSD project jointly funded by Airbus, ARA, TranscenData and the UK Technology Strategy Board. The medial object computation and automatic blocking technology was developed in the context of TranscenData Europe Ltd's CADfix product [9].

References

1. Blum, H.A.: Transformation for Extracting New Descriptors of Shape. In: Models for the Perception of Speech and Visual Form (1967)
2. Sheehy, D.S., Armstrong, C.G., Robinson, D.J.: Computing the medial surface of a solid from a domain Delaunay triangulation. In: Proc. ACM/IEEE Symposium on Solid Modeling and Applications, Salt Lake City (1995)
3. Price, M., Stops, C., Butlin, G.: A Medial Object Toolkit for Meshing and Other Applications. In: Proceedings of the 4th th International Meshing Roundtable (1995)
4. Robinson, T., Fairey, R., Armstrong, C., Ou, H., Butlin, G.: Automated Mixed Dimensional Modelling with the Medial Object. In: Proceedings of the 17th International Meshing Roundtable (2008)
5. Mitchell, S.A.: High fidelity interval assignment. In: Proceedings of the 6th International Meshing Roundtable (1997)
6. Beatty, K., Mukherjee, N.: A Transfinite Meshing Approach for Body-In-White Analyses. In: Proceedings of the of the 19th International Meshing Roundtable (2010)
7. Faux, I.D., Pratt, M.J.: Computational Geometry for Design and Manufacture. Ellis Horwood, Chichester (1979)
8. Ice Accretion Simulation Evaluation Test. Report of the Applied Vehicle Technology Panel (AVT) Task Group AVT-006, RTO Technical Report 38 (2001)
9. http://www.transcendata.com/cadfix.htm

QUAD MESHING
(Tuesday Afternoon Parallel Session)

A Frontal Delaunay Quad Mesh Generator Using the L^∞ Norm

J.-F. Remacle[2], F. Henrotte[2], T. Carrier Baudouin[2], C. Geuzaine[1],
E. Béchet[3], Thibaud Mouton[3], and E. Marchandise[2]

[1] Université de Liège, Department of Electrical Engineering and Computer Science, Montefiore Institute B28, Grande Traverse 10, 4000 Liège, Belgium
cgeuzaine@ulg.ac.be
[2] Institute of Mechanics, Materials and Civil Engineering, Universit catholique de Louvain, Avenue Georges-Lemaître 4, 1348 Louvain-la-Neuve, Belgium
{jean-francois.remacle,emilie.marchandise,
tristan.carrier,francois.henrotte}@uclouvain.be
[3] Université de Liège, Aerospace and Mechanical Engineering Department, Chemin des Chevreuils, 1,4000 Liège, Belgium
{eric.bechet,thibaud.mouton}@ulg.ac.be

Summary. A new indirect way of producing all-quad meshes is presented. The method takes advantage of a well known algorithm of the graph theory, namely the Blossom algorithm that computes the minimum cost perfect matching in a graph in polynomial time. Then, the triangulation itself is taylored with the aim of producing right triangles in the domain. This is done using the infinity norm to compute distances in the meshing process. The alignement of the triangles is controlled by a cross field that is defined on the domain. Meshes constructed this way have their points aligned with the cross field direction and their triangles are almost right everywhere. Then, recombination with our Blossom-based approach yields quadrilateral meshes of excellent quality.

Keywords: Quadrilateral mesh generation, graph theory, infinity norm.

1 Introduction

This paper describes a new methodology for generating meshes with quadrilateral elements (quad meshes). There exist so far essentially two approaches to generate automatically such meshes. With *direct methods*, quadrilaterals are constructed at once, using either advancing front techniques [1] or regular grid-based methods (quadtrees) [2]. *Indirect methods*, on the other hand, rely on an initial triangular mesh and apply merging techniques to recombine the triangles of the initial mesh into quadrangles [3, 4]. Other more sophisticated indirect methods use a mix of advancing front and recombination [5].

In order to motivate our work, let us first explain why standard indirect quadrilateralization methods fail to produce optimal quad meshes. Figure 1-(a) shows a uniform triangular mesh in R^2 with equilateral elements; all elements and all edges are of size a. This mesh can be deemed perfect in the sense that optimality criteria, both in size and shape, are fulfilled. In this case, the Voronoi cell of each vertex $\mathbf{x}$ is an hexagon of area $a^2\sqrt{3}/2$, and the number of points per unit of surface is $2/(a^2\sqrt{3})$. Comparing with a uniform mesh made of right triangles of size a, Figure 1-(b), one sees that the Voronoi cells are now squares of area a^2. Filling R^2 with equilateral triangles requires thus $2/\sqrt{3}$ times more vertices (i.e. about 15% more) than filling the same space with right triangles. So, although quad meshes can be obtained by recombination of any triangular meshes, conventional triangular meshes are not the most appropriate starting point because they are essentially made of (nearly) equilateral triangles and contain therefore about 15% too many vertices. The purpose of this paper is to introduce a method to generate triangular meshes suited for recombination into well-behaved quad meshes.

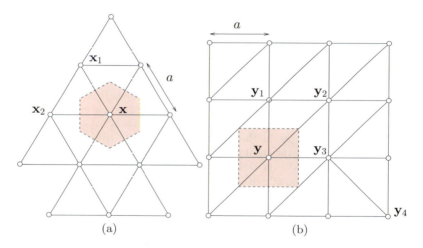

Fig. 1. Voronoi cells of one vertex that belongs either to mesh of a equilateral triangles (a) or of right triangles (b).

The mesh in Figure 1-(b) contains edges of different sizes. For example, $\|\mathbf{y} - \mathbf{y}_2\|_2 = a\sqrt{2}$ whereas $\|\mathbf{y} - \mathbf{y}_1\|_2 = a$. This mesh contains long edges at 45 degrees and short edges aligned with the axis. This explains why mesh (b) contains less points than mesh (a). Yet, long edges will be eliminated by the recombination procedure and the final mesh will be made of quadrilaterals with all edges of size a.

In devising a procedure to generate triangular meshes well-suited for recombination into quadrangles, one should recognize that the optimal size of an edge to be inserted at a point by the Delaunay algorithm depends on its orientation. A first possibility would be to encapsulate this directional information into some kind of anisotropic L^2 metric. This is however not possible as such a metric $\mathcal{M}$ would have to ensure that (See Figure 1-(b))

$$\begin{aligned}(\mathbf{y}_1-\mathbf{y}_2)^T\mathcal{M}(\mathbf{y}_1-\mathbf{y}_2) &= (\mathbf{y}_3-\mathbf{y}_2)^T\mathcal{M}(\mathbf{y}_3-\mathbf{y}_2)\\ &= (\mathbf{y}_3-\mathbf{y}_4)^T\mathcal{M}(\mathbf{y}_3-\mathbf{y}_4)\\ &= (\mathbf{y}-\mathbf{y}_2)^T\mathcal{M}(\mathbf{y}-\mathbf{y}_2),\end{aligned}$$

which is clearly impossible. This suffices to conclude that standard metric-based triangulation algorithms are unable to produce meshes made exclusively of right triangles.

The approach proposed in this paper is based on the following observation. If distances between points are measured in the L^∞-norm, the triangular elements of Figure 1-(a) are no longer equilateral: $\|\mathbf{x}-\mathbf{x}_2\|_\infty = a$ and $\|\mathbf{x}-\mathbf{x}_1\|_\infty = a\sqrt{3}/2$. On the other hand, the elements of Figure 1-(b), which are right triangles in the L^2 norm, are equilateral in the L^∞-norm: $\|\mathbf{y}-\mathbf{y}_1\|_\infty = \|\mathbf{y}-\mathbf{y}_2\|_\infty = a$.

On this basis, the frontal Delaunay algorithm can be adapted to work in the L^∞-norm so as to produce triangular meshes with the right number of nodes and triangles suitably shaped for producing high quality quadrilaterals after recombination.

The paper is divided in three parts. In the first part, we make use of a famous algorithm of the theory of graphs: the Blossom algorithm, proposed by Edmonds in 1965 [6, 7], which allows to find the minimal cost perfect matching of a given graph. While classical triangle merge procedures [3, 4] are based on some kind of heuristics that allow to find which pairs of triangles forming good quadrangles after recombination, the new method has some clear advantages: (i) it provides a mesh that is guaranteed to be quadrilateral only, (ii) it is optimal in a certain way and (iii) it is fast.

Then, a method to build the so-called "cross fields" [8] is presented. The cross fields represent at each point of the domain the preferred orientations of the quadrilateral mesh. In the finite element community, it is usually appreciated that quadrilateral elements have orientations parallel to the domain boundaries.

Finally, the mesh generation procedure is described in detail. A frontal Delaunay approach inspired by [9] is proposed for determining the successive position of new points. Frontal meshers usually insert a point in the mesh so as to form an equilateral triangle (in L^2-norm). Here, we also aim at generating an equilateral triangle, yet in the sense of the local L^∞-norm aligned with the cross field. Meshes constructed this way have their points aligned with the cross field direction and their triangles are almost right everywhere. Then, recombination with our Blossom-quad approach [10] yields quadrilateral meshes of excellent quality.

In all meshes that are presented as results, the quality of the quadrangular meshes are evaluated by computing the quality η of every quadrangle as follows:

$$\eta = \max\left(1 - \frac{2}{\pi}\max_k\left(\left|\frac{\pi}{2} - \alpha_k\right|\right), 0\right), \quad (1)$$

where $\alpha_k, k = 1, .., 4$ are the four angles of the quadrilateral. This quality measure is $\eta = 1$ if the element is a perfect quadrilateral and is $\eta = 0$ if one of those angles is either ≤ 0 or $\geq \pi$. The average quality of elements is noted $\bar{\eta}$ and the worst element quality is noted η_w.

2 Triangle Merging Using the Blossom Algorithm

The idea of the Blossom algorithms is to build a specific weighted graph $G(V, E, c)$ from the triangle adjacencies in a given mesh $\mathcal{T}_0$. Fig. 2 shows a simple triangular mesh with its graph G. Here, V is the set of graph vertices, E the set of graph edges and $c(E)$ an graph-edge-based cost function. As can be seen every vertex of the graph is a triangle t_i of the mesh and every edge of the graph is an internal edge e_{ij} of the mesh that connects two neighboring triangles t_i and t_j. We aim at finding a subset of edges that forms a perfect matching, i.e. that makes pairs of triangles, leaving no triangle alone.

2.1 Blossom: A Minimum Cost Perfect Matching Algorithm

Let us consider $G(V, E, c)$ now such an undirected weighted graph. A *matching* is a subset $E' \subseteq E$ such that each node of V has at most one incident edge in E'. A matching is said to be perfect if each node of V has exactly one incident edge in E'. As a consequence, a perfect matching contains exactly $|E'| = |V|/2$ edges. This means that a perfect matching can only be found for graphs with an even number of vertices. A matching is optimum if $c(E')$ is minimum among all possible perfect matchings.

In 1965, Edmonds [11, 6] invented the *Blossom algorithm* that solves the problem of optimum perfect matching in polynomial time. A straightforward implementation of Edmonds's algorithm requires $\mathcal{O}(|V|^2|E|)$ operations. Since then, the worst-case complexity of the Blossom algorithm has been steadily improving. Both Lawler [12] and Gabow [13] achieved a running time of $\mathcal{O}(|V|^3)$, Galil, Micali and Gabow [14] improved it to $\mathcal{O}(|V||E|\log(|V|))$. The current best known result in terms of $|V|$ and $|E|$ is $\mathcal{O}(|V|(|E| + \log|V|))$ [15].

There is also a long history of computer implementations of the Blossom algorithm, starting with the Blossom I code of Edmonds, Johnson and Lockhart [7]. In this paper, our implementation makes use of the Blossom IV code of Cook and Rohe [16][1] that has been considered for several years as the fastest available implementation of the Blossom algorithm.

---

[1] Computer code available at http://www2.isye.gatech.edu/ wcook/blossom4.

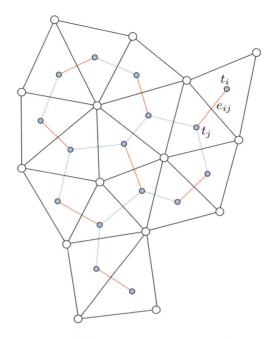

Fig. 2. A mesh (in black) and its graph (in cyan and red). The cyan points are the V graph vertices, the cyan and red lines the set of graph edges E and the subset of red edges forms a perfect matching.

2.2 Optimal Triangle Merging

In term of what has just been defined, the subset E' of edges that have been used for triangle merging in the approach of [4] is a matching that is very rarely a perfect matching. The one of [3] is usually a perfect matching, but not necessarily the optimal one.

Here, we propose a new indirect approach to quadrilateral meshing that takes advantage of the Blossom algorithm of Edmonds. To this end we apply the Blossom IV algorithm to the graph of the mesh. We intend to find the optimum perfect matching with respect to the following total cost function

$$c = \sum_{e \in E'} (1 - \eta(q_{ij})), \qquad (2)$$

that is, the sum of all elementary cost functions (or "badnesses") q_{ij} of the quadrilaterals that result in the merging of the edges of the perfect matching E'. Such a cost function can be related to any mesh quality measure [17].

An obvious requirement for the final mesh to be quadrilateral only is that the initial triangular mesh contains an even number of triangles (i.e., an even number of graph vertices). Euler's formula for planar triangulations states that the number of triangles in the mesh is

$$n_t = 2(n_v - 1) - n_v^b, \qquad (3)$$

where n_v^b is the number of mesh nodes on its boundary. So, the number of mesh points on the boundary n_v^b should be even.

If for some graphs it is possible to find different perfect matchings, there is in general no guarantee that even one single perfect matching exists in a given graph. The general problem of counting the number of perfect matchings in a general graph is #P-complete[2]. In other words, there is no hope to find the number of perfect matchings in a general graph (however, there is a way to find out, in polynomial time, wether a perfect matching exists by detecting a breakdown in the Blossom algorithm). More details about Blossom, matchings and graphs can be found in [10], where we describe a way to ensure that the graph of the mesh always has a perfect matching.

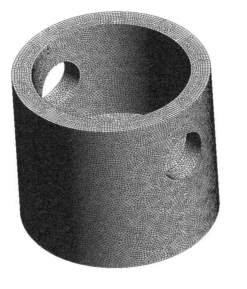

Fig. 3. Mesh of the Piston created from a combination of triangle using the Blossom-quad algorithm.

Fig. 3 shows an example of the use of the Blossom algorithm for recombining the triangles together with the optimization procedure that is described in [10]. The mesh is composed of 39,386 quadrilateral elements that are of average quality $\bar{\eta} = 0.78$, which is good but not great. It is indeed clear that elements have no preferred orientation. The mesh is composed of patches of

---

[2] Sharp p-complete, i.e. much harder than NP-complete.

quadrilaterals that have random orientations. In the next section we will develop a new method for inserting points in the domain in a way such that Blossom will be able to provide an optimal quad-mesh that is oriented correctly.

3 Cross Fields

In what follows, we consider that we have a first triangulation of a surface $\mathcal{T}_0$ in $\mathcal{R}^3$ and that we have build a parametrization $\mathbf{x}(\mathbf{u})$ that maps every point of the 3D surface mesh to a point in a parametric space in $\mathcal{R}^2$. The parametrized surface mesh is denotes $\mathcal{T}_0'$. Moreover, we are able to compute the derivatives of the mapping in order to build the mesh metric :

$$\mathbf{M} = \mathbf{x}_{,\mathbf{u}}^T \mathbf{x}_{,\mathbf{u}} = \begin{bmatrix} \mathbf{x}_{,u} \cdot \mathbf{x}_{,u} & \mathbf{x}_{,u} \cdot \mathbf{x}_{,v} \\ \mathbf{x}_{,v} \cdot \mathbf{x}_{,u} & \mathbf{x}_{,v} \cdot \mathbf{x}_{,v} \end{bmatrix}. \tag{4}$$

A cross field $\theta(\mathbf{u})$ is supposed to give enough information to orient a local system of axis at each point $\mathbf{u}$ in the parameter plane. The edges of the quadrilaterals generated around $\mathbf{u}$ should then be aligned with the cross field. Figure 4 shows an annular domain, the cross field and the resulting quad mesh. The computer graphics community has already been confronted to the issue of computing "cross fields" in the context of (global) surface parametrization [18, 19]. Cross fields can be based on principal directions of curvature of the surface [8].

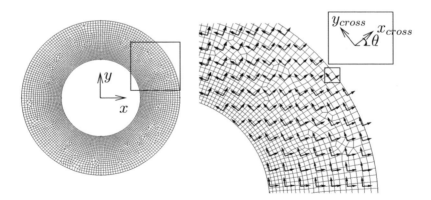

Fig. 4. Mesh of annular domain and a zoom on the cross field.

Here, we consider an *ad hoc* approach based on the following criteria:

- The cross field should be computed automatically;
- Mesh directions should be parallel to the boundaries of the domain at the vicinity of those boundaries;
- The cross field should be smooth.

In order to fulfill those constraints, we have chosen to compute θ using a boundary value problem. The value of θ is fixed at the boundary of the domain and is propagated inside the domain using an elliptic PDE.

The angular oriention of a cross being defined up to the multiples of $\pi/2$, it cannot be represented univocally by the orientation of one branch of the cross. The complex valued function

$$\alpha(\mathbf{u}) = a(\mathbf{u}) + ib(\mathbf{u}) = e^{4i\theta(\mathbf{u})}$$

however offers an univocal representation, as it takes one same value for the directions of the 4 branches of a local cross.

A first triangular mesh $\mathcal{T}_0$ is generated using any available algorithm. If a parametrization of the surface needs to be computed, then we use the same mesh as the one that has been used for computing the parametrization. Two Laplace equations with Dirichlet boundary conditions are then solved for each surface of the mesh in order to compute the real part $a(\mathbf{u}) = \cos 4\theta$ and the imaginary part $b(\mathbf{u}) = \sin 4\theta$ of α:

$$\begin{aligned} \nabla^2 a = 0, \quad \nabla^2 b = 0 \quad &\text{on } \mathcal{T}_0', \\ a = \cos(4\theta_b), \quad b = \sin(4\theta_b) \quad &\text{on } \partial\mathcal{T}_0', \end{aligned} \quad (5)$$

where θ_b is the angle between the normal to the boundary and the coordinate axis. Then, we have to supply the boundary conditions $\bar{a}(\mathbf{u})$ and $\bar{b}(\mathbf{u})$ ensuring that θ is aligned with $\partial \mathcal{S}'$. After solving, the cross field is represented by

$$\theta(\mathbf{u}) = \frac{1}{4}\text{atan2}(b(\mathbf{u}), a(\mathbf{u})),$$

where $\text{atan2}(b, a)$ is the 4-quadrant inverse tangent.

4 Triangulation in the L^∞-Norm

In this section, standard geometrical notions usually defined in the L^2-norm are extended in the L^∞ norm.

4.1 Distances and Norms

In the R^2 plane, the distance between two points $\mathbf{x}_1(x_1, y_1)$ and $\mathbf{x}_2(x_2, y_2)$ is usually based on the Euclidean norm (L^2-norm). Other distances can be defined however, based on other norms:

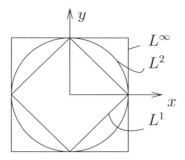

Fig. 5. Illustration of the unit circle in different norms $\|\mathbf{x}\|_p$

- The L^1-norm distance $\|\mathbf{x}_2 - \mathbf{x}_1\|_1 = |x_2 - x_1| + |y_2 - y_1|$,
- The L^2-norm distance $\|\mathbf{x}_2 - \mathbf{x}_1\|_2 = (|x_2 - x_1|^2 + |y_2 - y_1|^2)^{1/2}$,
- The L^p-norm distance $\|\mathbf{x}_2 - \mathbf{x}_1\|_p = (|x_2 - x_1|^p + |y_2 - y_1|^p)^{1/p}$,
- The L^∞-norm distance $\|\mathbf{x}_2 - \mathbf{x}_1\|_\infty = \lim_{p \to \infty} \|\mathbf{x}_2 - \mathbf{x}_1\|_p$
  $= \max(|x_2 - x_1|, |y_2 - y_1|)$.

Figure 5 shows the unit circle in different norms. One important thing to remark is that only the L^2-norm is rotation invariant. The L^∞-norm depends this on the local orientation of the coordinate axes.

In order to simplify the notations, we consider in what follows that (x, y) are the local coordinates aligned with the cross field (x_{cross}, y_{cross}) (see for example the cross field in Fig. 4).

4.2 Bisectors in the L^∞-Norm

The perpendicular bisector, or bisector of the segment delimited by the points $\mathbf{x}_1 = (-x_p, -y_p)$ and $\mathbf{x}_2 = (x_p, y_p)$ is by definition the set of points $\mathbf{x} = (x, y)$ equidistant to $\mathbf{x}_1$ and $\mathbf{x}_2$. In the L^2-norm, it is the union of all 2 by 2 intersections of circles centered at $\mathbf{x}_1$ and $\mathbf{x}_2$ and having the same radius. Those intersections are each times two points and their union form a straigth line. In the L^∞-norm, the circles have the geometric appearance of squares and their intersections 2 by 2 are either 2 points or a segment. The bisector is then a broken line (see Fig. 6) in general but it can also form a diabolo-shaped region whenever the considered segment is aligned with an axis ($x_1 = x_2$ or $y_1 = y_2$).

It is assumed, without loss of generality, that $x_p \geq y_p$. The bisector of the segment in the L^∞-norm is the set

$$\mathcal{L} = \{\mathbf{x} = (x, y), \max(|x - x_p|, |y - y_p|) = \max(|x + x_p|, |y + y_p|)\}.$$

The vertical segment of Figure 6 that passes through the origin $(0,0)$ is the intersection of the L^∞-circles of L^∞-radius x_p centered at $\mathbf{x}_1$ and $\mathbf{x}_2$.

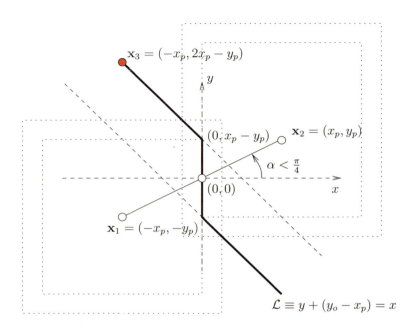

Fig. 6. Bisector of two points $x_1 = (-x_p, -y_p)$ and $x_2 = (x_p, y_p)$ using the L^∞-norm. The dottes squares are the L_∞ circles.

It belongs thus to the bisector. Increasing now the radius progressively, the intersection of the two L^∞-circles is a pair of points forming two half lines oriented at $3\pi/4$ and starting at $(0, x_p - y_p)$ and $(0, -x_p + y_p)$ respectively. The equation of the bisector $\mathcal{L}$ is then:

$$\mathcal{L} \equiv y + (y_p - x_p) = x \tag{6}$$

There exists an ambiguity when $y_p = 0$. In this case, the bisector contains 2D regions of the plane. It is assumed in what follows that points are in general position, i.e. that there does not exist two points that share either the same x or y coordinate.

4.3 The Equilateral Triangle Using the L^∞-Norm

A triangle $T(x_1, x_2, x_3)$ is equilateral in the L^∞-norm if

$$\|x_2 - x_1\|_\infty = \|x_3 - x_1\|_\infty = \|x_3 - x_2\|_\infty.$$

It is possible to build such a triangle starting from Figure 6. We take again $x_1 = (-x_p, -y_p)$ and $x_2 = (x_p, y_p)$ and look for a point on the bisector of $x_1 x_2$ located at a distance $2x_p$ from the endpoints of the segment. This point is $x_3 = (-x_p, 2x_p - y_p)$.

4.4 Circumcenter, Circumradius and Circumsquare in the L^∞-Norm

Consider a triangle $T_i(\mathbf{x}_1, \mathbf{x}_2, \mathbf{x}_3)$. Its circumcenter $\mathbf{x}_c = (x_c, y_c)$ in the L^∞-norm verifies

$$\|\mathbf{x}_1 - \mathbf{x}_c\|_\infty = \|\mathbf{x}_2 - \mathbf{x}_c\|_\infty = \|\mathbf{x}_3 - \mathbf{x}_c\|_\infty.$$

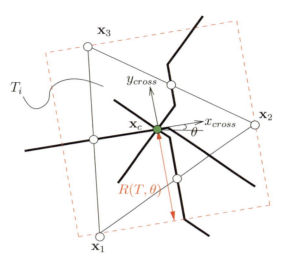

Fig. 7. Circumcenter $\mathbf{x}_c$ and circumradius $R_\infty(T, \theta)$ of a triangle T_i using the L^∞-norm. The circumsquare is the red dotted square.

The L^∞-circumcenter of the triangle is located at the intersection of the L^∞-perpendicular bisectors.

The circumcircle in the L^∞-norm (also called circumsquare), is the smallest square centered at the circumcenter that encloses the triangle, Figure 7. The circumradius $R_\infty(T, \theta)$ is the distance in the L^∞-norm between the circumcenter and anyone of the three vertices. It is given by:

$$R_\infty(T, \theta) = \frac{1}{2} \max\left((\max(x_1, x_2, x_3) - \min(x_1, x_2, x_3), (\max(y_1, y_2, y_3) - \min(y_1, y_2, y_3))\right). \quad (7)$$

One interesting fact is that the computation of circumcenters and circumradii in the L^∞-norm is a very stable numerical operation.

5 A Frontal-Delaunay Mesher in the L^∞-Norm

Let us recall briefly the pros and cons of the two main approaches for mesh generation. Advancing front techniques start from the discretization of the boundary (edges in 2D). The set of edges of the boundary discretization is called the front. A particular edge of this front is selected and a new triangle is formed with this edge as its base and the front is updated accordingly. The algorithm advances in the domain until the L^∞ front is emptied and the domain fully covered by triangles. The main advantage of advancing front techniques is that they generate points and triangles at the same time, which makes it possible to build optimum triangles, e.g. equilateral triangles in our case. The main drawback of the method is that parts of the front advance independently, leading to possible clashes when they meet.

Delaunay-based mesh generation techniques are more robust because a valid mesh exists at each stage of the mesh generation process. Yet, inserting a point using the Delaunay kernel [20] requires the creation and the deletion of a number of triangles, so that there is less control on the element shapes than in the case of advancing front techniques.

The frontal Delaunay approach makes the best of both techniques. As it is based on a Delaunay kernel, a valid mesh is maintained at each stage of the process. Yet, some kind of front is defined in the triangulation and points are inserted in a frontal manner. The process stops when every element of the mesh has the right size according to the size field $\delta(\mathbf{x})$.

The ideas of the new frontal-quad algorithm are inspired by the frontal Delaunay approach of [9].

In what follows, we consider parametric surfaces that have a conformal parametrization i.e parametrization that conserv angles. Withing this perspective, isotropic meshes on the parameter plane result in isotropic meshes in the 3D space. This hypothesis of conformality may seem over restrictive: for example, the usual parametrization of a spherical surface using spherical coordinates is not conformal. Nevertheless, in this paper, we use reparametrization techniques that allow to build conformal mappings [19, 21]. Moreover, the technique that is presented here can be extended to anisotropic quadrilateral mesh generation.

Consider a surface mesh $\mathcal{T}_0$ for which me have computed a discrete conformal parametrization $\mathbf{u}(\mathbf{x})$ (the parametrized mesh is $\mathcal{T}_0'$) and a given mesh size field $\delta(\mathbf{x})$. Consider also that we have computed on $\mathcal{T}_0'$ a cross field $\theta(\mathbf{u})$ from (5). An new *delquad* mesh $\mathcal{T}_1'$ is constructed in the parameter plane that contains the parametrized points of the boundary edges of $\mathcal{T}_0$. Let us define an adimensional L^∞-mesh size

$$h_i = \frac{R_\infty(T_i, \theta(\mathbf{u}))}{\delta(\mathbf{x}(\mathbf{u}))|\det M(\mathbf{u})|^{1/4}} = \frac{R_\infty(T_i, \theta(\mathbf{u}))}{\delta'(\mathbf{u})}$$

for each triangle T_i of $\mathcal{T}_0$. Quantities $\theta(\mathbf{u})$, $M(\mathbf{u})$ and the size function $\delta(\mathbf{x}(\mathbf{u}))$ are evaluated at the usual (L^2)-centroid of the triangle. Triangles are then classified into three categories

1. A triangle is *resolved* if $h_i \leq h_{\max}$;
2. A triangle is *active* if $h_i > h_{\max}$ and, either one of its three neighbors is resolved or one of its sides is on the boundary of the domain;

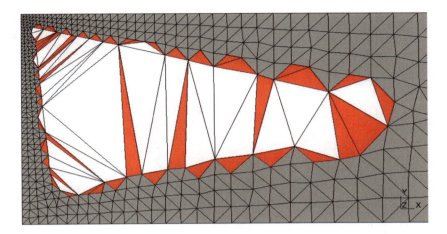

Fig. 8. Illustration of the frontal algorithm with resolved (grey), active (red) and waiting (white) triangles.

3. A triangle is *waiting* if it is neither resolved nor active.

We choose $h_{\max} = 4/3$. This choice is standard in the domain of mesh generation [22]. Figure 8 is an illustration of the way triangles are classified in the algorithm. The front is defined as the set of active triangles. Active triangles are sorted with respect to h_i. Front edges are therefore defined as those edges separating active and resolved triangles.

The frontal algorithm inserts a new point so as to form an optimal triangle with the edge corresponding to the largest active triangle. Consider the edge $\mathbf{x}_2\mathbf{x}_3$ in Figure 9 and assume it corresponds to the largest active triangle of the mesh (the red triangle on Figure 9). For the discussion, the coordinate system has been centered at the mid-edge point $\mathbf{x}_m = \frac{1}{2}(\mathbf{x}_2 + \mathbf{x}_3)$ and aligned with the local cross field, this can be achieved by a translation and a rotation of angle $\theta(\mathbf{x}_m)$.

We choose to position of the new point $\mathbf{x}_n$ along the L^∞-perpendicular bisector $\mathcal{L}$ of $\mathbf{x}_2\mathbf{x}_3$. The exact position of the new point $\mathbf{x}_n$ will be chosen in order to fullfill the size criterion $\delta(\mathbf{x}_m)$.

In order to create a new triangle $T_i(\mathbf{x}_2, \mathbf{x}_3, \mathbf{x}_n)$ of size $R_\infty(T_i, \theta) = \delta'(\mathbf{x}_m)$, we position $\mathbf{x}_n$ at the intersection of $\mathcal{L}$ with the square C_n of side $\delta'(\mathbf{x}_m)$ passing through points $\mathbf{x}_2$ and $\mathbf{x}_3$ (see Figure 9).

The following considerations should be made.

- The new point should not be beyond $\mathbf{x}_c$, the center of the circumsquare of the active triangle (see Figure 9) as this would create a triangle with

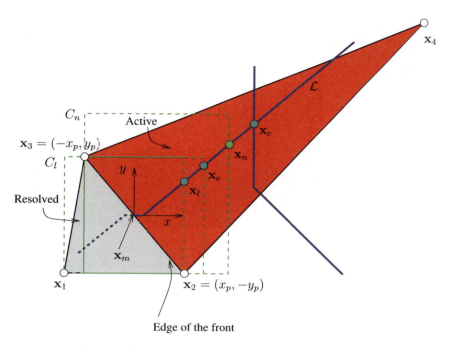

Fig. 9. Illustration of the point insertion algorithm.

a small edge $\mathbf{x}_n \mathbf{x}_4$. Note that this limit case corresponds to a classical point insertion scheme where new points are inserted at the center of the circumcircle of the worst triangle, yet in the L^∞-norm in this case.
- The new point should not be placed below $\mathbf{x}_l$ where $\mathbf{x}_l$ is the intersection of the ∞-perpendicular bisector of $\mathbf{x}_2\mathbf{x}_3$ and the circumsquare C_l of the resolved triangle $(\mathbf{x}_1, \mathbf{x}_2, \mathbf{x}_3)$. Inserting a point into C_l would make the resolved triangle invalid by means of the Delaunay criterion.
- If $\delta'(\mathbf{x}_m) = \|\mathbf{x}_3 - \mathbf{x}_2\|_\infty$, then the optimal point is $\mathbf{x}_n = \mathbf{x}_e$. It correspond to the largest triangle $T_i(\mathbf{x}_e, \mathbf{x}_2, \mathbf{x}_3)$ that verifies $R_\infty(T_i, \theta) = \delta'(\mathbf{x}_m)$.

The position of the optimal point is computed as follow:

$$\mathbf{x}_n = \mathbf{x}_e + t(\mathbf{x}_c - \mathbf{x}_e) \tag{8}$$

with

$$t = \min\left(\max\left(1, \frac{\|\mathbf{x}_3 - \mathbf{x}_2\|_\infty - \delta'(\mathbf{x}_m)}{\|\mathbf{x}_3 - \mathbf{x}_2\|_\infty - \|\mathbf{x}_c - \mathbf{x}_2\|_\infty}\right), -\frac{\|\mathbf{x}_l - \mathbf{x}_e\|_2}{\|\mathbf{x}_c - \mathbf{x}_e\|_2}\right), \tag{9}$$

and $\mathbf{x}_c, \mathbf{x}_e$ and $\mathbf{x}_l$ are computed from the equation of the bisector $\mathcal{L}$ (6) :

$$\mathbf{x}_c = \left(\frac{1}{2}(x_4 - x_p), \frac{1}{2}(x_4 + x_p) - y_p\right),$$

$$\mathbf{x}_e = (\delta'(\mathbf{x}_m) - x_p + y_p, \delta'(\mathbf{x}_m)) \quad \text{and} \quad \mathbf{x}_l = (\delta'(\mathbf{x}_m), \delta'(\mathbf{x}_m) + x_p - y_p).$$

Another important ingredient of the advancing front strategy is the fact that the fronts should be updated layer by layer. A initial front is created with the edges of the 1D discretization. The algorithm inserts points until every edge of the active front have been treated. Then other fronts are created and emptied until no active triangle is left in the mesh.

6 Examples

6.1 Piston

As a first example, let us apply the new algorithm to the geometry of the same piston of Figure 3. The result of the advancing front delaunay quad mesher is shown in Figure 10. The new mesh is close to be perfect. The quadrilateral mesh has been automatically generated with the new algorithm, the only control parameter being a constant mesh size field. The mesh is composed of $31,979$ quads and has been generated in about 15 seconds. Compared with the $39,386$ quads of the mesh shown in Figure 3, this mesh has about 19% less nodes. This number is close to the theoretical value $1 - \sqrt{3}/2 = 0.134$.

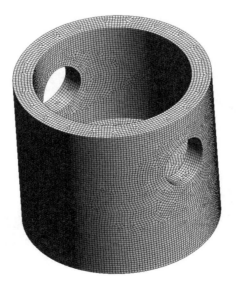

Fig. 10. Mesh of the Piston computed with the presented frontal *delquad* mesh algorithm.

The average element quality is now $\bar{\eta} = 0.93$ and the worst element is of quality 0.39, which can be considered as very good. Moreover, 92% of the nodes have 4 adjacent quadrangles, which is also very good.

6.2 Falcon Aircraft

As a second example, let us consider the Falcon aircraft of Figure 11. The mesh size field is composed of a uniform bulk size field $\delta_b = 0.1$ and of line and point sources positioned at strategic zones of the aircraft.

The resulting mesh is presented on Figure 11. The mesh is composed of 53,297 quadrangles. The total time for the surface meshing was 22 seconds. The average and worst quality of the mesh are $\bar{\eta} = 0.86$ and $\eta_w = 0.17$ which can be considered as excellent.

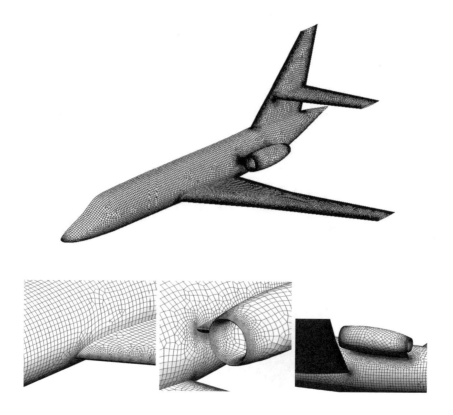

Fig. 11. Final quadrilateral surface mesh of the Falcon aircraft.

7 Conclusion

A new method for automatic quad-meshing of surfaces has been proposed. The new algorithm uses distances in the L^∞ norm as a base for the insertion of new points in the mesh and the generation of edges of the right size in this specific norm.

Perspectives of this approach are numerous. The automatic generation of hex-dominant meshes remains a challenge in the community of finite elements in general. The extension of the new Delquad approach to a Delhex algorithm that would generate 3D tetrahedral meshes that have the right number of points and the right orientation to be recombined optimally into hexaedra is a natural sequel to this work.

Acknowledgements

This work has been partially supported by the Belgian Walloon Region under WIST grants ONELAB 1017086 and DOMHEX 1017074.

References

1. Blacker, T.D., Stephenson, M.B.: Paving: A new approach to automated quadrilateral mesh generation. International Journal for Numerical Methods in Engineering 32, 811–847 (1991)
2. Frey, P.J., Marechal, L.: Fast adaptive quadtree mesh generation. In: Proceedings of the Seventh International Meshing Roundtable, Citeseer (1998)
3. Lee, C.K., Lo, S.H.: A new scheme for the generation of a graded quadrilateral mesh. Computers and Structures 52, 847–857 (1994)
4. Borouchaki, H., Frey, P.J.: Adaptive triangular–quadrilateral mesh generation. International Journal for Numerical Methods in Engineering 45(5), 915–934 (1998)
5. Owen, S.J., Staten, M.L., Canann, S.A., Saigal, S.: Q-morph: An indirect approach to advancing front quad meshing. International Journal for Numerical Methods in Engineering 9, 1317–1340 (1999)
6. Edmonds, J.: Maximum matching and a polyhedron with vertices. J. of Research at the National Bureau of Standards 69B, 125–130 (1965)
7. Edmonds, J., Johnson, E.L., Lockhart, S.C.: Blossom I: A computer code for the matching problem. In: Watson, T.J., Edmonds, J., Johnson, E.L., Lockhart, S.C. (eds.) IBM T. J. Watson Research Center, Yorktown Heights (1969)
8. Lévy, B., Liu, Y.: Lp centroidal voronoi tesselation and its applications. In: ACM Transactions on Graphics (SIGGRAPH Conference Proceedings) (2010)
9. Rebay, S.: Efficient unstructured mesh generation by means of delaunay triangulation and bowyer-watson algorithm. Journal of Computational Physics 106(1), 125–138 (1993)
10. Remacle, J.-F., Lambrechts, J., Seny, B., Marchandise, E., Johnen, A., Geuzaine, C.: Blossom-quad: a non-uniform quadrilateral mesh generator using a minimum cost perfect matching algorithm. International Journal for Numerical Methods in Engineering (submitted 2011)

11. Edmonds, J.: Paths, trees, and flowers. Canad. J. Math. 17, 449–467 (1965)
12. Lawler, E.L.: Combinatorial Optimization: Networks and Matroids. Holt, Rinehart, and Winston, New York, NY (1976)
13. Gabow, H.: Implementation of Algorithms for Maximum Matching on Nonbipartite Graphs. PhD thesis, Stanford University (1973)
14. Gabow, H., Galil, Z., Micali, S.: An o(ev log v) algorithm for finding a maximal weighted matching in general graphs. SIAM J. Computing 15, 120–130 (1986)
15. Gabow, H.N.: Data structures for weighted matching and nearest common ancestors with linking. In: Proceedings of the 1st Annual ACM-SIAM Symposium on Discrete Algorithms. pp. 434–443 (1990)
16. Cook, W., Rohe, A.: Computing minimum-weight perfect matchings. INFORMS Journal on Computing 11(2), 138–148 (1999)
17. Pébay, P.P.: Planar quadrangle quality measures. Engineering with Computers 20(2), 157–173 (2004)
18. Bommes, D., Zimmer, H., Kobbelt, L.: Mixed-integer quadrangulation. In: SIGGRAPH 2009: ACM SIGGRAPH 2009 Papers, pp. 1–10. ACM Press, New York (2009)
19. Lévy, B., Petitjean, S., Ray, N., Maillot, J.: Least squares conformal maps for automatic texture atlas generation. ACM Transactions on Graphics 21(3), 362–371 (2002)
20. Watson, D.F.: Computing the n-dimensional delaunay tessellation with application to voronoi polytopes. The Computer Journal 24(2), 167–172 (1981)
21. Marchandise, E., de Wiart, C.C., Vos, W.G., Geuzaine, C., Remacle, J.F.: High-quality surface remeshing using harmonic maps–part ii: Surfaces with high genus and of large aspect ratio. International Journal for Numerical Methods in Engineering 86, 1303–1321 (2011)
22. Frey, P.J., George, P.-L.: Mesh Generation - Application To Finite Elements. Wiley (2008)

L_p Lloyd's Energy Minimization for Quadrilateral Surface Mesh Generation

Tristan Carrier Baudouin[1], Jean-François Remacle[1], Emilie Marchandise[1], and Jonathan Lambrechts[1]

Institute of Mechanics, Materials and Civil Engineering, Université catholique de Louvain, Avenue Georges-Lemaître 4, 1348 Louvain-la-Neuve, Belgium
{tristan.carrier,jean-francois.remacle,emilie.marchandise, jonathan.lambrechts}@uclouvain.be

Summary. Indirect methods recombine the elements of triangular meshes to produce quadrilaterals. The resulting quadrilaterals are usually randomly oriented, which is not desirable. However, by aligning the vertices of the initial triangular mesh, precisely oriented quads can be produced. Levy's algorithm is a non-linear optimization procedure that can align points according to a locally defined metric. It minimizes an energy functional based on the L_p distance in the local metric. The triangulation of a set of vertices smoothed with Levy's algorithm is mainly composed of right-angled triangles, which is ideal for quad recombination. An implementation of Levy's algorithm for the purpose of finite element computation has been developed. The implementation can create quads of desired size and orientation. The algorithm has been tested on two-dimensional geometries as well as parametrized curved surfaces. The results show an improvement of the quads alignment.

Keywords: centroidal Voronoi tessellations, non-linear optimization, quad mesh generation.

1 Introduction

For finite element analysis, quad meshes are advantageous compared to triangular meshes [1]. For example, in computational fluid mechanics, they accelerate grid convergence [2][3][4] and they capture boundary layers with a higher precision [5]. They are also very useful in structural mechanics. They are not subject to numerical locking [6] and they allow schemes to remain stable under inexact integration [7][8][9]. In the context of high order methods, quad meshes can be curved more robustly [10].

However, quad mesh generation techniques are not as mature as triangular ones. The indirect approach looks like a promising solution. It consists of combining two by two the elements of a triangular mesh in order to create quads. Triangular mesh generators are usually designed to produce near-equilateral triangles. Combining these triangles yields randomly oriented quads, which is not ideal. However, if the vertices of the initial triangular mesh are well aligned, an indirect algorithm like Blossom-Quad [11] can create high quality, precisely oriented quads. Blossom-Quad optimizes the mean quality of the resulting quads. Unlike triangles, quads are always oriented. It is desirable to prescribe this orientation.

Levy and Liu developed a method based on centroidal Voronoi tessellations in L_p norm for aligning points in a rectangular manner [12]. The method minimizes an energy functional equal to the sum of the L_p moment of inertia of the Voronoi cells. The L_p distances are measured with respect to a locally defined metric. This metric field controls the orientation of the quads. Levy's algorithm uses the limited-memory Broyden-Fletcher-Goldfarb-Shanno optimization procedure (LBFGS) to minimize the energy functional. LBFGS has the advantage of only requiring the value of the functional and its gradient. Low energy solutions are sets of points having rectangular Voronoi cells. When the Voronoi cells are rectangular, the points are also aligned in a rectangular way. In other words, Levy's algorithm optimizes the shape of the Voronoi cells. This process is illustrated on Fig. 1 and 2. Fig. 1(a) is the initial triangular mesh and Fig. 1(b) is the Voronoi diagram of the vertices. Fig. 2(a) is the final quad mesh obtained with Levy's and Blossom-Quad algorithms. Fig. 2(b) is the corresponding Voronoi diagram.

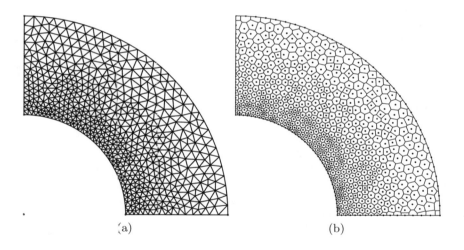

Fig. 1. The initial triangular mesh (a) and its corresponding Voronoi diagram (b).

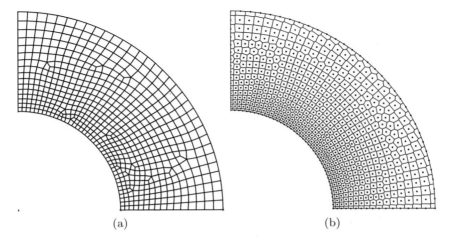

Fig. 2. The final quad mesh (a) and its corresponding Voronoi diagram (b) (A Laplacian smoothing has been applied, which explain the slight differences between the two sets of vertices).

A two-dimensional implementation of Levy's algorithm adapted to the field of finite element computation is presented in this paper. It can take into input metric and density fields. These metric and density fields define the orientation and the size of the quads. By using a parametrization, the algorithm can be applied to curved surfaces. The energy and the gradient are computed with Gauss integration techniques. The Alglib library[1] is used for the LBFGS optimization part.

2 Levy's Algorithm

The implementation of Levy's algorithm described in this article can only find local minimums. It cannot find global minimums even for simple geometries such as squares and rectangles. It can nevertheless align points in a very satisfactory manner.

The clipped Voronoi diagram is the part of the Voronoi diagram that is inside the domain, as shown in Fig. 1(b) and 2(b). It is an essential part of Levy's algorithm [12]. Without it, computing accurate values for the energy and the gradient is impossible. The method used here is inspired from [13].

Subsections 2.1 and 2.2 introduce the energy functional and the gradient. Subsection 2.3 shows how the domain is divided into triangular elements. Subsection 2.4 explains how to compute the energy and the gradient with Gauss integration techniques. The addition of a metric and a density is discussed in subsection 2.5.

---

[1] ALGLIB (www.alglib.net), Sergey Bochkanov and Vladimir Bystritsky.

The main difference between Levy's implementation and the one described in this article is the way of computing the energy and the gradient. Like it was said before, the present implementation uses Gauss integration techniques, while Levy employs analytical formulas.

2.1 Energy Functional

The energy functional minimized by Levy's algorithm uses the L_p distance defined here [12]:

$$||\boldsymbol{y} - \boldsymbol{x}||_p^p = |y_1 - x_1|^p + |y_2 - x_2|^p \qquad (1)$$

Each curve on Fig. 3 contains a set of points equidistant to the origin according to its particular L_p distance. The curves become more and more rectangular as p increases. For any p higher than two, the L_p distance become anisotropic.

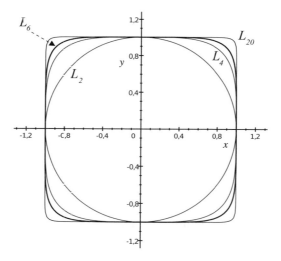

Fig. 3. Unit circles in various L_p distances.

The energy functional is defined as [12]:

$$F(\boldsymbol{x}_1, \boldsymbol{x}_2, ..., \boldsymbol{x}_N) = \underbrace{\sum_{i=1}^{N} \int_{R_i} ||\boldsymbol{y} - \boldsymbol{x}_i||_p^p d\boldsymbol{y}}_{I^{R_i}(\boldsymbol{x}_i)} \qquad (2)$$

The index i runs over all mesh vertices x_i. R_i is the domain corresponding to the i-th voronoi cell generated by mesh vertex x_i. Evidently, the Voronoi diagram is computed with the L_2 metric. In what follows, the energy integral will be denoted by $I^{R_i}(x_i)$.

2.2 Energy Gradient

The energy gradient is the total derivative of F with respect to the position of each non-boundary mesh vertex [12]. Boundary vertices are considered unmovable. It is assumed that $x_1, x_2,..., x_n$ are non-boundary vertices and that $x_{n+1}, x_{n+2},..., x_N$ are boundary vertices. The energy gradient is given by:

$$\frac{\mathrm{d}}{\mathrm{d}x_k} F(x_1, x_2, ..., x_N) = \frac{\mathrm{d}}{\mathrm{d}x_k} \sum_{i=1}^{N} I^{R_i}(x_i) = \sum_{i=1}^{N} \frac{\mathrm{d}I^{R_i}(x_i)}{\mathrm{d}x_k} \quad k=1,...,n$$

The total derivative on the right hand side can be rewritten in terms of partial derivatives.

$$\frac{\mathrm{d}}{\mathrm{d}x_k} F(x_1, x_2, ..., x_N) = \sum_{i=1}^{N} \frac{\partial I^{R_i}(x_i)}{\partial x_k} + \sum_{i=1}^{N} \frac{\partial I^{R_i}(x_i)}{\partial R_i} \frac{\mathrm{d}R_i}{\mathrm{d}x_k}$$

The partial derivative of $I^{R_i}(x_i)$ with respect to x_k is non-null only when $i = k$.

$$\frac{\mathrm{d}}{\mathrm{d}x_k} F(x_1, x_2, ..., x_N) = \frac{\partial I^{R_k}(x_k)}{\partial x_k} + \sum_{i=1}^{N} \frac{\partial I^{R_i}(x_i)}{\partial R_i} \frac{\mathrm{d}R_i}{\mathrm{d}x_k}$$

R_i can be expressed in terms of the Voronoi vertices $C_{i1}, C_{i2},...,C_{iM_i}$ of the i-th Voronoi cell. All Voronoi cells do not have the same number of vertices M_i. Fig. 4 shows a Voronoi cell with six Voronoi vertices.

The gradient can be rewritten in terms of the position of the Voronoi vertices.

$$\frac{\mathrm{d}}{\mathrm{d}x_k} F(x_1, x_2, ..., x_N) = \frac{\partial I^{R_k}(x_k)}{\partial x_k} + \sum_{i=1}^{N} \sum_{j=1}^{M_i} \frac{\partial I^{R_i}(x_i)}{\partial C_{ij}} \frac{\mathrm{d}C_{ij}}{\mathrm{d}x_k} \quad (3)$$

Most of the terms $\frac{\mathrm{d}C_{ij}}{\mathrm{d}x_k}$ are null. They are non-null only when the mesh vertex is very close to the Voronoi vertex, as detailed in the next section.

2.3 Gradient Assembly

In order to evaluate the integrals with Gauss techniques, the Voronoi cells are divided into triangular elements. Each triangular element is composed of a mesh vertex and two successive Voronoi vertices, as shown in Fig. 4. In Fig. 4 to 7, the black dots are assumed to be mesh vertices and the hollow dots are assumed to be Voronoi vertices. The dotted segments are Delaunay edges.

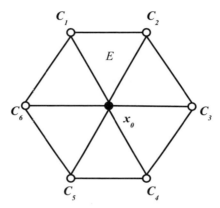

Fig. 4. A Voronoi cell divided into triangular elements.

Most of the terms inside the double summation found in equation (3) vanish. By considering the relationship between the mesh vertices and the Voronoi vertices, the relevant contributions can be identified. Each Voronoi vertex falls into one of three categories.

1. A Voronoi vertex can be the center of the circle circumscribing a Delaunay element. Voronoi vertex C_1 from Fig. 5 belongs to this category. As long as the displacements are infinitesimal, C_1 depends only on x_0, x_1 and x_2.

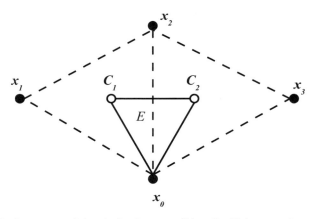

Fig. 5. C_1 is the center of the circle circumscribing the Delaunay element x_0-x_1-x_2.

2. A Voronoi vertex can be the intersection point between a Voronoi facet and a boundary line segment. Voronoi vertex C_2 from Fig. 6 belongs to this category. Again, as long as the displacements are infinitesimal, C_2 depends only on x_0, x_2 and the boundary line segment.

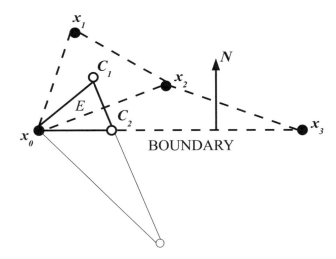

Fig. 6. C_2 is the intersection point between the bisector of the Delaunay edge x_0-x_2 and the boundary line segment x_0-x_3.

3. A Voronoi vertex can be the median point between two boundary mesh vertices. Voronoi vertex C_2 from Fig. 7 belongs to this category. It depends only on x_0 and x_2.

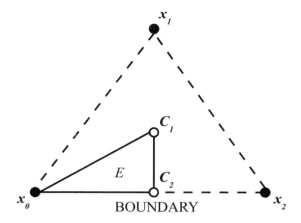

Fig. 7. C_2 is the median point between x_0 and x_2.

The following formulas were derived from equation (3). They apply to Fig. 5. It is assumed that x_0, x_1, x_2 and x_3 are non-boundary mesh vertices. A += sign is used because there will be contributions from other elements as well.

$$\frac{dF}{dx_0} += \frac{\partial I^E(x_0)}{\partial x_0} + \frac{\partial I^E(x_0)}{\partial C_1}\frac{dC_1}{dx_0} + \frac{\partial I^E(x_0)}{\partial C_2}\frac{dC_2}{dx_0} \tag{4}$$

$$\frac{dF}{dx_1} += \frac{\partial I^E(x_0)}{\partial C_1}\frac{dC_1}{dx_1} \tag{5}$$

$$\frac{dF}{dx_2} += \frac{\partial I^E(x_0)}{\partial C_1}\frac{dC_1}{dx_2} + \frac{\partial I^E(x_0)}{\partial C_2}\frac{dC_2}{dx_2} \tag{6}$$

$$\frac{dF}{dx_3} += \frac{\partial I^E(x_0)}{\partial C_2}\frac{dC_2}{dx_3} \tag{7}$$

2.4 Gauss Integration

The terms inside equations (4) to (7) are integrals on the element E. A linear transformation T will be used in order to go from the reference triangle E' to the triangle E, as shown in Fig. 8. It will then become possible to evaluate the various integrals with Gauss techniques.

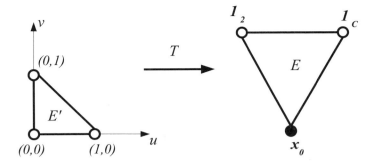

Fig. 8. The linear transformation T.

$$y = T(u,v) = x_0(1-u-v) + C_1 u + C_2 v$$

Equation (2) defines $I^E(x_0)$ as the energy contribution of element E. J is the Jacobian of the transformation T:

$$I^E(x_0) = \int_E \|y - x_0\|_p^p dy = \int_{E'} \|T(u,v) - x_0\|_p^p J du dv$$

The partial derivative of $I^E(x_0)$ with respect to the position of mesh vertex x_0 can be written as:

$$\frac{\partial I^E(x_0)}{\partial x_0} = \frac{\partial}{\partial x_0} \int_E \|y - x_0\|_p^p dy = \int_{E'} \frac{\partial \|T(u,v) - x_0\|_p^p}{\partial x_0} J du dv$$

The partial derivative of $I^E(x_0)$ with respect to the position of Voronoi vertex C_1 is given by:

$$\frac{\partial I^E(x_0)}{\partial C_1} = \frac{\partial}{\partial C_1} \int_E \|y - x_0\|_p^p dy = \frac{\partial}{\partial C_1} \int_{E'} \|T(u,v) - x_0\|_p^p J du dv$$

$$= \int_{E'} \frac{\partial \|T(u,v) - x_0\|_p^p}{\partial T(u,v)} \frac{\partial T(u,v)}{\partial C_1} J + \|T(u,v) - x_0\|_p^p \frac{\partial J}{\partial C_1} du dv$$

A Voronoi vertex can sometimes depend on three mesh vertices, as in Fig. 5. The following matrix is the derivative of Voronoi vertex C_1 with respect to mesh vertex x_0 [12]. The derivatives of C_1 with respect to mesh vertices x_1 and x_2 can be obtained by replacing x_0 with x_1 or x_2.

$$\frac{d\boldsymbol{C}_1}{d\boldsymbol{x}_0} = \begin{bmatrix} (\boldsymbol{x}_1 - \boldsymbol{x}_0)^T \\ (\boldsymbol{x}_2 - \boldsymbol{x}_0)^T \end{bmatrix}^{-1} \begin{bmatrix} (\boldsymbol{C}_1 - \boldsymbol{x}_0)^T \\ (\boldsymbol{C}_1 - \boldsymbol{x}_0)^T \end{bmatrix}$$

A Voronoi vertex can instead depend on two mesh vertices and one boundary line segment, as in Fig. 6. The following matrix needs to be used in this situation [12]. The derivative of $\boldsymbol{C}_1$ with respect to mesh vertex $\boldsymbol{x}_1$ can be obtained by replacing $\boldsymbol{x}_0$ with $\boldsymbol{x}_1$. $\boldsymbol{N}$ is the normal vector to the boundary line segment. It can be multiplied by any non-zero constant without affecting the value of the derivative.

$$\frac{d\boldsymbol{C}_1}{d\boldsymbol{x}_0} = \begin{bmatrix} (\boldsymbol{x}_1 - \boldsymbol{x}_0)^T \\ \boldsymbol{N}^T \end{bmatrix}^{-1} \begin{bmatrix} (\boldsymbol{C}_1 - \boldsymbol{x}_0)^T \\ \boldsymbol{0} \end{bmatrix}$$

More details about the procedure used to obtain these matrices can be found in Levy's article [12]. It is important to recall that it is only the derivatives with respect to non-boundary mesh vertices that need to be calculated.

2.5 Non-uniform Metric and Density

The addition of a metric M and a density ρ allows varying orientation and quad sizes. The following energy functional takes into account these two parameters [12][14].

$$F(\boldsymbol{x}_1, \boldsymbol{x}_2, ..., \boldsymbol{x}_N) = \sum_{i=1}^{N} \int_{R_i} \rho(\boldsymbol{y}) \|M(\boldsymbol{y} - \boldsymbol{x}_i)\|_p^p d\boldsymbol{y}$$

The following formula illustrates the relationship between the density ρ and the mesh size h. The procedure that has been used to obtain it is similar to the one described in another article [14].

$$\rho \sim \frac{1}{h^{p+2}}$$

The metric can be considered constant by element. It is preferable to use a continuous density however.

3 Results

In this section, different examples will be presented in order to show that Levy's algorithm can create high quality quad surface meshes that have well defined orientations.

Five steps are necessary to produce quad meshes :

1. Compute a conformal mapping that maps the 3D surface to a 2D parametric space [15].
2. Create a triangular mesh in the parametric space using a 2D mesh generator.
3. Use this mesh to compute an alignment direction. The alignment direction is already known at the boundary. By using the heat transfer equation, it can be obtained everywhere else inside the domain [16].

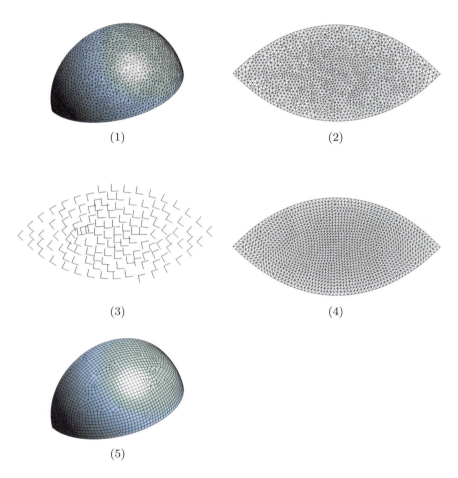

Fig. 9. The different steps necessary to produce quad meshes.

4. Use Levy's algorithm to optimize the location of the mesh vertices. L_6 is a good compromise between speed and squareness.
5. Apply the Blossom-Quad algorithm described in [11] to combine the triangles into quads. Blossom-Quad uses the well-known Blossom algorithm in order to find a perfect matching between triangles. The matching also optimizes the mean quality of the quads.

Fig. 9 illustrates the different steps of the global process. (1) is the triangular mesh of the geometry in the three-dimensional space. (2) is the triangular mesh of the geometry in the parametric space. (3) shows the cross-field determining the orientation of the quads. (4) is the triangular mesh in the parametric space after the application of Levy's algorithm. (5) is the final quad mesh.

A car hood is depicted on Fig. 10. (A) is the triangular mesh obtained with the *mesh adapt* algorithm from Gmsh [17]. It contains 1740 triangles. (AB) is the quad mesh obtained by applying Blossom-Quad directly on (A). (AB) contains randomly oriented quads. (ABL) is the quad mesh obtained by applying Levy's algorithm before the recombination. Most of the quads are now oriented in the specified direction.

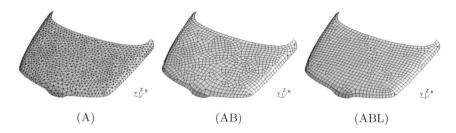

(A)   (AB)   (ABL)

Fig. 10. Different meshes of a car hood. (A) is a triangular mesh, (AB) is a quad mesh created only with Blossom-Quad algorithm and (ABL) is a quad mesh created with both Levy's and Blossom-Quad algorithms.

The (AB) mesh has a mean quality of $\bar{\eta} = 0.79$. The (ABL) mesh has a mean quality of $\bar{\eta} = 0.88$. The quality of a quadrilateral $\eta(q)$ is defined by the values of its four angles $\alpha_k, k = 1, 2, 3, 4$ [11]:

$$\eta(q) = \max\left(1 - \frac{2}{\pi}\max_k\left(\left|\frac{\pi}{2} - \alpha_k\right|\right), 0\right)$$

If the element is a perfect square, $\eta(q)$ is equal to one.

In an ideal quad mesh, each non-boundary mesh vertex is connected to four neighbors. In (AB), 62% of the non-boundary mesh vertices are 4-valent. In (ABL), this number reaches 77%.

Fig. 11 shows the decrease of the energy in function of the number of iterations for the car hood problem. Voronoi diagrams at various steps of the optimization process are also shown. As the number of iteration increases, the Voronoi cells become more and more rectangular.

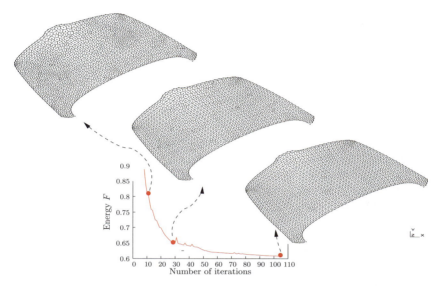

Fig. 11. Convergence curve of the car hood problem. The Voronoi diagrams in the parametric space for three different iterations (12, 26, 106) are shown.

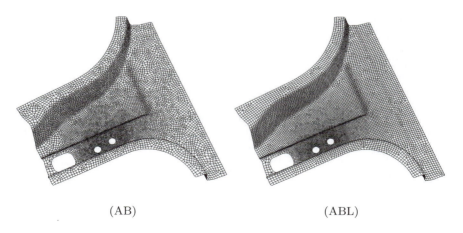

Fig. 12. (AB) is a quad mesh created only with Blossom-Quad algorithm and (ABL) is a quad mesh created with both Levy's and Blossom-Quad algorithms.

Fig. 12 shows another car body part. Again, it compares two quad meshes, one created with Levy's algorithm and one without. Both meshes contain 9954 quads. It took 6 minutes and 22 seconds to perform the 202 iterations of Levy's algorithm on a standard 2010 laptop.

(AB) has a mean quality of $\bar{\eta} = 0.79$ and (ABL) has a mean quality of $\bar{\eta} = 0.90$. The percentage of 4-valent vertices is equal to 71% in (AB). It is equal to 89% in (ABL).

4 Conclusion

A two-dimensional implementation of Levy's algorithm has been described in this article. When used in combination with an indirect algorithm like Blossom-Quad, it is able to create well-oriented quads of varying size. By taking advantage of parametrization techniques, curved surfaces can also be meshed. The only apparent drawback of this method is the execution time. It is much more complex than the traditional Lloyd's algorithm. Optimizing large meshes can take very long.

Indirect methods can also be used to create hexahedra. However, in order to create good quality hex meshes aligned in precise directions, a three-dimensional version of Levy's algorithm would be necessary. The implementation described in this article can be used as a starting point. Computing the energy and the gradient would not be particularly more difficult in three-dimension. Nevertheless, clipping a Voronoi diagram in three-dimension would be much more complex, but feasible.

Acknowledgement. This work has been partially supported by the Belgian Walloon Region under WIST grants ONELAB 1017086 and DOMHEX 1017074.

References

1. Remacle, J.-F., Marchandise, E., Geuzaine, C., Béchet, E.: The domhex proposal (2010)
2. Prakash, S., Ethier, C.R.: Requirements for mesh resolution in 3d computational hemodynamics. Journal of Biomechanical Engineering 123, 134–144 (2001)
3. Vinchurkar, S., Longest, P.W.: Effect of mesh style and grid convergence on particle deposition in bifurcating airway models with comparisons to experimental data. Medical Engineering and Physics 29, 350–366 (2007)
4. Vinchurkar, S., Longest, P.W.: Evaluation of hexahedral, prismatic and hybrid mesh styles for simulating respiratory aerosol dynamics. Computers and Fluids 37, 317–331 (2008)
5. Ferziger, J., Peric, M.: Computational Methods for Fluid Dynamics. Springer, Berlin (1999)
6. Roache, P.: Computational Fluid Dynamics. Hermosa, Albuquerque (1992)
7. Zienkiewicz, O.C., Taylor, R.L.: The Finite Element Method. The Basis, vol. 1. Butterworth-Heinemann, Oxford (2000)

8. Hughes, T.: The Finite Element Method - Linear Static and Dynamic Finite Element Analysis. Dover Publications, New York (2000)
9. Zienkiewicz, O.C., Rojek, J., Taylor, R.L., Pastor, M.: Triangles and tetrahedra in explicit dynamic codes for solids. International Journal for Numerical Methods in Engineering 43, 565–583 (1998)
10. Sherwin, S.J., Peiró, J.: Mesh generation in curvilinear domains using high-order elements. International Journal for Numerical Methods in Engineering 53, 207–223 (2002)
11. Remacle, J.-F., Lambrechts, J., Seny, B., Marchandise, E., Johnen, A., Geuzaine, C.: Blossom-quad: a non-uniform quadrilateral mesh generator using a minimum cost perfect matching algorithm. International Journal for Numerical Methods in Engineering (submitted 2011)
12. Lévy, B., Liu, Y.: l_p centroidal voronoi tessellation and its applications. In: Hoppe, H. (ed.) ACM Transactions on Graphics, Los Angeles, USA (2010)
13. Yan, D.-M., Wang, W., Lévy, B., Liu, Y.: Efficient Computation of 3D Clipped Voronoi Diagram. In: Mourrain, B., Schaefer, S., Xu, G. (eds.) GMP 2010. LNCS, vol. 6130, pp. 269–282. Springer, Heidelberg (2010)
14. Du, Q., Wang, D.: Tetrahedral mesh generation and optimization based on centroidal voronoi tessellations. International Journal for Numerical Methods in Engineering 56, 1355–1373 (2003)
15. Marchandise, E., Crosetto, P., Geuzaine, C., Remacle, J.F., Sauvage, E.: Quality open source mesh generation for cardiovascular flow simulations. Springer, Heidelberg (2011)
16. Remacle, J.-F., Henrotte, F., Carrier Baudouin, T., Béchet, E., Marchandise, E., Geuzaine, C., Mouton, T.: A frontal delaunay quad mesh generator using the l^∞ norm. International Journal for Numerical Methods in Engineering (in preparation, 2011)
17. Geuzaine, C., Remacle, J.-F.: Gmsh: a three-dimensional finite element mesh generator with built-in pre- and post-processing facilities. International Journal for Numerical Methods in Engineering 79, 1309–1331 (2009)

CSALF-Q: A Bricolage Algorithm for Anisotropic Quad Mesh Generation

Nilanjan Mukherjee

Meshing & Abstraction Group
Digital Simulation Solutions
Siemens PLM Software
SIEMENS
2000 Eastman Dr., Milford, Ohio 45150 USA
mukherjee.nilanjan@siemens.com

Summary. "Bricolage", a multidisciplinary term used extensively in visual arts, anthropology and cultural studies, refers to a creation that borrows elements from a diverse range of existing designs. The term is applied here to describe CSALF-Q, a new automatic 2D quad meshing algorithm that combines the strengths of recursive subdivision, loop-paving and transfinite interpolation to generate surface meshes that are anisotropic yet boundary structured, robust yet sensitive to size fields and competitive in performance. This algorithm is based on the understanding of boundary loops which are initially meshed recursively with a new, boundary based "loop-paving" technique that balances anisotropy and mesh quality. The remaining interior domain is filled out by a symbiotic process that shuttles between recursive subdivision, transfinite interpolation and loop-paving in an efficient manner. The mesh is finally cleaned and smoothed. Loop-fronts are classified and rule sets are defined for each, to aid optimum point placement. Stencils used for loop-closure are presented. Results are presented that compare both local and global element quality of meshes generated by the new algorithm with that of TQM (TriaQua Mesher) which is also known to handle variegated sizemaps.

Keywords: subdivision, tri-qua mesher, paving, transfinite, mapped, flattening, loop, loop-paving, parameterization, quadrilateral.

1 Introduction

Unstructured quadrilateral and hex mesh generation of a large gamut of industrial problems ranging from automotive, aerospace, electronic and appliance structures and thermal and fluid flow problems continue to be a key pre-processing step leading to the complex analyses performed for design validation. In the industry today, we continue to faithfully use just a handful of quad mesh generation algorithms. Added to the existing complexity of engineering demands and interaction problems there is a need to create hybrid meshes that capture quintessential properties of several mesh generation algorithms, thus enabling the application engineer or analyst to produce meshes that largely meet both diverse local and global requirements. This "hybridity", called for thus, is defined in two ways - first, a mix of elements (quad or hex-dominant meshes) and secondly quad/hex or quad/hex-dominant meshes confronted with stringent needs of variegated local and global size fields thus requiring challenging anisotropy.

2 Past Research

Owen [1] and Frey and George [2] present detailed overviews on unstructured meshing algorithms both from historical and practical perspectives. The basic work on unstructured mesh generation with quadrilaterals could be classified into the following:

1. Transfinite methods
2. Paving or Advancing front type methods
3. Quadtree methods
4. Subdivision or domain decomposition methods
5. Medial Axis methods
6. Sphere/ball/circle packing methods

This paper will only focus on paving, transfinite and subdivision algorithms.

2.1 Advancing Front Methods

The advancing front approach to domain triangulation was introduced by Marcum [3]. Ted Blacker and Stephenson [4] published their ground breaking work on a quadrilateral advancing front approach they called *Paving*. Paving and advancing front meshes guarantee a boundary conforming mesh with a very high quality. However, they do not guarantee a "controllable layered mesh". There may not exist a completely structured layer of elements around boundary loops - as a result properties of such layers cannot be user-controlled. Staten et al extended the unstructured paving approach to 3D which they called "plastering" [5]. Earlier White and Kinney [6] had attempted to progress a single element row until collision occurs. Although an interesting approach, the method does not help to improve quad element quality and nor guarantee controllable structured layers of high quality quads near interior boundaries. With triangles however, Pirzadeh [7] provided an algorithm to advance the boundary layers in a controlled and structured manner. Peraire et. al [8] made another important contribution in this class of problems by laying down the essential 3D Euler equations for boundary layer advancement. Very recently, Moreno et al [9] reported an improvisation of the paving algorithm that creates continuous and homogeneous curved triangular and quadrilateral meshes directly on NURBS geometry. Another strong limitation of the paving algorithm is its inability to handle complex sizemaps not making it the obvious choice for highly anisotropic quadrilateral meshes.

2.2 Subdivision Methods

Subdivision algorithms dissect a given polygon into either a simultaneous set of convex domains or a recursively split dynamic domain. The main idea here is to generate a best-split-line to subdivide the polygon into smaller areas that are predominantly convex. The polygonal postulate that states that a convex polygon, when linearly truncated, leads to only convex domains, serves as the basis for this approach. Since convex polygon domain yields a dominant convex mesh, the mesh quality produced by these algorithms is enviably high. It needs to be categorically pointed out that subdivision methods pre-existed advancing-front types and some of these provided the first automatic method of discretizing areas for the purpose of finite element analyses. There are two basic approaches with subdivision - (i) recursive subdivision and (ii) simultaneous subdivision. Each one has its strength and limitations. Sluiter and Hansen [10] and Schoofs et al. [11] made pioneering contributions in this area of recursive subdivision. Sluiter's work was later revisited and revised by many including Talbert and Parkinson [12], Sarrate and Huerta [13] and Cabello [14]. The strength of these algorithms lie in their robustness in handling complex shapes, constraints, sizemaps and efficiency. The meshes produced have reasonable good quality. Mezentsev et al [15] described a generic approach to unstructured mesh generation on subdivision geometry. Berzin et al [16] developed in recent years a subdivision technique based on Modified Butterfly interpolation scheme for triangular mesh generation. Barry Joe [17], in his public domain code GeomPack, proposes a method to proceed from a simultaneous set of convex sub-domains. A similar methodology was proposed by Nowottny [18].

3 CSALF-Q

Automotive and aerospace structural analysts from the engine and transmission areas have long voiced their need for layered boundary conforming meshes in generally unstructured domains. To be able to have suitable control on the parameters of these layered meshes is a key requirement. Another need has been transitioning meshes, well-adapted to surface curvature and local constraints. User control, especially in the boundary layers is also key to these analyses.

In a 1962 essay titled "The Savage Mind" [19] legendary French anthropologist Claude Lévi-Strauss used the word **bricolage** to mean "make creative and resourceful use of whatever materials are at hand". Since then, the term "bricolage" has been used extensively in a large range of disciplines including science, visual arts and engineering to suggest construction or creation of a work from a diverse range of existing designs or algorithms. In the present paper, a similar effort is made to create a new hybrid quadrilateral mesher in 2D that combines two mutually complimentary aspects of unstructured meshing. One of them is a recursive subdivision algorithm. The other is a new paving loop-front method. The original recursive subdivision algorithm **TriQuaMesh,** reported more than three decades back by Sluiter and Hansen [10] emphasizes on a recursive contour or loop-splitting algorithm that produces reasonably good quality quad meshes in an automatic mode. An industrial version of the algorithm has been improvised in the **I-DEAS** and **NX** softwares. The algorithm has had more than a quarter-century long industrial mileage and is well known as one of the most general purpose and robust surface meshing codes in the industry. However, this technique does not create boundary-structured quad meshes like paving. Paving, on the other hand, does not adapt to large size transitions very effectively and is known to be susceptible to the presence of a large number of interior point constraints – a dire necessity for body-in-white meshing in the car industry. In this paper, the traditional paving technique of Blacker and Stephenson [4] is improvised to create an advancing loop-paving-loop algorithm that is coupled to the modified recursive subdivision technique. This results in CSALF (Combined Subdivision And Loop-Front) mesher. A triangular version of the same algorithm has been recently reported elsewhere [20]. It is a *bricolage* mesher that can produce anisotropic unstructured meshes that are predominantly boundary structured. The user can optionally control layer advancement in terms of number and thickness. The mesh, however, has an overall unstructured nature and elegantly adapts to sharply varying size fields and can honor any number of valid interior point constraints.

Given a precisely discretized (isotropic or anisotropic) polygonal boundary, representing the area to be meshed, the proposed hybrid mesher algorithm initiates the meshing process with an advancing loop front method. The initial goal is to generate high-quality boundary conforming layered meshes. As the layer penetrates into the mesh area, new loops are created each time, reducing the mesh area. As the new loops begin overlapping, the advancing loop front algorithm is terminated and the subdivision algorithm is activated as shown in Fig. 1.

At this point, all loops are connected into a single continuous contour following procedures described by Cabello [14]. All interior point constraints are treated as a single-point loop. Next, the contour is split recursively by determining a best-splitting line. The splitting line dissects the compound contour into a left sub-contour and a right sub-contour (Fig 4.) Each left sub-contour is now recursively split into two children sub-contours - one on the right and a another one on the left. The split lines are so chosen that they make angles nearing 90 degrees with the contour boundaries.

Recursive splitting finally fills out the mesh area.

The overall algorithmic logic can be expressed in two phases represented by the following twin-algorithms –

ALGORITHM I: Generating a single continuous contour-loop

1. The 3D facetted surface is first flattened into a 2D domain following domain generation procedures as discussed recently by Beatty and myself [21].
2. Face-interior features like holes, scars (loops with repeated edges) and convex-shaped inner loops are identified.
3. A local and global size-map is developed over the 2D domain. Number of paving layers around inner loops are determined based on feature recognition, proximity to other boundaries, global size-map and local user-driven size criteria.
4. Nodes are generated on the boundary of the face accordingly and a triangular background mesh (based on the TQM algorithm, [15]) is created for size-sampling.
5. A mesh area is defined in a 2D domain with **N** meshed loops
6. The **N** meshed loops represent the area boundary and are unique and non-intersecting
7. While loops are non-intersecting and number of layers defined on various loops are not exhausted
   { 7a. Cycle all **N** loops, inner loops first, the outer loop last
      {
         7aa. Activate the loop-paving-front algorithm on the **i-th** loop
         7ab. Check if the new loop self-intersects. If it does, turn off the loop-paver for this loop and go to 7ac.
         7ac. Continue the cycle, go to the next loop
      }
   }
8. Connect all **N** loops into a continuous contour **C**

ALGORITHM II : The Symbiotic Triad

To fill the single continuous contour-loop C with a quad or quad-dominant mesh, three meshing algorithms are used in tandem in a symbiotic manner - Recursive Subdivision, Transfinite Interpolation and Loop-Paving following a stratagem described by the flowchart in Fig.1. As soon as the contourloop is split, it results in two child contourloops - a left loop C_L and a right loop C_R.

If the left contourloop C_L is convex, meets the sizemap limitations (as explained in section 7) and is map-meshable, the transfinite mesher is called. Otherwise it is loop-paved, provided all apriori and posteriori checks pass. The new loop produced by loop-paving, or the unaffected original loop is returned to the recursive subdivider. This symbiotic handling of the loops continues between the three meshing algorithms until there is no right loop and the left loop is completely filled. The right loop is now split into two loops again and the process is repeated until the complete mesh-area is filled out. Some regions of the 2D mesh generated is topologically cleaned by a mesh cleaner [22] and smoothed using a variational smoother [23].

4 Paving Loop-Front

The paving loop front approach differs from traditional advancing fronts in the sense that a series of "loop fronts" are considered for mesh advancement. When all loop fronts of a loop are advanced successfully, a new loop results that imitates and preserves the profile of the starting loop. This results in a perfectly layered mesh near inner and outer boundaries of the mesh area – a feature which is often preferred by structural and computational fluid dynamics analysts. Each loop front as shown in Fig 2., consists of a real segment BC and a virtual segment AB (represented by the previous element edge on the loop). The virtual segment represents a trailing edge of the real segment of the previous loop-front.

The virtual segment helps pave the loop front. The angle θ, between the real and virtual segments of the loop front is used to characterize the loop front. Table I lists the various loop fronts supported by the algorithm. **Q** denotes the new node that is created upstream of the loop-front, **P** denotes the node downstream of the loop-front that mostly pre-exists (created in certain special cases). All additional new nodes **R, S, T** and **U** are created between **P** and **Q**.

4.1 New Node Placement

When a paving loop-front is advanced, the new or optimal node placement depends on the type of the loop-front. For an flat paving loop-front, the position of the only new node Q can be expressed as

$$r_Q = r_B + V_t \cdot g \tag{1}$$

where r_B is the position vector of node B, V_t is the bisector of angle ABC and grading g is assumed to be the harmonic mean of the sizes at the nodes defining the loop-front.

$$g = 3g_A g_B g_C / \sin(\theta/2)(g_A g_B + g_B g_C + g_C g_A) \tag{2}$$

g_A, g_B, g_C are the grading values at nodes A, B & C.

CSALF-Q: A Bricolage Algorithm for Anisotropic Quad Mesh Generation

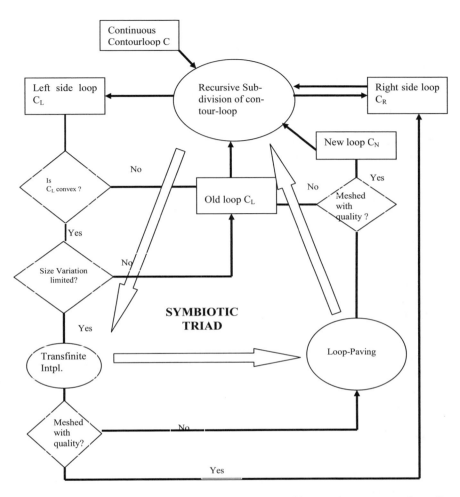

Fig. 1. Flowchart of Symbiotic Triad Algorithm for meshing continuous contourloop C

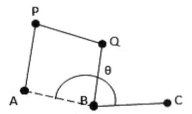

Fig. 2. A typical paving loop-front

Table 1 shows the formulas for positioning the new nodes during loop-paving. For a Right-Reversal paving loop-front, 3 new nodes (Q,R&S) are created. The angle bisector unit vector is transformed by ± θ/6 to obtain the translational unit vectors for nodes Q and R. For all nodes that are diagonally opposite to node B, the grading is multiplied by a factor of √2 to account for orthogonality. In a similar manner for a Reversal paving loop-front 5 nodes (Q,R,S,T,U) are created. The angle bisector unit vector is transformed by ± θ/4 and ± θ/8 to obtain the translational unit vectors for nodes Q,R and U,S respectively. Each new node is tested for proximity with its immediate neighbors in an attempt to avoid creating nearly-squished or hourglass-shaped quads. When such situations arise decision is taken either to merge proximate nodes or completely reject the loop for paving.

4.2 Loop Closure

Loop closing is a delicate affair. Exceptions need to be made to the forward creation theorems (as illustrated and explained in Table 1) for the various paving loop-fronts in this case. Table 2 lists the stencils used for each-pair of loop-front types. The terminal loop front is represented by XAB while the first loop-front is ABC. All nodes marked P",Q", R" etc. refer to the original nodes created during the advancement of the first front. Nodes P,Q,R,T etc. denote the last loop-fronts new advancing node positions created at the time of loop closure. To close the loop, 1,2 or 3 terminal elements need to be created based on the pairing type.

After each paving-loop is advanced or paved, the elements and nodes generated in the process are smoothed using an isoparametric smoother as described by Blacker and Stephenson [4] with a minor difference in that it is applied to a closed loop of elements and node rings and thus the Dirichlet and Neumann boundary conditions are placed accordingly.

4.3 Loop Front Evaluators

Loop closing is a delicate affair. Exceptions need to be made to the forward creation theorems (as illustrated and explained in the Appendix (Table I) for the various loop-fronts in this case. Table II lists the various stencils used for each-pair of loop-front types. The terminal loop front is represented by ABC while the first loop-front is BCE. Pt D represents the first new advancing point of the loop. Pt Q denotes the last new advancing point which, at the time of loop closure, has already been created. To close the loop, 1 or 2 terminal elements need to be created based on the pairing type.

Before and after a loop is paved a number of checks are performed, each one for a different purpose. These checks are explained below

4.3.1 Loop Area Shrinkage

With every successful loop-paving, the loop area shrinks. This check is a posteriori check that is performed on the paved loop to ensure the ratio of the loop areas of the new and original loops (A_n, A_o) do not shrink below an area tolerance (ε_A) -

$$A_n/A_o > \varepsilon_A \qquad (3)$$

4.3.2 Loop Perimeter Growth/Shrinkage

The loop perimeter can grow and shrink as the loop advances. This check is also posteriori and compares the ratio of the perimeters of the new and original loops (P_n, P_o) against a perimeter tolerance (ε_P) as described by eqn. (4).

$$|P_n/P_o| > \varepsilon_P \qquad (4)$$

4.3.3 Loop Length Variation

Since paving is more heuristic than recursive subdivision, the mesh pattern and number of elements tend to vary overtly for small variations of the global size [24]. Also, in traditional paving *wedging* and *tucking* are performed [4] to prevent fronts from shrinking and expanding beyond size limits. With wedges and tucks, paved meshes often produce diamond shaped quads in the middle of isotonic flow-lines. Certain types of analyses, especially crash analysis (of vehicles), are sensitive to that topology [25]. Therefore, a check is performed apriori where the rate of variation (**dL/ds**) of loop lengths (L) along its boundary (s) is kept below a certain limit (ε_S).

$$|dL/ds| < \varepsilon_S \qquad (5)$$

4.3.4 Loop-Front Relative Growth Rate

This posteriori check ensures that the ratio of the extremums (**|dL/ds|**$_{max}$, **|dL/ds|**$_{min}$) of the loop-front differential growth rates are within a specified limit (ε_G). Any irregular and abrupt loop-front size changes are avoided during paving so as to ensure that the paved mesh layers are good quality and regular.

$$|dL/ds|_{max} / |dL/ds|_{min} < \varepsilon_G \qquad (6)$$

5 Recursive Subdivision

The recursive subdivision algorithm consists of the following steps-

1. join all loops into a single continuous loop
2. recursively split the continuous loop by a best-split-line
3. determine the best split line
4. estimate the number of nodes to be generated on the split line
5. space the nodes on the split line
6. if no more nodes can be generated construct elements if loop has 3 or 4 nodes only
7. goto step 3 and continue until mesh is done

5.1 Connecting Loops

In order to connect all loops to a single continuous loop, a cartesian grid of the global element size is constructed in the background. Given a mesh area with **n** loops, these cells are used to identify a pair of nodes representing the shortest distance between outer loop l_o and any inner loop l_i along a line whose optimum angular deviation φ (discussed in sect. 5.2) is minimum. The connecting line is checked for intersection with any other loops. Once this connection is made the problem now is reduced to one connecting **n-1** loops. The process is repeated until it a single continuous loop results.

5.2 Recursion Algorithm

The recursive subdivision algorithm takes a single 2D contourloop defined by a sequence of nodes and recursively splits it to fill the region. The input contourloop must not be self-intersecting nor have coincident nodes but can be self-touching. Nodes can also have repeated entry in the loop. The subdivision logic is described by the following flow-logic

ALGORITHM III : The recursive subdivision logic

While the compound contour is not completely filled
(refer Fig. 3b)

{   1a. Get the best splitting line **A** that makes an
angle close to 90 degree with the contour.
The splitting line divides the contour into
a sub-contour on the left (C_L) and one on
the right(C_R).
   1b. while the left sub-contour C_L is unfilled
   {
     Repeat step 1a
   }
   1c. while the right sub-contour C_R stays unfilled
   {
     Repeat step 1a
   }
}

5.3 Selecting the Best Split Line

The split line functional Φ for a split line joining boundary nodes j and k, is expressed as a linear combination of normalized parameters, **L**, φ and ε (where A_1, A_2, A_3 are constants).

$$\Phi = A_1 L + A_2 \varphi + A_3 \varepsilon \qquad (7)$$

The normalized length parameter **L** is given by $L = l_{jk}/l_d$; l_{jk} is the length of the split line jk (as shown in Fig. 3b), l_d is the characteristic length, which is the diagonal of the rectangular box bounding the mesh area $B \equiv [(x_{min}, y_{min}), (x_{max}, y_{max})]$ and given by

$$l_d^2 = (x_{max} - x_{min})^2 + (y_{max} - y_{min})^2 \tag{8}$$

The normalized split angle φ is expressed as the normalized sum of the deviations of the 4 split angles (shown in Fig. 3b) from the ideal quadrilateral angle **π/2**.

$$\varphi = \frac{\sum_{i=1}^{4} |\varphi_i - \pi/2|}{2\pi} \tag{9}$$

The percentage of length error ε resulting from fitting n nodes on the split line based on the grading values of sample points is discussed elsewhere [14] in details.

The minimum value of the split line functional gives the best split line. However, many of the split line candidates in a concave loop are invalid as they fall outside the domain (as shown in Fig. 3a). A boundary visibility criterion is set up to eliminate these invalid candidates.

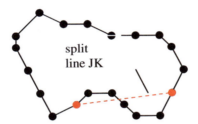

Fig. 3a. An invalid split line JK joining nodes j and k of the single continuous loop.

Based on experience with a large range of mesh areas, a range for the constants (A_1, A_2, A_3) are heuristically determined [14].

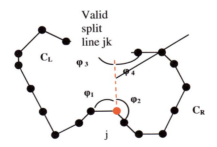

Fig. 3b. A valid split line joining nodes j and k.

5.4 Split Line Discretization

Split line discretization is extremely critical to the ability of the mesh to adapt to a given size field. The first step is to determine the number of nodes **n,** to be generated on a split line. A set of **s** sample points is first created on the split line with an uniform spacing. s is calculated as

$$s = l_{jk}(g_j + g_k)/2g_jg_k - 1 \tag{10}$$

The grading values (g_i) at these **s** sample points are determined from eqn. (11). The grading distribution along the split line is assumed to be an **s+2**-polynomial variation of the natural line coordinate ξ expressed as

$$g_i(\xi) = 1+C_1\xi_i+C_2\xi_i^2+C_3\xi_i^3+....C_{s+2}\xi_i^{s+2} \tag{11}$$

Substituting the grading values at these **s+2** interior sample points, the simultaneous equation family (11) is solved to determine the coefficients $C_1, C_2,...C_{s+2}$.

For computational efficiency, the grading function could be limited to a quintic order, i.e. $s < 4$.

$$\begin{aligned}
g_1 &= 1+C_1\xi_1+C_2\xi_1^2+C_3\xi_1^3+....C_{s+2}\xi_1^{s+2} \\
g_2 &= 1+C_1\xi_2+C_2\xi_2^2+C_3\xi_2^3+....C_{s+2}\xi_2^{s+2} \\
&\quad\quad\quad\quad .... \\
g_{s+2} &= 1+C_1\xi_{s+2}+C_2\xi_{s+2}^2+C_3\xi_{s+2}^3+....C_s\xi_{s+2}^{s+2}
\end{aligned} \tag{12}$$

The number of nodes **n** to be generated on the split line jk is estimated as

$$n = l_{jk}/g_l \text{ where } g_l = \int_{\xi=0}^{1} g_\xi \, d\xi \tag{13}$$

In the natural or parametric coordinates of the split line these n nodes will be equally spaced. The location of the i-th node on the split line can be expressed in terms of its grading value g_i, coordinates of the previous point p on the same line, and the coordinates of the two end nodes j and k. The following pair of equations need to be solved to evaluate the location of node i. The split line functional Φ for a split line joining boundary nodes j and k, is expressed as a linear combination of normalized parameters, L, φ and ε.

$$x_i^2(1+m^2)+2x_i[x_p + m(c-y_p)]+x_p^2+(c-y_p)^2+g_i^2 = 0 \tag{14}$$

and the equation of the split line

$$y_i = mx_i + c \tag{15}$$

where the slope of the split line is

$$m = (y_k-y_j)/(x_k-x_j) \text{ and} \tag{16}$$

$$c = (x_ky_j-x_jy_k)/(x_k-x_j) \tag{17}$$

6 Transfinite Interpolation

Transfinite interpolation (extensively discussed by Armstrong and Tam [26], Mitchell [27,28]) and more recently [25]) is optionally employed in convex subcontour loops only if the size-map variation within its domain is small. The advantage of TFIs is two fold - a) firstly it is fast and efficient and b) it is insensitive to small local variations in the size-map and is guaranteed to generate a structured quad mesh which neither loop-paving nor recursive subdivision can promise.

7 Local and Global Anisotropy

The detailed operation of the symbiotic triad explained in Fig. 1 depends on the driving size-map function defined by the local and global anisotropy requirement. Local anisotropy applies to all face-interior boundaries or constraints and is defined by two parameters - a) Number of layers desired (n_l); and b) A layer thickness function $L(n)$. Figures 4(a)-(b) depict the meshes around a hole for three different sizing functions (number of layer desired $n_l = 5$). Many applications, especially structural mechanics, require boundary-structured graded local meshes and even if the aspect ratio is higher than usual, two rings of well-shaped quads ($J_r < 1.2$) minimizes stress solution and smoothening errors. The Fibonacci function [$L_i = 1,1,2,3,5$; i =0-4;] thus assumes importance. The ramp function, however, is more popular for its simplicity.

Global anisotropy applies to the entire surface except for interior boundaries with local anisotropic definitions and is defined by a sizing function. The size or grading of the mesh at any interior node is evaluated from the size field represented by the background mesh as

$$g = \Gamma(x,y) \tag{18}$$

To evaluate the grading at any interior point **i** **(x,y)**, it's owner triangle **j** in the background mesh is identified by a space hashing mechanism. The grading at point **i** is thus expressed in terms of the field values at the vertices of triangle j as

$$g_i(x,y) = \sum_{k=1}^{3} N_k g_{jk}(x,y) \tag{19}$$

g_{jk} **(x,y)** denote the grading at the vertices of triangle **j**
N_k represent the shape functions of triangle **j**

As the mesher fills out the surface, for the unmeshed area **U** with N_s sampling points, an *Extremum Size Gradient* (g_{rU}) is constantly computed as the ratio of the maximum to the minimum grading averaged over the unmeshed area. It can expressed as -

$$g_{rU} = (\int_U (g_{max}(x,y)/g_{min}(x,y))\, du)/N_s \tag{20}$$

For all ESG > 1.2, transfinite interpolation is never attempted on sub-contour loops as it might render the mesh insensitive to the varying size field.

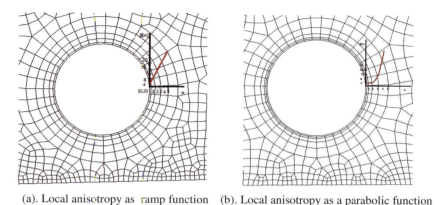

(a). Local anisotropy as ramp function   (b). Local anisotropy as a parabolic function

Fig. 4. Local and global anisotropy near a hole for various local sizing functions

8 Mesh Quality

Mesh quality is measured in both local and global domains. It is defined as the harmonic mean of the ratio of the extremum Jacobian's (J_r) and can be expressed, for a mesh of N elements as

$$\tau = N / \left(\sum_{i=1}^{N} 1.0/(J_{ri}) \right) \tag{21}$$

where $J_{ri} = J_{imax}/J_{imin}$ is the Jacobian ratio of the i-th element

It is measured both in the local paved layers (τ_l) and the entire mesh (τ_g). Both are used to compare meshes generated by the proposed algorithm with that generated by recursive subdivision [14].

9 Results and Discussion

Fig. 5 shows a typical example of a curved face with a central circular cut-out. The analysis demands a tenth of the global element size at the inner loop than the outer. Fig. 5a shows the TQM (recursive subdivision mesh) while Fig. 5b shows the mesh generated by CSALF-Q. While the TQM mesh adapts well to the acutely varying sizemap between the inner loop and the outer, element quality suffers immensely, especially near the inner boundary - the zone of analytical interest. Neither is the mesh along the outer boundary structured. The local mesh quality τ_l over a region defined by a circle with radius 1.5r (r = radius of the inner loop) is 0.7 while the global mesh quality τ_g is 0.78. The CSALF-Q mesh grows the paved ring around the inner loop gradually trying to compromise between the local sizemap gradient and the mesh quality desired. In comparison to TQM it reports admirably high local and global mesh qualities ($\tau_l = 0.98$, $\tau_g = 0.83$) and one layer of boundary structured elements along the outer loop. The insensitivity of loop-paving to the local sizemap allows for a good number of boundary structured inner layers followed by a smoother transition leading to an overall enhancement of mesh quality while honoring the sizemaps with reasonable accuracy. For the TQM mesh, the worst quad angle (163^0) is on the inner boundary while in the CSALF-Q mesh the worst quad ($152\ °$) is in between the boundaries in the zone of low importance.

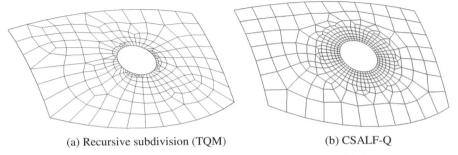

(a) Recursive subdivision (TQM) (b) CSALF-Q

Fig. 5. Mesh anisotropy on a curved face with a central hole.

Fig. 6 shows a swept hex mesh on a body with 4 rectangular slots where the wall faces are frozen with a very fine transfinite mesh build to expedite a local contact analysis. As a result, the source face of the body, although absolutely planar, gets a widely varying sizemap. Fig. 7 & 8 depict meshes on a highly curved face for TQM and CSALF-Q respectively. While the TQM and CSALF-Q meshes are close, one can easily notice a more boundary structured mesh and a smoother transition quad mesh [$\tau_g = 0.79$ (CSALF-Q), $\tau_g = 0.68$ (TQM)] resulting in relatively better quality mesh for CSALF-Q.

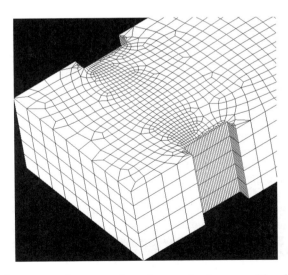

Fig. 6. Mesh anisotropy on the source face of a swept volume resulting from certain wall faces frozen with a finer transfinite mesh

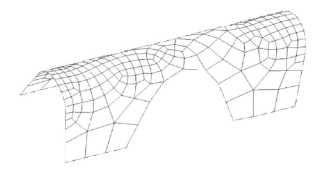

Fig. 7. Anisotropic quad mesh generated by TQM [recursive subdivision, [10,14]] alone.

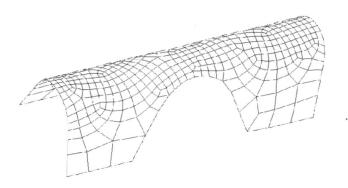

Fig. 8. Anisotropic quad mesh generated by CSALF-Q.

10 Conclusion

Paving is known to produce boundary structured quad meshes of admirable quality but fares poorly when challenged by a variegating global sizemap. Recursive subdivision, on the contrary, handles such size variations with elegance and robustness but produces relatively poor quality meshes that are rarely boundary-structured. In this paper, by adopting a *bricolage* technique the contrasting strengths of paving and recursive subdivision are fused and used in conjunction with subarea transfinite meshing to produce CSALF-Q - a new automatic 2D quad meshing algorithm that maintains element quality but is sensitive to both local and global sizemaps. To enable the bricolage algorithm to produce anisotropic yet high-quality boundary structured meshes, the loops are initially meshed recursively with a new "loop-paving" technique. The remaining interior domain is filled out by a symbiotic triad combining recursive subdivision, transfinite interpolation and loop-paving in a balanced and efficient manner. Loop-fronts are classified and rule sets are defined for each, to aid optimum point placement. Stencils used for loop-closure are presented. Results presented compare both local and global element quality of meshes generated by the new algorithm with that of

TQM. They clearly indicate that apart from offering user control on the number and thickness of paved layers, CSALF-Q tends to generate meshes that are more boundary structured and smoothens out the quad mesh transition in a manner that strikes a good compromise between quad mesh quality and anisotropy.

Acknowledgement

The author would like to dedicate the present paper to Thijs Sluiter and Ted Blacker, the founding fathers of TriQuaMesher and Paving respectively. He also extends sincere thanks to Michael Hancock, Jean Cabello and Kirk Beatty of the Meshing and Abstraction Team at Siemens PLM software for their continued support of this research.

References

1. Owen, S.: A Survey of Unstructured Mesh Generation Technology. In: Proceedings, 7th International Meshing Roundtable, Sandia National Lab, pp. 239–267 (1998)
2. George, P.L., Frey, P.J.: Mesh Generation. Hermes Science, UK (2000)
3. Marcum, D.L.: http://www.erc.msstate.edu/thrusts/grid/
4. Blacker, T., Stephenson, M.: Paving: A new approach to automated quadrilateral mesh generation. Int. Journal Num. Meth. Engg. 32, 811–847 (1991)
5. Staten, M.L., Kerr, R.A., Owen, S.J., Blacker, T.D.: Unconstrained Paving and Plastering: Progress Update. In: Proceedings, 15th International Meshing Roundtable, pp. 469–486. Springer, Heidelberg (2006)
6. White, D.R., Kinney, P.: Redesign of the Paving Algorithm: Robustness Enhancements through Element by Element Meshing. In: Proceedings, 6th International Meshing Roundtable, Sandia National Laboratories, pp. 323–335 (1997)
7. Pirzadeh, S.: Unstructured viscous grid generation by advancing-layers method. AIAA 93-3453-CP, AIAA, pp. 420-434 (1993)
8. Peraire, J., Peiro, J., Formaggia, L., Morgan, K., Zienkiewicz, O.C.: Finite Element Euler Computations in Three Dimensions. Int. Journal Num. Meth. Engg. 26, 2135–2159 (1988)
9. Moreno, J., Algar, M.J., Diego, I.G., Catedra, F.: A new mesh generator optimized for electromagnetic analysis. In: Proceedings, 5th European Conference on Antennas and Propagation (EUCAP), Rome, pp. 1734–1738 (2011)
10. Sluiter, M.L., Hansen, D.L.: A general purpose automatic mesh generator for shell and solid finite elements. Computers in Engineering, ASME, pp. 29–34 (1982)
11. Schoofs, L.H., Van Buekering, T.M., Sluiter, M.L.C.: TRIQUAMESH User's Guide, Report WE 78-01, Eindhoven University of Technology, Netherlands (1978)
12. Talbert, J.A., Parkinson, A.R.: Development of an automatic, two-dimensional finite element mesh generator using quadrilateral elements and bezier curve boundary definitions. Int. J. Num. Meth. Engg. 29, 1551–1567 (1991)
13. Sarrate, J., Huerta, A.: Efficient unstructured quadrilateral mesh generation elf-organizing mesh generation program. Int. Journal Num. Meth. Engg. 49, 1327–1350 (2000)
14. Cabello, J.: Towards Quality Surface Meshing. In: Proceedings, 12th International Meshing Roundtable, pp. 201–213 (2003)

15. Mezentsev, A.A., Munjiza, A., Latham, J.P.: Unstructured computationaal meshes for subdivision geometry of scanned geological objects. In: Proceedings. 14th International Meshing Roundtable, Sandia National Lab, pp. 73–89. Springer, Heidelberg (2005)
16. Berzin, D., Kojekine, N., Hagiwara, I.: Finite element mesh generation using subdivision technique. In: Nihon Kikai Gakkai Sekkei Kogaku, Japan, vol. 11, pp. 207–210 (2001)
17. Joe, B.: Quadrilateral mesh generation in polygonal regions. Computer Aided Design 27, 209–222 (1995)
18. Nowottny, D.: Quadrilateral mesh generation via geometrically optimized domain decomposition. In: Proceedings, 6th Int. Meshing Roundtable, pp. 309–320 (1997)
19. Lévi-Strauss, C.: La Pensée sauvage, Paris (1962); English translation as The Savage Mind, Chicago (1966), ISBN 0-226-47484-4
20. Mukherjee, N.: A Combined Subdivision and Advancing Loop-Front Surface Mesher (Triangular) for Automotive Structures. Int. J. Vehicle Structures & Systems 2(1), 28–37 (2010)
21. Beatty, K., Mukherjee, N.: Flattening 3D Triangulations for Quality Surface Mesh Generation. In: Proceedings, 17th International Meshing Roundtable, pp. 125–139 (2008)
22. Kinney, P.: CleanUp: Improving Quadrilateral Finite Element Meshes. In: Proceedings, 6th International Meshing Roundtable, pp. 437–447 (1997)
23. Mukherjee, N.: A hybrid, variational 3D smoother for orphaned shell meshes. In: Proceedings, 11th Int. Meshing Roundtable, pp. 379–390 (2002)
24. Knupp, P.: Applications of mesh smoothing: Copy, morph, and sweep on unstructured quadrilateral meshes. International Journal for Numerical Methods in Engineering (1999)
25. Mukherjee, N.: High Quality Bi-Linear Transfinite Meshing with Interior Point Constraints. In: Proceedings, 15th International Meshing Roundtable, pp. 309–323 (2006)
26. Tam, T.K.H., Armstrong, C.G.: Finite element mesh control by integer programming. International Journal for Numerical Methods in Engineering 36, 2581–2605 (1993)
27. Mitchell, S.: Choosing corners of rectangles for mapped meshing. In: Proceedings, 13th Annual Symposium on Computational Geometry, pp. 87–93 (1993)
28. Mitchell, S.: High Fidelity Interval Assignment. In: Proceedings, 6th International Meshing Roundtable, pp. 33–44 (1997)
29. Paving algorithm, UG/Scenario (1999-2002); NX3, Siemens PLM Software (2004)

Appendix

Table 1. Paving Loop Front Models and Front Advancement Rules

Paving Loop front Type	No. of new nodes	New Elements
Acute Loop Front $\theta \leq 70°$	None	None
Right/Obtuse Loop Front $70° < \theta \leq 110°$ $110° < \theta \leq 140°$	Pt P or Q if this is the first front. Else, none. Node P comes from the previous front, Q is same as P. The obtuse loop-front is differentiated from right, because sometimes based on quality criteria, the obtuse front may be treated as a flat-front.	$\square$PABC or $\square$QABC If element is created, next front must be skipped.
Flat Loop Front $140° < \theta \leq 220°$	1 Node Q Position vector of pt Q $r_Q = r_B + Tg_Q$ T = unit bisector vector of angle θ Grading at Q $g_Q = 3g_A g_B g_C / \sin(\theta/2) (g_A g_B + g_B g_C + g_C g_A)$	$\square$QPAB
Right Reversal Loop Front $220° < \theta \leq 310°$	3-Nodes, Q, R & S $r_Q = r_B + T_Q g_Q$ $r_R = r_B + Tg_R$ $r_S = r_B + T_S g_S$ T= unit bisector vector of angle θ T_Q = T transformed by $(-\theta/6)$ T_R = T transformed by $(\theta/6)$ $g_Q = g_R = 3g_A g_B g_C / \sin(\theta/3)$ $(g_A g_B + g_B g_C + g_C g_A)$ $g_S = 3\sqrt{2}g_A g_B g_C / \sin(\theta/3) (g_A g_B + g_B g_C + g_C g_A)$	$\square$BRPA $\square$QSRB
Reversal Loop Front $310° < \theta \leq 360°$	5- Nodes, Q, R, S, T & U $r_Q = r_B + T_Q g_Q$ $r_R = r_B + T_R g_R$ $r_S = r_B + Tg_S$ $r_T = r_B + T_T g_T$ $r_U = r_B + T_U g_U$ T = bisector vector of angle θ T_Q = T transformed by $(-\theta/4)$ T_U = T transformed by $(-\theta/8)$ T_S = T transformed by $(\theta/8)$ T_R = T transformed by $(\theta/4)$ $g_Q = g_T = g_R = 3g_A g_B g_C / \sin(\theta/4)$ $(g_A g_B + g_B g_C + g_C g_A)$ $g_S = g_U = 3\sqrt{2}g_A g_B g_C / \sin(\theta/4)$ $(g_A g_B + g_B g_C + g_C g_A)$	$\square$ABRP $\square$BTSR $\square$BQUT

Table 2. Paving Loop Front Closure Stencils

First Paving Front (ABC)	Last Paving Front (XAB)	Elements to create	Description
Acute/Obtuse/Right	Acute	None	Not applicable
	Obtuse/Right	☐ XYAC Node Q" is replaced by paving front node X	
	Flat	☐ PXAQ ☐ Q"ABC changes to QABC Node Q" of the first loop-front is merged with Q of the last front.	
	Right Reversal	☐s PXAR, SRAQ ☐ Q"ABC changes to QABC Pt Q" of the first loop-front is merged with Q of the last front.	
	Reversal	☐s PBAR, TSRA, UTAQ ☐ Q"ABC changes to QABC Node Q" ≡ Node Q (merged)	
Flat	Acute/Obtuse/Right	☐ XABQ	
	Flat	☐s Q"QAB	
	Right Reversal	☐s PXAR, AQSR, Q"QAB Node Q ≡ Node P" (merged)	
	Reversal		Similar to previous

CSALF-Q: A Bricolage Algorithm for Anisotropic Quad Mesh Generation 509

Right Reversal	Acute/Obtuse/Right	☐ R"P"AB is changed to R"XAB Node P" is replaced by Node X	
	Flat	☐QPXA ☐R"P"AB is changed to R"QAB as Node Q ≡ Node P" (merged)	
	Right Reversal	☐sRPXA, SRAQ ☐R"P"AB is changed to R"QAB as Node P" ≡ Node Q (merged)	
	Reversal	☐sRPXA, TSRA, UTAQ ☐R"P"AB is changed to R"QAB as Node P" ≡ Node Q (merged)	
Reversal	Acute/Obtuse/Right	☐R"P"AB is changed to R"XAB as Node P" is replaced by Node X	
	Flat	☐sQPXA, R"QAB ☐R"P"AB is changed to R"QAB as Node P" ≡ Node Q (merged)	
	Right Reversal	None	Not Applicable
	Reversal	None	Not Applicable

Jaal: Engineering a High Quality All-Quadrilateral Mesh Generator

Chaman Singh Verma[1] and Tim Tautges[2]

[1] Department of Computer Sciences, University of Wisconsin, Madison, 53706
csverma@cs.wisc.edu
[2] Argonne National Laboratory, Argonne IL, 60439
tautges@mcs.anl.gov

Summary. In this paper, we describe the implementation of an open source code (*Jaal*) for producing a high quality, isotropic all-quadrilateral mesh for an arbitrary complex surface geometry. Two basic steps in this process are: (1) Triangle to quad mesh conversion using Suneeta Ramaswamy's tree matching algorithm and (2) Global mesh cleanup operation using Guy Bunin's one-defect remeshing to reduce irregular nodes in the mesh.

These algorithms are fairly deterministic, very simple, require no input parameters, and fully automated yet produce an extremely high quality all-quadrilateral mesh (with very few 3 and 5 valence irregular nodes) for large class of problems.

1 Introduction

There are many applications where an all-quadrilateral and all-hexahedral mesh is preferred over a triangle mesh (non-linear structural mechanics, higher order spectral methods, texture mapping etc). While many provably robust, high-quality triangle mesh generators have been described and are available as open source software (e.g. the Triangle package from [9]), to our knowledge, there are no provably robust all-quadrilateral mesh generation algorithms available in open source form. In this paper, we describe a robust, high-quality quadrilateral mesh generation algorithm freely available in source code form.

We seek to implement a robust, high-quality all-quadrilateral mesh generation algorithm capable of meshing 3D, non-simply-connected surfaces. This algorithm must be theoretically sound, to support robustness for a wide class of input domains. The algorithm must be capable of working from a pre-defined list of bounding edges, with the only constraint being that the number of edges is even (note, we do not constrain the evenness of intervals on bounding loops, only on the total number of bounding edges). The resulting meshes should have relatively few "defects" (internal nodes with degree other than four), and where possible, should be boundary-sensitive (with lower-quality elements located relatively far from boundaries). The algorithm should operate on 3D surfaces, possibly with multiple bounding loops, without reliance on an underlying 2D parameterization. Finally, the algorithm should be robust even for large relative size transitions in the bounding edges.

Quadrilateral meshing algorithms (if not unstructured meshing algorithms in general) can be classified in two groups. Indirect methods generate some other intermediate decomposition of the domain, e.g. a triangle mesh or a medial surface decomposition, afterwards converting the pieces to quadrilaterals through recombination or further decomposition. In contrast, direct methods generate quads directly, often using advancing fronts. When deciding which approach to use, two key observations guide us. First, we observe that geometric computations often interfere with the initial generation of quadrilaterals. This occurs frequently in advancing front methods, usually by missing the intersections of the advancing fronts, especially for problems with rapid size transitions. Second, virtually all unstructured meshing algorithms finish with a cleanup phase, where local topological modifications are used to improve the regularity of the mesh, sometimes making extensive changes to the initial mesh. This reduces the impact of poor initial mesh quality, and guides us to focus more on "closing" the initial mesh, no matter the initial quality. Hence, the approach we take is to generate the initial quadrilateral mesh using an indirect method, by converting triangles to quads. As we show later, this is a provably reliable method for obtaining the initial mesh. This mesh is subsequently cleaned up using topological transformations. The result is an algorithm with provable robustness that in practice generates high-quality all-quadrilateral meshes.

This paper is organized as follows. In section 3.1, we describe a tree matching algorithm and in sections 4 and 5, we describe some local and global operations for mesh cleanup. In section 6, we show results for several input problems, followed by conclusions of this work.

2 Previous Work

The Q-Morph algorithm [22] transforms a given triangle mesh into a quadrilateral mesh using an *advancing front* method. In this approach, quadrilaterals are formed using existing edges in the triangulation, by inserting additional nodes, or by performing local transformations to the triangles. The final mesh quality is improved by topological clean-up and local smoothing operations. This approach has been implemented by several commercial meshing packages, but is not available in open-source form.

Marcelo's [23] *CQMesh* software converts a given triangulated mesh into an all-convex quadrilateral mesh, but the resulting mesh has many irregular nodes. QMPP is an extension to CQMesh that improves the quality at the boundaries, but topological improvements in the interior are limited. The first part of our work is greatly influenced by this approach.

Betul et al. [24] developed a direct quadrangulation algorithm with guaranteed angle bounds (18.43-171.86 degrees). A circle packing algorithm by Bern et al. [19] constructs quadrilaterals in a polygonal domain with an upper bound of 120 degrees in the interior, but provides no bounds on the smallest angle. These two theoretically proven algorithms are direct algorithms and work only for 2D planar surfaces. To the best of our knowledge, theoretically, nothing is proven for 3D surface quadrangulation.

Spectral methods[20] provide a novel approach for quadrangulating a manifold polygonal mesh using Laplacian eigenfunctions which are the natural harmonics of the surface. The surface Morse functions distribute their extrema evenly across a mesh, which connect via gradient flow into a quadrangular base mesh (known as a Morse-Smale Complex). An iterative relaxation algorithm refines this initial complex to produce a globally smooth parameterization of the surface. From this, well-shaped quadrilateral mesh with very few defects are generated. Although very elegant, this approach has many implementation problems: (1) The quality of mesh depends on an appropriate morse function, whose choice is often heuristic and, if poor, leads to high numbers of critical points; (2) calculations of the first few (about 40) eigenvectors are very expensive and sometimes impractical for a large mesh. (3) Tracing the curves from maxima to minima via saddle points may have geometric robustness issues.

Felix et al. [7] generate a high quality quadrilateral mesh using global parameterization which is guided by a user-defined frame field (often the principal curvature directions). These frame fields simplify to vector fields on the covering spaces, so that the problem of parameterization with frame fields reduces to the problem of finding a proper integrable vector field on the covering surface. Similarly, Bommes et al. [4] formulated the quadrangulation problem as two step process (cross field generation and global parameterization), both formulated as a mixed-integer problems. This scheme allows placement of singularities at geometrically meaningful locations, and produces meshes with favorable orientation and alignment properties. However, the method still requires the solution of a numerical problem, which can be expensive.

Bommes et al. [5] developed an algorithm that optimizes high-level structure of a given quadrilateral mesh, especially for helical structures. Their algorithm is able to detect helical structures and remove most of them by applying grid preserving operators.

Hormann et al. [10] presented an algorithm that converts an unstructured triangle mesh with boundaries into a regular quadrilateral mesh using global parameterization that minimizes geometric distortion. The second part of our work to reduce irregular nodes is also based on *remeshing patches*, but we do this on topologically convex patches instead of using global parameterization.

In the area of topological cleanup operations, many methods have been presented. Canann [2] and Kinney [18] present ever-growing numbers of operations, usually applied using an isomorphism approach. In contrast to those, Bunin [8] presents a very elegant technique for removing defect vertices based on patch replacement, whose results are better than those of the isomorphism approach while also being vastly simpler to implement. We make heavy use of this technique in the present work.

3 The JAAL All-Quadrilateral Meshing Algorithm

We describe an open source, all-quadrilateral mesh generator that uses an indirect approach. The input to the algorithm is any triangulated surface; the output is an all-quadrilateral mesh with the following features:

1. The mesh may have very few or no irregular nodes.
2. All the quad elements are convex.
3. All the interior nodes have degree range [3-5]. The boundary nodes have degree range [1-4].

The flowchart of this algorithm is shown in Figure 1. In the step-I, triangles are combined to form quadrilaterals and in the step-II and III, geometric and topological improvements are carried out on a topologically valid quadrilateral mesh.

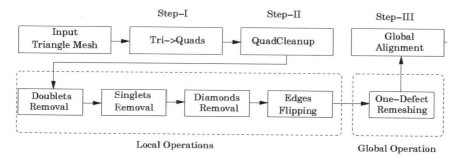

Fig. 1. Flowchart of Jaal Quadrilateral Mesh Generation

This algorithm relies on two fundamental ideas, tree matching and patch remeshing, which are described in following sections. When combined with other freely available components for triangle meshing, geometric evaluation, and mesh smoothing, the result is an open-source, all-quadrilateral meshing algorithm for 2D/3D surfaces of arbitrary topology. We refer to this algorithm using the name JAAL.

3.1 Tree Matching Algorithm (TMA)

Converting triangles into quadrilaterals essentially involves combining adjacent pairs of triangles to form a single quadrilateral. Greedy algorithms, although simple, may leave many unmatched triangles in the mesh, therefore some heuristics are used to maximize number and quality of the quads. For an example, *QMorph* uses advancing front method to combine triangles into quads[22].

One of the most important algorithms in combinatorial graph theory is *Edmond's perfect matching algorithm* [6]. Presently, the fastest known implementation of Edmond's algorithm is from Micali and Vazirani [17] that runs in $O(EV^{1/2})$ time complexity. For binary trees (derived from the dual graph of a triangulation), Suneeta [23] proposed the *Q-percolation algorithm* that has time complexity of $O(n)$. For many theoretical results, algorithms, and complete analysis, we recommend her original paper [23]. The simplicity in the Q-percolation is achieved by inserting *Steiner points* in the triangle mesh to allow perfect tree matching. In Table 1, we have compared the efficiency of general graph matching (available in the Boost library [1]) with our own implementation of the *Q-percolation* algorithm. These times clearly demonstrate that Q-percolation significantly outperforms the general algorithm for trees originating from a triangulated mesh. Table 2 shows that *Q-percolation* also introduces fewer than four percent steiner points.

Suneeta describes two algorithms for tree matching, a simple but non-optimal tree matching algorithm that considers local matching rules, and a theoretically optimal but complex algorithm that uses non-local patches to generate all convex quadrilaterals. We use the non-optimal approach for three reasons:

- Even with the optimal tree matching solution, in most cases, the resulting mesh will still be far from optimal in terms of defect vertices. For 3D surfaces, rules for obtaining a geometrically optimal quadrangulation may be difficult to pose, and expensive to evaluate.
- Experiments have shown that a non-optimal tree matching produces fewer than four percent steiner points, which is not a significant increase in the total number of vertices.
- The non-optimal solution is extremely fast, very easy to implement, and requires no geometric information (which is important for 3D surfaces).

Table 1. Boost Graph Matching v/s Jaal Q-Percolation Performance

# Input triangles	Graph Matching(sec)	Jaal Tree Matching (sec)
10^3	9.73E-03	5.95E-03
10^4	1.88E-01	9.8E-02
10^5	25.36	1.44
10^6	2423.23	25.64

Table 2. Experiments show that TMA introduces few steiner points

Nodes/Triangles	Nodes/Quads	# Steiner Points	% Nodes
542/1031	562/535	20	3.69
5106/10043	5306/5221	200	3.91
50433/100436	52346/52131	1913	3.79
501611/1001611	520419/519613	18808	3.74

4 Local Mesh Cleanup Operations

Although many different local mesh cleanup operations have been described [2, 18, 15, 13], we use only four local operations: *Singlet Removal, Doublet Removal, Diamond Removal*, and *Edge Flipping* (Figure 4). Unlike previous work, we do not attempt to order or prioritize these operations, as in our case, the main quality improver will be the patch-based defect removal described in the next section. We apply local mesh cleanup operations only as a means for removing well-known configurations that are always unacceptable, e.g. two quadrilaterals sharing two edges (a *doublet*). Although these local operations are simple, checks must be done so that (1) inverted elements are not created (2) doublets are not introduced. These operations are fairly inexpensive, as shown in Table 3. Among the four operations, *Edge flipping* is the most expensive for two reasons:

Table 3. Performance of local operations

Quads	Singlets	Doublets	Diamonds	Edge Flipping
65000	5	500	3000	3327
	0.128s	0.0725s	0.1772s	2.651s
701193	1	4450	32443	36034
	1.494s	0.5460s	2.2503s	32.44s

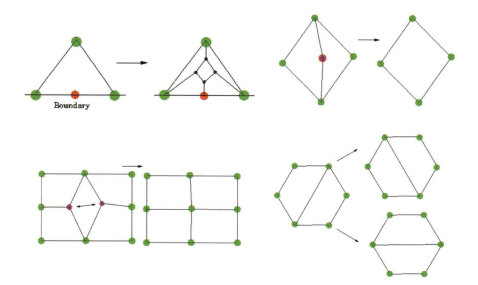

Fig. 2. Removal of a boundary singlet (top left), doublet (top right), diamond (bottom left), and edge flipping (bottom right).

- The operation requires checking for inverted elements.
- Number of successful operations is, in general, very large and many sweeps over the entire mesh are essential to exhaust all the possibilities of flipping.

Experiments have shown that these local operations are quite effective and in a quadrilateral mesh generated from the Delaunay triangulation, as much as half of the irregular nodes are removed from the mesh. Also with the edge flipping operations, we can ensure that the output mesh with the lower bound of vertex degree three and an upper bound of vertex degree five is achieved. For the boundary nodes, interior angle is used to enforce degree.

5 Defect Removal by Patch Replacement

Although Bunin described algorithms for a variety of defect types, we have found that considering of only 3-, 4-, and 5-sided patch regions to be sufficient, while also being straightforward to implement.

5.1 3- and 5-sided Patch Remeshing

3- and 5-sided patches are broken into three and five quadrilateral regions, respectively, as shown in Figure 3, with one irregular node of degree three or five at the intersection. Since the number of intervals on the patch boundary is fixed, the defects in a given patch can be replaced by a single defect if there is an everywhere-positive solution to the corresponding interval matching problem for the patch. For a 3-sided patch, the corresponding matrix problem is

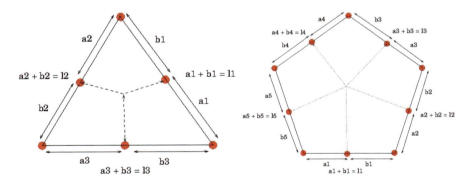

Fig. 3. Template of 3-sided (left) and 5-sided (right) patches.

$$\begin{bmatrix} 0 & 0 & -1 & 1 & 0 & 0 \\ -1 & 0 & 0 & 0 & 1 & 0 \\ 0 & -1 & 0 & 0 & 0 & 1 \\ 1 & 0 & 0 & 1 & 0 & 0 \\ 0 & 1 & 0 & 0 & 1 & 0 \\ 0 & 0 & 1 & 0 & 0 & 1 \end{bmatrix} \begin{Bmatrix} a_1 \\ a_2 \\ a_3 \\ b_1 \\ b_2 \\ b_3 \end{Bmatrix} = \begin{bmatrix} 0 \\ 0 \\ 0 \\ l_1 \\ l_2 \\ l_3 \end{bmatrix}$$

The problem for 5-sided patches is similar (see [8] for details). Note that the inverse matrix for this problem can be pre-computed, since it is always the same for 3- and 5-sided patches. Figure 4 shows examples of 3- and 5-sided patch replacement.

5.2 4-sided Patch Remeshing

For 4-sided patches, there are two possibilities, shown in Figure 5. In a regular patch, opposite sides have equal number of nodes, and the patch can be replaced with a simple structured mesh. In an irregular patch, sides have unequal number of nodes; the patch can be broken into two quadrilateral patches and one 3-sided patch. The resulting quadrilateral patches are remeshed using a regular 4-sided algorithm, with the 3-sided region replaced using the algorithm described in Section 5.1. 4-sided patches can be remeshed only if the interior 3-sided patch is remeshable, as described in Section 5.1. An example of this type of patch replacement is shown in Figure 6.

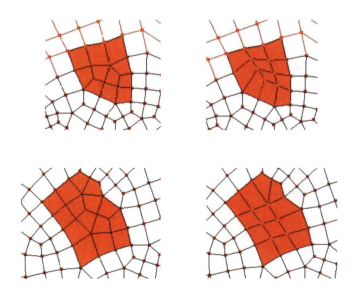

Fig. 4. Remeshing 3-sided (top) and 5-sided (bottom) patches.

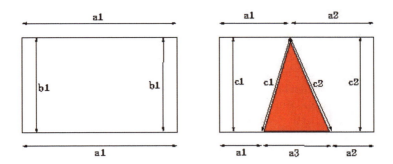

Fig. 5. Remeshing a 4-sided loop. Regular quad patch (left); irregular quad patch (right).

5.3 Patch Identification

Therefore, the crux of the problem is to identify all the remeshable patches. We now present pseudo-codes to explain the entire one-defect remeshing.

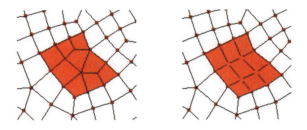

Fig. 6. Remeshing 4 Sided loop

void **remesh_defective_patches**(Mesh *mesh)
  begin
1.   **identify_boundary_nodes**(mesh);
2.   for(i = 4; i < 12; i++)
3.   max_faces_allowed = pow(2,i);
4.   while(1)
5.   ncount = 0;
6.   IR = **build_irregular_nodeset**(mesh);
7.   while(!IR.empty())
8.   Vertex *vertex = IR.top();
9.   IR = IR - vertex;
10.   patch = **build_one_defect_patch**(mesh, vertex)
11.   if(patch)
12.   if(patch.remesh() == 0)
13.   IR = IR - patch.get_all_irregular_nodes();
14.   IR = IR + patch.get_new_defective_node();
15.   ncount++;
16.   if(ncount == 0) break;
17.   end

Line 1: Here we identify all the boundary nodes and tag them as *boundary*. Global operations are valid only for the internal nodes.

Line 2: In this loop we set the maximum number of faces that a patch can have. We start with small number of faces, and once all the possibilities have exhausted, we increase the patch size. Many experiments have shown that 5000 elements are more than sufficient for reasonably good quality mesh. Large patches are generally avoided on highly curved 3D surfaces.

Line 4: We start the loop of global operation. This while loop exit only when no global operation is performed inside the loop. For this, ncount is initialized to zero at line 5 and checked at line 16.

Line 6: We build all the irregular nodes (internal) having valency of 3 or 5. (Assuming that the input nodes have valency 3,4 or 5).

Line 7: Loop starts for every irregular node in the set.

Line 10: Try to build a remeshable patch at the given irregular node. (This is explained in the next pseudo code).

Line 12: Try to remesh the patch. Remeshing may fail, if it results in any inverted element.

Line 13: A successful patch may contain many irregular nodes, remove all of them from the set.

Line 14: There will be at the most one irregular node in the remeshed region, insert it into the set and continue.

Patch **build_one_defect_patch**(Mesh *mesh, Vertex *v)
    begin
1.   initPath = **dijkstra_shortest_path**(mesh,v);
2.   faces = **initial_blob**(initPath);
3.   while(1)
4.   if(faces.size() > max_faces_allowed) return NULL;
5.   bn = **get_boundary_nodes**(faces);
6.   topo_convex_region = 1;
7.   for(i = 0; i < bn.size(); i++)
8.   if(is_internal_node(bn[i])
9.   topo_angle = **get_topological_outer_angle**(bn[i])
10.   if(topo_angle < 0)
11.   **expand_blob**(bn[i])
12.   topo_convex_region = 0
13.   if(topo_convex_region)
14.   if **patch345_meshable**(bn) == 0)
15.   return new Patch(faces, bn);
16.   for(i = 0; i < bn.size(); i++)
17.   **expand_blob**(bn[i])

Line 1: Starting from a given irregular node, search for another irregular node in the mesh using *Dijkstra's shortest path* algorithm and record all the nodes in the path. This initial path is the skeleton from where expansion will start.

Line 2: Collect all the faces around the skeleton, these form an initial *blob*.

Line 3: Blob expansion loop starts.

Line 4: If the blob size is more than specified, we just return with no patch found.

Line 5: Collect all the outer nodes of the boundary of the current patch.

Line 9: For each boundary node, topological outer angle (as defined in Guy Bunin's paper) is calculated, and if it is negative (line 10), more outer elements surrounding the vertex are added to the blob (line 11). The continues till all the boundary nodes of the patch have valid topological angle.

Line 14: Identify the corners of the valid patch. We support patches having 3, 4 or 5 sides only. If a valid patch is found, check if is remeshable by solving linear equations. If the patch is remeshable, return the patch with all the faces, boundary nodes and corners. This patch will be remeshed by the callee program.

Line 15: If the patch is not remeshable (i.e. no integer solution to the linear equations) then expand the blob by including more elements at the boundary and continue.

6 Results

All the development, testing, and evaluation were done on Ubuntu11.04 running on Intel Quad-Core processors with 4GM RAM. For mesh quality [11], we plotted Min-Max angles and aspect ratio distribution. We used the Delaunay triangulation as an input triangle mesh using *Triangle* [9] software. Experiments have shown that the Delaunay triangulation generate approximately half irregular nodes. In most of the cases, local operations are capable of reducing irregular nodes to almost half as shown in the table 6.2. After the global operation, very few irregular nodes remain in the mesh. Some experimental results are shown in 9, 10, 11, and 12.

6.1 Mesh Smoothing

Our implementation is conservative in the sense, that we do not allow creation of any inverted element and then use expensive untangling algorithms. Convex regions increases the chances of successful operations. Experiments have shown that standard Laplacian smoothing is less effective in correcting concave regions as shown in Figure 8, therefore we use *Quasi-Newton* non-linear mesh optimization supported by Mesquite software.

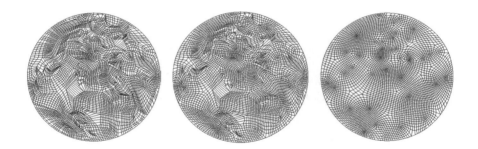

Fig. 7. (A) Mesh after one global operation (B) Standard Laplacian smoothing (C) Non-linear smoothing by Mesquite software

6.2 Power of Simple Global Operation

Converting a triangle into three quadrilateral is probably the simplest algorithm for quadrangulation (known as Berg's algorithm). Although this has complexity of $O(n)$, in literature, this approached has been shunned because it (1) increases the mesh complexity by three times (2) createas large number of irregular nodes. We used this algorithm for stress testing for both local and global operations.

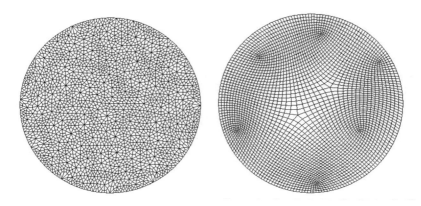

Fig. 8. (A) Quadrangulation using Berg's algorithm(B) Meshclean up using JAAL

Table 4. Performance results of Quadrangulation operations(F:#faces, I:# irregular nodes)

Operations	Dataset I	Dataset II	Dataset III
Quads/Irregular nodes	F:5088 I:2535	F:10316 I:5097	F:51323 I: 25317
Local Operations			
Doublet Removal	1.68E-03	9.993E-03	5.24E-02
Singlet Removal	3.44E-03	1.073E-02	9.71E-02
Diamonds Removal	1.91E-03	4.032E-03	2.05E-01
Edge Flipping	2.38E-01	3.014E-03	2.9296
Shape Optimization	0.2407	1.0	33.0
Quads/Irregular nodes	F:4732 I: 1091	F:9596 I:2280	F:48093 I: 11262
Global Operations			
# 3-Sided Patches	273	441	25875
# 4-Sided Patches	57	104	5507
# 5-Sided Patches	88	195	11305
Time for searching Patches	3.363	9.9462	800
Time for remeshing patches	8.80E-03	1.9E-02	30.0
Shape Optimization Time	2.92E-02	0.50	120.0
Quads/Irregular nodes	F:4144 I:4	F:9779 I:4	F:38036 I:4

7 Software Implementation

The entire software is written in C++ and part of the **MeshKit** software. (Toolkit for mesh generation) which can be freely downloaded from: https://trac.mcs.anl.gov/projects/fathom/wiki/MeshKit. In addition, we used the following modules (all source codes are freely available).

1. **MOAB:** This is a component for representing and evaluating mesh data. With this software, various relationships among the mesh entities can be stored, modified and queried very efficiently.
2. **CGMA:** This software provides all the geometric functionalities that are needed for mesh generation. We used this software for geometric queries and projecting vertices on the model.

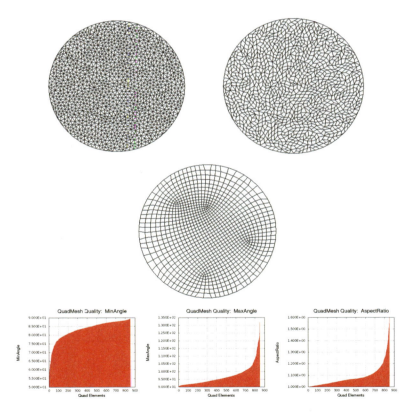

Fig. 9. Result-I: Quadrangulation in a disk produces 4 irregular nodes. (A) Initial triangulation (B) Quadrangulation after tree matching (C) Quadrilateral mesh after mesh cleanup.

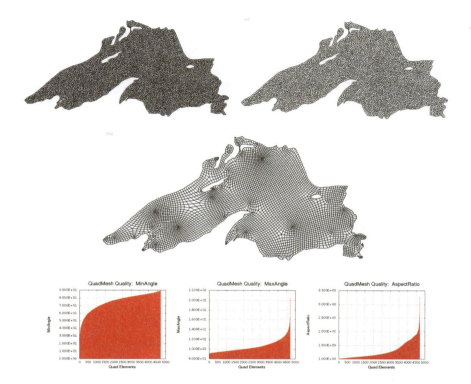

Fig. 10. Result-II: Quadrangulation of Lake Superior model. (A) Initial triangulation (B) Quadrilateral mesh after tree matching (C) Quadrilateral mesh after mesh cleanup.

3. **Mesquite:** This software provides sophisticated non-linear algorithms for mesh shape optimization. This is the most critical component in the Jaal software pipeline.
4. **Verdict:** This software is used for measuring various mesh elements qualities.

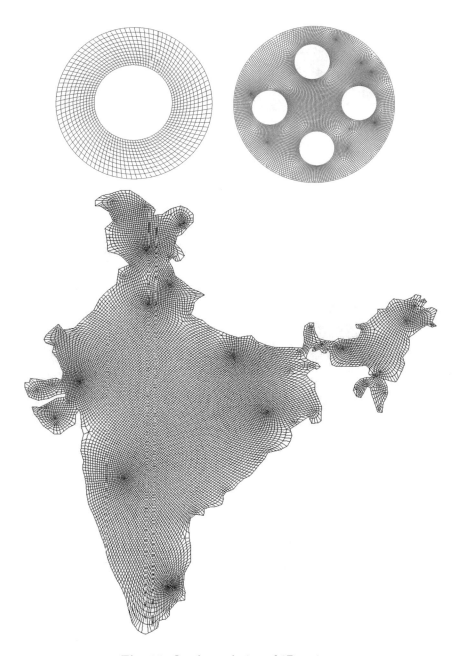

Fig. 11. Quadrangulation of 2D regions

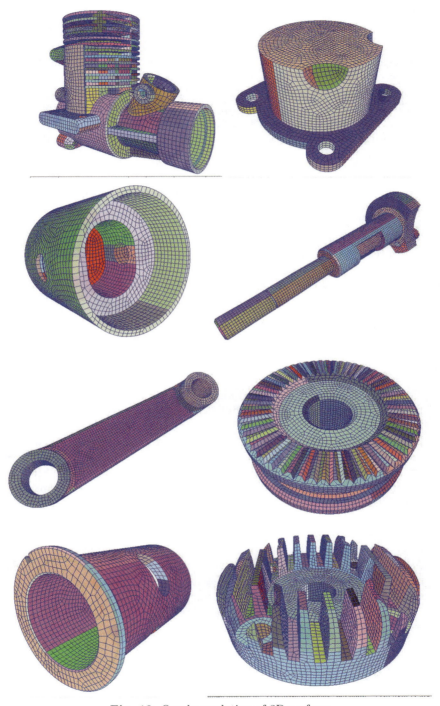

Fig. 12. Quadrangulation of 3D surfaces

8 Conclusion

We have engineered two algorithms to develop an open-source code for all-quadrilateral mesh for 2D/3D surfaces using indirect approach. We have done extensive experiments with the present software and conclude that a large class of surfaces can be quadrangulated with high quality using very simple operations (four local and one global). The entire pipeline is fully automatic and require no user's parameters. Both geometrically and topologically, the final mesh quality, in all cases, exceeded our initial expectations. Still, there are scopes for further improvements:

- The biggest performance bottleneck in the pipeline is *Patch Searching* which is purely heuristic. It is quite possible that other approaches and data structures may have less time complexity than what we have presented.
- The present code has no control over spatial distribution of irregular nodes. For good quality mesh, irregular nodes must be deep inside the domain and evenly distributed.
- At present, the code does not support constrained quadrangulation, which are essential for feature preservation.

9 Future Work

We plan to extend the current work in the following directions:

- *Parallelization :* The presented algorithm is inherently parallelizable on both shared memory and distributed memory architecture machines. In the immediate future, we will implement this algorithm using Intel Thread Building Blocks(TBB) library for multi-core machines and Message Passing Interface (MPI) for large scale problem on distributed memory machines. Transactional memory paradigm have also been successful in parallelization of incremental Delaunay triangulation(IDT) [16]. Since our global cleanup algorithm is very similar to IDT, therefore, scalable implementation will be quite interesting development.
- *Non-geometrical models :* The present method will be extended to handle arbitrary polygonal models where the underlying geometry is not known.
- *Feature preservation:* Not preserving user defined edges is one of the biggest weakness of the present algorithm. This can be done in either step-I or step-II, which we will explore.

References

1. Boost Graph Libary, http://www.boost.org/doc/libs/1_37_0/libs/graph/doc/maximum_matching.html
2. Canann, S., Muthukrishnan, S., Phillips, R.: Topological Improvement Procedures for Quadrilateral Finite Element Meshes. Engineering with Computers 14, 168–177 (1998)
3. Daniels, J., Lizier, M., Siqueira, M., Silva, C.T., Nonato, L.G.: Template Based Quadrilateral Meshing. Computers and Graphics 35(3), 471–482 (2011); Shape Modeling International (SMI) Conference 2011
4. Bommes, D., Zimmer, H., Kobbelt, L.: Mixed-Integer Quadrangulation. ACM Transactions on Graphics (TOG) 28(3) (2009), Article No. 77 (2009)
5. Bommes, D., Lempfer, T., Kobbelt, L.: Global Structure Optimization of Quadrilateral Meshes. Eurographics (2011)
6. Edmonds, J.: Paths, trees, and flowers. Canad. J. Math. 17, 449–467 (1965)
7. Klberer, F., Nieser, M., Polthier, K.: QuadCover Surface Parameterization using Branched Coverings. In: EUROGRAPHICS 2007, vol. 26(3) (2007)
8. Bunin, G.: Non-Local Topological Cleanup. In: 15th International Meshing Roundtable (2006)
9. Shewchuk, J.R.: Triangle: A Two dimensional quality mesh generator and Delaunay triangulator, http://www.cs.cmu.edu/~quake/triangle.html
10. Hormann, K., Greiner, G.: Quadrilateral Remeshing Proceedings of Vision, Modeling, and Visualization. In: Girod, B., Greiner, G., Niemann, H., Seidel, H.-P. (eds.) pp. 153–162 (2000)
11. Knupp, P.: Achieving Finite Element Mesh Quality via Optimization of the Jacobian Matrix Norm and Associated Quantities. International Journal for Numerical i Methods in Engineering 48(8), 1165–1185 (2000)
12. Mucha, M., Sankowski, P.: Maximum matchings in planar graphs via Gaussian elimination. In: Algorithmica (2004)
13. Dewey, M.W., Benzley, S.E., Shepherd, J.F., Staten, M.L.: Automated Quadrilateral Coarsening by Ring Collapse
14. Bern, M.: Quadrilateral meshing by circle packing. International Journal of Computational Geometry and Applications 7–20 (1997)
15. Staten, M., Canann, S.A.: Post Refinement Element Shape Improvement For Quadrilateral Meshes, 220 Trends in Unstructured Mesh Generation. ASME, 9–16 (1997)
16. Kulkarni, M., Paul Chew, L., Pingali, K.: Using Transactions in Delaunay Mesh Generation,
https://engineering.purdue.edu/~milind/docs/wtw06-slides.pdf
17. Micali, S., Vazirani, V.V.: An $O(\sqrt{|v|}\cdot|E|)$ algorithm for finding maximum matching in general graphs. In: 21st Annual Symposium on Foundations of Computer Science (1980)
18. Kinney, P.: CleanUp: Improving Quadrilateral Finite Element Meshes. In: 6th International Meshing Roundtable, pp. 449–461 (1997)
19. Bern, M., Eppstein, D.: Quadrilateral Meshing by Circle packing. International Journal of Computational Geometry and Applications, 7–20 (1997)
20. Dong, S., Bremer, P.-t., Garland, M.: Spectral surface quadrangulation. ACM Transaction on Graphics 25, 1057–1066 (2006)

21. Arnborg, S., Lagergren, J.: Easy Problems for Tree-Decomposable Graphs (1991),
    http://www.informatik.uni-trier.de/~ley/db/journals/jal/jal12.html#ArnborgLS91
22. Owen, S.J., Staten, M.L., Canann, S.A., Saigal, S.: Advancing Front Quadrilateral Meshing Using Triangle Transformations (1998)
23. Ramaswami, S., Ramos, P., Toussaint, G.: Converting triangulations to quadrangulations. Computational Geometry: Theory and Applications 9, 257–276 (1998)
24. Betual Atalay, F., Ramaswami, S., Xu, D.: Quadrilateral Meshes with Provable Angle Bounds (2011)
25. Zhou, T., Shimada, K.: An angle-based approach to two-dimensional mesh smoothing. In: Proceedings, 9th International Meshing Roundtable, pp. 373–384 (2000)

CROSS-CUTTING TOPICS
(Wednesday Morning Parallel Session)

Design, Implementation, and Evaluation of the Surface_mesh Data Structure

Daniel Sieger and Mario Botsch

Computer Graphics & Geometry Processing Group,
Bielefeld University, Germany
{dsieger,botsch}@techfak.uni-bielefeld.de

Summary. We present the design, implementation, and evaluation of an efficient and easy to use data structure for polygon surface meshes. The design choices that arise during development are systematically investigated and detailed reasons for choosing one alternative over another are given. We describe our implementation and compare it to other contemporary mesh data structures in terms of usability, computational performance, and memory consumption. Our evaluation demonstrates that our new Surface_mesh data structure is easier to use, offers higher performance, and consumes less memory than several state-of-the-art mesh data structures.

1 Introduction

Polygon meshes, or the more specialized triangle or quad meshes, are the standard discretization for two-manifold surfaces in 3D or solid structures in 2D. The design and implementation of mesh data structures therefore is of fundamental importance for research and development in as diverse fields as mesh generation and optimization, finite element analysis, computational geometry, computer graphics, and geometry processing.

Although the requirements on the mesh data structure vary from application to application, a generally useful and hence widely applicable data structure should be able to (i) represent vertices, edges, and triangular/quadrangular/polygonal faces, (ii) provide access to all incidence relations of these simplices, (iii) allow for modification of geometry (vertex positions) and topology (mesh connectivity), and (iv) allow to store any custom data with vertices, edges, and faces. In addition, the data structure should be easy to use, be computationally efficient, and have a low memory footprint.

Since it is hard to implement a mesh data structure that meets all these goals, many researchers and developers in both academia and industry rely on publicly available C++ libraries like CGAL [8] (computational geometry), Mesquite [6] (mesh optimization), and OpenMesh [5] (computer graphics).

However, even these highly successful data structures have their individual deficits and limitations, as we experienced during several years of research and teaching in geometry processing. In this paper we systematically derive the design choices for our new Surface_mesh data structure and provide an analysis and comparison to the widely used mesh data structures of CGAL, Mesquite, and OpenMesh. These comparisons demonstrate that Surface_mesh is easier to use than these implementations, while at the same time being superior in terms of computational performance and memory consumption.

2 Related Work

Due to their fundamental nature, a wide variety of data structures to represent polygon meshes have been proposed. Some are highly specialized to only represent a certain type of polygons, such as triangles or quadrilateral elements. Others are designed for specific applications, e.g. parallel processing of huge data sets. In general, mesh data structures can be classified as being either *face-based* or *edge-based*. We refer the reader to [18, 4] for a more comprehensive overview of mesh data structures for geometry processing.

In its most basic form a face-based data structure consists of a list of vertices and faces, where each face stores references to its defining vertices. However, such a simple representation does not provide efficient access to adjacency information of vertices or faces. Hence, many face-based approaches additionally store the neighboring faces of each face and/or the incident faces for each vertex. Examples for face-based mesh data structures include CGAL's 2D triangulation data structure [8], Shewchuck's Triangle [23], Mesquite [6], and VCGLib [29].

In contrast to face-based approaches, edge-based data structures store the main connectivity information in edges or halfedges [2, 13, 7, 20]. In general, edges store references to incident vertices/faces as well as neighboring edges. Kettner [18] gives a comparison of edge-based data structures and describes the design of CGAL's halfedge data structure. Botsch et al. [5] introduce OpenMesh, a halfedge-based data structure widely used in computer graphics. Alumbaugh and Jiao [1] describe a compact data structure for representing surface and volume meshes by halfedges and half-faces.

Furthermore, a fairly large number of publications describe more specialized mesh representations. For instance, Blandford et al. [3] introduce a compact and efficient representation of simplicial meshes containing triangles or tetrahedra. Other works focus on data structures for non-manifold meshes [10, 11], highly compact representations of static triangle meshes [14, 15], or mesh representations and databases for numerical simulation [12, 27, 22, 9].

3 Design Decisions

While virtually all of the publications cited above describe the specific design decisions made for a particular implementation, a comprehensive and systematic investigation of the design choices available is currently lacking. We therefore try to provide such an analysis in this section.

As mentioned in the introduction, the typical design goals for mesh data structures are computational performance, low memory consumption, high flexibility and genericity, as well as ease of use. Since these criteria are partly contradicting, one has to set priorities and make certain compromises.

Based on our experience in academic research and teaching as well as in industrial cooperations, our primary design goal is *ease of use*. An easy-to-use data structure is learned faster, allows to focus on the main problem (instead of on the details of the data structure), and fosters code exchange between academic or industrial research partners. The data structure should therefore be just as flexible and generic as needed, but should otherwise be free of unnecessary switches and parameters. At the same time, however, we have to make sure not to compromise computational performance and memory consumption. Otherwise the data structure would be easy to use, but not useful, and hence would probably not be used at all.

In the following we systematically analyze the typical design choices one is faced with when designing a mesh data structure. Driven by our design goals we argue for choosing one alternative over another for each individual design criterion. We start with high-level design choices and successively focus on more detailed questions.

3.1 Element Types

The most fundamental question is which types of elements or faces to support. While in computer graphics and geometry processing triangle meshes still are the predominant surface discretization [4], quad meshes are at least as important as triangle meshes for structural mechanics. For many applications, restricting to pure triangle or quad meshes is not an option, though. Polygonal finite element methods [26] decompose their simulation domain into arbitrary polygons. In discrete exterior calculus many computations are performed on the dual mesh [16]. In computational geometry, computations on Voronoi diagrams also need arbitrary polygon meshes [8]. Since we want our data structure to be suitable for an as wide as possible range of applications we choose to support *arbitrary polygonal elements*.

3.2 Connectivity Representation

As discussed in Section 2 there are two ways to represent the connectivity of a polygon mesh: a face-based or an edge-based representation.

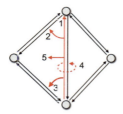

1. Target vertex
2. Next halfedge
3. Previous halfedge
4. Opposite halfedge
5. Adjacent face

Fig. 1. Connectivity relations within a halfedge data structure.

Face-based data structures store for each face the references to its defining vertices. While this is sufficient for, e.g., visualization or setting up a stiffness matrix, it is inefficient for mesh optimization, since vertex neighborhoods cannot be accessed easily. Some implementations therefore additionally store all incident faces per vertex (e.g., [6, 29]), but even then it is still inefficient to enumerate all incident *vertices* of a center vertex—a query frequently required for many algorithms, such as mesh smoothing, decimation, or remeshing. Furthermore, since for a general polygon mesh the number of vertices per face and the number of incident faces per vertex are not constant, they have to be stored using dynamically allocated arrays or lists, which further complicates the data structure. Edges are typically not represented at all.

In contrast, storing the main connectivity information in terms of edges or halfedges naturally handles arbitrary polygon meshes. The data types for vertices, (half-)edges, and faces all have constant size. The vertices and face-neighbors of a face can be efficiently enumerated, as well as the vertices or faces incident to a center vertex. Attaching additional data to vertices, halfedges, and faces is simple, since all entities are explicitly represented. Finally, a halfedge-based data structure allows for simple and efficient implementation of connectivity modifications as required by modern approaches to interleaving mesh generation and optimization [28, 25] or simulation [30]. We therefore choose a *halfedge data structure* to store the connectivity of a polygon mesh. The basic connectivity relations within a typical halfedge data structure are shown in Figure 1.

3.3 Storage

On an implementation level one has to decide whether to store the mesh entities in either doubly-linked lists or simple arrays.

Lists have the advantage that they allow for easy removal of individual vertices, edges, or faces, which is required, e.g., when collapsing edges or removing vertices in a mesh decimation algorithm. However, this flexibility comes at the price of higher memory consumption and less coherent memory layout compared to array-based storage, both resulting in considerable performance loss. We evaluated this on the halfedge data structure [17] of CGAL [8], which allows to switch between a list-based and an array-based implementation. Our benchmarks in Section 5 show the list-based implementation to be up to twice as slow as the array-based version.

Array-based storage on the one hand is more compact and faster, but on the other hand the removal of mesh entities is more difficult. Typically mesh entities are first marked as deleted and later removed by some form of garbage collection. However, the advantages in terms of performance and memory consumption clearly outweigh the additional effort needed to support removal. For these reasons we choose an *array-based storage* scheme.

3.4 Entity References

When using array-based storage for mesh entities, references (or handles) to entities can be represented either as pointers or indices.

Pointers have three important drawbacks: First, they become invalid upon a relocation of the array, which happens if the array has to allocate more memory (e.g., for refinement or subdivision algorithms). While the data structure can automatically update all *internally* stored pointers, references that are stored externally by the user will inevitably become invalid. Second, on 64-bit architectures pointers consume twice as much memory as 32-bit indices. For larger meshes, however, one has to use 64-bit addressing, since complex meshes easily exceed the 2GB limit for 32-bit architectures. Finally, pointers cannot be used to access additional properties of mesh entities that are stored in additionally "property arrays" (see the next section). We therefore choose *indices* as entity references.

3.5 Custom Properties

Additional information about the mesh entities can be stored either by extending the mesh entities themselves or by using additional arrays. For instance, vertex normals can be incorporated either by adding a member variable `normal` to the class `Vertex`, or by having an additional array `vertex_normals` where the i'th entry is the normal of vertex i.

The first approach, as e.g. chosen by CGAL, is more elegant from an object-oriented point of view, but has the following drawbacks: Since the class types of mesh entities are extended at compile-time, all custom properties are allocated over the whole running time of the application, even if the properties are used for a short time only. This does not only waste memory, it also slows down the algorithms due to a less compact memory layout: Just adding vertex and face normals to the CGAL mesh by extending the `Vertex` and `Facet` types slowed down our benchmarks (Section 5) by about 25% on average. This can be a significant drawback for larger mesh processing applications, where many individual algorithms need some custom data at some point in time.

In contrast, additional arrays can be dynamically allocated at run-time, such that custom properties are just allocated when needed and deleted afterwards (as implemented in OpenMesh and Mesquite). Keeping all property arrays synchronized upon resize and swap operations can easily be implemented. Furthermore, computations on the property arrays are also more cache-friendly, thereby increasing performance compared to extended mesh entities. Finally, if the model is meant to be visualized in an interactive application, property arrays can also be used in conjunction with OpenGL vertex arrays (normals, colors, texture coordinates), which speeds up rendering performance considerably. We therefore store custom properties in additional *synchronized arrays*.

3.6 Ease of Use

Up to this point, our previous mesh data structure OpenMesh [5], at least in its current version [21], follows most of the design decisions made so far. From our experience in research and teaching, however, the level of genericity offered by OpenMesh is not needed in practice. For instance, custom properties can be allocated both by extending mesh entities as well as using additional arrays, where due to the former the mesh entities (and hence the whole mesh) become template classes. Furthermore, the large (template-parametrized) inheritance hierarchy makes the code unnecessarily hard to document and understand. In terms of C++ sophistication, the polyhedral data structure of CGAL requires an even higher level of template expertise, which makes it hard to use this data structure with students or inexperienced programmers, too.

To reduce the negative effect that heavy use of templates and complicated inheritance hierarchies have on the ease of use of the data structure, we made our design *as simple as possible* while maintaining maximum applicability.

4 Implementation

In the following we highlight the most important aspects of our implementation. We first describe the fundamental organization of our new data structure and successively proceed to higher-level functionality.

Since OpenMesh already satisfies all design choices except simplicity, we started our implementation from a massively stripped-down and simplified version of OpenMesh. In contrast to other implementations, ours is concentrated within a single class, namely Surface_mesh. While the core of Surface_mesh (without file I/O) is implemented in three files using about 2250 lines of code, the part of OpenMesh that implements the same functionality requires 41 files or 8400 lines of code. In contrast to CGAL and OpenMesh, Surface_mesh is not a class template, i.e., it does not require so-called traits classes as template parameters. However, the fundamental types Scalar and Point can still be defined by simple typedefs.

Surface_mesh implements an array-based halfedge data structure. The basic entities of the mesh, i.e., vertices, (half-)edges, and faces are represented by the types Vertex, Halfedge, Edge, and Face, respectively, all of which are

```
Surface_mesh mesh;

// allocate property storing a point per edge
Surface_mesh::Edge_property<Point> edge_points
  = mesh.add_edge_property<Point>("property-name");

// access the edge property like an array
Surface_mesh::Edge e;
edge_points[e] = Point(x,y,z);

// remove property and free memory
mesh.remove_edge_property(edge_points);
```

Listing 1. Working with a custom edge property.

basically 32-bit indices. Edges are represented implicitly, since two opposite halfedges (laid out consecutively in memory) build an edge.

The connectivity information is stored in form of custom properties (i.e., synchronized arrays) of vertices, faces, and halfedges: Each vertex stores an outgoing halfedge, each face an incident halfedge. Each halfedge stores its incident face, its target vertex, and its previous and next halfedges within the face. Since opposite halfedges are laid out consecutively in memory, the opposite halfedge can be accessed by simple modulo operations on the `Halfedge` indices and therefore does not have to be stored explicitly.

Managing internal mesh data as well as dynamically allocated user-defined properties within the same framework for synchronized arrays on the one hand simplifies implementing and maintaining the data structure. On the other hand the performance of the data structure then crucially depends on efficient access to these properties. Our property mechanism deviates from Mesquite and OpenMesh in that it (i) avoids inefficient virtual function calls, (ii) does not require error-prone casting of `void`-pointers, (iii) avoids unnecessary indirections, and (iv) offers a cleaner interface. From a user's point of view, working with a custom property is as simple as shown in Listing 1.

In addition to access to all incidence relations and custom properties, `Surface_mesh` also offers higher-level topological operations, such as adding vertices and faces, performing edge flips, edge splits, face splits, or halfedge collapses. Based on these methods typical geometry processing algorithms (smoothing, decimation, subdivision, remeshing) can be implemented conveniently. Since `Surface_mesh` uses an array-based storage special care has to be taken when removing items from the mesh. Such operations do not delete mesh entities immediately, but instead mark them as being to be deleted. The function `garbage_collection()` eventually deletes those items from the arrays, while preserving the integrity of the data structure.

In order to sequentially access mesh entities we provide iterators for each entity type, namely `Vertex_iterator`, `Halfedge_iterator`, `Edge_iterator` and `Face_iterator`. Each iterator stores a reference to the current entity and to the mesh. The latter is used to automatically detect and skip deleted

Fig. 2. Traversal of one-ring neighbors of a center vertex. From left to right: 1. Start from center vertex. 2. Select outgoing halfedge (and access its target vertex). 3. Move to previous halfedge. 4. Move to opposite halfedge (access its target vertex). Steps 1 and 2 correspond to the initialization of a Vertex_around_vertex_circulator using mesh.vertices(Vertex), steps 3 and 4 to the ++-operator of the circulator.

entities, for instance when the user collapsed some edges but did not yet clean-up using garbage_collection(). We decided for these "safe iterators" despite their small performance penalty, since "unsafe" iterators turned out to be a frequent source of errors for novice OpenMesh users.

Similar to iterators, we also provide circulators for the ordered enumeration of all incident vertices, halfedges, or faces around a given face or vertex. Since there is no clear begin- and end-circulator, we follow the CGAL convention and use do-while loops for circulators. The traversal of the one-ring neighborhood of a vertex—which corresponds to a Vertex_around_vertex_circulator—is shown in Figure 2. An example usage of iterators and circulators is demonstrated in the smoothing example in Listing 2.

5 Evaluation and Comparison

In this section we evaluate our mesh data structure and compare it to three other widely used data structures: OpenMesh, CGAL, and Mesquite. Our evaluation criteria are ease of use, run-time performance, and memory usage.

All tests were performed on a Dell T7500 workstation with an Intel Xeon E5645 2.4 GHz CPU and 6GB RAM running Ubuntu Linux 10.04 x86_64. All libraries and tests were compiled with gcc version 4.4.3, optimization turned on (using -O3) and debugging checks disabled (-DNDEBUG).

For each of the mesh libraries in our comparison we used the latest version available, i.e., OpenMesh 2.0.1, CGAL 3.8, and Mesquite 2.1.4. To achieve comparable results, we chose double-precision floating point values for scalars, vertex coordinates, and normal vectors for all benchmarks and data structures. Since one benchmark requires vertex and face normals, all data structures allocate these properties, either by extending vertex and face types (CGAL) or using property arrays (Mesquite, OpenMesh, Surface_mesh).

```
#include <Surface_mesh.h>

int main(int argc, char** argv)
{
    Surface_mesh mesh;

    // read mesh from file
    mesh.read(argv[1]);

    // get (pre-defined) property storing vertex positions
    Surface_mesh::Vertex_property<Point> points
        = mesh.get_vertex_property<Point>("v:point");

    // iterators and circulators
    Surface_mesh::Vertex_iterator vit, vend = mesh.vertices_end();
    Surface_mesh::Vertex_around_vertex_circulator vc, vc_end;

    // loop over all vertices
    for (vit = mesh.vertices_begin(); vit != vend; ++vit)
    {
        if (!mesh.is_boundary(*vit))
        {
            // move vertex to barycenter of its neighbors
            Point  p(0,0,0);
            Scalar c(0);
            vc = vc_end = mesh.vertices(*vit);
            do
            {
                p += points[*vc];
                ++c;
            }
            while (++vc != vc_end);
            points[*vit] = p / c;
        }
    }

    // write mesh to file
    mesh.write(argv[2]);
}
```

Listing 2. A simple smoothing program implemented using Surface_mesh.

Note that regarding CGAL we compare to both the list-based and the vector-based version of the Polyhedron_3 mesh data structure, denoted as CGAL_list and CGAL_vector, respectively. Furthermore, following [24], we removed the storage for the plane equation from face entities in order to increase performance.

In contrast to CGAL, OpenMesh, and Surface_mesh, which are all halfedge data structures, Mesquite employs a face-based data structure that stores both downward adjacency (vertices of a face) and upward adjacency (all incident faces of a vertex).

5.1 Ease of Use

Being our primary design goal, we begin our evaluation by comparing the ease of use of Surface_mesh to the other libraries.

```
typedef CGAL::Simple_cartesian<double>   Kernel;
typedef Kernel::Point_3                  Point_3;

template <class Refs>
struct My_halfedge : public CGAL::HalfedgeDS_halfedge_base<Refs>
{
    Point_3 halfedge_point;
};

class Items : public CGAL::Polyhedron_items_3
{
public:
    template <class Refs, class Traits>
    struct Halfedge_wrapper
    {
        typedef My_halfedge<Refs> Halfedge;
    };
};

typedef CGAL::Polyhedron_3<Kernel, Items>   Mesh;
```

Listing 3. Declaring a custom halfedge property in CGAL.

Simplicity

As already outlined in Section 3.6, simplicity is a key criterion for the ease of use of a software library. By design, Surface_mesh is as simple as possible while maintaining high applicability. In contrast, both OpenMesh and CGAL offer a higher level of genericity. While this enables the customization of the mesh data structure for specialized applications, it also makes the library less accessible for students and inexperienced programmers.

The differences in complexity are demonstrated best by example. Listing 3 shows how to declare a custom halfedge property in CGAL, which is roughly equivalent to Listing 1 showing the usage of properties in Surface_mesh.

Compared to OpenMesh, our increased simplicity (and decreased genericity) is due to the definition of basic types (e.g., use float or double as scalar type, 2D or 3D vertex coordinates) through typedefs instead of through template parameters. While this allows Surface_mesh not to be a class template, it restricts each application to use a single Surface_mesh definition. In contrast, OpenMesh and CGAL allow for several custom-tailored template instances in a single application.

Properties

Comparing Listings 1 and 3 not only serves as an example for evaluating simplicity, but also demonstrates the differences between CGAL's extended entities and and Surface_mesh's synchronized arrays for property handling. While the declaration of the former is rather involved and bound to compile-time properties, the latter is easy to use and dynamically allocated at runtime. Both OpenMesh and Mesquite also support dynamic property arrays. In case of OpenMesh however, the interface is slightly more complicated.

Table 1. Compilation times (in seconds) of our benchmark program for the different mesh data structures. The chart shows timings relative to Surface_mesh.

Mesquite	1.57
CGAL_list	2.83
CGAL_vector	2.75
OpenMesh	3.37
Surface_mesh	1.13

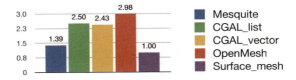

Mesquite's implementation of properties relies on casting void-pointers, a practice generally discouraged and also relevant to our next evaluation criterion.

Safety

Especially for inexperienced programmers protection against common sources of errors is a crucial aspect of usability. The use of void pointers in Mesquite mentioned above can be considered harmful in this context, since this practice essentially circumvents the static type-safety of the programming language. The use of pointers as entity references for CGAL's array-based mesh data structure is prone to errors, since the pointers (and iterators) become invalid upon resizing. While OpenMesh uses safe, index-based entity references, its iterators by default do not skip deleted items, which turned out to be a common source of errors. In contrast, Surface_mesh's implementation of safe iterators protects the user from iterating over deleted entities.

Compilation Time

Finally, compilation time is a usability factor frequently overlooked. While the times to compile the individual programs in our test suite are relatively short, compilation time becomes a significant factor for the speed and efficiency of the development process in more complex projects. As can be seen from Table 1, Surface_mesh offers the fastest compilation times, mostly due to minimizing the use of templates.

User Study

We evaluated the usability of Surface_mesh in a user study among the participants of a two-day course of mesh processing (involving lectures and programming exercises) held at the Symposium on Geometry Processing 2011. The attendees had a varying degree of programming experience and exposure to other mesh libraries. After the two-days the participants were asked anonymously if Surface_mesh was easy to use and understand for them. Out of 18 participants seven strongly agreed to this statement (5/5 points), another seven agreed (4/5 points). On average, Surface_mesh received 4.1/5 points. While this is not a representative survey, the results are still encouraging.

5.2 Performance

In order to compare the efficiency of our implementation with other mesh data structures we designed several benchmarks, which either evaluate a fundamental functionality of a data structure (e.g., iterators or adjacency queries) or test the performance in common application domains (e.g., mesh smoothing or subdivision). The benchmark tests are described below and their pseudo-code is shown in Algorithms 1–6:

1. Circulator Test: For each vertex enumerate its incident faces. For each face enumerate its vertices. This test measures the efficiency of iterators and circulators.
2. Barycenter Test: Center the mesh at the origin by first computing the barycenter of all vertex positions and then subtracting it from each vertex. This test evaluates the performance of iterators and of the access to and basic computations on the vertex coordinates.
3. Normal Test: First compute (and store) face normals, then compute vertex normals as the average of the incident faces' normals. This test measures the performance of iterators, circulators, vertex computations, and custom properties (storing face and vertex normals).
4. Smoothing Test: Perform Laplacian smoothing by moving each (non-boundary) vertex to the barycenter of its neighboring vertices. This test requires (and evaluates) the enumeration of incident vertices of a vertex.
5. Subdivision Test: Perform one step of $\sqrt{3}$-subdivision [19] by first splitting all faces at their centers, smoothing the old vertices, and then flipping all the old edges. This test mainly evaluates the performance of the face split and edge flip operators.
6. Edge Collapse Test: First split all faces at their center and then collapse each newly introduced vertex into one of its (old) neighbors, thereby restoring the original connectivity. This test evaluates the operators face split and halfedge collapse.

These benchmarks were performed on the Imp model, consisting of 300k vertices and 600k triangles, and the Dual Dragon model, a dualized triangle mesh consisting of 100k vertices and 50k polygonal faces. The models are shown in Figure 3. All tests were iterated sufficiently many times in order to get more reliable accumulated timings. The results are listed in Tables 2 and 3. Note that we also performed the tests with other models and setups (CPU, compiler version, and operating system). While the results quantitatively vary to a certain extent, they were qualitatively equivalent to the ones shown here.

It can be observed that for some tests the performance varies significantly between different libraries. While it is hard to track down the reasons in detail, we point out the most important issues we identified.

For Mesquite, a significant performance penalty comes from the large number of virtual functions (e.g., to access incidences or vertex coordinates), as

Imp Model	Lucy Model	Dual Dragon Model
300k vertices, 600k triangles	10M vertices, 20M triangles	100k vertices, 50k polygons

Fig. 3. The three models used in the evaluation.

Algorithm 1. Circulator Test

Initialize $counter = 0$;
for *each vertex v* do
    for *each face f incident to v* do
        | $counter = counter + 1$;
    end
end
for *each face f* do
    for *each vertex v incident to f* do
        | $counter = counter - 1$;
    end
end

Algorithm 2. Barycenter Test

Initialize $\mathbf{p} = (0, 0, 0)$;
for *each vertex v* do
    | $\mathbf{p} = \mathbf{p} + \texttt{point}(v)$;
end
$\mathbf{p} = \mathbf{p}/\texttt{number\_of\_vertices}()$;
for *each vertex v* do
    | $\texttt{point}(v) = \texttt{point}(v) - \mathbf{p}$;
end

Algorithm 3. Normal Test

for *each face f* do
    | Compute the face normal of f;
end
for *each vertex v* do
    $\mathbf{n} = (0, 0, 0)$;
    for *each face f incident to v* do
        | $\mathbf{n} = \mathbf{n} + \texttt{face\_normal}(f)$;
    end
    $\texttt{vertex\_normal}(v) = \texttt{normalize}(\mathbf{n})$;
end

Algorithm 4. Smoothing Test

for *each vertex v* do
    if *v is not a boundary vertex* then
        $\mathbf{p} = (0, 0, 0)$;
        $c = 0$;
        for *each vertex w incident to v* do
            | $\mathbf{p} = \mathbf{p} + \texttt{point}(w)$;
            | $c = c + 1$;
        end
        $\texttt{point}(v) = \mathbf{p}/c$;
    end
end

Algorithm 5. Subdivision Test

for *each face f* do
    Compute centroid c;
    Split f at centroid c;
end
for *each old vertex v* do
    | Smooth vertex position;
end
for *each old edge e* do
    | Flip e;
end

Algorithm 6. Collapse Test

for *each face f* do
    | Split f;
end
for *each new vertex v* do
    | Collapse v into one of its neighbors;
end

Fig. 4. The six benchmark tests used to evaluate and compare the run-time performance of Surface_mesh to Mesquite, CGAL, and OpenMesh.

Table 2. Timings for performing Algorithms 1–6 on the Imp model of 300k vertices and 600k triangles. The table lists timings in milliseconds, the chart visualizes the performance relative to `Surface_mesh`.

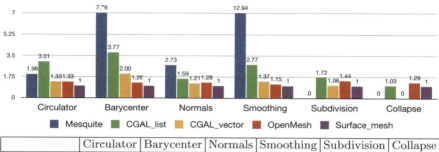

	Circulator	Barycenter	Normals	Smoothing	Subdivision	Collapse
Mesquite	3479.57	15039.9	11406.4	23228.9	—	—
CGAL_list	5329.89	7298.91	6642.29	4976.79	506.158	1582.94
CGAL_vector	2358.51	3879.86	5064.38	2467.66	312.607	—
OpenMesh	2359.36	2443.59	5356.68	2071.79	423.925	1987.44
Surface_mesh	1673.34	1412.28	4181.92	1757.07	294.24	1547.53

Table 3. Timings for performing Algorithms 1–4 and 6 on the Dual Dragon model consisting of 100k vertices and 50k arbitrary polygonal faces. The table lists timings in milliseconds, the chart visualizes the performance relative to `Surface_mesh`.

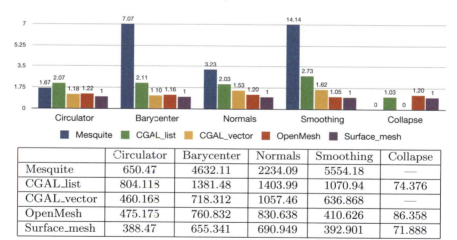

	Circulator	Barycenter	Normals	Smoothing	Collapse
Mesquite	650.47	4632.11	2234.09	5554.18	—
CGAL_list	804.118	1381.48	1403.99	1070.94	74.376
CGAL_vector	460.168	718.312	1057.46	636.868	—
OpenMesh	475.175	760.832	830.638	410.626	86.358
Surface_mesh	388.47	655.341	690.949	392.901	71.888

Table 4. Memory usage for the Imp, Lucy, and Dual Dragon models. The table lists resident size memory usage after reading the meshes, without performing any further tests or processing. The chart visualizes the relative difference to Surface_mesh.

	Imp	Dragon	Lucy
Mesquite	88M	16M	2.8G
CGAL_list	172M	30M	5.5G
CGAL_vector	105M	19M	3.4G
OpenMesh	67M	14M	2.2G
Surface_mesh	60M	12M	1.9G

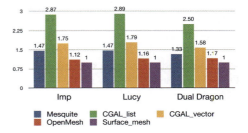

well as from memory fragmentation due to dynamically allocated arrays for storing per-vertex and per-face incidences. Moreover, enumerating incident *vertices* of a center vertex is not directly supported by this face-based data structure and therefore has to be implemented less efficiently by looping over the vertices of the incident faces. Since Mesquite does not support connectivity modifications, the subdivision and collapse test were not implemented.

The performance difference between CGAL_list and CGAL_vector is due to the higher memory consumption and memory fragmentation of the list-based version. Both CGAL mesh data structures store 64-bit references, vertex positions, and normal vectors in extended mesh entities, leading to a less compact memory layout, which in turn results in performance penalties. Note that the array-based version does not support removal of entities, so that the collapse test could be implemented with the slower list-based version only.

Since OpenMesh is closest to Surface_mesh in terms of design and implementation, it also is close in terms of performance. The differences of about 20%–30% are due to our more efficient mechanism for accessing custom properties, which requires fewer indirections. Furthermore, our `do-while` circulators are slightly more efficient than the `for` circulators of OpenMesh, which use a rather complex test for detecting the end of the loop.

The results clearly demonstrate the performance of Surface_mesh to be (in most cases) superior to or at least on par with the other data structures.

5.3 Memory Efficiency

Besides run-time performance, memory consumption is a key criterion to measure the efficiency of a library, especially when it comes to applications dealing with highly complex data sets. We compare the memory consumption of the data structures on three different models: the Imp model (300k vertices, 600k triangles) and the Dual Dragon (100k vertices, 50k polygons) already used in the performance comparison and the complex Lucy model (10M vertices, 20M triangles). The results are shown in Table 4.

Although a face-based data structure in general consumes less memory than a halfedge data structure, Mesquite requires more memory than Surface_mesh because (i) of the overhead of the dynamic arrays used to store incidences, (ii) the use of 64-bit references, and (iii) the storage of several helper data per face and vertex.

In addition to the memory overhead due to the doubly-linked of the CGAL list-kernel, both CGAL data structures use 64-bit pointers as references, which consume twice as much memory than the 32-bit indices employed by OpenMesh and Surface_mesh.

Our slight performance advantage with respect to OpenMesh comes from the different storage of the information whether a vertex, edge, or face is deleted. We store this information in custom `bool` property arrays, which in a `std::vector<bool>` require approximately 1 bit per entity. In contrast, OpenMesh uses one status byte per entity, similar to Mesquite.

These results show that Surface_mesh is superior to the other data structures in terms of memory consumption.

6 Conclusion and Future Work

Our results show that the design decisions made during the development of a mesh data structure have a crucial impact on both the usability and the efficiency of the library. By systematically analyzing the design questions we derived design decisions that—if carefully implemented—result in a mesh data structure that is more usable, offers higher performance and consumes less memory than several other mesh data structures publicly available.

Considering the sometimes drastic differences in performance and memory consumption between the individual libraries, it is important to keep in mind that some of them have originally been designed and implemented with a strong focus on a given application domain, such as computational geometry in case of CGAL and mesh optimization in case of Mesquite. As a consequence, both libraries provide significantly more functionality that goes beyond a pure surface mesh data structure. For example, Mesquite supports the optimization of surface and volume meshes within a single framework.

While we are confident with the tests and results achieved thus far, we feel that our benchmark tests should be expanded to a wider variety of different setups (i.e. different hardware, operating systems, compilers and mesh models). Furthermore, additional algorithms and additional mesh data structures, for instance VCGLib, could be included in future evaluations.

Our performance and memory benchmarks can be a first step towards a general benchmark for mesh data structures. We will therefore make the source code and the results of the benchmarks publicly available. Furthermore, in order to facilitate wide adoption of our new data structure, we will also make Surface_mesh freely available under an Open Source license allowing for both academic and commercial usage.

While our current work is focused on surface meshes only, we are aware that applications such as physical simulations often require volumetric meshes. We feel that a systematic approach as presented in this paper might also be beneficial for the design and implementation of a volumetric mesh data structure. In particular, design decisions such as array-based storage, indices as entity references and custom properties as synchronized arrays should carry over to such a data structure seamlessly.

Acknowledgments

The authors are grateful to the participants of the SGP 2011 graduate school for their valuable feedback. Daniel Sieger and Mario Botsch are supported by the Deutsche Forschungsgemeinschaft (Center of Excellence in "Cognitive Interaction Technology", CITEC). The Lucy and Dragon models are courtesy of the Stanford University Computer Graphics Laboratory.

References

1. Alumbaugh, T., Jiao, X.: Compact array-based mesh data structures. In: Proceedings of the 14th International Meshing Roundtable, pp. 485–504 (2005)
2. Baumgart, B.G.: Winged-edge polyhedron representation. Technical Report STAN-CS320, Computer Science Department, Stanford University (1972)
3. Blandford, D., Blelloch, G., Cardoze, D., Kadow, C.: Compact representations of simplicial meshes in two and three dimensions. In: Proceedings of the 12th International Meshing Roundtable, pp. 135–146 (2003)
4. Botsch, M., Kobbelt, L., Pauly, M., Alliez, P., Lévy, B.: Polygon Mesh Processing. AK Peters (2010)
5. Botsch, M., Steinberg, S., Bischoff, S., Kobbelt, L.: Openmesh: A generic and efficient polygon mesh data structure. In: Proc. of OpenSG Symposium (2002)
6. Brewer, M., Freitag Diachin, L., Knupp, P., Leurent, T., Melander, D.: The Mesquite mesh quality improvement toolkit. In: Proceedings of the 12th International Meshing Roundtable, pp. 239–250 (2003)
7. Campagna, S., Kobbelt, L., Seidel, H.-P.: Directed edges: A scalable representation for triangle meshes. Journal of Graphics, GPU, and Game Tools 3(4), 1–12 (1998)
8. CGAL. Computational Geometry Algorithms Library (2011), http://www.cgal.org
9. Edwards, H.C., Williams, A.B., Sjaardema, G.D., Baur, D.G., Cochran, W.K.: SIERRA toolkit computational mesh conceptual model. Technical Report SAND2010-1192, Sandia National Laboratories (2010)
10. De Floriani, L., Hui, A.: Data structures for simplicial complexes: An analysis and a comparison. In: Proc. of Eurographics Symposium on Geometry Processing, Berlin, pp. 119–128 (2005)
11. De Floriani, L., Hui, A., Panozzo, D., Canino, D.: A dimension-independent data structure for simplicial complexes. In: Proceedings of the 19th International Meshing Roundtable, pp. 403–420 (2010)

12. Garimella, R.: MSTK - a flexible infrastructure library for developing mesh based applications. In: Proceedings of the 13th International Meshing Roundtable, pp. 203–212 (2004)
13. Guibas, L., Stolfi, J.: Primitives for the manipulation of general subdivisions and computation of Voronoi diagrams. ACM Transaction on Graphics 4(2), 74–123 (1985)
14. Gurung, T., Laney, D., Lindstrom, P., Rossignac, J.: SQuad: Compact representation for triangle meshes. Computer Graphics Forum 30(2), 355–364
15. Gurung, T., Luffel, M., Lindstrom, P., Rossignac, J.: LR: Compact connectivity representation for triangle meshes. ACM Trans. Graph. 30(3) (2011)
16. Hirani, A.N.: Discrete Exterior Calculus. PhD thesis, California Institute of Technology (2003)
17. Kettner, L.: Designing a data structure for polyhedral surfaces. In: Proceedings of 14th Symposium on Computational Geometry, pp. 146–154 (1998)
18. Kettner, L.: Using generic programming for designing a data structure for polyhedral surfaces. Computational Geometry – Theory and Applications 13(1), 65–90 (1999)
19. Kobbelt, L.: $\sqrt{3}$ subdivision. In: Proceedings of ACM SIGGRAPH 2000, pp. 103–112 (2000)
20. Mantyla, M.: An Introduction to Solid Modeling. Computer Science Press, New York (1988)
21. OpenMesh (2011), http://www.openmesh.org
22. Seegyoung Seol, E., Shephard, M.S.: Efficient distributed mesh data structure for parallel automated adaptive analysis. Engineering with Computers 22(3), 197–213 (2006)
23. Shewchuk, J.R.: Triangle: Engineering a 2D Quality Mesh Generator and Delaunay Triangulator. In: Applied Computational Geometry: Towards Geometric Engineering, vol. 1148, pp. 203–222 (1996)
24. Shiue, L.-J., Alliez, P., Ursu, R., Kettner, L.: A tutorial on CGAL Polyhedron for subdivision algorithms. In: Symp. on Geometry Processing Course Notes (2004)
25. Sieger, D., Alliez, P., Botsch, M.: Optimizing Voronoi diagrams for polygonal finite element computations. In: Proceedings of the 19th International Meshing Roundtable, pp. 335–350 (2010)
26. Sukumar, N., Malsch, E.A.: Recent advances in the construction of polygonal finite element interpolants. Archives of Computational Methods in Engineering 13(1), 129–163 (2006)
27. Tautges, T.J., Meyers, R., Merkley, K., Stimpson, C., Ernst, C.: MOAB: A mesh-oriented database. Technical Report SAND2004-1592, Sandia National Laboratories (2004)
28. Tournois, J., Alliez, P., Devillers, O.: Interleaving Delaunay refinement and optimization for 2D triangle mesh generation. In: Proceedings of the 16th International Meshing Roundtable, pp. 83–101 (2007)
29. VCGLib (2011), http://vcg.sourceforge.net/
30. Wicke, M., Ritchie, D., Klingner, B.M., Burke, S., Shewchuk, J.R., O'Brien, J.F.: Dynamic local remeshing for elastoplastic simulation. ACM Transaction on Graphics 29, 49:1–49:11 (2010)

The Meccano Method for Isogeometric Solid Modeling

José María Escobar[1], José Manuel Cascón[2], Eduardo Rodríguez[1], and Rafael Montenegro[1]

[1] University Institute for Intelligent Systems and Numerical Applications in Engineering, SIANI, University of Las Palmas de Gran Canaria, Spain
{jmescobar,erodriguez,rmontenegro}@siani.es
http://www.dca.iusiani.ulpgc.es/proyecto2008-2011
[2] Department of Economics and History of Economics,
Faculty of Economics and Management, University of Salamanca, Spain
casbar@usal.es

Summary. We present a new method to construct a trivariate T-spline representation of complex solids for the application of isogeometric analysis. The proposed technique only demands the surface of the solid as input data. The key of this method lies in obtaining a volumetric parameterization between the solid and a simple parametric domain. To do that, an adaptive tetrahedral mesh of the parametric domain is isomorphically transformed onto the solid by applying the meccano method. The control points of the trivariate T-spline are calculated by imposing the interpolation conditions on points situated both on the inner and on the surface of the solid. The distribution of the interpolating points is adapted to the singularities of the domain in order to preserve the features of the surface triangulation.

Keywords: Trivariate T-spline, isogeometric analysis, volumetric parameterization, mesh optimization and meccano method.

1 Introduction

CAD models usually define only the boundary of a solid, but the application of isogeometric analysis [2, 3, 10] requires a fully volumetric representation. An open problem in the context of isogeometric analysis is how to generate a trivariate spline representation of a solid starting from the CAD description of its boundary. As it is pointed by Cotrell et al. in [10], "the most significant challenge facing isogeometric analysis is developing three-dimensional spline parameterizations from surfaces".

There are only a few works addressing this problem, and they all have in common the use of harmonic functions to establish the volumetric parameterization [20, 22, 23, 24, 31].

For example, Li et al. [20] construct a harmonic volumetric mapping through a meshless procedure by using a boundary method. The algorithm can be applied to any genus data but it is complex and requires placing some source and collocation points on an offset surface. Optimal results of source positions are unknown, and in practice they are chosen in a trial-and-error manner or with the help of human experience. Therefore, the problem is ill-conditioned and regular system solvers often fail.

Martin et al. [23, 24] present a methodology based on discrete harmonic functions to parameterize a solid. They solve several Laplace's equations, first on the surface and then on the complete 3-D domain with FEM, and use a Laplacian smoothing to remove irregularities. During the process, new vertices are inserted in the mesh and retriangulations (in 2-D and 3-D) are applied in order to introduce the new vertex set in the mesh. The user has to make an initial choice of two critical points to establish the surface parameterization and to fix a seed for generating the skeleton. The parameterization has degeneracy along the skeleton. The extension to genus greater than zero [24] requires finding suitable midsurfaces.

We propose a different approach in which the volumetric parameterization is accomplished by transforming a tetrahedral mesh from the parametric domain to the physical domain. This is a special feature of our procedure; we do not have to give the tetrahedral mesh of the solid as input, as it is a result of the parameterization process. Another characteristic of our work is that we use an interpolation scheme to fit a trivariate B-spline to the data, instead of an approximation, as other authors do. This performs a more accurate adaptation of the T-spline to the input data.

One of the main drawbacks of NURBS (see for example [27]) is that they are defined on a parametric space with a tensor product structure, making the representation of detailed local features inefficient. This problem is solved by the T-splines, a generalization of NURBS conceived by Sederberg [28] that enables the local refinement. The T-splines are a set of functions defined on a T-mesh, a tiling of a rectangular prism in $\mathbb{R}^3$ allowing T-junctions (see [2] and [28]).

In this paper we present a new method [13] for constructing volumetric T-meshes of genus-zero solids whose boundaries are defined by surface triangulations. Our procedure can be summarized in two stages. In the first one, a volumetric parameterization of the solid is developed by using the meccano method [7, 8, 25, 26]. Broadly speaking, we can consider that the construction of a volumetric parameterization is a process in which an adaptive tetrahedral mesh, initially defined in the unitary cube $\mathcal{C} = [0, 1]^3$, is deformed until it achieves the shape of the solid (the physical domain). This deformation only affects the positions of the nodes, that is, there is not any change in their connectivities: we say that both meshes are isomorphic. Given that a point is fully determined by the barycentric coordinates relative to the tetrahedron in which it is contained, we can define a one-to-one mapping between $\mathcal{C}$ and the solid assuming that the barycentric coordinates are the same in both spaces.

In the second stage, the modeling of the solid by trivariate T-splines is carried out. The control points of the T-splines are calculated enforcing the

T-splines to verify the interpolation conditions. Here is where the volumetric parametrization plays its part, mapping the interpolation points from the parametric domain, the T-mesh, onto the solid. In our case, the T-mesh is an octree partition of $\mathcal{C}$ with a similar resolution than the tetrahedral mesh defined in $\mathcal{C}$.

Our technique is simple and it automatically produces a T-spline adapted to the geometry with a low computational complexity and low user intervention.

In [25], we introduced the meccano method to construct volumetric parameterizations of solids of genus greater than zero, where the surface parameterization is explicitly given. The method is based on the construction of a rough approximation of the solid joining cuboids, i.e. the meccano. In this case, the construction of a T-spline representation also demands a T-mesh adapted to the discretization of the meccano. In this paper, we introduce a way to undertake this task by an octree subdivision of a cube enclosing a polycube decomposition of the meccano. Possibly, the most complex question, from a technical point of view, is the automatic generation of a meccano by using a specific CAD system and the corresponding surface parameterization. This last topic is widely discussed in the literature and we could proceed as in PolyCube-Maps [21, 30, 31].

The paper is organized as follows. In the next Section we describe the main steps to parameterize a genus-zero solid onto a cube. Some parts of this Section are taken from our previous works on mesh untangling and smoothing and the meccano method [7, 8, 11, 13, 25, 26], but they have been adapted to the requirements of the present work. The representation of the solid by means of trivariate T-splines is developed in Section 3. In Section 4 we show a test problem and several applications that highlight the ability of our method for modeling complex objects. Finally, in Section 5 we present the conclusions and set out some challenges.

2 Volumetric Parameterization

2.1 Boundary Mapping

The first step to construct a volumetric parameterization of a genus-zero solid consists of establishing a bijective correspondence between the boundary of the cube and the solid. To do that, the given surface triangulation of the solid, $\mathcal{T}_S$, is divided in six patches or *connected subtriangulations*, $\mathcal{T}_S^i$ ($i = 1, 2, \ldots, 6$), having the same connectivities as the cube faces. Specifically, if we consider that each subtriangulation corresponds to a vertex of a graph and two vertices of the graph are connected if their corresponding subtriangulations have at least a common edge, then, the graphs corresponding to the solid and the graph of the cube must be isomorphic (see [8, 26] for details).

Once $\mathcal{T}_S$ is decomposed into six patches, we map each $\mathcal{T}_S^i$ to the corresponding cube face by using the parameterization of surface triangulations proposed by M. Floater in [14, 15]. This is a well-known method to transform a surface triangulation onto a plane triangulation defined in a convex domain, that is, the cube faces in our case. Many and more recent alternative solutions have been proposed to solve the surface parameterization (see for example the surveys [16, 17]), but in most of them the plane triangulation is not defined in a convex set, which is a restriction for us. Thus, if τ_F^i is the resulting triangulation on the i-th face of the cube, the parameterization $\Pi_F^i : \tau_F^i \to \mathcal{T}_S^i$ is a piece-wise linear function that maps a point p inside triangle $T \in \tau_F^i$ onto a point q belonging to triangle $\Pi_F^i(T) \in \mathcal{T}_S^i$ with identical barycentric coordinates.

In order to ensure the compatibility of $\{\Pi_F^i\}_{i=1}^6$, the boundary nodes of $\{\tau_F^i\}_{i=1}^6$ must coincide on common cube edges. The six transformations $\{\Pi_F^i\}_{i=1}^6$ define a global parameterization between $\tau_F = \bigcup_{i=1}^6 \tau_F^i$ and $\mathcal{T}_S$ given by

$$\Pi_F : \tau_F \to \mathcal{T}_S \qquad (1)$$

The parameterization Π_F is used in the following step of the algorithm to map a new triangulation defined over the boundary of $\mathcal{C}$ onto the boundary of the solid.

2.2 Generation of an Adapted Tetrahedral Mesh of the Cube

Let us consider $\mathcal{C}_K$ is a tetrahedral mesh of $\mathcal{C}$ resulting after applying several local bisections of the Kossaczky algorithm [19] to an initial mesh formed by six tetrahedra (see Fig. 1(a)). Three consecutive global bisections are presented in Figures 1(b), 1(c) and 1(d). The mesh of Fig. 1(d) contains 8 cubes similar to the one shown in Fig. 1(a). Therefore, the successive refinement of this mesh produces similar tetrahedra to those of Figures 1(a), 1(b) and 1(c).

If $\tau_K = \partial \mathcal{C}_K$ is the new triangulation defined on the boundary of $\mathcal{C}$, then we define a new parameterization

$$\Pi_K : \tau_K \to \mathcal{T}_S^* \qquad (2)$$

where $\mathcal{T}_S^*$ is the surface triangulation obtained after Π_F-mapping the nodes of τ_K. The points of τ_K are mapped to $\mathcal{T}_S^*$ by preserving their barycentric coordinates. Note that $\mathcal{T}_S^*$ is an approximation of $\mathcal{T}_S$. In order to improve this approximation we must refine the tetrahedra of $\mathcal{C}_K$ in contact with the surface of the cube in such a way that the *distance* between $\mathcal{T}_S^*$ and $\mathcal{T}_S$ decreases until reaching a prescribed tolerance ε. The concept of *distance* between two triangulations can be defined and implemented in several ways. In our case, it is as follows:

Let T be a triangle of τ_K, where a, b and c are their vertices and let $p_k \in \{p_i\}_{i=1}^{N_q}$ be a Gauss quadrature point of T, then, the distance, $d(T)$, between

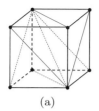 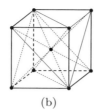 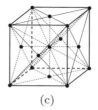 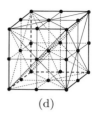

(a) (b) (c) (d)

Fig. 1. Refinement of a cube by using Kossaczky's algorithm: (a) cube subdivision into six tetrahedra, (b) bisection of all tetrahedra by inserting a new node in the cube main diagonal, (c) new nodes in diagonals of cube faces and (d) global refinement with new nodes in cube edges.

$\Pi_K(T)$ and the underlaying triangulation $\mathcal{T}_S$ is defined as the maximum of the volumes of the tetrahedra formed by $\Pi_F(a)$, $\Pi_F(b)$, $\Pi_F(c)$ and $\Pi_F(p_k)$. If we considerer the distance between $\mathcal{T}_S^*$ and $\mathcal{T}_S$ as the maximum of all $d(T)$, the local refinement stops when $d(T) < \epsilon$ for all $T \in \tau_K$. A more accurate approach based on Hausdorff distance can be found in [4].

Once the adapted tetrahedral mesh $\mathcal{C}_K$ has been constructed by using the proposed method, the nodes of τ_K are mapped to the surface of the solid giving the triangulation $\mathcal{T}_S^*$, which is the final approximation of $\mathcal{T}_S$. Note that inner nodes of $\mathcal{C}_K$ stay in their initial positions, so the current tetrahedral mesh of the solid will most likely be tangled. The following step plays a crucial roll in our procedure. We have to relocate the inner nodes in suitable positions such that this tetrahedral mesh gets untangled and the distortion introduced by the associated parameterization is as small as possible.

2.3 Relocation of Inner Nodes

Usual techniques to improve the quality of a *valid* mesh, that is, one that does not have inverted elements, are based upon local smoothing. In short, these techniques consist of finding the new positions that the mesh nodes must hold, in such a way that they optimize an objective function. Such a function is based on a certain measurement of the quality of the *local submesh* $N(q)$, formed by the set of tetrahedra connected to the *free node* q. Usually, objective functions are appropriate to improve the quality of a valid mesh, but they do not work properly when there are inverted elements. This is because they present singularities (barriers) when any tetrahedron of $N(q)$ changes the sign of its Jacobian.

Most of what is stated below is taken from [11], where we developed a procedure for untangling and smoothing meshes simultaneously. For that purpose, we use a suitable modification of the objective function such that it is regular all over $\mathbb{R}^3$. When a feasible region (subset of $\mathbb{R}^3$ where q could be placed, being $N(q)$ a valid submesh) exists, the minima of both the original and the modified objective functions are very close, and when this region does not exist, the minimum of the modified objective function is located in such a way that it tends to untangle $N(q)$. The latter occurs, for example, when the fixed boundary of $N(q)$ is tangled. With this approach, we can use any standard and efficient unconstrained optimization method to find the minimum of the modified objective function, see for example [1].

If we name $\mathcal{T}$ to the tetrahedral mesh of the solid once the inner nodes have been relocated, the corresponding volumetric parameterization is

$$\Pi : \mathcal{C}_K \to \mathcal{T} \qquad (3)$$

A point p included in a tetrahedron of $\mathcal{C}_K$ is mapped, preserving barycentric coordinates, into a point q belonging to the transformed tetrahedron of $\mathcal{T}$.

Objective Functions

Several tetrahedron shape measures could be used to construct an objective function. Nevertheless, those obtained by algebraic operations [18] are specially indicated for our purpose because they can be computed very efficiently and they allow us to choose the shape of the tetrahedra to optimize. Our objective is to relocate the nodes of $\mathcal{T}$ in positions where not only the mesh gets untangled, but also the distortion introduced by the parameterization is minimized.

Let T be a tetrahedral element of $\mathcal{T}$ whose vertices are given by $\mathbf{x}_k = (x_k, y_k, z_k)^T \in \mathbb{R}^3$, $k = 0, 1, 2, 3$ and T_R be the reference tetrahedron with vertices $\mathbf{u}_0 = (0,0,0)^T$, $\mathbf{u}_1 = (1,0,0)^T$, $\mathbf{u}_2 = (0,1,0)^T$ and $\mathbf{u}_3 = (0,0,1)^T$. If we choose $\mathbf{x}_0$ as the translation vector, the affine map that takes T_R to T is $\mathbf{x} = A\mathbf{u} + \mathbf{x}_0$, where A is the Jacobian matrix of the affine map referenced to node $\mathbf{x}_0$, and expressed as $A = (\mathbf{x}_1 - \mathbf{x}_0, \mathbf{x}_2 - \mathbf{x}_0, \mathbf{x}_3 - \mathbf{x}_0)$.

Let us consider that T_I is our ideal or target tetrahedron whose vertices are $\mathbf{v}_0, \mathbf{v}_1, \mathbf{v}_2$ and $\mathbf{v}_3$. If we take $\mathbf{v}_0 = (0,0,0)^T$ the linear map that takes T_R to T_I is $\mathbf{v} = W\mathbf{u}$, where $W = (\mathbf{v}_1 - \mathbf{v}_0, \mathbf{v}_2 - \mathbf{v}_0, \mathbf{v}_3 - \mathbf{v}_0)$ is its Jacobian matrix. As the parametric and real meshes are topologically identical, each tetrahedron of $\mathcal{T}$ has its counterpart in $\mathcal{C}_K$. Thus, in order to reduce the distortion in the volumetric parameterization we will fix the target tetrahedra of $N(q)$ as their counterparts of the local mesh in the parametric space.

The affine map that takes T_I to T is given by $\mathbf{x} = AW^{-1}\mathbf{v} + \mathbf{x}_0$, and its Jacobian matrix is $S = AW^{-1}$. Note that this weighted matrix S depends on the node chosen as reference, so this node must be the same for T and T_I. We can use matrix norms, determinant or trace of S to construct algebraic quality metrics of T. For example, the *mean ratio*, $Q = \frac{3\sigma^{\frac{2}{3}}}{|S|^2}$, is an easily computable algebraic quality metric of T, where $\sigma = \det(S)$ and $|S|$ is the Frobenius norm of S. The maximum value of Q is the unity, and it is reached when $A = \mu RW$, where μ is a scalar and R is a rotation matrix. In other words, Q is maximum if and only if T and T_I are similar. Besides, any flat tetrahedron

The Meccano Method for Isogeometric Solid Modeling

has quality measure zero. We can derive an optimization function from this quality metric. Thus, let $\mathbf{x} = (x, y, z)^T$ be the position of the free node, and let S_m be the weighted Jacobian matrix of the m-th tetrahedron of $N(q)$. We define the objective function of $\mathbf{x}$, associated to an m-th tetrahedron as

$$\eta_m = \frac{|S_m|^2}{3\sigma_m^{\frac{2}{3}}} \tag{4}$$

Then, the corresponding objective function for $N(q)$ is constructed by using the p-norm of $(\eta_1, \eta_2, \ldots, \eta_M)$ as

$$|K_\eta|_p(\mathbf{x}) = \left[\sum_{m=1}^{M} \eta_m^p(\mathbf{x}) \right]^{\frac{1}{p}} \tag{5}$$

where M is the number of tetrahedra in $N(q)$.

Although this optimization function is smooth in those points where $N(q)$ is a valid submesh, it becomes discontinuous when the volume of any tetrahedron of $N(q)$ goes to zero. It is due to the fact that η_m approaches infinity when σ_m tends to zero and its numerator is bounded below. In fact, it is possible to prove that $|S_m|$ reaches its minimum, with strictly positive value, when q is placed in the geometric center of the fixed face of the m-th tetrahedron. The positions where q must be located to get $N(q)$ to be valid, i.e., the feasible region, is the interior of the polyhedral set P defined as $P = \bigcap_{m=1}^{M} H_m$, where H_m are the half-spaces defined by $\sigma_m(\mathbf{x}) \geqslant 0$. This set can occasionally be empty, for example, when the fixed boundary of $N(q)$ is tangled. In this situation, function $|K_\eta|_p$ stops being useful as an optimization function. Moreover, when the feasible region exists, that is $int\ P \neq \emptyset$, the objective function tends to infinity as q approaches the boundary of P. Due to these singularities, it is formed a barrier which avoids reaching the appropriate minimum when using gradient-based algorithms, and when these start from a free node outside the feasible region. In other words, with these algorithms we can not optimize a tangled mesh $N(q)$ with the above objective function.

Modified Objective Functions

We proposed in [11] a modification in the previous objective function (5), so that the barrier associated with its singularities will be eliminated and the new function will be smooth all over $\mathbb{R}^3$. An essential requirement is that the minima of the original and modified functions are nearly identical when $int\ P \neq \emptyset$. Our modification consists of substituting σ in (5) by the positive and increasing function

$$h(\sigma) = \frac{1}{2}(\sigma + \sqrt{\sigma^2 + 4\delta^2}) \tag{6}$$

being the parameter $\delta = h(0)$. Thus, the new objective function here proposed is given by

$$\left|K_\eta^*\right|_p(\mathbf{x}) = \left[\sum_{m=1}^{M} (\eta_m^*)^p(\mathbf{x})\right]^{\frac{1}{p}} \tag{7}$$

where

$$\eta_m^* = \frac{|S_m|^2}{3h^{\frac{2}{3}}(\sigma_m)} \tag{8}$$

is the modified objective function for the m-th tetrahedron. With this modification, we can untangle the mesh and, at the same time, improve its quality. An implementation of the simultaneous untangling and smoothing procedure for an equilateral reference tetrahedron is freely available in [12].

Rearrangement of the Inner Nodes

The computational effort to optimize a mesh depends on the initial position of the nodes. An arrangement of the nodes close to their optimal positions significantly reduces the number of iterations (and the CPU time) required by the untangling and smoothing algorithm. Therefore, an interesting idea is to construct a rough approximation of the solid and to use the corresponding parametrization to relocate interior nodes of more accurate subsequent approximations.

Taking into account that the grade of refinement attained by the tetrahedral mesh depends on the maximum allowed distance, ε, between $\mathcal{T}_S^*$ and $\mathcal{T}_S$, we will write $\mathcal{C}_K(\varepsilon)$, $\mathcal{T}(\varepsilon)$, $\mathcal{T}_K(\varepsilon)$ and $\mathcal{T}_S^*(\varepsilon)$ to express this dependence.

Let suppose that $\Pi_{\varepsilon_i} : \mathcal{C}_K(\varepsilon_i) \to \mathcal{T}(\varepsilon_i)$ is the volumetric parameterization for a given tolerance ε_i. We want to find the approximate location of the nodes of a more accurate mesh $\mathcal{T}(\varepsilon_{i+1})$, assuming that $\varepsilon_i > \varepsilon_{i+1}$. Firstly, the mesh $\mathcal{C}_K(\varepsilon_i)$ is locally refined until the distance between $\mathcal{T}_S^*(\varepsilon_i)$ and $\mathcal{T}_S$ is below ε_{i+1}. In that moment we have the new mesh of the cube $\mathcal{C}_K(\varepsilon_{i+1})$. Afterward, their inner nodes are mapped by using the previous parameterization, that is, we construct the new tetrahedral mesh $\mathcal{T}^*(\varepsilon_{i+1})$ after Π_{ε_i}-mapping the nodes of $\mathcal{C}_K(\varepsilon_{i+1})$. Note that $\mathcal{T}^*(\varepsilon_{i+1})$ has the same topology as $\mathcal{T}(\varepsilon_{i+1})$, but their nodes are not located at optimal positions. Although $\mathcal{T}^*(\varepsilon_{i+1})$ could be tangled, their interior nodes are close to their final positions. Therefore, the computational effort to optimize the mesh is drastically reduced. The last step of this iteration consists on relocating the inner nodes of $\mathcal{T}^*(\varepsilon_{i+1})$ in their optimal position following the mesh smoothing and untangling procedure above described. This sequence is repeated several times until we achieve the desired tolerance. In Fig. 2 it is shown a sequence of gradual approximations to the mesh of a horse. The initial surface triangulation $\mathcal{T}_S$ has been obtained from the Large Geometric Model Archives at Georgia Institute of Technology.

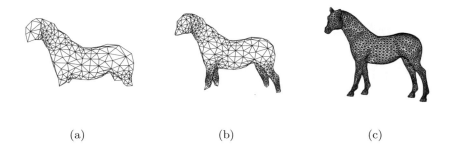

(a) (b) (c)

Fig. 2. Gradual approximations: from a coarse mesh to the final accurate mesh.

3 Representation of the Solid by T-Splines

3.1 Construction of an Adapted Volumetric T-Mesh

We will start this Section with a short introduction on T-splines. A detailed report about T-splines and their relationship with isogeometric analysis can be found in [2].

The T-mesh is the control grid of the T-splines. In 3-D it is a division of a rectangular prism forming a grid in which the T-juntions are allowed. In 2-D T-junctions are inner vertices of the grid connecting 3 edges. T-junctions in 3-D are inner vertices shared by one edge in some direction and two edges in other directions at the same time [29]. T-splines are rational spline functions defined by local knot vectors, which are inferred from certain points of the T-mesh known as *anchors* [2]. The anchors of the odd-degree T-splines are situated on the vertices of the T-mesh and the anchors of the even-degree T-splines are located in the center of each prism. We will focus on odd-degree T-splines and, in particular, on cubic T-splines because they are the ones implemented in the present work. Cubic T-splines have 5 knots in each parametric direction. Let us consider the 2-D example of Fig. 3 to understand how the knot vectors are deduced from the anchor. The parametric coordinates of the anchor $\mathbf{t}_\alpha$ in Fig. 3(a) are given by $\left(\xi_4^1, \xi_4^2\right)$, then, by examining the intersections of horizontal and vertical lines (red lines in the Figure) with the edges of the T-mesh, we deduce that the the knot vector in ξ^1 direction is $\Xi_1^\alpha = \left(\xi_1^1, \xi_2^1, \xi_4^1, \xi_5^1, \xi_6^1\right)$ and, the knot vector in ξ^2 direction is $\Xi_2^\alpha = \left(\xi_2^2, \xi_3^2, \xi_4^2, \xi_5^2, \xi_6^2\right)$. In the case of Fig. 3(b) only one edge is found when marching horizontally from $\mathbf{t}_\beta$ to the right. In such situations we have two possibilities: repeat knots in order to form a clamped local knot vector or, as we have implemented in our work, add phantom knots and form an unclamped one. These phantom knots are placed following the pattern shown in Fig. 3(b). The construction of knot vector in 3-D is analogous but we must examine the intersections with T-mesh faces encountered when marching in each space direction. The points of the parametric domain are written as $\boldsymbol{\xi} = \left(\xi^1, \xi^2, \xi^3\right)$.

A T-spline is a rational function from the parametric domain to the physical space given by

$$\mathbf{S}(\boldsymbol{\xi}) = \sum_{\alpha \in A} \mathbf{P}_\alpha R_\alpha (\boldsymbol{\xi}) \qquad (9)$$

where $\mathbf{P}_\alpha$ is the control point corresponding to the α-th blending function

$$R_\alpha (\boldsymbol{\xi}) = \frac{w_\alpha B_\alpha (\boldsymbol{\xi})}{\sum_{\beta \in A} w_\beta B_\beta (\boldsymbol{\xi})} \qquad (10)$$

being w_α its weight and $B_\alpha(\boldsymbol{\xi}) = N_\alpha^1(\xi^1) N_\alpha^2(\xi^2) N_\alpha^3(\xi^3)$ the product of univariate B-splines. In these expressions $A \subset \mathbb{Z}^3$ represents the index set containing every α such that $\mathbf{t}_\alpha$ is an anchor.

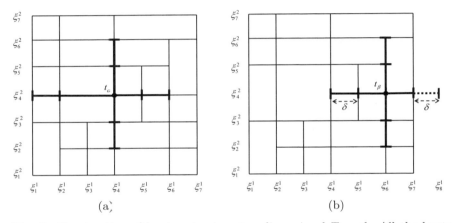

Fig. 3. Construction of knot vector in a two-dimensional T-mesh. All the knots associated to the anchor $\mathbf{t}_\alpha$ lie inside the T-mesh (a). The phantom knot ξ_8^1 has been added to construct an unclamped local knot vector (b).

The T-spline $\mathbf{S}(\boldsymbol{\xi})$ is the sum of rational C^2 blending functions, so it is also a C^2 function. Nevertheless, as the surface of the solid is the union of six patches obtained by mapping the six faces of the cube, and these faces match with C^0 continuity, we only can assure the C^0 continuity for the surface of the solid.

Our objective is to get a representation of the solid suitable for isogeometric analysis by means of trivariate T-splines. This representation, $\mathcal{V}$, must preserve the features and details of the input data, the triangulation $\mathcal{T}_S$. To do that, we construct an adapted T-mesh by partitioning the parametric domain $\mathcal{C}$ in cells by using an octree subdivision. The unitary cube $\mathcal{C}$ is divided in 8 identical cells and, each cell is, in turn, divided in other 8 cells and so on, until all the cells of the octree do not contain any node of $\mathcal{C}_K$

in their inner. This last is possible due to the particular characteristics of the Kossaczky subdivision scheme, in which the edges of $\mathcal{C}_K$ are the result of successive division of the edges of $\mathcal{C}$ by two. The octree partition defines a T-mesh, $\mathcal{C}_T$, that is used to determine the local knot vector and the anchors of the T-splines. Note that all the nodes of $\mathcal{C}_K$ are vertices of $\mathcal{C}_T$, so it is to be hoped that the surface of $\mathcal{V}$ achieves the same resolution than the input triangulation $\mathcal{T}_S$. Another consequence of the proposed octree subdivision is that the cell faces of $\mathcal{C}_T$ contain no more than one inner T-junction.

3.2 Interpolation

Basically there are two ways of fitting splines to a set points: interpolation and approximation. We have adopted the first one because it is more appropriate for reducing all features of the input triangulation. Assuming that the set of blending functions are linearly independent, we need as many interpolation points as blending functions.

Recently Buffa et al. [5] have analyzed the linear independence of the bicubic T-spline blending functions corresponding to some particular T-meshes. They prove linear independence of hierarchical 2-D T-meshes generated as the refinement of a coarse and uniform T-mesh (this is the 2-D counterpart to our case). However, the extension of these results to 3-D is not straightforward.

We have chosen the images of the anchors as interpolation points, and all the weights have been taken equal to 1. Thus, the control points, $\mathbf{P}_\alpha$, are obtained by solving the linear system of equations

$$\Pi\left(\mathbf{t}_\beta\right) = \mathbf{S}\left(\mathbf{t}_\beta\right) = \sum_{\alpha \in A} \mathbf{P}_\alpha R_\alpha\left(\mathbf{t}_\beta\right), \ \forall \mathbf{t}_\beta, \ \beta \in A \qquad (11)$$

where the images $\Pi\left(\mathbf{t}_\beta\right)$ have been calculated through the volumetric parameterization (3).

The linear independence has become evident in all the applications considered until now, as the resolution of (11) is only possible if the blending functions are linearly independent.

4 Results

4.1 Test Example

We have chosen a 2-D domain as first example in order to dicuss how the proposed technique works.

At present, there are no quality metrics for isogeometric analysis analogous to the ones for traditional FEA to help us characterize the impact of the mesh on analysis, as it is indicated in [9]. Xu et al. [32, 33] give sufficient conditions for getting both an injective parameterization for planar splines without self-intersections and an isoparametric net of good uniformity and orthogonality, but there are not similar studies for T-splines.

One of the factors to take into account is the variation of the Jacobian in the elements. Usually, a large variation leads to poor accuracy in the numerical approximation, so we can explore the suitability of a T-spline for isogeometric simulations by analyzing the scaled Jacobian in the quadrature points of the cells. The scaled Jacobian, given by

$$J_s(\xi^1, \xi^2, \xi^3) = \frac{\det(\mathbf{S}_{\xi^1}, \mathbf{S}_{\xi^2}, \mathbf{S}_{\xi^3})}{\|\mathbf{S}_{\xi^1}\| \|\mathbf{S}_{\xi^2}\| \|\mathbf{S}_{\xi^3}\|} \qquad (12)$$

where $\mathbf{S}_{\xi^i}$ is the derivative of the trivariate T-spline (9) with respect to ξ^i, has been evaluated in the eight Gaussian quadrature points (see for example [6]) of each cell in the real domain. For doing that, we set the eight quadrature points in the hexahedra of the parametric domain and calculate their transformation to the real domain by applying (9). We can get an idea about whether the distortion introduced by the spline is or is not too large by plotting the average, minimum and maximum of the scaled Jacobian in the quadrature points. The following test model shows a procedure, based on local mesh refinement, to improve the scaled Jacobian values. The goal is to reach values of the scaled Jacobian close to one in most parts of the solid.

The test model (see Fig. 4) is a T-spline representation of a deformed unitary square in which the corner $(1, 1)$ has been displaced toward position $\left(\frac{3}{4} - \frac{1}{10}, \frac{3}{4} - \frac{1}{10}\right)$, producing a degenerate cell. This displacement makes the new optimal position for the central node to become $(0.38, 0.38)$. The same model is approximated by two T-meshes with 9 (Fig. 4(a)) and 14 (Fig. 4(d)) interpolating points. The corresponding T-spline representations are shown in Figures 4(b) and 4(e), respectively. Note that the representation of Fig. 4(b) has a wide folded region around the corner in which the Jacobian is negative. However, this region has been remarkably reduced in the refined version (Fig. 4(e)). This example indicates that, although the refinement of the T-mesh around the corners (and edges in 3-D) does not completely solve the problem of degenerate cells, it tends to diminish the region in which the Jacobians become negative. It can be more clearly seen in Figures 4(c) and 4(f), where the scaled Jacobian has been represented by a color map. The dark colors correspond to the regions in which the Jacobian is negative.

4.2 Solid with Surface of Genus Zero

In Fig. 5 we have shown a tetrahedral and T-spline representation of the Stanford bunny. Note how similar discretization of the respective parametric domains give rise to similar grade of detail in the physical domains. It can be seen how the isoparametric curves are nearly orthogonal in most parts of the solid, which entails low distortion and values of scaled Jacobian close to one. Nevertheless, the distortion becomes high in some regions of the surface. We have computed 39 cells out of 9696 in which at least one of the eight Gaussian quadrature points has a negative Jacobian.

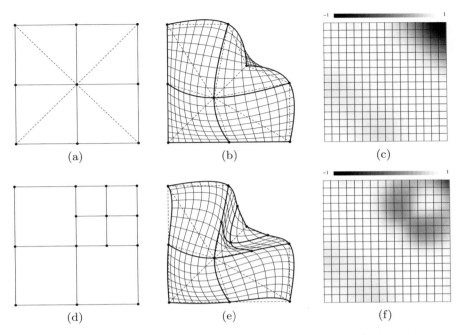

Fig. 4. Initial T-mesh in the parametric domain (continuous line) and the underlying triangular mesh (dashed line) (a). T-spline of a deformed square with a reentrant corner (b). Scaled Jacobian representation in the parametric domain (c). Corresponding representations for a refined version (d), (e) and (f).

As we have mentioned, the T-spline is enforced to interpolate all the nodes of the tetrahedral mesh $\mathcal{T}$ and this mesh is as close as we want to the input surface $\mathcal{T}_S$. Moreover, the interpolating points are exactly situated on the input surface. These reasons suggest a good accuracy between the surface of the T-spline and $\mathcal{T}_S$. In order to estimate the gap between both surfaces we have analyzed the differences between the volumes enclosed by $\mathcal{T}_S$ and the T-spline, $\mathcal{V}$. The first volume is measured by applying the divergence theorem and the second one is calculated integrating $\det\left(\mathbf{S}_{\xi^1}, \mathbf{S}_{\xi^2}, \mathbf{S}_{\xi^3}\right)$ in the unitary cube $\mathcal{C}$ with 8 Gaussian quadrature points in each cell. The quadrature points with negative Jacobians have been rejected from the calculations. The results for the bunny application are: the volume enclosed by $\mathcal{T}_S$ is 754.9; the volume of $\mathcal{T}$ is 750.9 (a difference of 0.5% in relation to $\mathcal{T}_S$) and the volume of $\mathcal{V}$ is 757.4 (a difference of 0.3% in relation to $\mathcal{T}_S$).

Guided by the results of the test example of Fig. 4, we are interested in knowing the effect of refining the cells with worst quality. To do that, we develop an iterative procedure in which the scaled Jacobian is evaluated in the center of each cell and, if it is negative, we store the point in a list of vertices to be included in the T-mesh of the subsequent iteration. If the

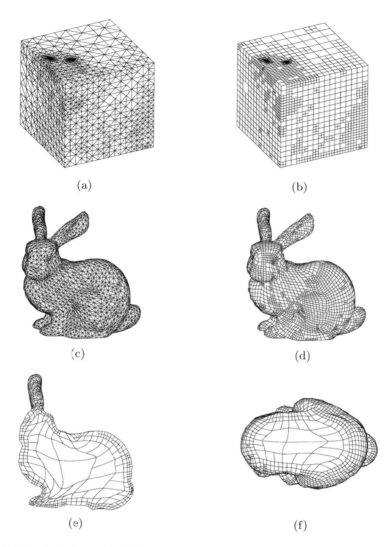

Fig. 5. Tetrahedral mesh of the parametric domain $\mathcal{C}_K$ (a), T-mesh $\mathcal{C}_T$ (b), tetrahedral mesh $\mathcal{T}$ (c) and T-spline representation $\mathcal{V}$ (d) of the Stanford bunny. Two transversal sections of $\mathcal{V}$ (e) and (f).

impact of such refinement is similar to the one of the test example, it should be hoped a reduction of the region in which the Jacobian had negative values. In fact, the number of cells with negative scaled Jacobian evaluated in their centers have been: 5 in the first iteration, 4 in the second, 2 in the third and 0 in the fourth. Moreover, only 6% of the cells have a scaled Jacobian less than 0.5.

We remark that in this application we have obtained positive Jacobians in all the centers of the cells of $\mathcal{C}_T$. Therefore, the most distorted cells are susceptible of being integrated with at least one Gaussian quadrature point. Obviously, a better numerical approximation is possible in most of the cells.

4.3 Solid with Surface of Genus Greater Than Zero

We now consider the extension of the proposed isogeometric modeling to solids with surface of genus greater than zero. In this case, the volumetric parameterization of the solid is constructed by using the meccano method [7, 8, 25, 26]. The method is based on the composition of a meccano, joining cuboid pieces in order to get a rough approximation of the solid. Afterward, we use a parameterization to map the boundary of the solid to the meccano faces. To obtain a T-mesh adapted to the discretization of the meccano, we apply an octree subdivision of a cube enclosing the initial polycube decomposition of the meccano. This subdivision produces vertices both inside and outside the meccano, but only the inner vertices must be considered as anchors. The external vertices will be used to complete the unclamped knot vectors. As an example, we present the modeling of a solid with a genus-one surface that is explicitly given. The main stages of the process are shown in Fig. 6. In this case, the meccano is formed by four cuboids. We remark that we have also obtained positive Jacobians in all the centers of the cells of the T-mesh.

5 Conclusions and Challenges

Focused on the application of isogeometric analysis, this work is a new approach to the automatic generation of trivariate T-splines representation of solids. Our procedure has been presented with detail for genus-zero solids and has been introduced for genus greater than zero. The key lies on having a volumetric parameterization of the solid by using the meccano method [25, 26]. In this paper we have considered a genus-one surface parameterization explicitly given, but we think that this handicap could be overcome by applying a technique similar to PolyCube-Maps [21, 30, 31].

Furthermore, the input data of the solid boundary is generally described by CAD. Such information could be used to map the points lying on the surface of the parametric mesh to the surface of the solid, making unnecessary the stage of surface parameterization in the meccano method.

In general, the distortion introduced by the proposed volumetric parameterization is low, but the existence of critical points where the Jacobian of the T-spline may become negative constitutes an inconvenience for isogeometric simulations. Just as we have shown in Section 4, the selective refinement of the most degenerate cells palliates the problem.

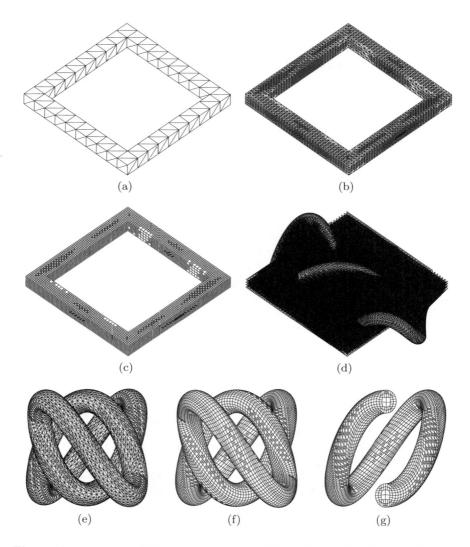

Fig. 6. Main stages of the isogeometric modeling of a solid with a surface of genus one: (a) Coarse tetrahedral mesh of the meccano, (b) refined tetrahedral mesh of the meccano, (c) T-mesh of the meccano, (d) tangled tetrahedral mesh after the mapping on solid surface, (e) resulting tetrahedral mesh after inner node relocation and mesh optimization, (f) T-spline representation of the solid and (g) two transversal sections.

Acknowledgements

This work has been supported by the Spanish Government, "Secretaría de Estado de Universidades e Investigación", "Ministerio de Ciencia e Innovación", and FEDER, grant contracts: CGL2008-06003-C03 and UNLP08-3E-010. The authors are also grateful to the "Junta de Castilla y León", grant number SA124A08, for their support.

References

1. Bazaraa, M.S., Sherali, H.D., Shetty, C.M.: Nonlinear programing: Theory and algorithms. John Wiley and Sons Inc., New York (1993)
2. Bazilevs, Y., Calo, V.M., Cottrell, J.A., Evans, J., Hughes, T.J.R., Lipton, S., Scott, M.A., Sederberg, T.W.: Isogeometric analysis using T-splines. Comput. Meth. Appl. Mech. Eng. 199, 229–263 (2010)
3. Bazilevs, Y., Calo, V.M., Cottrell, J.A., Evans, J., Hughes, T.J.R., Lipton, S., Scott, M.A., Sederberg, T.W.: Isogeometric analysis: Toward unification of computer aided design and finite element analysis. In: Trends in Engineering Computational Technology, pp. 1–16. Saxe-Coburg Publications, Stirling (2008)
4. Borouchaki, H., Frey, P.J.: Simplification of surface mesh using Hausdorff envelope. Comput. Meth. Appl. Mech. Eng. 194, 4864–4884 (2005)
5. Buffa, A., Cho, D., Sangalli, G.: Linear independence of the T-spline blending functions associated with some particular T-meshes. Comput. Meth. Appl. Mech. Eng. 199, 1437–1445 (2010)
6. Carey, G.F., Oden, J.T.: Finite elements, a second course. Prentice-Hall, New Jersey (1982)
7. Cascón, J.M., Montenegro, R., Escobar, J.M., Rodríguez, E., Montero, G.: A new meccano technique for adaptive 3-D triangulations. In: Proc. 16th Int. Meshing Roundtable, pp. 103–120. Springer, Berlin (2007)
8. Cascón, J.M., Montenegro, R., Escobar, J.M., Rodríguez, E., Montero, G.: The meccano method for automatic tetrahedral mesh generation of complex genus-zero solids. In: Proc. 18th Int. Meshing Roundtable, pp. 463–480. Springer, Berlin (2009)
9. Cohen, E., Martin, T., Kirby, R.M., Lyche, T., Riesenfeld, R.F.: Analysis-aware modeling: Understanding quality considerations in modeling for isogeometric analysis. Comput. Meth. Appl. Mech. Eng. 199, 334–356 (2010)
10. Cottrell, J.A., Hughes, T.J.R., Bazilevs, Y.: Isogeometric Analysis: Toward Integration of CAD and FEA. John Wiley & Sons, Chichester (2009)
11. Escobar, J.M., Rodríguez, E., Montenegro, R., Montero, G., González-Yuste, J.M.: Simultaneous untangling and smoothing of tetrahedral meshes. Comput. Meth. Appl. Mech. Eng. 192, 2775–2787 (2003)
12. Escobar, J.M., Rodríguez, E., Montenegro, R., Montero, G., González-Yuste, J.M.: SUS Code – Simultaneous mesh untangling and smoothing code (2010), http://www.dca.iusiani.ulpgc.es/proyecto2008-2011
13. Escobar, J.M., Cascón, J.M., Rodríguez, E., Montenegro, R.: A new approach to solid modeling with trivariate T-splines based on mesh optimization. Comput. Meth. Appl. Mech. Eng. (2011), doi:10.1016/j.cma, 07.004
14. Floater, M.S.: Parametrization and smooth approximation of surface triangulations. Comput. Aid. Geom. Design 14, 231–250 (1997)
15. Floater, M.S.: Mean Value Coordinates. Comput. Aid. Geom. Design 20, 19–27 (2003)
16. Floater, M.S., Hormann, K.: Surface parameterization: A tutorial and survey. In: Advances in Multiresolution for Geometric Modelling, Mathematics and Visualization, pp. 157–186. Springer, Berlin (2005)

17. Hormann, K., Lévy, B., Sheffer, A.: Mesh parameterization: Theory and practice. In: SIGGRAPH 2007: ACM SIGGRAPH, Courses. ACM Press, New York (2007)
18. Knupp, P.M.: Algebraic mesh quality metrics. SIAM J. Sci. Comput. 23, 193–218 (2001)
19. Kossaczky, I.: A recursive approach to local mesh refinement in two and three dimensions. J. Comput. Appl. Math. 55, 275–288 (1994)
20. Li, X., Guo, X., Wang, H., He, Y., Gu, X., Qin, H.: Harmonic Volumetric Mapping for Solid Modeling Applications. In: Proc. of ACM Solid and Physical Modeling Symposium, pp. 109–120. Association for Computing Machinery, Inc. (2007)
21. Lin, J., Jin, X., Fan, Z., Wang, C.C.L.: Automatic PolyCube-Maps. In: Chen, F., Jüttler, B. (eds.) GMP 2008. LNCS, vol. 4975, pp. 3–16. Springer, Heidelberg (2008)
22. Li, B., Li, X., Wang, K.: Generalized PolyCube trivariate splines. In: SMI 2010 - Int. Conf. Shape Modeling and Applications, pp. 261–265 (2010)
23. Martin, T., Cohen, E., Kirby, R.M.: Volumetric parameterization and trivariate B-spline fitting using harmonic functions. Comput. Aid. Geom. Design 26, 648–664 (2009)
24. Martin, T., Cohen, E.: Volumetric parameterization of complex objects by respecting multiple materials. Computers and Graphics 34, 187–197 (2010)
25. Montenegro, R., Cascón, J.M., Escobar, J.M., Rodríguez, E., Montero, G.: An automatic strategy for adaptive tetrahedral mesh generation. Appl. Num. Math. 59, 2203–2217 (2009)
26. Montenegro, R., Cascón, J.M., Rodríguez, E., Escobar, J.M., Montero, G.: The meccano method for automatic 3-D triangulation and volume parametrization of complex solids. Computational Science, Engineering and Technology Series 26, 19–48 (2010)
27. Piegl, L., Tiller, W.: The NURBS book. Springer, New York (1997)
28. Sederberg, T.W., Zheng, J., Bakenov, A., Nasri, A.: T-splines and T-NURCCSs. ACM Trans. Graph. 22, 477–484 (2003)
29. Song, W., Yang, X.: Free-form deformation with weighted T-spline. The Visual Computer 21, 139–155 (2005)
30. Tarini, M., Hormann, K., Cignoni, P., Montani, C.: Polycube-maps. ACM Trans. Graph. 23, 853–860 (2004)
31. Wang, H., He, Y., Li, X., Gu, X., Qin, H.: Polycube splines. Comput. Aid. Geom. Design 40, 721–733 (2008)
32. Xu, G., Mourrain, B., Duvigneau, R., Galligo, A.: Optimal Analysis-Aware Parameterization of Computational Domain in Isogeometric Analysis. In: Mourrain, B., Schaefer, S., Xu, G. (eds.) GMP 2010. LNCS, vol. 6130, pp. 236–254. Springer, Heidelberg (2010)
33. Xu, G., Mourrain, B., Duvigneau, R., Galligo, A.: Parametrization of computational domain in isogeometric analysis: Methods and comparison. INRIA-00530758, 1–29 (2010)

A Volume Flattening Methodology for Geostatistical Properties Estimation

Mathieu Poudret, Chakib Bennis, Jean-François Rainaud, and Houman Borouchaki

IFP Energies nouvelles, 1-4 avenue de Bois-Préau, 92852 Rueil-Malmaison, France
{mathieu.poudret,chakib.bennis,j-francois.rainaud}@ifpen.fr
Charles Delaunay Institute, FRE CNRS 2848, 10010 Troyes, France
houman.borouchaki@utt.fr

Summary. In the domain of oil exploration, geostatistical methods aim at simulating petrophysical properties in a 3D grid model of reservoir. Generally, only a small amount of cells are populated with properties. Roughly speaking, the question is: which properties to give to cell c, knowing the properties of n cells at a given distance from c? Obviously, the population of the whole reservoir must be computed while respecting the spatial correlation distances of properties. Thus, computing of these correlation distances is a key feature of the geostatistical simulations.

In the classical geostatistical simulation workflow, the evaluation of the correlation distance is imprecise. Indeed, they are computed in a Cartesian simulation space which is not representative of the geometry of the reservoir. This induces major deformations in the final generated petrophysical properties.

We propose a new methodology based on isometric flattening of sub-surface models. Thanks to the flattening, we accurately reposition the initial populated cells in the simulation space, before computing the correlation distances. In this paper, we introduce our different flattening algorithms depending on the deposit mode of the sub-surface model and present some results.

1 Introduction

The method presented in this article is concerned with the domain of oil exploration. More particularly, we focus on an important phase of oil reservoir characterization: populating lithostratigraphic units represented by a fine stratigraphic grid (see Fig. 1) with rock properties, for instance lithofacies (types of rock), porosity or permeability. Properties are assigned to the center of the cells.

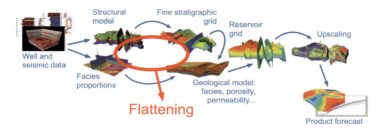

Fig. 1. Reservoir characterization workflow

An oil reservoir is a portion of the sub-surface of the geographic space which looks like a multilayered rock plate more or less bent and torn into pieces (see Fig. 2). The usual dimensions are:

- Total thickness of the reservoir: from tens to a few hundred meters;
- Total thickness of a lithostratigraphic unit: some tens of meter (typically 30 m);
- Horizontal extent: a few tens of kilometers (typically 10 x 10 km²).

After the sedimentation period, the lithostratigraphic units are broken and distorted by tectonic events. This geological process creates geological "faults" which are tearing these stratigraphic units and their upper and lower limits.

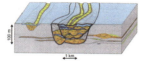 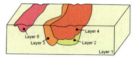

Fig. 2. A reservoir represented in the geographical space composed of 5 lithostratigraphic units. Each unit has its own geometry and deposit mode

Our work is part of the geological modeling of the reservoir. The geometric description of the reservoir in the geographic space is embedded within a 3D grid composed of different layers. Each lithostratigraphic unit is a subset of these gridded layers (which upper and respectively lower limits are called top and bottom horizons respectively). The average cell dimension is about 25 m x 25 m wide and 0,3 m thick.

Another input is coming from drilled wells data in the geographic space. Lithofacies and petrophysical properties as porosity and permeability are measured along the well trajectory. These data are assigned to every cell of the grid which intersects a well trajectory. At this step, just a very small number of cells are populated. Geostatistical simulations are used in order to populate all remaining cells (representing the major part of the grid) with respect to the geological constraints.

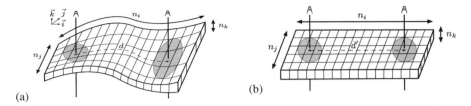

Fig. 3. Classical workflow. (a) The geographic space. (b) The simulation space

The classical geostatistical simulation workflow is illustrated in Fig. 3. In the geographic layer of Fig. 3(a), two cells are crossed by a well. These ones are the only populated cells. The petrophysical properties are computed in the regular Cartesian grid of Fig. 3(b). In this simulation space, the dimensions of the cells are the average dimensions d_i, d_j, and d_k of the geographic cells, and the number of cells remains same: $n_i \times n_j \times n_k$ of cells.

The computing of the correlation distances between wells (the influence distances of geological measures between each other) is a decisive simulation step. Classically, the wells are discretized in the geographic space. Each cell crossed by a well becomes a well cell populated with the corresponding well measure. Nevertheless, the correlation distances are computed in the simulation space. Consequently, they are not accurate. In our example, because of the fold in the geographic grid, we have $d >> d'$.

Moreover, the simulated properties are deformed when, at the end, they are transferred cell to cell from the simulation space grid into the geographic space grid. Here, the spherical properties on the simulation space grid become elliptic when mapped back into the geographic space grid.

Thus, depending on the deformation degree of the lithostratigraphic units in the geographic space, significant errors may be introduced in the geostatistical simulation. This lack of accuracy has prompted us to work on and devise a new methodology in order to increase the reliability of the parameters required by the geostatistical simulators.

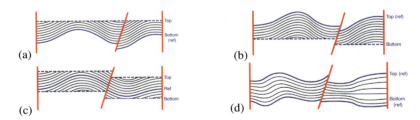

Fig. 4. Deposit modes. (a) Parallel to bottom. (b) Parallel to top. (c) Parallel to one inner limit. (d) Proportional

Our methodology is based on an "isometric" flattening of the studied lithostratigraphic units in order to have a better estimation of the correlation distances between wells. Our flattening process is optimal in the reasonable assumption of thin lithostratigraphic units. Thanks to this flattening process, the well trajectories are accurately repositioned in the simulation space. In other respects, it is commonly known that there is not a unique way to flatten a non developable layer isometrically. We propose in this paper two flattening methods driven by the deposit modes of sedimentation: the parallel deposit mode and the proportional deposit mode (see Fig. 4).

Our paper is organized as follows. Our whole flattening methodology is presented in section 2. In section 3 and 4, we respectively detail the flattening process in the case of the parallel and proportional deposit modes. Before concluding, to illustrate our flattening-based methodology, we present in section 5 some results of actual lithostratigraphic unit populating.

2 Our Flattening Methodology

Our flattening methodology improves the precision of the geostatistical methods. Indeed, the simulation of petrophysical properties is computed in a flat simulation space where the well trajectories are repositioned. Thus, accurate correlation distances between wells are easily computed.

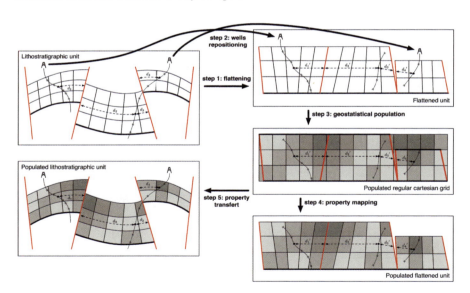

Fig. 5. The flattening methodology

Our flattening methodology is illustrated in Fig. 5. The inputs consist of one thin lithostratigraphic unit in the geographic space (represented with a grid) equipped with some well trajectories. Five steps are required to populate this unit with some geological properties:

- step 1: isometric flattening of the lithostratigraphic unit;
- step 2: repositioning of the well trajectory in the simulation space;
- step 3: constitution of a bounding regular Cartesian grid and geostatistical simulation of petrophysical properties;
- step 4: mapping of the simulated properties on the flattened unit;
- step 5: transfer of the properties in the geographic space.

The steps are detailed in the followings. In Fig. 5, we supposed that the layers are parallel to the bottom of the unit (see Fig. 4(a)). The methodology is identical in the case of a proportional deposit mode.

a) Step 1: flattening of the lithostratigraphic unit

The input lithostratigraphic unit is modeled with a 3D line support grid (see Fig. 10(a)). Such a grid is constituted with one surface (which is generally the bottom of the grid) where each vertex is attached to a coordline (illustrated with arrows). Some nodes are distributed along each coordline and every coordlines have the same number of nodes. The combination of the surface and coordlines represents an (i,j,k) cell partition of the space.

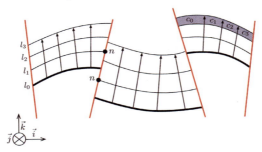

Fig. 6. Vertical cut of a grid

A line support representation of the example unit is illustrated in Fig. 6. Here, only a 2D cut is represented (all of the cells have the same j index). Along faults (represented in red), a same node can be split in several geometric vertices. For instance, the two black dots of the figure represent two vertices which are associated to a same node n. The line support structure allows ones to represent such split nodes.

In our 3D grids, each layer has the same number of cells. Some of them are not related to any geological data, for instance in an eroded layer. In order to preserve the regularity, a so-called Actnum property is associated to each grid cell. This property indicates whether a cell is active. For example, as cells c_0, c_1, c_2 and c_3 are not related to geological data, they are inactive.

The volume flattening of the unit is based on the isometric unfolding of one are two reference surfaces (one for the parallel deposit mode or two for the proportional one). First, we have to extract the reference surfaces from the grid. In our example, it is the l_0 limit of the unit. For that, we use the extraction algorithm previously introduced in [ECMOR]. This algorithm preserves the structure of the unit limits. Thus, there exists a correspondence between the grid nodes and the associated geometric vertices in the extracted surfaces.

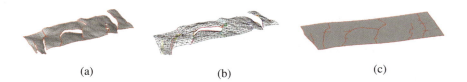

(a)   (b)   (c)

Fig. 7. Unfolding of a surface. (a) A folded surface. (b) Fault lips correspondence. (c) Corresponding unfolded surface with fault lips sticking

The unfolding is computed with an extension of the [APLAT] software. A specific code associates horizon and fault contacts separated during tectonic periods. Intuitively, the isometric unfolding works in the following way: from a 3D folded surface, we compute an unfolded plane surface in which the deformations are minimal. The criterion chosen is a minimization of the elastic deformation tensor. During the deformation, all z coordinates are computed to obtain a plane surface. Horizons and faults contacts are then put in correspondences and smoothed. The Fig. 7 illustrates the result of the isometric unfolding of a surface. We are preparing another publication dedicated to the isometric unfolding of extracted surfaces.

After the unfolding of the reference surface, the whole unit is isometrically deformed by conserving the topology of the initial geographic unit. Moreover, the isometric unfolding algorithm ensures that deformations between the geometric space and the simulation space are minimal. The deformation algorithms for both parallel and proportional deposit modes are detailed in section 3 and 4.

b) Step 2: wells trajectory repositioning

There exists a complete correspondence between the structure of the initial lithostratigraphic grid and the structure of the flattened grid. Indeed, our flattening process preserves the topology. Starting from the trajectory of the well, we can calculate the barycentric coordinates of the well measurements in each crossed cell of the original lithostratigraphic grid. Then, thanks to the correspondence, we replace these well measurements in each corresponding cell of the flattened grid and thus obtain there coordinates in the simulation space.

c) Step 3: geostatistical population of the regular Cartesian grid

Thanks to the repositioning of wells in the simulation space, the correlation distances between wells are easily computed. They allow ones to constitute the variogram models required by the geostatistical simulation. We then create a regular Cartesian grid which bound the flattened grid. The petrophysical properties are simulated in this new grid. Here, the discretizations along the i, j and k axes are regular. The number of cells along the axes is not necessary the same than in the geographic lithostratigraphic unit. The result of the geostatistical simulation is illustrated with the grey tone attributed to each cell of the regular Cartesian grid.

d) Step 4 and 5: mapping and transfer of properties

In the last two steps, we plug back the simulated petrophysical properties in the geographic space. In step 4, we map the properties from the regular Cartesian grid to the flattened grid. For each cell of the flattened grid, we compute its center and then compute in which cell of the Cartesian grid it is located. Then, the value of the regular Cartesian grid cell is copied in the flattened cell. Let us notice that when several Cartesian grid cells correspond to a unique flattened cell, an interpolation of these Cartesian cell values may be use to obtain a more accurate results.

The step 5 consists in copying the properties of the flattened grid in the lithostratigraphic unit. A simple cell by cell copy is sufficient as the flattened and geographic grids have the same topology.

3 Parallel Volume Flattening Method

In this section, we present our parallel volume flattening method (a more detailed presentation is proposed in [ECMOR]). This case corresponds to the parallel deposit mode of Fig. 4: every limits of the considered unit are parallel to one reference surface. This reference surface may be the bottom one, the top one or an inner limit of the unit.

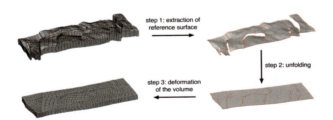

Fig. 8. The parallel volume flattening algorithm

The isometric unfolding of the reference surface plays a decisive role in the flattening process. Roughly speaking, it provides a geometric transformation which allows one to straighten every limit and then to deform the whole lithostratigraphic unit. Here is the general algorithm of the parallel volume flattening (the different steps are illustrated in Fig. 8):

- step 1: extraction and triangulation of the reference surface (see [ECMOR]);
- step 2: isometric unfolding of the reference surface (see section 2.a);
- step 3: deformation of the whole lithostratigraphic volume.

Deformation of the Lithostratigraphic Volume

We need to flatten all layers included in the stratigraphic unit. For that, layers included between bottom and top limits are straightened by considering the isometric unfolding of the reference surface.

Fig. 9. Computing of a deformation for each vertex of the reference surface. (a) A folded reference surface. (b) Corresponding unfolded surface.

A deformation is computed for each vertices of the reference surface. For each vertex, two coordinate bases are defined. The one with regard to the folded surface (represented with plain arrows in Fig. 9(a)) and the other one with regard to the unfolded surface (see the doted arrows). A deformation is the operation which transforms a base from the geographic space to the simulation space.

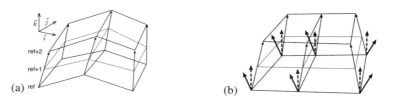

Fig. 10. Application of the deformation on each coordline. (a) A parallel unit in the geographic space. (b) Corresponding unit in the simulation space.

As shown in Fig. 10, the deformations are then applied iteratively along all of the coordlines. Our algorithm uses the correspondence between the reference surface and each coordline. It is thus necessary to ensure that every coordline corresponds strictly to one vertex of the reference surface. For this purpose, the horizons and faults contacts must correctly coincide and the removing of surface edges (for instance to simplify triangular meshes) must not remove required coordinates. The surface extraction and triangulation of surface proposed in [ECMOR] satisfy these consistency constraints.

Along faults, our method can produce discontinuities during the volume deformation. In [ECMOR], a process is provided in order to put fault lips in correspondence when the deformation is applied on the coordlines.

4 Proportional Volume Flattening Method

In this section, we introduce our proportional volume flattening method. It corresponds to the proportional deposit mode of one lithostratigraphic unit. Here, the geometry of every inner limit of the unit is given by interpolating the geometry of top and bottom limits. Thus, two reference surfaces are required: the top and bottom limits of the unit.

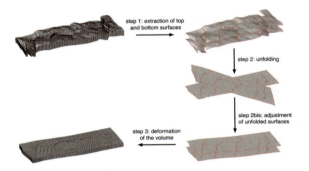

Fig. 11. The proportional volume flattening algorithm

The general algorithm of the proportional volume flattening is analogous to the parallel one. Nevertheless, an additional step is required in order to adjust the unfolded reference surfaces in space:

- step 1: extraction and triangulation of the top and bottom reference surfaces (see [ECMOR]);
- step 2: isometric unfolding of the reference surfaces (see section 2.a);
- step 2bis: adjustment of the resulting unfolded reference surfaces;
- step 3: deformation of the whole lithostratigraphic volume.

a) Adjustment of the unfolded reference surfaces

The flatten volume resulting from the proportional method is obtained by interpolating the geometry of the reference surfaces. Thus, the position of the two unfolded surfaces greatly affects the result.

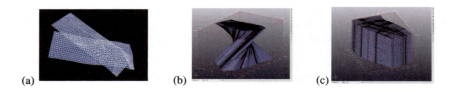

Fig. 12. Unfolded surfaces displacement. (a) Displacement of unfolded surfaces. (b) Twisted flattened volume. (c) Straight flatten volume

The [APLAT] isometric unfolding algorithm does not provide any control on the position of the output surfaces. For example, in Fig. 12(a), a displacement exists between two unfolded surfaces. The Fig. 12(b) displays a result of the proportional flattening method when the adjustment step 2bis is not processed. We observe a twist of the deposit model.

We propose an algorithm to avoid this symptomatic behaviour. For that, we compute the translation and rotation making the two unfolded surfaces superposable. Intuitively, we obtain these transformations by minimizing the distance between each pair of unfolded surfaces.

Choosing minimization support nodes

First, we choose two sets of support nodes: one in the bottom and another one in the top. These sets are such that for each node of a given set, there exists a corresponding node in the other set, a node of the opposite limit which comes from the same coordline. The choice of these support nodes is decisive for the robustness of our algorithm. A support node must satisfy the following conditions:

- condition 1: neither itself nor its corresponding node is adjacent to an inactivated cell (see section 2.a);
- condition 2: it is not located on fault.

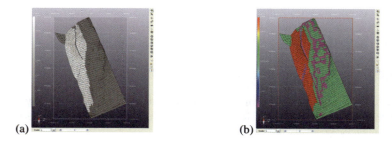

Fig. 13. Support nodes conditions. (a) A top limit. (b) The support nodes

These two conditions ensure that each support node has exactly one geometric position. Indeed, only the fault nodes and nodes adjacent to an inactivated cell may be associated to several geometric positions. In practice, it is not mandatory to take into account every node satisfying the previous conditions and sets of about ten support nodes are sufficient.

An example is given in Fig. 13. The top limit of one lithostratigraphic unit is displayed in Fig. 13(a). In this figure, the activated cells are represented in white. In Fig. 13(b) the red region represents the corresponding support nodes. The nodes of the green regions do not satisfy condition 1 while the ones of the purple regions do not satisfy condition 2.

Readjustment algorithm

We first adjust the unfolded top on the unfolded bottom (the unfolded bottom remains in position). We start by computing the two sets of corresponding support nodes. By using these nodes, we then minimize the distances between the two unfolded surfaces.

For that, we use the parallel flattening method to make sure that the angles between the bottom limit and the coordlines are well repositioned in the simulation space. By not considering these angles, the result of a proportional flattening is a vertical straight volume where the initial orientation of the grid is completely forgotten (see Fig. 12(c)).

Finally, to minimize the error, we repeat the adjustment by considering that the top remains in position (with the same sets of support nodes).

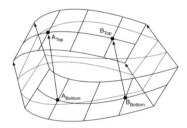

Fig. 14. A lithostratigraphic unit in the proportional deposit mode

Let us detail the steps of the algorithm by using an example:

1. Let us consider the geographic lithostratigraphic unit of Fig. 14 (only some of the coordlines are represented). In the bottom, we choose the following set of support nodes: $\{A_{Bottom}, B_{Bottom}\}$. Let $\{A_{Top}, B_{Top}\}$ be the corresponding set of support nodes in the top. Here, A_{Bottom} corresponds to A_{Top} and B_{Bottom} corresponds to B_{Top};
2. *UnfoldedBottom* and *UnfoldedTop* are respectively the unfolded bottom and top of the unit (see Fig. 15(a)). Thanks to the correspondence between grids and extracted surfaces, we get the vertices $A_{UnfoldedBottom}$ and $B_{UnfoldedBottom}$ that respectively correspond to A_{Bottom} and B_{Bottom}. We then use the parallel method in order

to reposition A_{Top} and B_{Top} in the simulation space and obtain vertices A_{Top}' and B_{Top}'. The parallel method ensure that the angles between limits and coordlines are respected;

3. $A_{UnfoldedTop}$ and $B_{UnfoldedTop}$ are the vertices which respectively correspond to A_{Top} and B_{Top} in *UnfoldedTop*. We compute a 2D minimization between the two sets $\{A_{Top}', B_{Top}'\}$ and $\{A_{UnfoldedTop}, B_{UnfoldedTop}\}$. This computation consists in minimizing a square error by using the Lagrange multiplier. We obtain a rotation angle and a translation vector;

4. We apply the inverse displacement on the whole set of *UnfoldedTop* vertices. Thus, *UnfoldedTop* is adjusted on *UnfoldedBottom* in the 2D space (see Fig. 15(b));

5. To minimize the resulting error, the steps 2 to 4 are repeated. For now, *UnfoldedTop* remains in position and *UnfoldedBottom* is adjusted;

6. The last step consists in elevating the resulting adjusted surface. We compute the average z position *ZMoyB* of all nodes of the folded bottom surface which satisfy the support nodes condition 1. Analogously, we compute *ZMoyT*. We finally assigned to each vertex of *UnfoldedBottom* and *UnfoldedTop* respectively *ZMoyB* and *ZMoyT*.

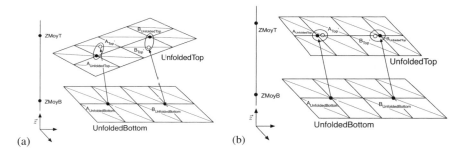

Fig. 15. Unfolded surfaces adjustment. (a) Using the parallel method to preserve angles. (b) UnfoldedTop is adjusted on UnfoldedBottom

b) Deformation of the lithostratigraphic volume

The flattening of the whole lithostratigraphic unit (step 4 of the proportional algorithm) is based on the inner limits computation by interpolating the unfolded reference surfaces.

For each couple of corresponding unfolded vertices, a vector is computed and then divided into N equal intervals (see Fig. 16). For each vector, we have $N+1$ nodes (the inner nodes are represented in white) which correspond to the limits of the flattened grid. The coordlines of the flattened grid are given by these vectors equipped with their corresponding nodes. Let us remark that as no nodes are split along the coordlines, no particular fault treatment exists.

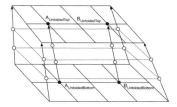

Fig. 16. A flatten unit in the proportional deposit mode

5 Results

In this section, we sum up our different results in the case of the Alwyn field. We present both different results of the flattening methods and some preliminary results of geostatistical population and mapping.

a) The Alwyn field

Fig. 17. The Alwyn field. (a) Alwyn grid. (b) Corresponding lithostratigraphy

The geographic Alwyn field (see Fig. 17(a)) is composed of 18 layers of 1500 cells. Its geological interpretation is represented in Fig. 17(b). Alwyn is composed of 3 Tarbert units, 2 Ness units and two additional units in the lower part. In the following, we focus first on N2 and T3.

b) Parallel flattening of the N2 unit

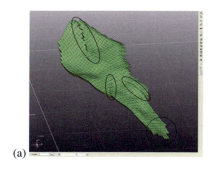 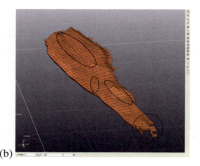

(a) (b)

Fig. 18. Parallel flattening. (a) The geographic N2 unit. (b) The flattened N2 unit

In Fig. 18(a), we have extracted the N2 unit of the Alwyn grid. We see that this unit is crossed with 3 major faults (circled with plain lines). The result of the parallel flattening process, which have been computed in a few seconds on an Intel Core 2 Quad CPU at 2.83GHz, is illustrated in Fig. 18(b). The chosen reference limit was the bottom one. We remark that the 3 fault have been correctly closed.

Two observations deserve to be pointed out. First, in the result, the top limit is not completely flat. This comes from the behavior of the parallel algorithm. Indeed, only the reference limit is explicitly unfolded. This feature is insignificant in the case of thin lithostratigraphic unit. Secondly, some differences exist between the shape of the unit in the geographic space and corresponding flattened unit (see doted circle in the bottom region). The footprint of the reference surface is often different from the footprint of other limits. Here, some cells of these limits do not have corresponding cells in the reference surface. For now, we choose to deactivate these "orphan cells".

c) Proportional flattening of the T3 unit and application to geostatistical population

The T3 unit of the Alwyn grid is illustrated in Fig. 19(a). In this figure, the wells trajectories are also represented. The result of the proportional flattening process and the repositioned wells (obtained in a few seconds with the same computer configuration) are illustrated in Fig. 19(b). Now, in the simulation space, the volume is completely flat. Indeed, both the top and bottom reference surfaces have been unfolded.

In Fig. 19(c), the geographic and simulation spaces have been superposed with an enhancement of our readjustment algorithm. This superposition provides us a visual quality control which allows one to evaluate the well trajectory repositioning. We see that the initial wells (in black) have been correctly repositioned (see white wells in the simulation space).

A Volume Flattening Methodology for Geostatistical Properties Estimation 583

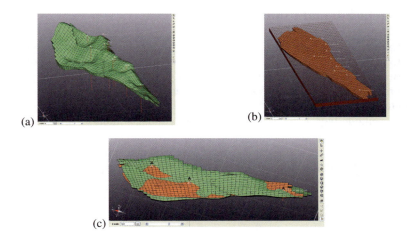

Fig. 19. Proportional flattening. (a) The T3 unit with wells. (b) The flattened T3 unit. (c) Superposition of the geographic and simulation spaces

In section 2, we have seen that geostatistical population requires a Cartesian simulation grid and corresponding wells properties logs. In Fig. 19(b), we generate a bounding Cartesian grid around the flattened lithostratigraphic unit. Thanks to it and to the repositioning of the wells trajectories, we are now able to generate the geological properties.

Fig. 20. Geostatistical population and mapping. (a) Lithofacies population. (b) Porosity population. (c) Mapping of the porosity property in the geographic space

In Fig. 20(a), we have successfully applied the geostatistical framework to populate the flattened Cartesian grid with lithofacies. In the same way, a porosity property has been associated to the Cartesian grid in Fig. 20(b). Finally, the generated properties have to be plugged back in the geographic space. In Fig. 20(c), a mapping and a copy of the porosity have been applied from the Cartesian grid to the geographic T3 unit (see section 2.d).

6 Conclusion and Future Works

In the scope of the reservoir characterization, the geostatistical simulations allow ones to populate the geographic lithostratigraphic units with some rock properties: for instance lithofacies, porosity or permeability. In this paper, we have proposed a new methodology which aims at making more accurate these geostatistical simulation methods.

Our methodology is based on an isometric volume flattening process, applied on thin lithostratigraphic units. By transporting the studied geographic lithostratigraphic unit into a flat "simulation space", we obtain a transformation which allows one to also transport the wells trajectories. In this flat space, it becomes easy to accurately compute the wells correlation distances playing a decisive role in the geostatistical simulation. Moreover, we ensure that the deformations between the geographic space and the simulation space are minimal.

In this paper, we have proposed two isometric flattening methods that depend on the deposit mode of the considered lithostratigraphic unit: the parallel one and the proportional one. Finally, we have illustrated our flattening-based methodology by populating an actual lithostratigraphic unit.

Surface extension and hole filling

The footprint of the reference limits often differs to the one of the other limits. See Fig. 21(a) where a whole unit is represented in green while the border of the reference limit is drawn in black. We can see that the reference surface contains some holes. Moreover, this particular surface is smaller than the other ones. Indeed, on the border of the unit, some cells do not have corresponding cells in the reference surface. These particular features of the reference surface can decisively weaken the precision of our methodology if some wells are supposed to cross the missing regions.

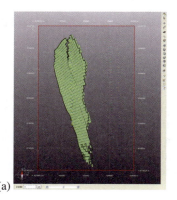
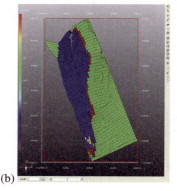

Fig. 21. Surface extension and hole filling. (a) Comparison between the shape of one reference and the whole unit. (b) Property map resulting from this comparison

We have developed a specific method to detect the limit of the extension of the volume and holes into the reference surface. The result of this method is a property map associated to the reference limit (see Fig. 21(b)). The regions to extend are represented in red, the holes are represented in yellow and the exterior inactivated cells, which have to remain inactive, are represented in green. The blue cells are correct.

For now, we simply make inactive the border regions which do not exist in the reference limit. In the future, the reference limit will have to be extended in order to cover the shape of the whole unit. By the same way, the hole cells of the reference surface are simply reactivated. This straightforward filling solution supposes that the geometry of the inactivated cells is correct. In the future, we will have to develop a true geometric hole filling.

External reference surface

In this paper, we always suppose that the reference surfaces of considered lithostratigraphic units exist among the limits of the unit. This hypothesis cannot be generalized to every unit. In Fig. 22, the bottom and top limit of the unit are represented with thick black lines while inner limits are represented with thin black lines. According to geologist knowledge, the deposit mode of such a unit is the parallel one. Nevertheless, neither the bottom limit nor the top one can be considered as a reference surface. Here, the reference limit is external: for instance the Ref z or Ref 1 one. This kind of unit cannot be handled in our current methodology and have to be prospected in the future. A solution may consist in using a geological modeler in order to provide us with the accurate reference limits.

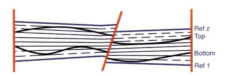

Fig. 22. External reference surfaces

References

[CG] Galera, C., Bennis, C., Moretti, I., Mallet, J.-L.: Construction of coherent 3d geological blocks. Computer and Geosciences 29, 971–984 (2003)

[JSG] Gibbs, A.: Balanced cross section from seismic sections in area of extensional tectonics. Journal of Structural Geology 5(2), 153–160 (1983)

[ECMOR] Horna, S., Bennis, C., Borouchaki, H., Delage, C., Rainaud, J.-F.: Isometric unfolding of stratigraphic grid units for accurate property populating -mathematical concepts. In: 12th European Conference on the Mathematics of Oil Recovery (2010)

[EAGE] Horna, S., Bennis, C., Crabie, T., Peltier, S., Rainaud, J.-F.: Extracting and unfolding a stratigraphic unit to update property population. In: 72nd EAGE Conference and Exhibition (2010)

[APLAT] APLAT3D Surface Flattening Toolkit, University of Technology of Troyes (2008)

Zipper Layer Method

Ning Qin[1], Yibin Wang[2], Greg Carnie[3], and Shahrokh Shahpar[4]

[1] Department of Mechanical Engineering, University of Sheffield
n.qin@shef.ac.uk
[2] Department of Mechanical Engineering, University of Sheffield
mep07yw@shef.ac.uk
[3] Department of Mechanical Engineering, University of Sheffield
g.carnie@shef.ac.uk
[4] Aerothermal Design System, Rolls Royce plc.,
Shahrokh.Shahpar@rolls-royce.com

Summary. Based on the buffer layer method, a new method called zipper layer method was developed. This method maintains the capability of the buffer layer which can link two topologically different multi-block structured meshes together, while significantly improving the robustness. This method can both locally and globally connect two dissimilar structured meshes with tetrahedrons and pyramids to form a conformal mesh. The NASA Rotor 37 and an open rotor are used as test cases here to generate the zipper layer meshes. The numerical results on these zipper layer meshes compared well with those on the multi-block structured meshes.

Keywords: mesh generation, hybrid mesh, zipper layer mesh.

1 Introduction

It is well known that the hybrid mesh is a promising and useful alternative to structured and unstructured meshes. The original hybrid mesh uses prisms near the solid wall to maintain the orthogonality while using tetrahedrons to fill the rest of the volume. This method acquires some advantages of structured and unstructured meshes, however, hexahedrons are more preferable for most of the CFD applications. Therefore, it is natural to generate the hexahedrons near the solid wall while using unstructured cells such as tetrahedrons and/or prisms to fill the gap. The dragon mesh1 is one of the generic hybrid mesh which generates structured meshes for different components and then assemblies them together by using tetrahedrons. This method guarantees the high quality meshes in the structured mesh region and simplifies the mesh generation process. However, it causes hanging nodes in the meshes which is not suitable for all the flow solvers. In order to overcome this problem, Qin *et al*.1 use the combination of unstructured mesh layers to link the topologically different structured mesh together and named these layers as buffer layer. This buffer layer method successfully applied on the numerical simulation of turbomachinery flow field. It avoids the hanging nodes in the mesh by using pyramids layers, prismatic layers and tetrahedral layers to bridge the two structured meshes. This method further enlarges the scope of application of the generic hybrid mesh. Though the buffer layer method can link the structured meshes, it is still not robust enough when the linkage is in a small gap. Hence a new method called zipper layer method 3 is proposed in this paper to overcome the issues. Based on the general idea of the buffer layer method, the zipper layer method introduces an interface and then meshes the interface with unstructured cells to link two structured meshes.

2 Zipper Layer Mesh Generation Method

The generation process of the zipper layer method is totally dissimilar from that of the buffer layer method, though it still uses some of the ideas generated in the buffer layer method. The whole process can be divided into the following steps:

1. Identify the interface of the two multi-block structured meshes, find intersecting points of the two surface mesh edges, and merge the nodes or project the nodes to the edge when it is necessary;
2. Generate an unstructured surface mesh (triangles and quadrilaterals) including all the mesh points from both sides and the intersection points;
3. Insert nodes at the geometrical centres of the hexahedrons which need to be split on both sides of the interface, then generate unstructured volume cells on both sides of the interface, including tetrahedrons, pyramids and hexahedrons;
4. Check the cell quality, fix the negative volume cells.

3 Dual Fast March Method and Node Movement

In order to indentify the node relation between the two meshes in Step 1 quickly, a method called dual fast march method is developed to locate the position in Mesh B for the nodes in another Mesh A. Based on the essence of the fast march method 4 applied on the single mesh, the dual fast march method is applied on the two different meshes at one time. Let Cell B contain Node A. Since the neighbouring nodes of Node A must be in the cells which are near the cell containing Node A, when locating these neighbouring nodes of Node A, the search region can be narrowed down to those cells near Cell B. As this method only searches the nearby cells, it can significantly decrease the search time than the brute force search.

In order to eliminate the sliver cells, node movement is adopted before the triangles on the interface are generated. The node movement is imbedded into the dual fast march method, so when locating the node position, the nodes are moved or projected to the edge. As shown in Fig.1, Node B is near Edge CD, so Edge CD becomes CBD to eliminate the small triangle which may be generated in the next procedure.

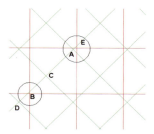

Fig. 1. Node movement

4 Generation of Interface Mesh and Volume Mesh

Before generating the unstructured meshes on the interface to form the interface mesh, some of the edges on both sides of the meshes are split by the intersection nodes. As shown in Fig.2, the cell's four edges (the red edges in Fig.2) are split by the intersection with another mesh (in green in Fig.2).

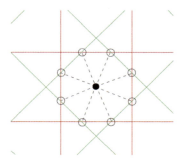

Fig. 2. Triangulation of the polygon

The interface mesh is formed by the newly generated edges and the original edges which have not been spilt. In Fig.2, the triangulation method is as follows:

1. Find the polygon which has more than four sides;
2. Insert a point into the geometry centre of the polygon;
3. Link the point with the two end nodes of each edge of the polygon to form triangles.

Fig.3 illustrates how a 2D interface mesh is generated. Fig.3 (a) and (b) are two topologically different structured meshes, whereas Fig.3 (c) is the interface mesh, and Fig.3 (d) is the magnificent view of Fig.3 (c).

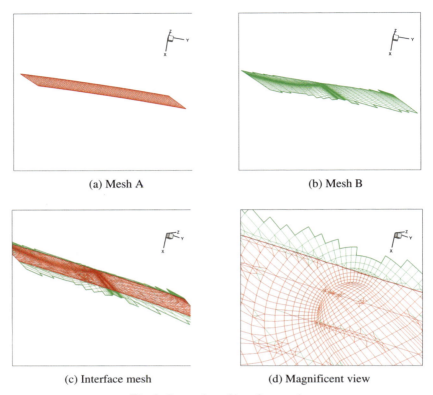

Fig. 3. Generation of interface mesh

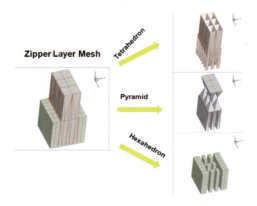

Fig. 4. Cell splitting required for zipper layer method to connect two non-matching structured meshes

The zipper layer approach adopts a similar cell splitting method as used in the buffer layer method, which is the major common ground between the two methods. However, in the buffer layer method, points are only introduced into each zone, while in zipper layer method, each split cell is treated as a single zone, and points are introduced into each cell. The general method is shown in Fig.4. First insert a point into the geometry centre of the cell, and then link the points of the triangles or quads to form tetrahedrons or pyramids.

5 Simple Test Case

This is a simple test case, as shown in Fig.5 to demonstrate the basic connection and capability of the method. On the left there is a uniformly distributed structured mesh; on the right there is a non-uniformly distributed mesh. The red surface highlights the interface surface where an interface mesh will be generated to link the top and bottom meshes, as shown in Fig.5. With the nodes on the interface surface and the intersection nodes, an interface mesh is then generated, as shown in Fig.6. Since the bottom mesh is smaller than the top one, all the top cells are split, while some of the cells at the bottom remain unchanged because they are 'contained' in the cells on the top. In order to form the volume mesh, the hexahedral cells are split based on the interface mesh in Fig.7. As can be seen from the figure, all the triangles on the interface mesh form the tetrahedrons, while the quadrilaterals formed by intersection nodes become pyramids.

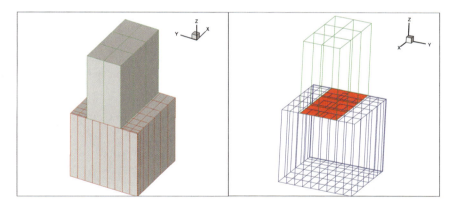

Fig. 5. Structured mesh(left) and interface (right, red plane highligts the interface of two different structured meshes)

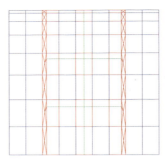

Fig. 6. Interface mesh (blue quads indicate the hexhedrons remain unchanged in the third step; while the rest of the quads form the pyramids)

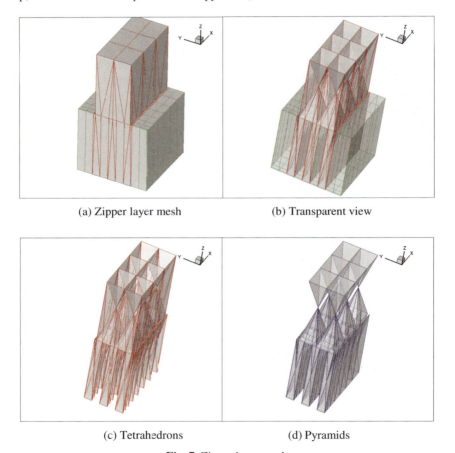

(a) Zipper layer mesh (b) Transparent view

(c) Tetrahedrons (d) Pyramids

Fig. 7. Zipper layer mesh

6 NASA Rotor 37 Case

The NASA Rotor 37 case is used here as a test case to verify the method. Several grooves need to be added to the casing wall of the NASA Rotor 37 case for the casing design. However, it is difficult to achieve this for the structured mesh, since the tip gap is 0.356mm. Hence, the most efficient and simple way is to locally graft the groove mesh and the casing mesh for the NASA Rotor 37 case, which is called local zipper layer mesh here.

Fig.8 shows the local zipper layer mesh. As can be seen from the picture, the zipper layer mesh is introduced directly on the casing near the middle chord, and then the groove part is removed for the numerical comparison with the structured mesh. Both meshes use the same flow solver with 4 levels of multi-grid and CFL=2.

In Fig.9, the pressure ratios calculated on different meshes are compared. In order to verify that the local zipper layer mesh does not introduce too many numerical errors to the flow field, several zipper layer meshes are generated for comparison. The results on the local zipper layer meshes show the consistency to those on the multi-block structured mesh and the experimental data, while the result on the buffer layer mesh slightly under-estimates the pressure ratio. It indicates that the local zipper layer mesh introduces fewer numerical errors than the buffer layer mesh for the NASA Rotor 37 case. Fig.10 shows the comparison of the total pressure along the span. Both results on the zipper layer mesh and the multi-block structured mesh are comparable to the experimental data, while the result on the buffer layer mesh shows more discrepancy to the above two results and the experimental data.

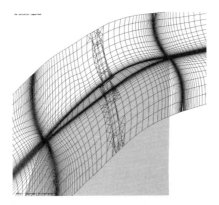

Fig. 8. Local zipper layer mesh for NASA rotor 37 case without groove

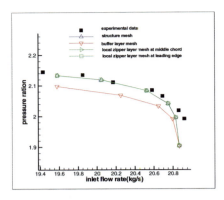 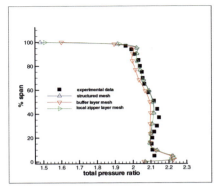

Fig. 9. Comparison of the total pressure ratios

Fig. 10. Total pressure ratios along the span

Different from the previous local zipper layer mesh which locally grafts the groove mesh to the casing, a more general way is to link the whole new casing mesh to the mesh from the blade side which is called the global zipper layer mesh here. As the global zipper layer creates more unstructured interface cells, it is therefore less efficient than the local zipper layer method. However, it makes the treatment of multiple components such as multiple grooves much more straightforward in programming.

To verify the global zipper layer mesh, the results on the multi-block structured mesh and the global zipper layer mesh were compared. Fig.11 shows the multi-block structured mesh and the global zipper layer mesh. Being different from the buffer layer method, the zipper layer method only affects the two layer meshes which can be seen in Fig.11. The original multi-block structured mesh consists of 5 H-type mesh blocks and 1 O-type mesh block for the blade. The upstream H-block is 40x59x84, the downstream H-block is 50x59x84, both the passage H-blocks are 74x16x84, the H-block mesh above the blade tip is 86x17x16 and finally the O-mesh block representing the blade is 203x12x84. The resulting mesh when all blocks are merged gives a mesh consisting of 810,228 hexahedral elements, which are then linked to 200x60x6 casing block. The zipper layer method effectively splits two hexahedral layers in the tip gap with unstructured cells, thus the total number of elements for this mesh increases to 1,070,609. Each mesh shown in Fig.11 was run from the choke to numerical stall condition and compared to the experimental data. Flow solver was run as a steady state calculation with a 4 level multi-grid approach and a CFL=2.0. The wall function and the SA turbulence model were used. All the solid walls were treated as non-slip adiabatic boundary. A subsonic inflow condition was specified at the inlet where the total pressure and temperature profile were set according to the data from the AGARD report [5]. A radial equilibrium subsonic boundary was used at the exit, which allows for a single pressure to be specified at a given radial point from which the exit pressure is calculated.

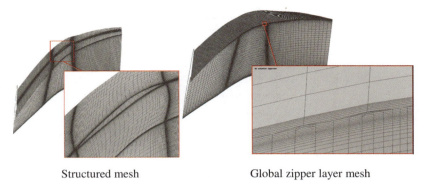

Structured mesh   Global zipper layer mesh

Fig. 11. Original Multi-block structured mesh and global zipper layer mesh

Fig.12 shows the comparison of the total pressure ratios calculated on the zipper layer meshes, the multi-block structured mesh and the experimental data. The result on the zipper layer mesh matches well with those on the multi-block structured mesh and the experimental data. As compared in this figure, the local zipper layer mesh and the global zipper layer mesh show slight differences, which is probably due to the distinction of the casing block. In Fig.13 the total pressure ratios along the span at peak efficiency condition were compared. Though the result on the global zipper layer mesh shows a slight deviation from the structured mesh at 90%-95% span, two results are identical at the rest of the region. This indicates that the global zipper layer mesh only slightly changes the flow field locally. The uniformly distributed structured mesh on the casing side may account for the mismatch, which improves the flow resolution at that region and makes the result on the zipper layer mesh closer to the experimental result.

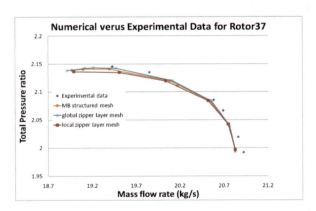

Fig. 12. Comparison of total pressure ratios

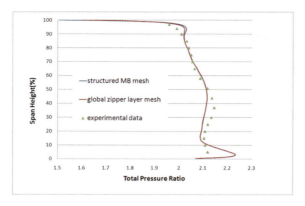

Fig. 13. Total pressure ratios along the span

In Fig.14, the stream line shows the path of the tip leakage vortex which is critical to the stall. The results given by different meshes are similar to each other. The static pressure at 98% span is also compared in Fig.15. The small mismatch is only observed at the leading edge, while most of the lines are identical.

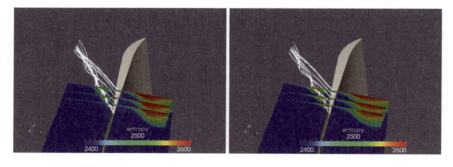

(a) Multi-block structured mesh   (b) Zipper layer mesh

Fig. 14. Entropy and tip leakage vortex

NASA Rotor 37 case with five grooves is generated by using the global zipper layer method. The five grooves have the same height, and the first groove starts form the place above leading edge. Fig.16 shows the zipper layer mesh of this configuration. As shown in Fig.16, the zipper layer mesh is in the tip gap which links two different multi-block structured meshes together. So far these types of configuration are deemed to be impossible to generate a merely multi-block structured mesh due to its complexity of geometry. The most challenging part is that the tip gap is very small, and the two meshes change dramatically from one topology to the other. However, by using the zipper layer mesh method, two meshes are smoothly linked together.

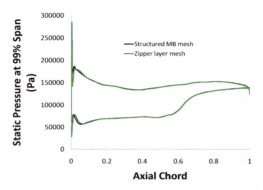

Fig. 15. Comparison of static pressure at 98% span

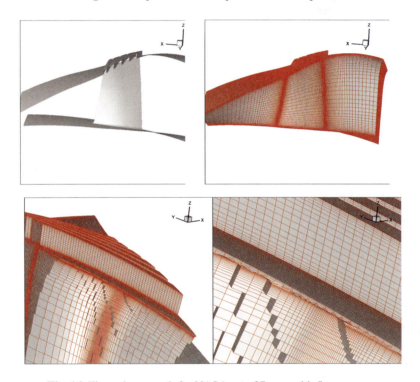

Fig. 16. Zipper layer mesh for NASA rotor37 case with 5 grooves

7 Open Rotor Case

The global zipper layer mesh method was also successfully applied on the open rotor case. As can be seen from Fig.17, the uniformly distributed mesh is smoothly linked to the clustered mesh. In Fig.18 the convergence histories are compared.

The zipper layer mesh shows a superior convergence than the multi-block structured mesh, as the uniformly distributed mesh may help the solver converge more quickly.

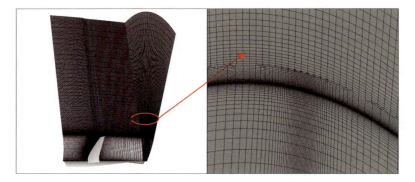

Fig. 17. Global zipper layer mesh for open rotor case

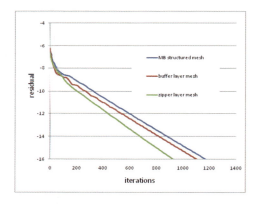

Fig. 18. Convergence history

8 Conclusion

A novel method for linking two dissimilar structured meshes is developed. This method maintains the structured mesh near the solid wall and links the different parts together. According to the numerical test, the addition of the zipper layer to a structured mesh has not degraded the quality of the flow solution, but rather gives as good as convergence and accuracy as the structured mesh.

References

1. Kao, K.H., Liou, M.S.: Advance in overset grid scheme: from Chimera to Dragon grids. AIAA Journal 33, 1809–1815 (1995)
2. Qin, N., Carnie, G., Moigne, A.L., Liu, X., Shahpar, S.: Buffer layer method for linking two non-matching multi-block structured grids. In: 47th AIAA Aerospace Sciences Meeting including The New Horizons Forum and Aerospace Exposition, Orlando, Florida, January 5-8, AIAA 2009-1361 (2009)
3. Shahpar, S., Qin, N., Wang, Y., Carnie, G.: A Method of Connecting Meshes. GB Patent number, 1101810.8
4. Sethian, J.A.: Fast marching method. SIAM Review 41(2), 199–235 (1999)
5. Dunham, J.: CFD Validation for Propulsion System Component. AGARD Report, AGARD-AR 355 (1998)

Fitting Polynomial Surfaces to Triangular Meshes with Voronoi Squared Distance Minimization

Vincent Nivoliers[1,2], Dong-Ming Yan[1,3], and Bruno Lévy[1]

[1] Project ALICE / Institut National de Recherche en Informatique et en Automatique (INRIA) Nancy Grand Est, LORIA
[2] Institut National Polytechnique de Lorraine (INPL)
[3] Geometric Modeling and Scientific Visualization Center,
King Abdullah University of Science and Technology (KAUST)
Vincent.Nivoliers@loria.fr, Dongming.Yan@inria.fr,
Bruno.Levy@inria.fr

This paper introduces Voronoi Squared Distance Minimization (VSDM), an algorithm that fits a surface to an input mesh. VSDM minimizes an objective function that corresponds to a Voronoi-based approximation of the overall squared distance function between the surface and the input mesh (SDM). This objective function is a generalization of Centroidal Voronoi Tesselation (CVT), and can be minimized by a quasi-Newton solver. VSDM naturally adapts the orientation of the mesh to best approximate the input, without estimating any differential quantities. Therefore it can be applied to triangle soups or surfaces with degenerate triangles, topological noise and sharp features. Applications of fitting quad meshes and polynomial surfaces to input triangular meshes are demonstrated.

1 Introduction

We focus on the problem of fitting a surface $\mathcal{S}$ to an input mesh $\mathcal{T}$, under the following assumptions:

- An initial template $\mathcal{S}^{(0)}$ is available. For instance, if $\mathcal{T}$ is a topological sphere, $\mathcal{S}^{(0)}$ can be initialized as the bounding box of $\mathcal{T}$ (see Figure 1-center). For higher genus, some existing automatic or interactive methods may be used to construct the template $\mathcal{S}^{(0)}$ (see Section 4);
- the reconstructed surface $\mathcal{S}$ can be a polygon mesh or a polynomial surface;
- the input mesh $\mathcal{T}$ may have degenerate triangles and/or topological degeneracies such as T-junctions, holes or topological noise.

We introduce VSDM (Voronoi Squared Distance Minimization), an algorithm that fits the template $\mathcal{S}$ to the input mesh $\mathcal{T}$ by minimizing an objective function $\tilde{F}$ of the set of coordinates that determines $\mathcal{S}$, i.e. the vertices of a polygon mesh or

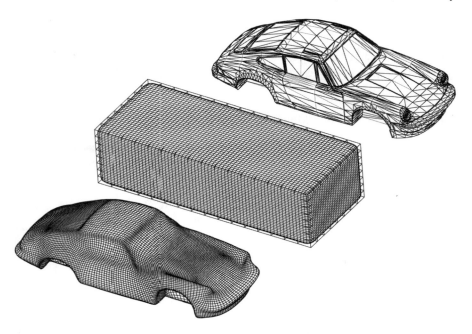

Fig. 1. Given an input mesh $\mathcal{T}$ (top, 2065 vertices and 4114 facets) and a control mesh (898 vertices and 896 quads) in an initial position (center), VSDM minimizes the squared distance between $\mathcal{T}$ and the polynomial surface $\mathcal{S}$ defined by the control mesh. The Hausdorff distance between the result (bottom) and the input mesh (top) is 0.554% of the bounding box diagonal. Other views of the same data are shown further in the paper.

the control points of a polynomial surface. Figure 1 shows an example of fitting a polynomial surface to an input triangulated mesh.

This paper makes the following contributions:

- definition of $\tilde{F}$ (Section 3.1), and proof that it converges to the integrated squared distance (Appendix A);
- solution mechanism to minimize $\tilde{F}$ (Sections 3.3, 3.4);
- some applications to quad mesh fitting and polynomial surface fitting (Section 3.5).

Advantages:

1. VSDM can fit a surface $\mathcal{S}$ to an input mesh $\mathcal{T}$ even if the initialization $\mathcal{S}^{(0)}$ is far away from $\mathcal{T}$ (typically a bounding box);
2. unlike methods based on parameterization, VSDM can process meshes with sharp angles and skinny triangles;
3. VSDM adapts the orientation of the control mesh in a way that best approximates the input surface, without requiring computation of its curvature tensor.

Limitations/uncovered aspects:

1. VSDM may generate pinchouts or overlaps, for instance when trying to fit a simple template $\mathcal{S}^{(0)}$ to a surface that has long protrusions / high Gaussian curvature. This can be fixed in most cases by designing a better template $\mathcal{S}^{(0)}$;
2. we do not prove the C^2 continuity of the objective function. However, in our empirical studies, the numerical optimization behaves well (see discussion in Section 3.4);
3. Dynamically modifying the topology of $\mathcal{S}$ is not adressed here. These topics will be studied in future works.

2 Background and Previous Work

Methods based on parameterization

Fitting splines was the motivation of early works in mesh parameterization for objects homeomorphic to a disc [8]. For fitting splines to objects of arbitrary genus, it is possible to use a parameterization defined over a base complex [7, 20], polycube maps [21, 22] or global parameterization methods [19, 12]. The relations between the curvature of the surface and the metric defined by the parameterization is studied in [10] and used to compute an anisotropic mesh that minimizes the approximation error. Since they require the estimates of differential quantities (gradients, curvature, shape operator ...), the methods above cannot be applied to meshes with degeneracies (skinny triangles, multiple components, holes, sharp creases). Our methods that directly minimizes the squared distance does not suffer from this limitation.

Methods based on point-to-point distances

To remesh surfaces, "shrink-wrap" methods [5, 9] iteratively project the template onto the input mesh while minimizing a regularization criterion. A similar idea can be applied to subdivision surfaces [15, 16], using an exact algorithm to find closest points on the subdivision surface and the exact evaluation of the subdivision surface. The "dual domain relaxation" method [25] uses some variants of Laplace surface editing to fit a template to the input mesh. Since they are based on point-to-point distances, the methods above can mostly do small corrections on the geometry, and have difficulties converging when the initialization is far away from the target surface. In contrast, VSDM can successfully fit a control mesh to a surface.

Squared distance minimization (SDM)

SDM was proposed by Pottmann et al. [18] for curve and surface fitting. The SDM framework fits a surface $\mathcal{S}$ to an input mesh $\mathcal{T}$ by minimizing an approximation of the objective function $E(\mathbf{X})$:

$$E(\mathbf{X}) = F_{\mathcal{S} \rightarrow \mathcal{T}}(\mathbf{X}) + \lambda R(\mathbf{X})$$
where:
$$F_{\mathcal{S} \rightarrow \mathcal{T}}(\mathbf{X}) = \int_{\mathcal{S}(\mathbf{X})} \| \mathbf{x} - \Pi_{\mathcal{T}}(\mathbf{x}) \|^2 \, d\mathbf{x} \qquad (1)$$
$$R(\mathbf{X}) = \|\mathbf{L}\mathbf{X}\|^2$$

In this equation, $\mathbf{X} = [\mathbf{x}_i]_{i=0}^n$ denotes the coordinates that determine $\mathcal{S}$ and $\Pi_{\mathcal{T}}(\mathbf{x})$ denotes the projection of $\mathbf{x}$ onto $\mathcal{T}$, i.e. the point of $\mathcal{T}$ nearest to $\mathbf{x}$. The term $R(\mathbf{X})$ is a quadratic regularization energy and $\mathbf{L}$ a discretization of the Laplacian. The regularization factor λ lets the user choose a tradeoff between the smoothness of $\mathcal{S}$ and the fitting criterion.

Wang et al. [23] showed that SDM can be characterized as a quasi-Newton method and they applied it to B-spline curve fitting. Cheng et al. [2, 3] proposed a subdivision surface fitting algorithm based on SDM. In the methods above, the approximation of the integral is based on both a point-sampling $\mathbf{X} = [\mathbf{x}_i]_{i=1}^n$ of $\mathcal{S}$ and a point-sampling $\mathbf{Y} = [\mathbf{y}_j]_{j=0}^m$ of $\mathcal{T}$. The approximation has several variants that correspond to Taylor expansions of different orders (see Figure 2).

The squared distance between $\mathbf{x}_i$ and $\mathcal{T}$ can be replaced by:

- Point Distance (PD): the squared distance to the nearest sample $\mathbf{y}_j$ (order 0 approximation);
- Tangent Distance (TD): the squared distance to the nearest point on the tangent plane of the nearest sample (order 1 approximation);
- Squared Distance (SD): the order 2 approximation of the squared distance around $\mathbf{y}_j$.

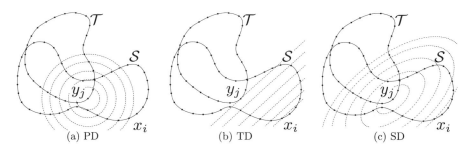

(a) PD (b) TD (c) SD

Fig. 2. Illustration of different approximations of SDM. The dashed lines represent iso lines of the distance function to the sample

3 Voronoi Squared Distance Minimization

SDM requires an accurate estimation of the curvature tensor on $\mathcal{T}$, which may be not available if $\mathcal{T}$ is a triangle soup or a CAD mesh with many skinny triangles. Therefore, to allow processing such degenerate input meshes, VSDM directly uses the local geometry of $\mathcal{T}$ around $\mathbf{y}_j$ by integrating the squared distance function over a small patch (a restricted Voronoi cell), as explained in the next subsection.

In addition, we note that $F_{\mathcal{S} \to \mathcal{T}}$ vanishes whenever $\mathcal{S}$ matches a subset of $\mathcal{T}$ (instead of the totality of $\mathcal{T}$). Therefore, to avoid degenerate minimizers that partially match $\mathcal{T}$, we propose to minimize a symmetrized version of SDM given by $F_{\mathcal{S} \to \mathcal{T}} + F_{\mathcal{T} \to \mathcal{S}}$. The benefit of the symmetrized formulation is demonstrated later (subsection 3.3).

3.1 Definition

VSDM minimizes an approximation of the following objective function:

$$F(\mathbf{X}) = F_{\mathcal{T} \to \mathcal{S}}(\mathbf{X}) + F_{\mathcal{S} \to \mathcal{T}}(\mathbf{X}) + \lambda R(\mathbf{X}) \quad (2)$$

Let us consider the first term $F_{\mathcal{T} \to \mathcal{S}}$. Using a sampling $\mathbf{X}$ of $\mathcal{S}$, we make the following approximation $\|\mathbf{y} - \Pi_{\mathcal{S}}(\mathbf{y})\| \simeq \min_i \|\mathbf{y} - \mathbf{x}_i\|$. Replacing the integrand of $F_{\mathcal{T} \to \mathcal{S}}$ gives:

$$\begin{aligned} F_{\mathcal{T} \to \mathcal{S}} &= \int_{\mathcal{T}} \|\mathbf{y} - \Pi_{\mathcal{S}}(\mathbf{y})\|^2 d\mathbf{y} \\ &\simeq \int_{\mathcal{T}} \min_i \|\mathbf{y} - \mathbf{x}_i\|^2 d\mathbf{y} \\ &= \sum_i \int_{\Omega_i \cap \mathcal{T}} \|\mathbf{y} - \mathbf{x}_i\|^2 d\mathbf{y} \end{aligned}$$

where Ω_i denotes the 3D Voronoi cell of $\mathbf{x}_i$. We shall now give the definition of the approximation $\tilde{F}$ of F minimized by VSDM:

$$\tilde{F}(\mathbf{X}) = \tilde{F}_{\mathcal{T} \to \mathcal{S}}(\mathbf{X}) + \tilde{F}_{\mathcal{S} \to \mathcal{T}}(\mathbf{X}) + \lambda \underbrace{\mathbf{X}^t \mathbf{L}^2 \mathbf{X}}_{R(\mathbf{X})}$$

where:

$$\begin{aligned} \tilde{F}_{\mathcal{T} \to \mathcal{S}} &= \sum_{\mathbf{x}_i \in \mathbf{X}} \int_{\mathcal{T} \cap \Omega_i} \|\mathbf{y} - \mathbf{x}_i\|^2 \, d\mathbf{y} \\ \tilde{F}_{\mathcal{S} \to \mathcal{T}} &= \sum_{\mathbf{y}_j \in \mathbf{Y}} \int_{\mathcal{S} \cap \Omega_j} \|\mathbf{x} - \mathbf{y}_j\|^2 \, d\mathbf{x} \end{aligned} \quad (3)$$

The matrix $\mathbf{L}$ is the uniform graph Laplacian of $\mathcal{S}$. The influence of the regularization factor λ is illustrated in Figure 3. Ω_i denotes the Voronoi cell of $\mathbf{x}_i$ in the Voronoi diagram of $\mathbf{X}$, and Ω_j the Voronoi cell of $\mathbf{y}_j$ in the Voronoi diagram of $\mathbf{Y}$ (see Figure 4).

3.2 Convergence to the Continuous Objective Function

The VSDM approximation replaces the nearest point on $\mathcal{S}$ with the nearest sample of $\mathbf{X}$ (in the term $F_{\mathcal{T} \to \mathcal{S}}$) and the nearest point on $\mathcal{T}$ with the nearest sample of $\mathbf{Y}$ (in the term $F_{\mathcal{S} \to \mathcal{T}}$). The accuracy of the approximation depends on the *density* of the point sets $\mathbf{X}$ and $\mathbf{Y}$ used to sample $\mathcal{S}$ and $\mathcal{T}$ respectively. The density of a sampling is formalized by the notion of ε-sampling [1]. A point set $\mathbf{X}$ is an ε-sampling of a surface $\mathcal{S}$ if for any point $\mathbf{x}$ of $\mathcal{S}$ there is a point $\mathbf{x}_i$ in $\mathbf{X}$ such that $\|\mathbf{x}_i - \mathbf{x}\| < \varepsilon \, \text{lfs}(\mathbf{x})$ where $\text{lfs}(\mathbf{x})$ denotes local feature size (distance to medial axis of $\mathcal{S}$). $\tilde{F}_{\mathcal{T} \to \mathcal{S}}$ satisfies the following property (proved in Appendix A).

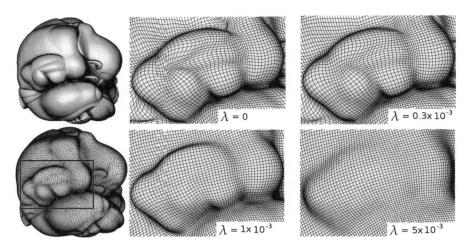

Fig. 3. Influence of the regularization factor λ on subdivision surface fitting.

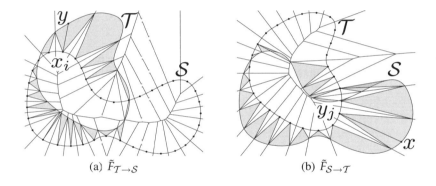

Fig. 4. Illustration of the terms of the VSDM objective function. The shaded regions represent for each sample $\mathbf{x}_i$ (resp. $\mathbf{y}_j$) the portion $\mathcal{T} \cap \Omega_i$ (resp. $\mathcal{S} \cap \Omega_j$) of the other surface whose squared distance with respect to the sample is integrated.

Property 1. Given $\mathbf{X}$, an ε-sampling of $\mathcal{S}$, we have:

$$\lim_{\varepsilon \to 0} \tilde{F}_{\mathcal{T} \to \mathcal{S}}(\mathbf{X}) = F_{\mathcal{T} \to \mathcal{S}}(\mathbf{X})$$

The same property is satisfied by the symmetric term $\tilde{F}_{\mathcal{S} \to \mathcal{T}}$ if $\mathbf{Y}$ is an ε-sampling of $\mathcal{T}$. Therefore, if $\mathbf{X}$ and $\mathbf{Y}$ are dense enough, $\tilde{F}$ is a good approximation of F. Note that ε-sampling is not defined for non-smooth surfaces. In all our experiments, we took $\mathbf{X}$ as the vertices of $\mathcal{S}$ and $\mathbf{Y}$ as a sampling of $\mathcal{T}$ with the same number of vertices as $\mathbf{X}$, optimized by CVT [24].

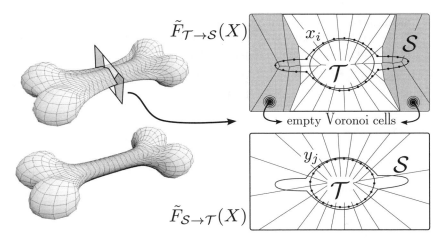

Fig. 5. Top: the minimizer of $\tilde{F}_{\mathcal{T}\to\mathcal{S}}$ has spurious parts that cannot be eliminated since their Voronoi cells do not intersect the input mesh $\mathcal{T}$. Bottom: the symmetrized $\tilde{F} = \tilde{F}_{\mathcal{T}\to\mathcal{S}} + \tilde{F}_{\mathcal{S}\to\mathcal{T}}$ detects and eliminates them.

3.3 Need for the Symmetrized Objective Function

The term $\tilde{F}_{\mathcal{T}\to\mathcal{S}}$ of $\tilde{F}$ corresponds to the objective function minimized by Restricted CVT. Therefore, omitting the term $\tilde{F}_{\mathcal{S}\to\mathcal{T}}$ results in the objective function $\tilde{F}_{\mathcal{T}\to\mathcal{S}} + \lambda R(\mathbf{X})$, that can be minimized by a straightforward modification of the CVT quasi-Newton algorithm used in [14, 24], i.e. by adding the term $\lambda \mathbf{X}^t \mathbf{L}^2 \mathbf{X}$ to the objective function and $2\lambda \mathbf{L}^2 \mathbf{X}$ to the gradient. However, as noted before, the function $F_{\mathcal{T}\to\mathcal{S}}$ reaches a minimum whenever $\mathcal{S}$ is a superset of $\mathcal{T}$. Therefore, a minimizer of $F_{\mathcal{T}\to\mathcal{S}}$ may have spurious parts, as shown in Figure 5. These spurious parts correspond to the set $\mathcal{S} - \Pi_{\mathcal{S}}(\mathcal{T})$, that does not yield any term in $F_{\mathcal{T}\to\mathcal{S}}$. In terms of the discretization $\tilde{F}_{\mathcal{S}\to\mathcal{T}}$, they correspond to Voronoi cells that have an empty intersection with $\mathcal{T}$.

3.4 Solution Mechanism

To minimize the function $\tilde{F} = \tilde{F}_{\mathcal{T}\to\mathcal{S}} + \tilde{F}_{\mathcal{S}\to\mathcal{T}} + \lambda R(\mathbf{X})$ in Equation 3, VSDM uses the L-BFGS algorithm [17, 13]. L-BFGS is a Newton-type algorithm, that uses successive evaluation of the function and its gradient to compute an approximation of the inverse of the Hessian. Although only the gradient is required in the computation, the objective function needs to be C^2 to ensure the proper convergence of the L-BFGS algorithms. We discuss here about the continuity of the three terms of $\tilde{F}$:

- The term $R(\mathbf{X})$ is a quadratic form (C^∞);
- the term $\tilde{F}_{\mathcal{T}\to\mathcal{S}}$ corresponds to the *quantization noise power*, which is the objective function minimized by a centroidal Voronoi tesselation. It is of class C^2, except in some rarely encountered degenerate configurations (see [14] for a proof);

- the term $\tilde{F}_{\mathcal{S}\to\mathcal{T}}$ is obtained by permuting the roles of the constant and variables in $\tilde{F}_{\mathcal{T}\to\mathcal{S}}$. We will study its continuity in future work. Experimentally, it is regular enough for obtaining a stable behavior of L-BFGS.

In practice, implementing L-BFGS requires to evaluate $\tilde{F}(\mathbf{X}^{(k)})$ and $\nabla\tilde{F}(\mathbf{X}^{(k)})$ for a series of iterates $\mathbf{X}^{(k)}$ (see Algorithm 1):

(1) $\mathbf{X}^{(0)} \leftarrow$ vertices of $\mathcal{S}^{(0)}$
(2) $\mathbf{Y} \leftarrow \varepsilon$-sampling of $\mathcal{T}$; Compute $Vor(\mathbf{Y})$
while *minimum not reached* **do**
 (3) Compute $Vor(\mathbf{X}^{(k)})$, $Vor(\mathbf{X}^{(k)})|_{\mathcal{T}}$ and $Vor(\mathbf{Y})|_{\mathcal{S}}$
 (4) Compute $\tilde{F}_{\mathcal{T}\to\mathcal{S}}(\mathbf{X}^{(k)})$ and $\nabla\tilde{F}_{\mathcal{T}\to\mathcal{S}}(\mathbf{X}^{(k)})$
 (5) Compute $R(\mathbf{X}^{(k)})$ and $\nabla R(\mathbf{X}^{(k)})$
 (6) Compute $\tilde{F}_{\mathcal{S}\to\mathcal{T}}(\mathbf{X}^{(k)})$ and $\nabla\tilde{F}_{\mathcal{S}\to\mathcal{T}}(\mathbf{X}^{(k)})$
 (7) Compute $\tilde{F}(\mathbf{X})$ and $\nabla\tilde{F}(\mathbf{X})$
 (8) $\mathbf{X}^{(k+1)} \leftarrow \mathbf{X}^{(k)} + \mathbf{p}^{(k)}$; Update $\mathcal{S}$ from $\mathbf{X}^{(k+1)}$
end

Algorithm 1. Fitting a polygon mesh using VSDM.

In order to make our work reproducible, we further detail each step:

(2): the sampling $\mathbf{Y}$ of $\mathcal{T}$, used by $\tilde{F}_{\mathcal{S}\to\mathcal{T}}$, is computed by the CVT algorithm in [24], with the same number of vertices as in $\mathbf{X}$;

(3): the Restricted Voronoi Diagrams $Vor(\mathbf{Y})|_{\mathcal{S}}$, $Vor(\mathbf{X}^{(k)})|_{\mathcal{T}}$ are computed as in [24] ;

(4),(5): $F_{\mathcal{T}\to\mathcal{S}}$ is the CVT objective function and R the regularization energy. The gradients are given by Equation 4:

$$\begin{aligned}\nabla|_{\mathbf{x}_i}\tilde{F}_{\mathcal{T}\to\mathcal{S}} &= 2m_i(\mathbf{x}_i - \mathbf{g}_i)\\ \nabla R(\mathbf{X}) &= \nabla\mathbf{X}^t\mathbf{L}^2\mathbf{X} = 2\mathbf{L}^2\mathbf{X}\end{aligned} \quad (4)$$

where m_i and $\mathbf{g}_i$ denote the volume and the centroid of the restricted Voronoi cell $\Omega_i \cap \mathcal{T}$ [6] ;

(6): the term $\tilde{F}_{\mathcal{S}\to\mathcal{T}}$ is obtained by exchanging the roles of $\mathcal{S}$ and $\mathcal{T}$ in $\tilde{F}_{\mathcal{T}\to\mathcal{S}}$ and using the point set $\mathbf{Y}$ instead of $\mathbf{X}$. The computation of this term and its gradient are explained in the next paragraph ;

(8): $\mathbf{p}^{(k)}$ denotes the step vector computed by L-BFGS.

The function $\tilde{F}_{\mathcal{S}\to\mathcal{T}}$ depends on the Voronoi diagram of $\mathbf{Y}$ restricted to $\mathcal{S}$ (see Figure 6). Each restricted Voronoi cell $\Omega_j \cap \mathcal{S}$ (colored polygons) is decomposed into a set of triangles. One of them $T = (\mathbf{c}_1, \mathbf{c}_2, \mathbf{c}_3)$ is highlighted. Each triangle T of the decomposition of $\Omega_j \cap \mathcal{S}$ contributes the following terms to $\tilde{F}_{\mathcal{S}\to\mathcal{T}}$ and $\nabla\tilde{F}_{\mathcal{S}\to\mathcal{T}}$:

$$\tilde{F}^T_{\mathcal{S}\to\mathcal{T}} = \frac{|T|}{6}\sum_{1\leq k\leq l\leq 3}(\mathbf{c}_k - \mathbf{y}_j)\cdot(\mathbf{c}_l - \mathbf{y}_j),\quad \frac{d\tilde{F}^T_{\mathcal{S}\to\mathcal{T}}}{d\mathbf{X}} = \sum_{k=1}^{3}\frac{d\tilde{F}^T_{\mathcal{S}\to\mathcal{T}}}{d\mathbf{c}_k}\frac{d\mathbf{c}_k}{d\mathbf{X}} \quad (5)$$

where $d\mathbf{A}/d\mathbf{B} = (\partial a_i/\partial b_j)_{i,j}$ denotes the Jacobian matrix of $\mathbf{A}$.

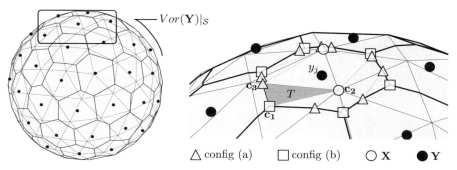

Fig. 6. Computing the gradient of the symmetric term $\nabla \tilde{F}_{\mathcal{S} \to \mathcal{T}}$: configurations of the vertices of $Vor(\mathbf{Y})|_{\mathcal{S}}$.

The set of possible configurations for a vertex $\mathbf{c}_k$ is similar to the combinatorial structure of the L_p-CVT function [11], with the exception that the roles of the variables and constants are exchanged. Each configuration yields a Jacobian matrix that propagates the derivatives of $\tilde{F}_{\mathcal{S} \to \mathcal{T}}^T$ from the $\mathbf{c}_k$'s to the $\mathbf{x}_i$'s. There are 3 possible configurations (see overview in Figure 6):

→ ○ $\mathbf{c}$ is a vertex $\mathbf{x}_i$ of $\mathcal{S}$ (then $d\mathbf{c}/d\mathbf{x}_i = \mathbf{I}_{3\times 3}$) ;
→ △ $\mathbf{c}$ has configuration (a):

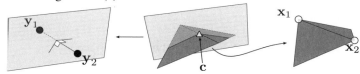

$\mathbf{c}$ corresponds to the intersection between the bisector of $[\mathbf{y}_1, \mathbf{y}_2]$ (left, plane shown in blue) and an edge $[\mathbf{x}_1, \mathbf{x}_2]$ of $\mathcal{S}$ (right). The Jacobian matrices $d\mathbf{c}/d\mathbf{x}_1$ and $d\mathbf{c}/d\mathbf{x}_2$ are given in Appendix B, Equation 13;

→ □ $\mathbf{c}$ has configuration (b):

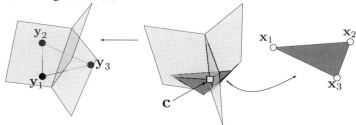

$\mathbf{c}$ corresponds to the intersection between the three bisectors of $[\mathbf{y}_1, \mathbf{y}_2]$, $[\mathbf{y}_2, \mathbf{y}_3]$, $[\mathbf{y}_3, \mathbf{y}_1]$ (left) and a facet $(\mathbf{x}_1, \mathbf{x}_2, \mathbf{x}_3)$ of $\mathcal{S}$ (right). The Jacobian matrices $d\mathbf{c}/d\mathbf{x}_1$, $d\mathbf{c}/d\mathbf{x}_2$ and $d\mathbf{c}/d\mathbf{x}_3$ are given in Appendix B, Equation 14.

3.5 Fitting Polynomial Surfaces

We consider now the problem of fitting a polynomial surface defined by its control mesh $\mathcal{C}$. At each iteration, we compute a polygonal approximation $\mathcal{S}$ of the polynomial surface. The vertices $\mathbf{X}$ of $\mathcal{S}$ are given as linear combinations of the control points $\mathbf{P}$:

$$\mathbf{X} = \mathbf{MP}$$

where $\mathbf{X} = [x_1 y_1 z_1 \ldots x_n y_n z_n]^t$ denotes the coordinates at the vertices of $\mathcal{S}$, $\mathbf{P}$ denotes the coordinates at the control points. One may use the exact evaluation of the polynomial surface, or simply use the approximation obtained by subdividing the control mesh several times with De Casteljau's rule.

Polynomial surface fitting is done by minimizing the function $G(\mathbf{P}) = \tilde{F}(\mathbf{MP})$. This can implemented with a change of variable in the VSDM algorithm (see Algorithm 2):

$\mathbf{P}^{(0)} \leftarrow$ vertices of $\mathcal{C}^{(0)}$
$\mathbf{Y} \leftarrow \varepsilon$-sampling of $\mathcal{T}$; Compute $Vor(\mathbf{Y})$
while *minimum not reached* **do**
  $\mathbf{X} \leftarrow \mathbf{MP}^{(k)}$; Update $\mathcal{S}$ from $\mathbf{X}$
  Compute $\tilde{F}(\mathbf{X})$ and $\nabla \tilde{F}(\mathbf{X})$ as in Algo. 1, steps (3) to (7)
  Compute $\nabla G(\mathbf{P}) = \mathbf{M}^t \nabla \tilde{F}(\mathbf{X})$
  $\mathbf{P}^{(k+1)} \leftarrow \mathbf{P}^{(k)} + \mathbf{p}^{(k)}$
end

Algorithm 2. Fitting a polynomial surface.

3.6 Feature-Sensitive Fitting

Using the algorithm above for fitting polynomial surfaces may result in over-smoothing sharp creases (Figure 7 center). However, this can be improved by injecting *normal anisotropy* [11] into the objective function $\tilde{F}$ (Figure 7 right). This changes the terms $\tilde{F}_{\mathcal{T} \to \mathcal{S}}$ and $\tilde{F}_{\mathcal{S} \to \mathcal{T}}$ as follows:

$$\tilde{F}^s_{\mathcal{T} \to \mathcal{S}} = \sum_{x_i \in X} \sum_{T \subset \mathcal{T} \cap \Omega_i} \int_T \| \mathbf{A}_s(\mathbf{N}_T)(\mathbf{y} - \mathbf{x}_i) \|^2 \, d\mathbf{y}$$
$$\tilde{F}^s_{\mathcal{S} \to \mathcal{T}} = \sum_{y_j \in Y} \sum_{T \subset \mathcal{S} \cap \Omega_j} \int_T \| \mathbf{A}_s(\mathbf{N}_j)(\mathbf{x} - \mathbf{y}_j) \|^2 \, d\mathbf{x}$$

where : (6)

$$\mathbf{A}_s(\mathbf{N}) = (s-1) \begin{pmatrix} \mathbf{N}_x[\mathbf{N}]^t \\ \mathbf{N}_y[\mathbf{N}]^t \\ \mathbf{N}_z[\mathbf{N}]^t \end{pmatrix} + \mathbf{I}_{3\times 3}$$

where the parameter $s \in (0, +\infty)$ specifies the importance of normal anisotropy. The normals are sampled from the input surface $\mathcal{T}$ in both terms, $\mathbf{N}_T$ is the normal of the triangle T, and $\mathbf{N}_j$ the normal to $\mathcal{T}$ at $\mathbf{y}_j$. Normal anisotropy is used in all the examples shown below.

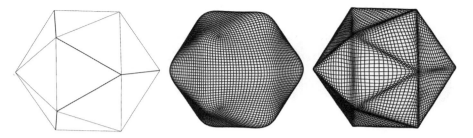

Fig. 7. Influence of the feature-sensitive fitting on meshes with sharp creases (from left to right: original mesh, result without and with normal anisotropy).

3.7 Implementation

For the Delaunay triangulation, we use CGAL (www.cgal.org). For the Restricted Voronoi Diagram computation (Section 3.4) and the normal anisotropy (previous subsection), we use the implementation provided with [11].

4 Results

We shall now show some results obtained with VSDM. In the results herein, the regularization term is set to $\lambda = 0.2 \times 10^{-3}$, the normal anisotropy is set to $s = 50$ and subdivision surfaces are approximated by subdividing the control mesh twice. Figures 8, 9, and 10 show the result obtained with an initial toroidal grid. Note on Figure 12 how the spacing of the iso-parameter line adapts to the features. Scanned meshes from AimAtShape can also be efficiently processed (see Figure 13). For each model, the result was obtained in less than 3 minutes on a 2 GHz machine.

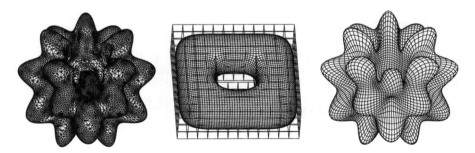

Fig. 8. Fitting a polynomial surface to an object with toroidal topology. Left: input mesh (16.8k vertices, 33.6k facets); Center: initial control mesh (512 vertices and 512 quads) and surface; Right: result. The Hausdorff distance between the resulting surface and the input mesh is 1.221% (relative to the diagonal length of the bounding box, measured by Metro [4]).

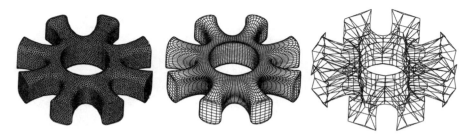

Fig. 9. Another example, using the same initial toroidal control mesh as in Figure 8. Left: input mesh (10k vertices, 20k facets); Center: result; Right: result (control mesh with 512 vertices and 512 quads). Hausdorff distance is 0.473% bbox. diag.

Fig. 10. Another example, still using the same initial toroidal control mesh. Left: input mesh (5.2k vertices and 10.4k facets); Center: result; Right: result (control mesh with 512 vertices and 512 quads). The Hausdorff distance is 0.699% bbox. diag.

Fig. 11. Other examples with geometrical shapes. Initialization from bounding box (386 vertices and 384 quads). Left: input mesh of sharp sphere (10.4k vertices, 20.9k facets) and result. Right: input mesh of octa-flower (7.9k vertics and 15.8k facets) and result. Hausdorff distances are 1.072% and 0.706% bbox. diag., respectively.

Voronoi Squared Distance Minimization 613

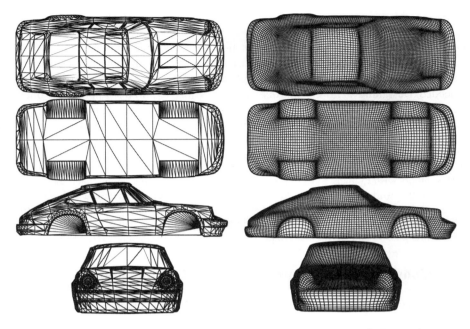

Fig. 12. Different views of the example shown in Figure 1. The Hausdorff distance between the input and result is 0.554% of the bounding box diagonal.

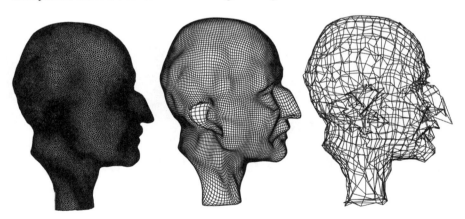

Fig. 13. Fitting a Catmull-Clark subdivision surface to the statue of Max Planck (52.8k vertices and 105.6 facets). Initialization from bounding box (6257 vertices and 6255 quads). The Hausdorff distance is 0.386% of the bounding box diagonal.

Discussion and Future Work

The examples shown in the previous section were obtained automatically, by using the bounding box (or a toroidal mesh) as the initial control mesh. However, for shapes with an arbitrary genus or a complicated geometry, an initial control mesh is needed. Designing an initial control mesh may be also required to improve the quality of the surface. In future work, we will study the generation of an initial control mesh and/or the dynamic modification of the control mesh during the optimization.

Acknowledgements

The authors wish to thank Sylvain Lefebvre for a discussion (about an unrelated topic) that inspired this work, Rhaleb Zayer, Xavier Goaoc, Tamy Boubekeur, Yang Liu and Wenping Wang for many discussions, Loic Marechal, Marc Loriot and the AimAtShape repository for data. This project is partly supported by the European Research Council grant GOODSHAPE ERC-StG-205693 and ANR/NSFC (60625202,60911130368) Program (SHAN Project).

A Convergence to Squared Distance – Error Bound

The following section proves that if $\mathbf{X}$ is an ε-sampling of $\mathcal{S}$ then:

$$\lim_{\varepsilon \to 0} \tilde{F}_{\mathcal{T} \to \mathcal{S}}(\mathbf{X}) = F_{\mathcal{T} \to \mathcal{S}}(\mathbf{X}) \tag{7}$$

Lemma 1. *Let $\mathbf{y}$ be a point of $\mathcal{T}$ and $\mathbf{x}_i$ its nearest point in $\mathbf{X}$ (see Figure 14). Let $d = \| \mathbf{y} - \Pi_{\mathcal{S}}(\mathbf{y}) \|$ and $\tilde{d} = \| \mathbf{y} - \mathbf{x}_i \|$. Then for $\varepsilon < 2$ the following bound is sharp:*

$$\tilde{d}^2 - d^2 \leq \varepsilon^2 \mathit{lfs}(\Pi_{\mathcal{S}}(\mathbf{y}))(\mathit{lfs}(\Pi_{\mathcal{S}}(\mathbf{y})) + d)$$

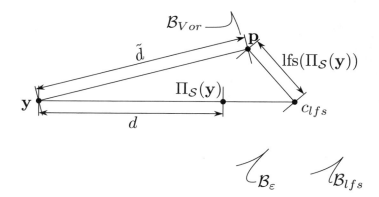

Fig. 14. Configuration of the nearest point of $\Pi_{\mathcal{S}}(\mathbf{y})$.

Proof. Let $\mathcal{B}_{Vor}$ be the ball centered at **y** passing through $\mathbf{x}_i$. This ball contains no point of **X**.

Let $\mathcal{B}_{lfs}$ be the ball tangent to $\mathcal{S}$ at $\Pi_\mathcal{S}(\mathbf{y})$ on the opposite side of **y** and of radius $\mathrm{lfs}(\Pi_\mathcal{S}(\mathbf{y}))$ and c_{lfs} its center. Since **X** is an ε-sampling this ball also contains no point of **X**.

Finally let $\mathcal{B}_\varepsilon$ be the ball centered at $\Pi_\mathcal{S}(\mathbf{y})$ of radius $\varepsilon\mathrm{lfs}(\Pi_\mathcal{S}(\mathbf{y}))$. Since **X** is an ε sampling this ball contains no point of $\mathcal{S}$, and therefore no point of **X**.

Since $\Pi_\mathcal{S}(\mathbf{y})$ is the nearest point of **y** on $\mathcal{S}$, **y**, $\Pi_\mathcal{S}(\mathbf{y})$ and c_{lfs} are aligned and the problem is completely symmetric around the line joining them. Figure 14 shows a cut containing this axis.

The bound follows from the fact that $\mathcal{B}_\varepsilon \not\subset \mathcal{B}_{Vor} \cup \mathcal{B}_{lfs}$. Let p be a point of $\mathcal{B}_{Vor} \cap \mathcal{B}_{lfs}$. This point exists since $\varepsilon < 2$ and $\mathcal{B}_\varepsilon$ is not included in $\mathcal{B}_{Vor}$. Using this point the previous condition can be reformulated as $p \in \mathcal{B}_\varepsilon$.

Using triangular identities in $(\Pi_\mathcal{S}(\mathbf{y}), c_{lfs}, \mathbf{p})$, we have:

$$\| \Pi_\mathcal{S}(\mathbf{y}) - \mathbf{p} \|^2 = 2\mathrm{lfs}(\Pi_\mathcal{S}(\mathbf{y}))^2(1 - \cos\alpha) \tag{8}$$

with α being the $(\mathbf{p}, c_{lfs}, \Pi_\mathcal{S}(\mathbf{y}))$ angle. Using the same identities in $(\mathbf{x}, c_{lfs}, \mathbf{p})$ we obtain:

$$\tilde{d}^2 = (d + \mathrm{lfs}(\Pi_\mathcal{S}(\mathbf{y})))^2 + \mathrm{lfs}(\Pi_\mathcal{S}(\mathbf{y}))^2$$
$$- 2(d + \mathrm{lfs}(\Pi_\mathcal{S}(\mathbf{y})))\mathrm{lfs}(\Pi_\mathcal{S}(\mathbf{y}))\cos\alpha$$
$$\tilde{d}^2 - d^2 = 2(d + \mathrm{lfs}(\Pi_\mathcal{S}(\mathbf{y})))\mathrm{lfs}(\Pi_\mathcal{S}(\mathbf{y}))(1 - \cos\alpha)$$

Using equation 8, $(1 - \cos\alpha)$ can be replaced:

$$\tilde{d}^2 - d^2 = (d + \mathrm{lfs}(\Pi_\mathcal{S}(\mathbf{y})))\frac{\| \Pi_\mathcal{S}(\mathbf{y}) - \mathbf{p} \|^2}{\mathrm{lfs}(\Pi_\mathcal{S}(\mathbf{y}))} \tag{9}$$

Finally since **p** is inside $\mathcal{B}_\varepsilon$, we have:

$$\| \Pi_\mathcal{S}(\mathbf{y}) - \mathbf{p} \| \leq \varepsilon\mathrm{lfs}(\Pi_\mathcal{S}(\mathbf{y})) \tag{10}$$

This finally provides the result:

$$\tilde{d}^2 - d^2 \leq \varepsilon^2 \mathrm{lfs}(\Pi_\mathcal{S}(\mathbf{y}))(\mathrm{lfs}(\Pi_\mathcal{S}(\mathbf{y})) + d) \tag{11}$$

This bound is sharp since it is reached whenever $\mathcal{S}$ is exactly $\mathcal{B}_{lfs}$ and $\mathbf{x}_i$ is located at **p**.

This lemma leads to a global bound:

Proposition 1. *If $\mathcal{S}$ is different from a plane and bounded, then:*

$$\tilde{F}_{\mathcal{T} \to \mathcal{S}}(\mathbf{X}) - F_{\mathcal{T} \to \mathcal{S}}(\mathbf{X}) \leq \varepsilon^2 |\mathcal{T}| \sigma_\mathcal{S}(\sigma_\mathcal{S} + d_\mathcal{H}(\mathcal{T}, \mathcal{S}))$$

where $\sigma_\mathcal{S} = \sup\{lfs(\mathbf{x}), \mathbf{x} \in \mathcal{S}\}$

Proof. Since $\mathcal{S}$ is not a plane and bounded, σ exists. In addition, the definition of the Hausdorff distance gives us $d \leq d_\mathcal{H}(\mathcal{T}, \mathcal{S})$.

$$\begin{aligned}
e &= \tilde{F}_{\mathcal{T} \to \mathcal{S}}(\mathbf{X}) - F_{\mathcal{T} \to \mathcal{S}}(\mathbf{X}) \\
&= \int_\mathcal{T} \min_i \| \mathbf{y} - \mathbf{x}_i \|^2 \, d\mathbf{y} - \int_\mathcal{T} \| \mathbf{y} - \Pi_\mathcal{S}(\mathbf{y}) \|^2 \, d\mathbf{y} \\
&= \int_\mathcal{T} \min_i \| \mathbf{y} - \mathbf{x}_i \|^2 - \| \mathbf{y} - \Pi_\mathcal{S}(\mathbf{y}) \|^2 \, d\mathbf{y} \\
&\leq \int_\mathcal{T} \varepsilon^2 \sigma_\mathcal{S}(\sigma_\mathcal{S} + d_\mathcal{H}(\mathcal{T}, \mathcal{S})) d\mathbf{y} \\
&\leq \varepsilon^2 |\mathcal{T}| \sigma_\mathcal{S}(\sigma_\mathcal{S} + d_\mathcal{H}(\mathcal{T}, \mathcal{S}))
\end{aligned} \qquad (12)$$

B Gradients of the Symmetric Term $\nabla \tilde{F}_{\mathcal{S} \to \mathcal{T}}$

Configuration (a):

$$\frac{d\mathbf{c}}{d\mathbf{x}_1} = \mathbf{e}\mathbf{w}_1^t + (1-u)\mathbf{I}_{3\times 3}$$
$$\frac{d\mathbf{c}}{d\mathbf{x}_2} = \mathbf{e}\mathbf{w}_2^t + u\mathbf{I}_{3\times 3}$$

where:
$$\begin{cases}
\mathbf{e} = (\mathbf{x}_2 - \mathbf{x}_1) \\
\mathbf{n} = (\mathbf{y}_2 - \mathbf{y}_1) \\
k = \mathbf{n} \cdot \mathbf{e} \\
h = \frac{1}{2} \mathbf{n} \cdot (\mathbf{y}_1 + \mathbf{y}_2) \\
u = \frac{1}{k}(h - \mathbf{n} \cdot \mathbf{x}_1) \\
\mathbf{w}_1 = -\frac{1}{k^2}(h - \mathbf{n} \cdot \mathbf{x}_2)\mathbf{n} \\
\mathbf{w}_2 = \frac{1}{k^2}(h - \mathbf{n} \cdot \mathbf{x}_1)\mathbf{n}
\end{cases} \qquad (13)$$

Configuration (b):

$$\frac{d\mathbf{c}}{d\mathbf{x}_1} = \mathbf{e}\mathbf{w}_1^t$$
$$\frac{d\mathbf{c}}{d\mathbf{x}_2} = \mathbf{e}\mathbf{w}_2^t$$
$$\frac{d\mathbf{c}}{d\mathbf{x}_3} = \mathbf{e}\mathbf{w}_3^t$$

where:
$$\begin{cases}
\mathbf{e} = (\mathbf{y}_1 - \mathbf{y}_2) \times (\mathbf{y}_1 - \mathbf{y}_3) \\
\mathbf{n} = (\mathbf{x}_1 - \mathbf{x}_2) \times (\mathbf{x}_1 - \mathbf{x}_3) \\
k = \mathbf{n} \cdot \mathbf{e} \\
\mathbf{w}_1 = ((\mathbf{x}_2 - \mathbf{x}_3) \times (\mathbf{x}_1 - \mathbf{c}) + \mathbf{n})/k \\
\mathbf{w}_2 = ((\mathbf{x}_3 - \mathbf{x}_1) \times (\mathbf{x}_1 - \mathbf{c}))/k \\
\mathbf{w}_3 = ((\mathbf{x}_1 - \mathbf{x}_2) \times (\mathbf{x}_1 - \mathbf{c}))/k
\end{cases} \qquad (14)$$

References

1. Amenta, N., Bern, M.: Surface reconstruction by Voronoi filtering. Discrete and Computational Geometry 22(4), 481–504 (1999)
2. Cheng, K.-S.D., Wang, W., Qin, H., Wong, K.-Y.K., Yang, H.-P., Liu, Y.: Fitting subdivision surfaces using SDM. In: Pacific Graphics Conf. Proc., pp. 16–24 (2004)
3. Cheng, K.-S.D., Wang, W., Qin, H., Wong, K.-Y.K., Yang, H.-P., Liu, Y.: Design and analysis of methods for subdivision surface fitting. IEEE TVCG 13(5) (2007)
4. Cignoni, P., Rocchini, C., Scopigno, R.: Metro: measuring error on simplified surfaces. Comp. Graphics Forum 17(2), 167–174 (1998)
5. Delingette, H.: General object reconstruction based on simplex meshes. IJCV 32(2), 111–146 (1999)
6. Du, Q., Faber, V., Gunzburger, M.: Centroidal Voronoi tessellations: applications and algorithms. SIAM Review 41(4), 637–676 (1999)

7. Eck, M., Hoppe, H.: Automatic reconstruction of B-spline surfaces of arbitrary topological type. In: Proc. ACM SIGGRAPH, pp. 325–334 (1996)
8. Floater, M.S.: Parametrization and smooth approximation of surface triangulations. Computer Aided Geometric Design 14(3), 231–250 (1997)
9. Kobbelt, L., Vorsatz, J., Labsik, U., Seidel, H.-P.: A shrink wrapping approach to remeshing polygonal surfaces. Comp. Graphics Forum 18(3) (2001)
10. Kovacs, D., Myles, A., Zorin, D.: Anisotropic quadrangulation. In: Proceedings of the 14th ACM Symposium on Solid and Physical Modeling, pp. 137–146 (2010)
11. Lévy, B., Liu, Y.: L_p centroidal Voronoi tessellation and its applications. ACM TOG (Proc. SIGGRAPH) 29(4), article no. 119 (2010)
12. Li, W.-C., Ray, N., Lévy, B.: Automatic and interactive mesh to T-spline conversion. In: Symposium on Geometry Processing, pp. 191–200 (2006)
13. Liu, D.C., Nocedal, J.: On the limited memory BFGS method for large scale optimization. Mathematical Programming: Series A and B 45(3), 503–528 (1989)
14. Liu, Y., Wang, W., Lévy, B., Sun, F., Yan, D.-M., Lu, L., Yang, C.: On centroidal Voronoi tessellation — energy smoothness and fast computation. ACM Trans. on Graphics 28(4), article no. 101 (2009)
15. Ma, W., Ma, X., Tso, S.-K., Pan, Z.: A direct approach for subdivision surface fitting. Comp. Aided Design 36(6) (2004)
16. Marinov, M., Kobbelt, L.: Optimization methods for scattered data approximation with subdivision surfaces. Graphical Models 67(5), 452–473 (2005)
17. Nocedal, J., Wright, S.J.: Numerical Optimization. Springer, Heidelberg (2006)
18. Pottmann, H., Leopoldseder, S.: A concept for parametric surface fitting which avoids the parametrization problem. Comp. Aided Geom. Design 20(6) (2003)
19. Ray, N., Li, W.C., Lévy, B., Scheffer, A., Alliez, P.: Periodic global parameterization. ACM Trans. on Graphics 25(4), 1460–1485 (2006)
20. Schreiner, J., Asirvatham, A., Praun, E., Hoppe, H.: Inter-surface mapping. ACM TOG (Proc. SIGGRAPH) 23(3), 870–877 (2004)
21. Tarini, M., Hormann, K., Cignoni, P., Montani, C.: Polycube-maps. ACM TOG (Proc. SIGGRAPH) 23(3), 853–860 (2004)
22. Wang, H., He, Y., Li, X., Gu, X., Qin, H.: Polycube splines. Comp. Aided Design 40(6), 721–733 (2008)
23. Wang, W., Pottmann, H., Liu, Y.: Fitting B-spline curves to point clouds by curvature-based sdm. ACM Trans. on Graphics 25(2), 214–238 (2006)
24. Yan, D.-M., Lévy, B., Liu, Y., Sun, F., Wang, W.: Isotropic remeshing with fast and exact computation of restricted Voronoi diagram. Comp. Graphics Forum (Proc. SGP) 28(5), 1445–1454 (2009)
25. Yeh, I.-C., Lin, C.-H., Sorkine, O., Lee, T.-Y.: Template-based 3D model fitting using dual-domain relaxation. IEEE TVCG 17(8), 1178–1190 (2011)

Dendritic Meshing*
LA-UR 11-04075

Brian A. Jean, Rodney W. Douglass, Guy R. McNamara, and Frank A. Ortega

Los Alamos National Laboratory
MS T085
P.O. Box 1663
Los Alamos, NM 87545
baj@lanl.gov

1 Introduction

A mesh is said to be dendritic if it contains elements with mid-side (edge) nodes when the predominant element topology has only corner nodes. A dendritic mesh is illustrated in Figure 1 where the predominant element is a four-node quadrilateral, but has also several five-node quadrilateral elements each with one mid-edge node plus four corner nodes. Such meshes arise when an approximately uniform element size is required across a mesh domain in cases, for example, where domain geometry changes would otherwise cause a significant variation in element size or in an Adaptive Mesh Refinement (AMR) context. In meshes created for multi-physics applications with explicit time-stepping, the maximum time-step size is intimately tied to element size through the Courant-Friedrichs-Lewy condition[6]:

$$\frac{u\,\Delta t}{\Delta x} \leq C \qquad (1)$$

where u is a representative speed, Δx the element size, Δt is the time-step size, and C is a constant appropriate for the physics being modeled. This implies $\Delta t \leq C'\Delta x$, C' a constant, and the smaller the element, the smaller the time-step size. The smallest element in a mesh therefore limits the time-step size providing motivation to equalize element size over a domain.

An additional concern is that the mesh adhere as nearly as possible to the domain boundary geometry. There are two general classes of boundary-fitted mesh generation methods (*e.g.,* [14]): block-structured and unstructured. Both of these methods have inherent strengths and weaknesses. The various unstructured methods are highly automated and tend to produce meshes with relatively uniform zone size. However unstructured techniques

---

* This work was performed at Los Alamos National Laboratory under the auspices of U.S. Department of Energy, under contract DE-AC52-06NA25396 and has been reviewed for general release as LA-UR 11-04075.

generally produce meshes with irregular zone connectivity and there is little or no control over zone orientation or zone aspect ratio. Structured methods enable control of zone orientation, have regular connectivity, and can produce very high aspect ratio zones (desired in some problem domains). However structured techniques are labor-intensive and control of zone-size is limited by the geometry of the problem and the domain decomposition into logical blocks. The discussion herein focuses on boundary-fitted meshes with a relatively uniform zone size in most regions, the option to produce high aspect ratio zones in other regions, and the ability to align the mesh with the predominant direction of shock propogation and/or material flow. Neither of the existing meshing methods mentioned above satisfies these criteria.

To meet these meshing requirements, a hybrid method is presented which is termed *dendritc meshing*. Dendritic meshing is a modified block-structured technique that allows logical edges within a structured mesh block to be "deactivated" as needed to control zone size. This method combines many of the advantages of both structured and unstructured methods. With dendritic meshing, zone size can be kept relatively uniform when needed, zone aspect ratio can be controlled, zone orientation can be controlled, and nodes of irregular connectivity within the mesh are minimized.

The starting point for a dendritic mesh is a standard block-structured mesh; a mesh consisting of perhaps multiple blocks each having a logical ij-structure. To produce a dendritic mesh, selected segments of constant-i or constant-j lines are removed. The "dendrites" are the nodes at which an i or j line segment becomes inactive. After removing the selected mesh segments, the mesh is re-interpolated using a specialized dendrite-aware transfinite interpolation algorithm, and is then smoothed with a dendrite-aware smoother. Figures 1 and 2 show the active and removed edges of a structured mesh in physical and logical space, respectively.

The following sections highlight the significant issues in the dendritic meshing process. These issues are: data structures, feathering of a structured mesh, how to build the initial mesh, trans-finite interpolation (TFI) with dendrites, smoothing of dendritic meshes, and a brief conclusion.

2 Data Structures

The mesh is composed of a set of logically rectangular mesh blocks with the individual blocks connected together in an unstructured fashion[14] creating a block-structured mesh data structure. Mesh nodes for the entire problem are stored in a single master node array. The array index of a node in this master array is its *id*. The array of nodes for a mesh block stores the id's of the nodes for the block and not the nodes themselves. Block-to-block connectivity is determined implicitly by shared nodes, not by an explicit block topology data structure.

Dendritic Meshing

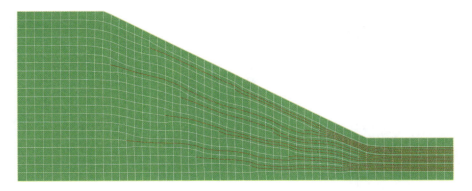

Fig. 1. A simple dendritic mesh with inactive edges shown in red.

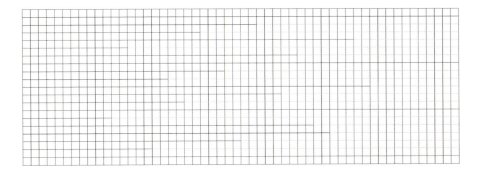

Fig. 2. Dendritic Mesh in Logical Space. The logical coordinate system origin is in the lower-left corner. The i-axis is horizontal and the j-axis is vertical.

Active Mesh Lines and Vertices

In addition to the array of mesh node id's, each block also maintains an array of Boolean ij-edge flags which indicate whether a zone mesh edge is active or inactive. A zone is identified by its lower left corner node indices, i, j. Figure 3 shows the edge flags associated with node (i, j).

Activity of a node is determined by whether or not its edges are active. A node is active if and only if at least one active i-edge and at least one active j-edge connect to the node. Figure 4 shows an inactive edge (i, j) and the associated inactive node at $(i + 1, j)$. The $(i + 1, j)$ node is inactive because no active i-edges touch it.

Node Slaving

Dendritic nodes (nodes at which ij-lines terminate) are also called *slave* nodes. Slave nodes are constrained to lie at the mid-point of the line connecting the slave's neighbors in the perpendicular logical direction of the

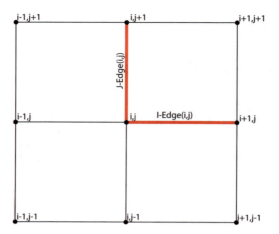

Fig. 3. Edge flags for node (i,j).

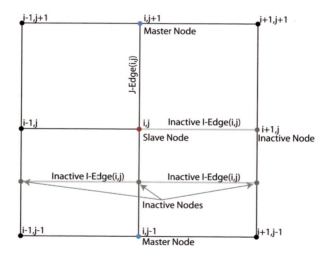

Fig. 4. Inactive edges/nodes and node slaving.

terminating line. The slaving constraint for nodes on terminating i-lines and terminating j-lines are given in Equations 2 and 3, respectively.

$$\mathbf{r}_{i,j} = \frac{1}{2}\left(\mathbf{r}_{i,j+1} + \mathbf{r}_{i,j-1}\right) \qquad (2)$$

$$\mathbf{r}_{i,j} = \frac{1}{2}\left(\mathbf{r}_{i+1,j} + \mathbf{r}_{i-1,j}\right) \qquad (3)$$

Note that in the above equations, the $\pm$ indexing on the nodes denotes the next/previous *active* node in the indicated direction and not simply the adjacent node in the original structured mesh (see Figure 4).

2.1 Conversion to an Unstructured Mesh

An unstructured representation of the mesh is used for smoothing (see section 5) and for output to some physics codes. The unstructured mesh is defined as a collection of arbitrary polyhedral zones and is built by generating polyhedral zones from the active nodes. This is done by marching around the active edges in the block structured mesh where a change in direction is made when the next active edge in the opposite logical direction, for example when changing from an i-edge to a j-edge. When a change in direction is found, the active vertex at that change then becomes a corner vertex of the unstructured zone. Any vertices in the zone that is not a corner vertex then becomes a slave vertex and the two corner vertices on each side of the save vertex are the master vertices for that slave.

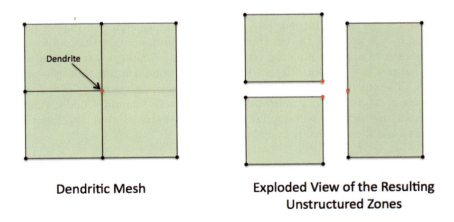

Dendritic Mesh   **Exploded View of the Resulting Unstructured Zones**

Fig. 5. Dendrite treatment during conversion to an unstructured mesh.

2.1.1 Arbitrary Polyhedral Topology

Due to the hanging vertices produced by dendritic mesh lines, an arbitrary polyhedral mesh topology is used for the unstructured mesh. This topology defines a zone by its faces and faces are defined by their vertices. The vertices in the face are ordered such that they define a right-handed normal relative to the zone center. In 2-D space, a face is an edge and the order of the vertices for each edges and the order of the edges are specified such that they define a right-handed normal relative to the X-Y axis plane. Because of the right-handedness of the arbitrary polyhedral topology, each zone has its own unique set of faces. Every polygonal zone in the mesh has it's own set of unique faces, however the faces between two zones have the same vertices but with a different sense. To provide mesh connectivity, the two faces with the same vertices are linked in the data structure.

2.2 Mesh Connectivity

Special consideration must be given to boundary nodes of connected dendritic mesh blocks. Figure 6 illustrates the cases that must be considered. These cases are:

1. **Vertex-Glue:** An active boundary node shared among blocks
2. **Edge-Glue:** A boundary node active in one block, but absent in the adjoing block
3. **Inactive Node:** An inactive boundary node in one block, but absent in the adjoing block

Vertex-glue is treated in the obvious way – the node is present in both blocks in the final mesh. Edge-glue represents a boundary dendrite. This requires identifying the edge-glue node as a dendrite and indentifying its master nodes to ensure it is properly slaved. During conversion to an unstructured format, treatement of zones containing edge-glue nodes is analogous to that of dendritic zones (see section 2.1) with the added complication that the zones are in different blocks. Inactive boundary nodes are the end points of inactive edges extending to a block boundary. Inactive boundary nodes exist to maintain the logically rectangular structure of the block, but are ignored in block-to-block connectivity calculations.

3 Feathering

Feathering is the process used to convert a structured mesh block into a dendritic mesh block. Currently, feathering is limited to a single logical direction within any given mesh block. Multi-directional feathering capability is in development.

Dendritic Meshing

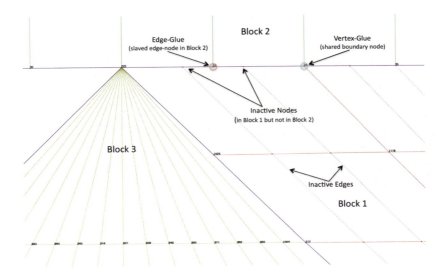

Fig. 6. Block-to-block connectivity: Vertex glue, edge glue, and inactive vertices.

Validity Constraints

A valid dendritic block-structured mesh must obey the following constraints:

1. The set of mesh zones is a disjoint partition of rectangles in the logical space of a mesh block. This basically means that zones must be convex in logical space (e.g., no L-shaped zones) and that there can be no dangling mesh edges (i.e. a vertex must have two active mesh edges).
2. *Slave* nodes cannot also be *Master* nodes (i.e., no recursive slaving)

Base Mesh Resolution

Unlike a structured mesh, a dendritic mesh block does not have constraints on the resolution of its boundaries in the direction of feathering. The boundary resolutions in the logical direction(s) of feathering need not match one another and may each be different than the resolution of the block in logical space. In some cases, the geometry of a mesh block may dictate that the maximum active edge count in the block occur in the interior of the block to maintain a desired zone size (*e.g.*, meshing a football). Figure 7 shows an example of a boundary resolution mismatch and an interior resolution that is larger than either of the boundary resolutions. The mesh block in Figure 7 is generated

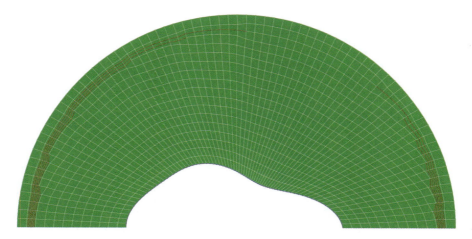

Fig. 7. Feathering with an interior and boundary resolution mismatch (red lines are inactive edges).

with the constraint that a uniform zone size be maintained. Note that for the case in Figure 7 the i-resolution of the block is determined by the resolution needed in the interior of the block to maintain uniform zoning after feathering.

Feather Schemes

Feather schemes determine the order in which mesh edges are removed for a given logical sheet. Note however, that the scheme does not determine *how many* edges are removed. There are three feathering schemes (patterns) currently supported. They are *boundary feathering* where edges are removed near one of the logical boundaries of the block, *centered feathering* where edges are removed near the center of the block logical space, and *distributed feathering* where edges are removed in a distributed manner throught the logical space of the block. The *boundary* and *distributed* schemes have the option to prevent edges from being removed immediately adjacent to a boundary by specifying an integer *offset* to indicate how many layers of un-featherd zones to leave next to either or both boundaries. Figures 7 and 8 show examples of the *boundary* and *centered* schemes. The *distributed* scheme is shown in Section 1, Figure 1.

Distributed feathering requires that edges to be removed from the mesh be more-or-less uniformly distributed across the width of the mesh. We do not want the removed edges to cluster to one side or the other, or to be preferentially located near the block's center or near the block boundaries. This outcome is achieved by means of a repulsive "potential" acting between removed edges. The potential is calculated in logical coordinate (index) space, and periodic boundary conditions are employed to prevent removed edges from clustering near the block boundaries. We arbitrarily use a $1/r^2$ potential

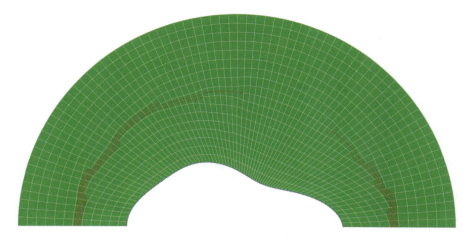

Fig. 8. Feathering with the *center* scheme.

(a $1/r$ potential would serve equally well) to select the next edge to be removed based on minimization of the potential produced by existing removed edges.

Removing Edges

Given a base mesh resolution and a desired feathering scheme, the only piece of information missing is the number of edges to be removed at each logical sheet in the feather direction. The count of edges to be removed is determined by calculating the arclength of the TFI projector in the logical direction perpendicular to the feather direction then calculating the number of edges which must be removed to give the desired zone size on the sheet. Once the number of edges is determined, the feather scheme determines which edges are removed.

4 Trans-Finite Interpolation with Dendrites

Trans-Finite Interpolation (TFI) is a method of positioning the internal vertices of logically-retangular mesh blocks based on the locations of the block's boundary vertices. Internal-vertex positions are computed as the vector sum of two linear interpolations between opposing-face pairs, minus a bilinear interpolation based on the corner vertices. For the special case of a block with two opposing uniformly-distributed straight-line faces, the bilinear interpolation exactly cancels the linear interpolation between the straight-line faces, leaving the linear interpolation between the remaining two (curvelinear) faces.

The simplest possible TFI uses the interior vertex's normalized logical coordinates ($i = I/(N_I - 1), j = J/(N_J - 1)$) as the interpolation parameters. Such a TFI will not accurately propagate non-uniform boundary vertex distributions into the interior. This defect may be overcome by replacing normalized logical coordinates (i,j) with normalized curvelinear coordinates (u,v), with u and v linear in arc length from 0 to 1 along opposing mesh block faces; u increases with increasing i, and v increases with j. Curvelinear coordinates for the (I,J)-th interior vertex are given by:

$$u = (1-v)u_l(I) + vu_h(I) \qquad (4)$$

$$v = (1-u)v_l(J) + uv_h(J) \qquad (5)$$

where $u_l(I)$ and $u_h(I)$ are the u coordinates of the I-th vertex on the $v=0$ and $v=1$ boundaries, respectively. A similar definition holds for $v_l(J)$ and $v_h(J)$.

To extend TFI to dendrited mesh blocks we introduce Normalized Dendrite-Logical (NDL) coordinates (i_d, j_d). NDL coordinates for the (I,J)-th vertex are calculated using active-vertex counts along the I and J mesh lines:

$$i_d = N_I A_{const-J}(I,J)/A_{const-J}(N_I - 1, J) \qquad (6)$$

$$j_d = N_J A_{const-I}(J,I)/A_{const-I}(N_J - 1, I) \qquad (7)$$

where $A_{const-J}(I,J)$ is the running count of active vertices along the Jth constant-J mesh line starting with $A_{const-J}(0,J) = 0$. $A_{constI}(J,I)$ is defined similarly. Note that NDL coordinates are normalized to the dimensions of the unfeathered mesh block. NDL coordinates are, strictly speaking, only required for active vertices, but it is useful for purposes of mesh visualization to apply the TFI to inactive vertices as well. For this purpose we linearly interpolate i_d values for succesive inactive vertices on a constant-J line from the i_d values of the bracketing active vertices (and like-wise for j_d values on constant-I lines).

TFI on a dendrited block is acomplished by replacing logical coordinates (I, J) in equations 4 and 5 with NDL coordinates (i_d, j_d). The block-boundary coordinates u_l, u_h, v_l and v_h are now determined by linear interpolation between successive boundary vertices, e.g.:

$$f(i_d) = i_d - \lfloor i_d \rfloor$$
$$u_{l,\text{dendrite}}(i_d) = (1 - f(i_d))\, u_l(\lfloor i_d \rfloor) + f(i_d) * u_l(\lceil i_d \rceil). \qquad (8)$$

Figure 9 shows the effect of a dendrite aware TFI on a feathered fan mesh. Note that the active vertices are uniformly distributed along each of the block's interior arcs.

5 Dendritic Mesh Smoothing

Historically, there are many methods proposed for mesh enhancement such as those documented in [5], [9],[11], [13], and [14], for example. The presence of dendritic zones (elements) within the mesh to be enhanced present challenges

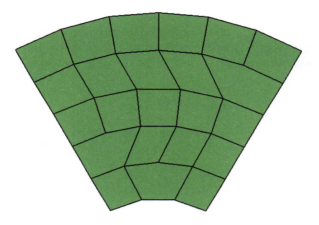

Fig. 9. Application of TFI to a dendritic mesh block.

to algorithms designed for smoothing non-dendritic meshes. There are two fundamental reasons why this is so: 1.) the presence of extra *hanging* nodes in an element and 2.) applying constraints to those hanging nodes.

In the first case, the presence of hanging nodes means that for the dendritic element the representation of physics is different than it is for the non-dendritic elements. For example, consider a finite element context wherein it is desired to solve a partial differential equation within a domain. Suppose a mesh consists (in two-dimensions) of primarily simplex triangular elements, the presence of a dendritic triangular element in the mesh will cause that element to no longer be a simplex element. It may have up to three mid-side (hanging) nodes. Consequently, dendritic elements are non-conforming (*c.f.*, Section 5.5 of [4]) requiring modifications to the element basis function set.

In the second case, it is often required that the hanging nodes be placed according to a constraint. A common constraint on node j, the local elemental index of a hanging node, in an unstructured mesh is

$$x_j^i = \frac{1}{2}\left(x_{j+1}^i + x_{j-1}^i\right), \tag{9}$$

where x_j^i are the $i = 1, \ldots, D$ coordinate components in D-dimensional space for the $j \in 1, \ldots, N^e$ (counter-clockwise numbered) nodes in element e.

The Laplace-Beltrami method of mesh enhancement (*e.g.*, Chapters 5 – 9, [11]) is used in the following discussion as a means of illustrating how these two issues impact a mesh enhancement algorithm. The discussion draws extensively from Douglass [7].

5.1 Laplace-Beltrami Mesh Enhancement

Consider a domain, Ω, in three-dimensional Euclidean space, $x^i = (x, y)$, $i = 1, 2, 3$, which is described locally with parametric coordinates, $u^i = (u, v, w)$. Then, harmonic coordinates, x^i, are defined by the system

$$\frac{1}{\sqrt{g}} \frac{\partial}{\partial u^\alpha} \left(\sqrt{g} g^{\alpha\beta} \frac{\partial x^i}{\partial u^\beta} \right) = 0, \tag{10}$$

where, for the two-dimensional problems discussed here, $i, \alpha, \beta = 1, 2$. The contravariant metric tensor components $g^{\alpha\beta}$ are related to the covariant components through $g^{\alpha\gamma} g_{\gamma\beta} = \delta^\alpha_\beta$ and $g = \det(g_{\alpha\beta})$, where

$$g_{\alpha\beta} = \frac{\partial x^i}{\partial u^\alpha} \frac{\partial x^i}{\partial u^\beta}, \tag{11}$$

and where the usual Einstein summation convention is applied to repeated sub- or superscripts within a term and δ^α_β is the Kronecker delta.

To solve these equations for x^i, the finite element method of weighted residuals (MWR) ([1], [8]) is used to find an approximate solution. The partial differential equation is converted to an equivalent weak form

$$I = \sum_{e=1}^{E} I_e = 0$$

$$I_e = \int_{\Omega_e} \frac{\partial w}{\partial u^\alpha} g^{\alpha\beta} \frac{\partial x^i}{\partial u^\beta} \sqrt{g} d^2 u - \int_{\partial \Omega_e} w \sqrt{g} g^{\alpha\beta} \frac{\partial x^i}{\partial u^\beta} \, ds_\alpha \tag{12}$$

where w is an appropriate weight function. The last term, the integral over the boundary of Ω_e, cancels for all interior element boundaries and is taken to be zero for those element boundaries on the domain boundary subject to natural boundary conditions.

The dependent variables within each element use a nodal interpolation of the form

$$x^i = \sum_{k=1}^{N_e} \phi_k(u, v) X_k^i \tag{13}$$

where $k = 1, \ldots, N_e$ and N_e is the number of element nodes with $X_k^i = (X_k, Y_k)$ the coordinates of the kth element node and ϕ_k are the N_e basis functions. The element in the physical (x, y)-plane is the image of a unit element in the (u, v)-plane through the mapping $x^i = x^i(u, v)$. It is common to define a basis function set for a master element having coordinates $\xi^i = (\xi, \eta)$ so that the local u^i-coordinates are mapped to the master element via, usually, an isoparametric mapping

$$u^i = \sum_{m=1}^{M} U_m^i \psi_m(\xi, \eta), \tag{14}$$

where U^i are the coordinates of the element nodes in u^i-space. Using isoparametric mapping, the number of basis functions in the map is equal to the number of basis functions used in the approximate solution, $M = N_e$. Equation 12 is invariant when the integration variables are changed from u^i to ξ^i resulting in

$$I_e = K^e_{ij} X^i_j, \text{ where} \tag{15}$$

$$K^e_{jk} = \int_{\Omega_e} \frac{\partial \psi_j}{\partial \xi^\alpha} g^{\alpha\beta} \frac{\partial \psi^i_k}{\partial \xi^\beta} \sqrt{g}\, d\xi d\eta. \tag{16}$$

5.1.1 Hanging Nodes and Element Basis Functions

The element basis functions address the first issue discussed above, that of the presence of hanging nodes in dendritic elements. Consider either linear triangles or bilinear quadrilateral elements as in Becker *et al.* [1], for example, whose basis functions are given in the upper part of Table 1.

Table 1. Basis functions for triangle and quadrilateral elements.

Linear Triangle Basis	Bilinear Quadrilateral Basis
$\psi_1 = 1 - \xi - \eta$	$\psi_1 = \frac{1}{4}(1-\xi)(1-\eta)$
$\psi_2 = \xi$	$\psi_2 = \frac{1}{4}(1+\xi)(1-\eta)$
$\psi_3 = \eta$	$\psi_3 = \frac{1}{4}(1+\xi)(1+\eta)$
	$\psi_4 = \frac{1}{4}(1-\xi)(1+\eta)$
6-Node Triangle Basis	**8-Node Quadrilateral Basis**
$\psi_1 = \bar{\psi}_1 - \frac{1}{2}(\Psi_6 + \Psi_4)$	$\psi_1 = \bar{\psi}_1 - \frac{1}{2}(\Psi_8 + \Psi_5)$
$\psi_2 = \Psi_4$	$\psi_2 = \Psi_5$
$\psi_3 = \bar{\psi}_2 - \frac{1}{2}(\Psi_4 + \Psi_5)$	$\psi_3 = \bar{\psi}_2 - \frac{1}{2}(\Psi_5 + \Psi_6)$
$\psi_4 = \Psi_5$	$\psi_4 = \Psi_6$
$\psi_5 = \bar{\psi}_3 - \frac{1}{2}(\Psi_5 + \Psi_6)$	$\psi_5 = \bar{\psi}_3 - \frac{1}{2}(\Psi_6 + \Psi_7)$
$\psi_6 = \Psi_6$	$\psi_6 = \Psi_7$
$\Psi_4 = 4(1-\xi-\eta)\,\xi\,\delta_{s1}$	$\psi_7 = \bar{\psi}_4 - \frac{1}{2}(\Psi_7 + \Psi_8)$
$\Psi_5 = 4\,\xi\,\eta\,\delta_{s2}$	$\psi_8 = \Psi_8$
$\Psi_6 = 4\,\eta\,(1-\xi-\eta)\,\delta_{s3}$	$\Psi_5 = \frac{3}{8}(1-\eta)(1-\xi\eta)\,\delta_{s1}$
	$\Psi_6 = \frac{3}{8}(1+\xi)(1-\eta^2)\,\delta_{s2}$
	$\Psi_7 = \frac{3}{8}(1+\eta)(1-\xi^2)\,\delta_{s3}$
	$\Psi_8 = \frac{3}{8}(1-\xi)(1-\eta^2)\,\delta_{s4}$

Dendritic elements, in the context of the finite element method, cannot be treated as though they are simple 3-node triangles or 4-node quadrilateral elements. The presence of mid-side (*i.e.*, hanging) nodes must be accounted for in the basis function set for the element. Such elements are considered transition elements as in Huang and Xie [12] who present a transition 5-node quadrilateral finite element addressing this refinement issue, which forms the foundation of the treatment of dendritic element interpolation used here. Dendritic triangular elements are handled in an analogous fashion.

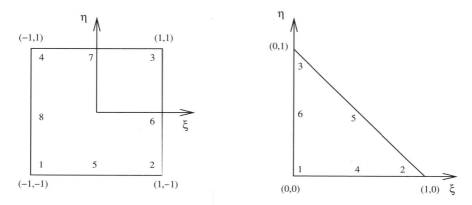

Fig. 10. Master transition quadrilateral (left) and triangular (right) elements.

In general, a transition triangular element may have up to 3 hanging nodes giving as many as 6 nodes per element while a 4-node quadrilateral transition element may have up to 4 hanging nodes leading to as many as 8 nodes in the element. For the most general cases, these elements are shown in Figure 10. If the linear/bilinear basis functions of the upper part of Table 1 are designated with an overbar ($\bar{\psi}_1$, for example), then it can be shown that the basis functions for general transition (*i.e.*, dendritic) elements are as given in the lower part of Table 1. The Ψ_i contain a multiplier, δ_{st}, the Kronecker delta function. The subscripts refer to the sides of the element containing a hanging node.

5.1.2 Hanging Node Constraints in Dendritic Elements

The second issue described above, hanging node constraints, can be addressed in the following manner. We have seen that a typical constraint for hanging node X_{HN}^i in an element is given in Equation 9. This constraint may be imposed by an appropriate penalty method or by explicitly imposing it after each iterative solution of the assembled problem, thereby lagging the solution. In Carey [3], a method is presented which applies a penalty method for inter-element constraints such as for hanging nodes. If the governing system admits

Dendritic Meshing

to a variational principle, as does Equation 10, and if the constraint within the element is of the form $\mathbf{G}\,\mathbf{v} = 0$, then the penalty functional becomes

$$J_\epsilon = \frac{1}{2}\mathbf{v}^t\,K\,\mathbf{v} - \mathbf{v}^t\,\mathbf{F} + \frac{\epsilon^{-1}}{2}\mathbf{v}^t\,\mathbf{G}^t\mathbf{G}\,\mathbf{v}, \tag{17}$$

where ϵ^{-1} is the penalty factor and from which the stationary condition gives

$$\left(K + \frac{1}{\epsilon}\mathbf{G}^t\mathbf{G}\right)\mathbf{v} = \mathbf{F}. \tag{18}$$

Since Equation 10 has a variational principle and the constraints are expressible in the form $\mathbf{G}\,\mathbf{v} = 0$, then Equation 18 applies. For an element with N^e nodes the constraint condition vector, $\mathbf{G}$, has components

$$G_i = \begin{cases} 1 & \text{if } i = j \\ -\frac{1}{2} & \text{if } i = j \pm 1 \\ 0 & \text{otherwise} \end{cases} \tag{19}$$

for node j a hanging node. For an element with 5-nodes and node 4 the hanging node, the constraint condition vector would be

$$\begin{bmatrix} 0, 0, -\frac{1}{2}, 1, -\frac{1}{2} \end{bmatrix} \begin{bmatrix} X_1^i \\ X_2^i \\ X_3^i \\ X_4^i \\ X_5^i \end{bmatrix} = 0, \tag{20}$$

and the element matrix becomes

$$K^e + \frac{1}{\epsilon}\begin{bmatrix} 0 & 0 & 0 & 0 & 0 \\ 0 & 0 & 0 & 0 & 0 \\ 0 & 0 & \frac{1}{4} & \frac{-1}{2} & \frac{1}{4} \\ 0 & 0 & \frac{-1}{2} & 1 & \frac{-1}{2} \\ 0 & 0 & \frac{1}{4} & \frac{-1}{2} & \frac{1}{4} \end{bmatrix}. \tag{21}$$

Should there be $N_{HN}^e \leq N^e$ hanging nodes within the element, N_{HN}^e constraint matrices sum together as in Equation 21 for 1 hanging node, using Equation 19 as before.

5.1.3 Metric Tensor Components

I_e is a non-linear function of the nodal coordinates, X^i, through the metric tensor components. In terms of their interpolants, the covariant metric tensor components are

$$g_{11} = \frac{\partial \psi_k}{\partial \xi} \frac{\partial \psi_l}{\partial \xi} X_k^i X_l^i, \qquad (22)$$

$$g_{12} = \frac{\partial \psi_k}{\partial \xi} \frac{\partial \psi_l}{\partial \eta} X_k^i X_l^i, \qquad (23)$$

$$g_{22} = \frac{\partial \psi_k}{\partial \eta} \frac{\partial \psi_l}{\partial \eta} X_k^i X_l^i. \qquad (24)$$

Mesh motion relies upon the metric components to "enhance" the mesh from an initial state to a "better" final state by solving the assembled global problem. The metric components are known only within an element. Consequently, if the metric tensor components are calculated using the initial mesh coordinates, then the metric tensor components are consistent with the initial mesh and the mesh does not "move." To circumvent this problem, a target mesh defined by the centroid of a composite region defined by the elements in contact with a node is used. In general, this point will not coincide with the node's coordinates, but will converge to it in the final solution. For node n, let M elements be in contact with it. Each surrounding element has an area, A_m, and centroidal coordinates, $C_m^i = (X_{C_m}, Y_{C_m})$ (e.g., [2]) so that the centroid coordinates of the composite region are then

$$C_n^i = \frac{\sum_{m=1}^M A_m C_m^i}{\sum_{m=1}^M A_m}. \qquad (25)$$

C_n^i is used in place of X_n^i in the metric tensor component calculations. The centroid of each boundary node that is also a fixed (*i.e.*, a Dirichlet) boundary condition node is set to its current coordinates.

Since the metric tensor components are a nonlinear function of the nodal coordinates, an iterative solution for the coordinates is required. Here a simple fixed point scheme is used, while others have used a Jacobian-free Newton-Krylov method to solve the fully nonlinear problem (*e.g.*, [10]).

5.1.4 Example Results

As an example illustrating enhancement of a composite triangular-quadrilateral element mesh, consider the initial mesh in Figure 11 (left side) having both triangles and quadrilateral elements. The triangle elements are dendritic, while only some of the quadrilateral elements are dendritic. In particular, the quadrilateral dendritic elements have two hanging nodes per element, a more challenging test of the algorithm. This example also illustrates boundary node movement. The right side of the Figure shows the mesh after enhancement, allowing the boundary nodes on the top and bottom boundaries to move, requiring 57 iterations to converge ($\epsilon_{\text{converged}} = 1.\%$).

A mesh quality metric [11](pg. 466) is defined to be E/E_0, where

$$E = \frac{1}{2} \int_{\hat{\Omega}} \left(g^{11} + g^{22} \right) \sqrt{g} \, d\xi d\eta \qquad (26)$$

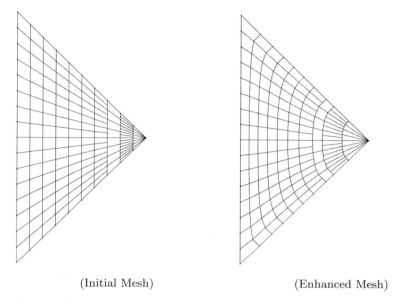

(Initial Mesh)   (Enhanced Mesh)

Fig. 11. An example two dimensional initial mesh (left) having both triangular and quadrilateral elements and also "hanging node" or "dendritic" elements. The quadrilateral dendritic elements have two hanging nodes per element and the triangular elements have one hanging node. The enhanced mesh with moving boundary nodes (right) with $E/E_0 = 0.822993$.

and E_0 is E evaluated on the initial mesh. E is the energy functional (variational principle) for the mesh. Consequently, when viewed relative to its initial value it gives a measure of the fractional energy in the mesh with the goal being to minimize the ratio, E/E_0. The final mesh minimizes the energy functional producing a minimum value of E/E_0, giving the solution to Equation 10.

Iterations cease when the largest node movement on the mesh is less than 1% of the maximum node movement of the first iteration.

6 Conclusion

We present a method for constructing dendritic meshes for two-dimensional domains. The resulting meshes consist of predominantly quadrilateral elements with triangles occurring when two quadrilateral nodes occupy the same position, that is, are degenerate quadrilaterals. A detailed discussions of relevant data structure concerns, feathering structured meshes to produce dendritic elements, trans-finite interpolation on dendritic meshes, and smoothing dendritic meshes are presented. Applying these ideas to construct three-dimensional dendritic meshes is, in principle, straightforward, but has not yet been done.

Acknowledgement. The authors wish to acknowledge these additional members of the ASC Setup Team at Los Alamos National Laboratory without whose contribution this article and the software used for dendritic meshing could not have been completed: S. Davis Herring, Laura M. Lang, and Joseph H. Schmidt.

References

1. Becker, E.B., Carey, G.F., Oden, J.T.: Finite Elements: An Introduction, vol. I. Prentice-Hall International, London (1981)
2. Bourke, P.: Calculating the area and centroid of a polygon (July 1988), http://paulbourke.net/geometry/polyarea
3. Carey, G.F., Kabaila, A., Utku, M.: On penalty methods for interelement constraints. Computer Methods in Applied Mechanics and Engineering 30, 151–171 (1982)
4. Carey, G.F., Oden, J.T.: Finite Elements: An Introduction, vol. II. Prentice-Hall International, London (1983)
5. Carey, G.F.: Computational Grids: Generation, Adaptation, and Solution Strategies. Taylor & Francis, Washington (1997)
6. Courant, R., Friedrichs, K., Lewy, H.: Über die partiellen differenzengleichungen der mathematischen physik. Mathematische Annalen 100(1), 32–74 (1928); English translation: IBM Journal, 215–234 (March 1967)
7. Douglass, R.W.: Laplace-Beltrami enhancement for unstructured two-dimensional meshes having dendritic elements and boundary node movement. Journal of Computational and Applied Mathematics (in review, 2011)
8. Finlayson, B.A.: The Method of Weighted Residuals and Variational Principles. Academic Press, New York (1972)
9. Frey, P.J., George, P.-L.: Mesh Generation: Application to Finite Elements. Hermes Science Publishing, Oxford (2000)
10. Hansen, G., Zardecki, A., Greening, D., Bos, R.: A finite element method for three-dimensional unstructured grid smoothing. J. Comput. Phys. 202(1), 281–297 (2005)
11. Hansen, G.A., Douglass, R.W., Zardecki, A.: Mesh Enhancement: Selected Elliptic Methods, Foundations, and Applications. Imperial College Press (2005)
12. Huang, F., Xie, X.: A modified nonconforming 5-node quadrilateral transition finite element. Adv. Appl. Math. Mech. 2(6), 784–797 (2010)
13. Knupp, P.M., Steinberg, S.: Fundamentals of Grid Generation. CRC Press, Boca Raton (1994)
14. Thompson, J.F., Soni, B.K., Weatherill, N.P.: Handbook of Grid Generation. CRC Press, Boca Raton (1999)

Author Index

Agouzal, A.   313
Alauzet, Frederic   99
Alliez, Pierre   81
Armstrong, Cecil G.   199

Baudouin, T. Carrier   455, 473
Béchet, E.   455
Belme, Anca   99
Bennis, Chakib   569
Benzley, Steven E.   143
Borouchaki, Houman   63, 569
Botsch, Mario   533

Carnie, Greg   587
Cascón, José Manuel   551
Chrisochoides, Nikos   3
Coffey, Todd S.   293

Dervieux, Alain   99
Desbrun, Mathieu   237
Douglass, Rodney W.   619

Ebeida, Mohamed S.   273
Escobar, José María   551

Fairey, Robin   437
Foteinos, Panagiotis   3

Gargallo-Peiró, Abel   365
Geuzaine, Christophe   21, 255, 455
Gould, Jeremy   437
Gu, Xianfeng   217

Han, Wei   217
Henrotte, F.   455

Jean, Brian A.   619
Johnen, Amaury   255
Juretić, Franjo   405

Kumar, Amitesh   387

Lambrechts, Jonathan   473
Laug, Patrick   63
Leng, Juelin   347
Lévy, Bruno   601
Lipnikov, K.   313
Lu, Jean Hsiang-Chun   179

Makem, Jonathan E.   199
Marchandise, Emilie   21, 419, 455, 473
Martineau, David   437
McNamara, Guy R.   619
Memari, Pooran   237
Mitchell, Scott A.   273
Montenegro, Rafael   551
Morvan, Jean-Marie   81
Mouton, Thibaud   455
Mukherjee, Nilanjan   489
Mullen, Patrick   237

Nivoliers, Vincent   601

Ortega, Frank A.   619
Owen, Steven J.   143, 161, 293

Paudel, Gaurab   143
Pellenard, Bertrand   81
Piret, C.   419

Poudret, Mathieu   569
Putz, Norbert   405

Qian, Jin   41
Qin, Ning   587
Quadros, William Roshan   179

Rainaud, Jean-François   21, 255, 419, 455, 473, 569
Robinson, Trevor T.   199
Roca, Xevi   365
Rodríguez, Eduardo   551

Salinger, Andrew G.   293
Sarrate, Josep   365
Sastry, Shankar P.   329
Shahpar, Shahrokh   587
Shih, Alan M.   387
Shimada, Kenji   125, 179
Shontz, Suzanne M.   293, 329
Sieger, Daniel   533

Song, Inho   179
Sorensen, Marguerite C.   161
Staten, Matthew L.   161, 293

Tautges, Tim   511

Vassilevski, Y.   313
Vavasis, Stephen A.   329
Verma, Chaman Singh   511

Wang, Yibin   587

Xu, Guoliang   347

Yamakawa, Soji   125
Yan, Dong-Ming   601
Yau, Shing-Tung   217
Yin, Xiaotian   217

Zhang, Yongjie   41, 347

Printing: Ten Brink, Meppel, The Netherlands
Binding: Stürtz, Würzburg, Germany